Innovative Design and Application of Modern Agricultural Machinery Systems in Cropping Systems

Innovative Design and Application of Modern Agricultural Machinery Systems in Cropping Systems

Guest Editors

Mustafa Ucgul
Chung-Liang Chang

MDPI

Basel • Beijing • Wuhan • Barcelona • Belgrade • Novi Sad • Cluj • Manchester

Guest Editors
Mustafa Ucgul
Faculty of Science and Engineering
Southern Cross University
Lismore
Australia

Chung-Liang Chang
Department of Biomechatronics Engineering
National Pingtung University of Science and Technology
Pingtung County
Taiwan

Editorial Office
MDPI AG
Grosspeteranlage 5
4052 Basel, Switzerland

This is a reprint of the Special Issue, published open access by the journal *Agriculture* (ISSN 2077-0472), freely accessible at: https://www.mdpi.com/journal/agriculture/special_issues/0907QFQ318.

For citation purposes, cite each article independently as indicated on the article page online and as indicated below:

Lastname, A.A.; Lastname, B.B. Article Title. *Journal Name* **Year**, *Volume Number*, Page Range.

ISBN 978-3-7258-6165-1 (Hbk)
ISBN 978-3-7258-6166-8 (PDF)
https://doi.org/10.3390/books978-3-7258-6166-8

Contents

About the Editors . . . . **vii**

Preface . . . . **ix**

Chung-Liang Chang and Mustafa Ucgul
Innovative Design and Application of Modern Agricultural Machinery Systems in Cropping Systems
Reprinted from: *Agriculture* **2025**, *15*, 2371, https://doi.org/10.3390/agriculture15222371 . . . . **1**

Akeem Shokanbi, Dhruvin Jasoliya and Costin Untaroiu
Parameter Identification of Soil Material Model for Soil Compaction Under Tire Loading: Laboratory vs. In-Situ Cone Penetrometer Test Data
Reprinted from: *Agriculture* **2025**, *15*, 2142, https://doi.org/10.3390/agriculture15202142 . . . . **6**

Jiaoli Jiang, Xueting Han, Qingting Liu, Hai Xu, Tao Wu, Jiamo Feng, et al.
Structural Improvement of Sugarcane Harvester for Reducing Field Loss When Harvesting Lodged Canes
Reprinted from: *Agriculture* **2025**, *15*, 1759, https://doi.org/10.3390/agriculture15161759 . . . . **38**

Yunxiang Li, Yinsong Qu, Yuan Fang, Jie Yang and Yanfeng Lu
Design and Experiment of Autonomous Shield-Cutting End-Effector for Dual-Zone Maize Field Weeding
Reprinted from: *Agriculture* **2025**, *15*, 1549, https://doi.org/10.3390/agriculture15141549 . . . . **68**

Zhikang Peng, Fengying Xu, Pan Xie, Jinpeng Chen, Tao Wu and Zhen Chen
Evaluation of a Pre-Cut Sugarcane Planter for Seeding Performance
Reprinted from: *Agriculture* **2025**, *15*, 1429, https://doi.org/10.3390/agriculture15131429 . . . . **90**

Sili Zhou, Fengguang He, Ganran Deng, Ye Dai, Xilin Wang, Bin Yan, et al.
Design and Experimentation of a Height-Adjustable Management Platform for Pineapple Fields
Reprinted from: *Agriculture* **2025**, *15*, 1420, https://doi.org/10.3390/agriculture15131420 . . . . **106**

Huanyu Liu, Zhihang Han, Junwei Bao, Jiahao Luo, Hao Yu, Shuang Wang and Xiangnan Liu
Intelligent Path Tracking for Single-Track Agricultural Machinery Based on Variable Universe Fuzzy Control and PSO-SVR Steering Compensation
Reprinted from: *Agriculture* **2025**, *15*, 1136, https://doi.org/10.3390/agriculture15111136 . . . . **123**

Wenjie Li, Zhun Cheng and Mengchen Yang
Configurational Comparison of a Binary Logic Transmission Unit Applicable to Agricultural Tractor Hydro-Mechanical Continuously Variable Transmissions and Its Wet Clutch Optimization Design Based on an Improved General Regression Neural Network
Reprinted from: *Agriculture* **2025**, *15*, 877, https://doi.org/10.3390/agriculture15080877 . . . . . **150**

Moxian Li, Xueliang Chang, Yaqing Gu, Ping Wang and Shuqi Shang
Design of Dynamic Deep Sowing System for Peanut Planter with Double-Loop Feedback Fuzzy PID Control
Reprinted from: *Agriculture* **2025**, *15*, 808, https://doi.org/10.3390/agriculture15080808 . . . . . **184**

Wancheng Dong, Xiongye Zhang, Zhen Jiang, Xue Hu, Yun Ge and Lixin Zhang
Study on Structure Design and Parameter Optimization of Diversion Rifled Feeder Based on CFD-DEM
Reprinted from: *Agriculture* **2025**, *15*, 351, https://doi.org/10.3390/agriculture15030351 . . . . . **203**

Dariusz Jaskulski, Iwona Jaskulska, Emilian Różniak, Maja Radziemska, Barbara Klik and Martin Brtnický
Smart Strip-Till One-Pass Machine: Winter Wheat Sowing Accuracy Assessment
Reprinted from: *Agriculture* **2025**, *15*, 411, https://doi.org/10.3390/agriculture15040411 . . . . . **223**

Zhao Li, Can Li, Ye Zeng, Chaodong Mai, Runpeng Jiang and Jun Li
Design and Realization of an Orchard Operation-Aid Platform: Based on Planting Patterns and Topography
Reprinted from: *Agriculture* **2025**, *15*, 48, https://doi.org/10.3390/agriculture15010048 . . . . . . **244**

Mingsheng Wang, Wuxuan Lai, Peng Sun, Hong Li and Qiang Song
Severity Estimation of Inter-Turn Short-Circuit Fault in PMSM for Agricultural Machinery Using Bayesian Optimization and Enhanced Convolutional Neural Network Architecture
Reprinted from: *Agriculture* **2024**, *14*, 2214, https://doi.org/10.3390/agriculture14122214 . . . . **268**

Chung-Liang Chang and Cheng-Chieh Huang
Design and Implementation of an AI-Based Robotic Arm for Strawberry Harvesting
Reprinted from: *Agriculture* **2024**, *14*, 2057, https://doi.org/10.3390/agriculture14112057 . . . . **298**

Ernest Owusu-Sekyere and Ying Chen
The Effect of Varying Compaction Levels on Soil Dynamic Properties and the Growth of Canola (*Brassica napus* L.)
Reprinted from: *Agriculture* **2024**, *14*, 1976, https://doi.org/10.3390/agriculture14111976 . . . . **319**

Seung-Jun Kim, Moon-Kyeong Jang, Seok-Joon Hwang, Won Suk Lee and Ju-Seok Nam
Development of a Prediction Model for Specific Fuel Consumption in Rotary Tillage Based on Actual Operation
Reprinted from: *Agriculture* **2024**, *14*, 1993, https://doi.org/10.3390/agriculture14111993 . . . . **335**

Yun-Jeong Yang, Moon-Kyeong Jang and Ju-Seok Nam
Dynamic Simulation Model of Miniature Tracked Forestry Tractor for Overturning and Rollover Safety Evaluation
Reprinted from: *Agriculture* **2024**, *14*, 1991, https://doi.org/10.3390/agriculture14111991 . . . . **354**

Gang Wang, Chengqian Jin, Min Zhang, Chongyou Wu, Qing Tang and Yao Yang
Reducing Grain Damage in Moist Corn Threshing via Corncob Division
Reprinted from: *Agriculture* **2024**, *14*, 1648, https://doi.org/10.3390/agriculture14091648 . . . . **373**

Jianjun Yin, Zefu Chen, Shiyu Lv, Han Wu, Yansu Gao and Luning Wu
Design and Fatigue Life Analysis of the Rope-Clamping Drive Mechanism in a Knotter
Reprinted from: *Agriculture* **2024**, *14*, 1254, https://doi.org/10.3390/agriculture14081254 . . . . **387**

Jeong-Hun Kim, Dong-Hyeon Gim and Ju-Seok Nam
Experimental Structural Safety Analysis of Front-End Loader of Agricultural Tractor
Reprinted from: *Agriculture* **2024**, *14*, 947, https://doi.org/10.3390/agriculture14060947 . . . . . **400**

About the Editors

Mustafa Ucgul

Mustafa Ucgul is a mechanical engineer with extensive expertise in computational mechanics and its applications in agricultural machinery design. With 21 years of research experience across Turkey and Australia, Dr. Ucgul has specialized in discrete element modeling of soil-engaging tools, the design of agricultural machinery, and the thermal analysis of agricultural greenhouses and desalination systems. Since completing his PhD, he has contributed to multiple industry-linked projects.

Chung-Liang Chang

Chung-Liang Chang received his Ph.D. in Electrical Engineering from National Cheng Kung University, Tainan, Taiwan, in 2008. He is currently a Distinguished Professor in the Department of Biomechatronics Engineering at National Pingtung University of Science and Technology. His research focuses on smart agricultural technologies and mechatronic system integration, with contributions spanning agricultural robotics, guidance and navigation control, intelligent environmental management, and plant factory engineering. Dr. Chang serves on editorial boards and as a guest editor for several international journals, and he has been actively involved as a technical program committee member and reviewer for numerous international conferences. He is a Senior Member of IEEE and a member of several professional societies. His current research interests include intelligent navigation and satellite positioning, mobile agricultural robot control, smart cultivation and environmental management systems, and next-generation GNSS receiver design. He remains committed to advancing technological innovation and talent cultivation in smart agriculture.

Preface

Agricultural machinery plays a central role in meeting the increasing demand for food production, especially as farmers face rising costs, labor shortages, and the pressure to operate more sustainably. Across the world, growers rely on machinery systems that are efficient, reliable, and suited to different soil and crop conditions. Improving these systems is essential if we want farming to remain productive, profitable, and environmentally responsible.

This Special Issue brings together a collection of studies that focus on how modern machinery and technologies can support better farming outcomes. The papers cover a wide range of practical and technical topics that are important for both researchers and industry. Each contribution provides useful insights into how machinery performance can be improved and how new ideas can be applied in real farming operations.

We would like to acknowledge all authors who contributed to this Special Issue. Their work represents valuable progress for the agricultural engineering community and will support ongoing improvements in machinery development and farm practices. We hope that this collection will be a useful resource for researchers, students, and anyone interested in advancing the machinery systems that support modern agriculture.

Mustafa Ucgul and Chung-Liang Chang
Guest Editors

agriculture

MDPI

Editorial

Innovative Design and Application of Modern Agricultural Machinery Systems in Cropping Systems

Chung-Liang Chang [1,*] and Mustafa Ucgul [2,*]

1 Department of Biomechatronics Engineering, National Pingtung University of Science and Technology, Neipu, Pingtung 91201, Taiwan
2 Faculty of Science and Engineering, Southern Cross University, Lismore, NSW 2084, Australia
* Correspondence: chungliang@mail.npust.edu.tw (C.-L.C.); mustafa.ucgul@scu.edu.au (M.U.)

The recent increase in extreme weather events and the variety of crop types and planting patterns have shifted agricultural machinery research toward achieving consistent, high-quality performance in the field [1]. The emphasis is now on integrating systems as a whole rather than making small adjustments to individual mechanisms [2]. This includes determining machine size, configuring specialized components, implementing closed-loop control systems, employing sensor-driven decision making, and rigorously assessing field work quality [3–6]. Standardized testing procedures are increasingly employed to allow for valid comparisons to be made across different sites and to assess the performance of complete functional modules under realistic conditions.

Intelligent field operation platforms are usually designed to fit row spacing and terrain, and then mapped to an accessible workspace, with their mechanical strength verified through structural stress analysis [7]. In pineapple cultivation, a height-adjustable platform employs a gantry chassis with hydraulic elevation to regulate ground clearance. Finite element analysis identifies stress concentrations at beam-to-plate connections, informing targeted reinforcement and mass reduction. Field trial results demonstrate stable operation on twenty-degree slopes and plug-and-play integration with multiple task modules [8]. In litchi and longan orchards, a mechanized platform is designed based on tree and site characteristics, taking into account row spacing, plant spacing, tree height, canopy width, and local slope. This enables multi-step tasks to be performed and allows for the operating parameters of the platform to be quickly adjusted across different cultivation systems and terrains [9,10]. In complex farmland, autonomous navigation for tracked-chassis platforms has advanced to a model-assisted and learning-based approach, allowing for perception-driven decisions to be made under challenging field conditions [6,11]. In agricultural field work, accurately applying fertilizer and planting seeds improves efficiency and reduces input usage when important factors are identified and tracked. These factors include the depth of seeding, individual seed placement, consistency of material flow, and spacing between rows [12]. Utilizing sensors and smart control systems can enhance machinery efficiency and reduce costs. This approach makes it easier to reliably and quantifiably link crop growth with planting results [13]. In peanut farming, a dynamic depth system uses surface pressure sensors to assess slope and reconstruct bed geometry in real time. A dual-loop fuzzy controller with velocity feedforward can adjust for variations in the beds and changes in travel speed, ensuring a consistent planting depth during field trials [14]. Similarly, field assessments of pre-cut sugarcane planters show that the efficiency of these machines is influenced by the length of the segments, the width of the outlet, the sprocket speed of the metering system, and the travel speed [15,16].

To address the problems of poor conveying stability and inadequate air–fertilizer mixing uniformity in traditional feed machines, new designs now employ guide chamber blades that enhance both the uniformity of air–fertilizer mixing and the stability of fertilizer transportation [17]. Examining such systems using coupled computational fluid dynamics and discrete element models can enhance the reliability and reproducibility of performance outcomes [18]. Additional innovations include a seedling-protection weeding end-effector for maize that couples deep learning detection with sliding-mode control to enforce a minimum tool-to-crop clearance. Field trials report an approximately eighty-eight percent weed-removal rate with only minimal crop injury [19]. In line with the design concept of agricultural machinery that integrates multiple operational functions, one-pass seeders have been developed that can loosen soil between rows, adjust seeding rate and depth, and apply fertilizer simultaneously [20]. Electrohydraulic depth control and flow monitoring ensure accurate in-row and between-row seeding under field conditions [21]. Improved harvesters and material-handling machines further reduce mechanical damage to crops [22,23]. In the threshing process of high-moisture maize, comparing whole maize ears with pre-split maize ears shows that dividing the ears into smaller sections changes stress paths and support conditions, reducing kernel breakage rates by around 70% when moisture content ranges between 25% and 37% [24]. This Special Issue also presents work on improved agricultural knotters featuring a double-fluted disk design, where redesigning to double synclastic fluted disks through combined simulation and experimental evaluation has enhanced the bundling performance of rice straw [25].

Another notable development is the design of modified harvesters for severely lodged sugarcane. For canes that become lodged, a structurally improved segmental harvester reduces field loss to about one to two percent by stabilizing pickup, transfer, and roller conveyance. Prototype testing through orthogonal experiments demonstrates its effectiveness for severely lodged cane, showing a significant improvement compared with conventional harvesters, which typically experience losses between 15% and 20% [26]. When traditional farming equipment is used to harvest large volumes of fruit, the fruits frequently bump into each other, causing damage to their appearance. This is particularly problematic for delicate fruits, which are easily harmed and can lose market value as a result [27]. This Special Issue also introduces techniques for picking fragile fruits in greenhouse environments, including a compliant cable-driven arm with a two-stage fuzzy controller and visual servoing system that synchronizes cable extension and bending to match peduncle geometry, thereby improving harvesting success [28]. This Special Issue further covers medium-to-large farm machinery, focusing on operational safety, energy efficiency, powertrain systems, soil compaction management, and electrification. Safety assessments for agricultural front-end loaders include both powertrain and structural aspects. Transmission shift quality, fatigue risk, and operational stability are evaluated, and reinforcement of mounting structures is recommended to maintain durability during lifting and dumping operations. A dynamic model developed using measured center of mass and vehicle geometry helps to reduce the risk and expense of rollover testing for small tracked vehicles [29,30]. Validation against critical side-tilt angle and minimum turning radius shows very small errors, enabling practical limits on slope and turning to be specified for safe operation [31]. Field-based predictive models of specific fuel consumption in rotary tillage operations provide contour charts that guide the selection of travel speed, power-take-off settings, power levels, and step lengths. This approach achieves fuel savings without compromising tillage quality [32]. Within power-split drivetrains for agricultural tractors, binary logic transmission units are integrated into hydromechanical continuously variable transmissions. Simulation models, verified through wet clutch test bench data, confirm accurate oil pressure behavior and realistic transmission response. Generalized regression neural networks optimized through advanced algorithms are used to calibrate clutch parameters, resulting in smooth shifts, controlled thermal load, and repeatable parameterization workflows [33]. Because tractor power consumption and soil compaction are interrelated [34], emergence rates for crops such as canola and overall growth performance have been quantified against soil penetration resistance, shear strength, water content, and bulk density. Cross-site trials across sandy loam, silt clay, and clay soils have established soil-specific thresholds, indicating that zero or minimal compaction is preferable [35]. Higher compaction increases draft and fuel demand and inhibits biomass accumulation. Moreover, low-pressure compaction equipment can reduce yield losses caused by soil degradation. Developing reliable soil–tire interaction prediction models is therefore essential [36]. Laboratory methods are used to determine initial soil-compaction parameters, which are then refined with field data to improve prediction accuracy. The final calibrated model supports field assessments under different moisture and load conditions [37]. In the area of electrified agricultural machinery, operating at low speed with high torque, along with fluctuating loads and frequent starts, increases the likelihood of inter-turn short circuits. A refined fault-severity estimation method tuned by probabilistic optimization provides accurate results under both constant and variable speed and load, thereby reducing false shutdowns and enabling effective preventive maintenance [38].

Overall, the research presented in this Special Issue reflects a continuous process from design to verification in modern agricultural machinery systems. It begins with mechanism design informed by field geometry and crop diversity, advances to sensing and control methods validated under practical field conditions, and culminates in performance indicators that evaluate the feasibility of field operations. Applied across cropping systems with diverse crops, terrains, and operational scenarios, these studies demonstrate reproducible engineering methods that can be transferred to different cultivation targets and environments. Collectively, these contributions advance the standardization of on-site equipment operations and lay the foundation for data sharing and the integration of heterogeneous modules, thereby enabling comparable, reproducible designs and verifiable outcomes. The collected studies provide a valuable reference for both researchers and industry professionals, guiding future innovation, comparison, and application in this rapidly advancing area of agricultural machinery engineering.

Funding: No external funding was received for this research.

Conflicts of Interest: The authors declare no conflicts of interest.

References

1. Ucgul, M.; Chang, C.L. Design and application of agricultural equipment in tillage systems. *Agriculture* **2023**, *13*, 790. [CrossRef]
2. Wang, B.; Du, X.; Wang, Y.; Mao, H. Multi-machine collaboration realization conditions and precise and efficient production mode of intelligent agricultural machinery. *Int. J. Agric. Biol. Eng.* **2024**, *17*, 27–36. [CrossRef]
3. Saiz-Rubio, V.; Rovira-Más, F. From smart farming towards agriculture 5.0: A review on crop data management. *Agronomy* **2020**, *10*, 207. [CrossRef]
4. Reddy, R. Innovations in agricultural machinery: Assessing the impact of advanced technologies on farm efficiency. *J. Artif. Intell. Big Data* **2022**, *2*, 1. [CrossRef]
5. Xie, D.; Chen, L.; Liu, L.; Chen, L.; Wang, H. Actuators and sensors for application in agricultural robots: A review. *Machines* **2022**, *10*, 913. [CrossRef]
6. Chang, C.L.; Chen, H.W.; Ke, J.Y. Robust guidance and selective spraying based on deep learning for an advanced four-wheeled farming robot. *Agriculture* **2023**, *14*, 57. [CrossRef]
7. Gao, J.; Jin, Z.; Ai, A. The optimized design of soil-touching parts of a greenhouse humanoid weeding shovel based on strain sensing and DEM–ADAMS coupling simulation. *Sensors* **2024**, *24*, 868. [CrossRef] [PubMed]
8. Zhou, S.; He, F.; Deng, G.; Dai, Y.; Wang, X.; Yan, B.; Pan, D. Design and experimentation of a height-adjustable management platform for pineapple fields. *Agriculture* **2025**, *15*, 1420. [CrossRef]

9. Li, J.; Zhong Lu, H.; Yang, Z.; Lü, E. Orchard facilities and equipment for litchi and longan in China. In Proceedings of the 2011 ASABE Annual International Meeting, Louisville, KY, USA, 7–10 August 2011; American Society of Agricultural and Biological Engineers: St. Joseph, MO, USA, 2011; p. 1.
10. Li, Z.; Li, C.; Zeng, Y.; Mai, C.; Jiang, R.; Li, J. Design and realization of an orchard operation-aid platform: Based on planting patterns and topography. *Agriculture* **2024**, *15*, 48. [CrossRef]
11. Wang, S.; Song, J.; Qi, P.; Yuan, C.; Wu, H.; Zhang, L.; He, X. Design and development of orchard autonomous navigation spray system. *Front. Plant Sci.* **2022**, *13*, 960686. [CrossRef]
12. Yu, H.; Ding, Y.; Fu, X.; Liu, H.; Jin, M.; Yang, C.; Dou, X. A solid fertilizer and seed application rate measuring system for a seed-fertilizer drill machine. *Comput. Electron. Agric.* **2019**, *162*, 836–844. [CrossRef]
13. Farooqui, N.A.; Haleem, M.; Khan, W.; Ishrat, M. Precision agriculture and predictive analytics: Enhancing agricultural efficiency and yield. In *Intelligent Techniques for Predictive Data Analytics*; Wiley-IEEE Press: Hoboken, NJ, USA; pp. 171–188.
14. Li, M.; Chang, X.; Gu, Y.; Wang, P.; Shang, S. Design of dynamic deep sowing system for peanut planter with double-loop feedback fuzzy PID control. *Agriculture* **2025**, *15*, 808. [CrossRef]
15. Wu, F.; Li, S.; He, Y.; Song, M.; Ma, F.; Teng, X.; Liang, Y. Optimized design and performance of pre-cutting vibrating sugarcane sett metering device. *Sugar Tech.* **2023**, *25*, 210–222. [CrossRef]
16. Yan, B.; Liu, H.; He, F.; Deng, G.; Zheng, S.; Cui, Z.; Li, B. Analysis and testing of pre-cut sugarcane seed stalk sawing performance parameters. *Agriculture* **2024**, *14*, 953. [CrossRef]
17. Zhang, L.; Yuan, W.; Jin, C.; Feng, Y.; Liu, G.; Hu, Y. Research progress on key mechanical components of the pneumatic centralized fertilizer discharge system. *Appl. Sci.* **2024**, *14*, 3884. [CrossRef]
18. Dong, W.; Zhang, X.; Jiang, Z.; Hu, X.; Ge, Y.; Zhang, L. Study on structure design and parameter optimization of diversion rifled feeder based on CFD–DEM. *Agriculture* **2025**, *15*, 351. [CrossRef]
19. Li, Y.; Qu, Y.; Fang, Y.; Yang, J.; Lu, Y. Design and experiment of autonomous shield-cutting end-effector for dual-zone maize field weeding. *Agriculture* **2025**, *15*, 1549. [CrossRef]
20. Squires, N.R.W.; Haggar, R.J.; Elliott, J.G. A one-pass seeder for introducing grasses, legumes and fodder crops into swards. *J. Agric. Eng. Res.* **1979**, *24*, 199–208. [CrossRef]
21. Zhou, L.; Ma, Y.; Zhou, H.; Niu, K.; Zhao, B.; Wei, L.; Zhang, W. Design and test of sowing depth measurement and control system for no-till corn seeder based on integrated electro-hydraulic drive. *Appl. Sci.* **2023**, *13*, 5823. [CrossRef]
22. Ampatzidis, Y.G.; Vougioukas, S.G.; Whiting, M.D.; Zhang, Q. Applying the machine repair model to improve efficiency of harvesting fruit. *Biosyst. Eng.* **2014**, *120*, 25–33. [CrossRef]
23. Zhou, L.; Jin, S.; Wang, J.; Zhang, H.; Shi, M.; Zhou, H. 3D positioning of Camellia oleifera fruit-grabbing points for robotic harvesting. *Biosyst. Eng.* **2024**, *246*, 110–121. [CrossRef]
24. Wang, G.; Jin, C.; Zhang, M.; Wu, C.; Tang, Q.; Yang, Y. Reducing grain damage in moist corn threshing via corncob division. *Agriculture* **2024**, *14*, 1648. [CrossRef]
25. Yin, J.; Chen, Z.; Lv, S.; Wu, H.; Gao, Y.; Wu, L. Design and fatigue life analysis of the rope-clamping drive mechanism in a knotter. *Agriculture* **2024**, *14*, 1254. [CrossRef]
26. Jiang, J.; Han, X.; Liu, Q.; Xu, H.; Wu, T.; Feng, J.; Zou, X.; Li, Y. Structural improvement of sugarcane harvester for reducing field loss when harvesting lodged canes. *Agriculture* **2025**, *15*, 1759. [CrossRef]
27. Li, Z.; Thomas, C. Quantitative evaluation of mechanical damage to fresh fruits. *Trends Food Sci. Technol.* **2014**, *35*, 138–150. [CrossRef]
28. Chang, C.L.; Huang, C.C. Design and implementation of an AI-based robotic arm for strawberry harvesting. *Agriculture* **2024**, *14*, 2057. [CrossRef]
29. Cameron, J.T.; Brennan, S.N. Vehicle Dynamic Modeling for the Prediction and Prevention of Vehicle Rollover. Ph.D. Thesis, Pennsylvania State University, University Park, PA, USA, 2005.
30. Wang, F.; Chen, Y. Vehicle rollover propensity detection based on a mass-center-position metric: A continuous and completed method. *IEEE Trans. Veh. Technol.* **2019**, *68*, 8652–8662. [CrossRef]
31. Yang, Y.J.; Jang, M.K.; Nam, J.S. Dynamic simulation model of miniature tracked forestry tractor for overturning and rollover safety evaluation. *Agriculture* **2024**, *14*, 1991. [CrossRef]
32. Kim, S.-J.; Jang, M.-K.; Hwang, S.-J.; Lee, W.S.; Nam, J.-S. Development of a prediction model for specific fuel consumption in rotary tillage based on actual operation. *Agriculture* **2024**, *14*, 1993. [CrossRef]
33. Li, W.; Cheng, Z.; Yang, M. Configurational comparison of a binary logic transmission unit applicable to agricultural tractor hydro-mechanical continuously variable transmissions and its wet clutch optimization design based on an improved general regression neural network. *Agriculture* **2025**, *15*, 877. [CrossRef]
34. Shaheb, M.R.; Venkatesh, R.; Shearer, S.A. A review on the effect of soil compaction and its management for sustainable crop production. *J. Biosyst. Eng.* **2021**, *46*, 417–439. [CrossRef]

35. Owusu-Sekyere, E.; Chen, Y. The effect of varying compaction levels on soil dynamic properties and the growth of canola (*Brassica napus* L.). *Agriculture* **2024**, *14*, 1976. [CrossRef]
36. Zeng, H.; Zhao, C.; Chen, S.; Xu, W.; Zang, M. Numerical simulations of tire–soil interactions: A comprehensive review. *Arch. Comput. Methods Eng.* **2023**, *30*, 4801–4829. [CrossRef]
37. Shokanbi, A.; Jasoliya, D.; Untaroiu, C. Parameter identification of soil material model for soil compaction under tire loading: Laboratory vs. in-situ cone penetrometer test data. *Agriculture* **2025**, *15*, 2142. [CrossRef]
38. Wang, M.; Lai, W.; Sun, P.; Li, H.; Song, Q. Severity estimation of inter-turn short-circuit fault in PMSM for agricultural machinery using Bayesian optimization and enhanced convolutional neural network architecture. *Agriculture* **2024**, *14*, 2214. [CrossRef]

agriculture

MDPI

Article

Parameter Identification of Soil Material Model for Soil Compaction Under Tire Loading: Laboratory vs. In-Situ Cone Penetrometer Test Data

Akeem Shokanbi [1], Dhruvin Jasoliya [1] and Costin Untaroiu [1,2,*]

[1] Department of Mechanical Engineering, Virginia Tech, Blacksburg, VA 24060, USA; akeems@vt.edu (A.S.); dhruvinj@vt.edu (D.J.)
[2] Department of Biomedical Engineering & Mechanics, Virginia Tech, Blacksburg, VA 24060, USA
* Correspondence: costin@vt.edu; Tel.: +1-540-505-0138

Abstract: Accurate numerical simulations of soil-tire interactions are essential for optimizing agricultural machinery to minimize soil compaction and enhance crop yield. This study developed and compared two approaches for identifying and validating parameters of a LS-Dyna soil model. The laboratory-based approach derives parameters from triaxial, consolidation, and cone penetrometer tests (CPT), while the optimization-based method refines them using in-situ CPT data via LS-OPT to better capture field variability. Simulations employing Multi-Material Arbitrary Lagrangian–Eulerian (MM-ALE), Smoothed Particle Hydrodynamics (SPH), and Hybrid-SPH methods demonstrate that Hybrid-SPH achieves the optimal balance of accuracy (2% error post-optimization) and efficiency (14-h runtime vs. 22 h for SPH). Optimized parameters improve soil–tire interaction predictions, including net traction and tire sinkage across slip ratios from −10% to 30% (e.g., sinkage of 12.5 mm vs. 11.1 mm experimental at 30% slip, with overall mean-absolute percentage error (MAPE) reduced to 3.5% for sinkage and 4.2% for traction) and rut profiles, outperforming lab-derived values. This framework highlights the value of field-calibrated optimization for sustainable agriculture, offering a cost-effective alternative to field trials for designing low-compaction equipment and reducing yield losses from soil degradation. While sandy loam soil at 0.4% moisture content was used in this study, future extensions to different soil types with varied moisture are recommended.

Keywords: soil-tire/tool interaction; SPH; hybrid-SPH; sandy loam soil; CPT; parameter optimization; granular behavior; soil compaction; agricultural engineering

1. Introduction

Soil compaction, a critical factor in sustainable agriculture, can significantly reduce crop yield, and accurate numerical simulations are essential for predicting and mitigating compaction caused by heavy agricultural machinery. This has been a key area of research, as demonstrated by recent studies that focus on the numerical simulation of soil–tire interactions and the effect of tire parameters on soil compaction [1,2]. The reliability of such a simulation hinges on the accuracy of the soil constitutive material model and the numerical methods employed [3]. However, modeling soil behavior is complex due to its non-linear, plastic, time-dependent, and anisotropic properties, which are further influenced by environmental factors like moisture content [4].

In recent years, several soil constitutive models have been developed to capture these complex behaviors, ranging from simple elastic-plastic models like Mohr–Coulomb

(MC) and Drucker–Prager (DP) to more advanced models like the Cap Plasticity (CP) model, which accounts for strain hardening in the cap region of the yield surface [5]. Parameter identification for these models has traditionally relied on laboratory tests, such as triaxial compression, hydrostatic compression, and oedometer tests, where parameters like cohesion, friction angle, and moduli are directly fitted from stress-strain data [3,4,6]. For instance, a comprehensive study on sandy loam soil [6] was performed in Abaqus software (2022HF2) (Dassault Systemes, Vélizy-Villacoublay, France), selecting a CP material model and identifying/validating parameters using experimental data from triaxial, consolidation, and cone penetrometer tests (CPT). That study compared the performance of mesh-based (Coupled Eulerian–Lagrangian, CEL) and meshless (Smoothed Particle Hydrodynamics, SPH) methods, finding that SPH better captured the granular nature of soil, achieving a 2% error in CPT simulations compared to CEL's 20% error.

Despite these advances, the current state of parameter identification methods reveals significant limitations. Lab-based approaches, while foundational, are often expensive, time-consuming, and may not fully represent in situ soil conditions due to sample disturbance, scale effects, and environmental variability [3,7,8]. Inverse methods, such as optimization algorithms (e.g., least-squares fitting or genetic algorithms integrated with tools like LS-OPT [9–12]), have emerged to refine parameters by minimizing discrepancies between simulations and experimental data. However, just a few applications were published in the identification of soil parameters, and lack integration with field-specific tests like CPT for direct in situ calibration [3,6]. In addition, few studies have explored various non-Lagrangian approaches (e.g., meshless methods) developed recently in robust platforms like LS-Dyna 2025 R16 (ANSYS, Canonsburg, PA, USA) [13], which excels in non-linear finite element field.

To address these gaps, this study compares a (1) direct derivation for parameter identification in the MAT_005 soil model from lab tests (triaxial and consolidation) to (2) an innovative optimization-based refinement using in-situ CPT data to capture field realism better, reducing reliance on costly lab procedures [13]. In addition, soil behavior under wheel loads is simulated and compared using three LS-Dyna computational approaches (MM-ALE, SPH, and Hybrid-SPH), with Hybrid-SPH offering the best balance of accuracy (errors reduced to 2–5%) and efficiency (14-h runtime vs. 22 h for SPH). By bridging lab and field data through optimization, this work enhances innovation in sustainable farming simulations, providing a cost-effective alternative to field trials for designing low-compaction machinery.

The paper is organized as follows: Section 2 outlines the materials and methods, detailing the agricultural relevance of soil parameters, laboratory-based parameter identification (via triaxial, consolidation, and CPT tests, with numerical CPT setups using MM-ALE, SPH, and Hybrid-SPH), optimization-based in situ parameter identification, and a case study on soil-tire compaction using Hybrid FEM-SPH, building on methodologies established in [6]. Section 3 discusses results, including lab-derived parameters, triaxial and CPT simulation outcomes compared to findings in [6], optimization results, sensitivity analysis, soil–tire interaction findings, and limitations. Section 4 concludes with key findings, agricultural implications, and future research directions.

2. Materials and Methods

Material models that include stress-strain constitutive equations describe the under-loading soil's behavior. Currently, in Ls-Dyna, a few material models are available for analyzing soil under loadings, including the geologic cap model [14,15], hysteretic soil [15], and the FHWA (Federal Highway Administration) soil model [16]. These models consider pressure hardening, strain hardening, and strain rate dependency of soil material, but not all are suitable for new modeling approach elements, such as ALE and SPH [15]. Therefore, in this study, the material soil and foam (MAT_005 in LS-Dyna [15]), a robust and straightforward linearly elastic perfectly plastic model [17] based on isotropic plasticity [7], was used with a focus on simulating soil compaction under wheel loads to support agricultural engineering applications. Limitations of this model include the absence of strain hardening and softening features, but it is compatible with all Lagrangian, ALE, and SPH elements. The methodology aims to enable accurate numerical simulations that reduce the reliance on costly and time-consuming field tests, which can cost USD 5000–10,000 per trial [18], for designing low-compaction agricultural machinery and addressing soil compaction, a critical factor in sustainable farming. Two complementary approaches are employed (Figure 1). The sandy loam soil at 0.4% moisture content is used as an example. The process begins with extracting experimental data from triaxial and consolidation tests [6] and estimating parameters for MAT_005 (Figure 1a).

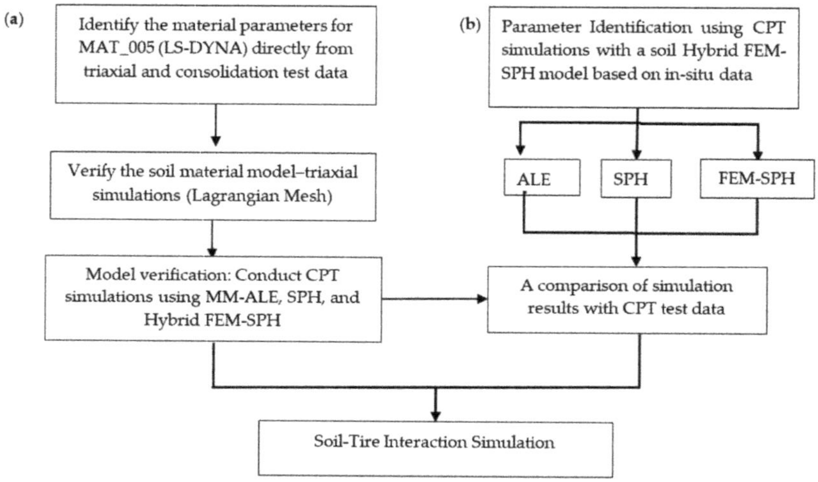

Figure 1. Methodology. (**a**) Direct Material Identification from lab material testing. (**b**) Material identification from in situ CPT tests.

Numerical simulations of the triaxial test are then performed to verify the model parameters using the Lagrangian meshing method. Subsequently, CPT simulations involving extensive mesh deformation are carried out using advanced meshing techniques such as MM-ALE, SPH, and Hybrid-SPH to validate the parameters against experimental data. The second approach focuses on parameter identification directly from in situ CPT test data (Figure 1b). This approach uses an optimization-based method to determine the model parameters using FE simulation with a Hybrid FEM-SPH model. Finally, the material models with optimized parameters are used in CPT simulations using different soil modelling approaches. They were applied in a practical agricultural case study to simulate soil compaction under tractor wheel loads, demonstrating their utility in replacing field trials.

2.1. Agricultural Relevance of the Soil Parameters

Key soil parameters for the MAT_005 model (Table 1), critical for modeling soil compaction under wheel loads, are determined through experimental tests and optimized via simulations to enhance machinery performance and soil health. Cohesion (c) governs soil shear strength, influencing resistance to compaction. The compaction depth is reduced by enhancing soil cohesion, critical for minimizing soil damage under wheel loads [3].

Friction Angle (φ) affects load-bearing capacity, impacting the soil's ability to support heavy machinery without excessive compaction [3]. Shear Modulus (G) controls soil stiffness under shear, influencing tire traction and compaction extent. Higher G increases resistance to deformation but may increase compaction under heavy loads [2]. Bulk Modulus (K) dictates soil compressibility directly affecting compaction depth and bulk density increase under wheel loads. Lower K reduces compaction depth by up to 15% [1]. Volumetric Strain Hardening (ε, P) is derived from consolidation tests this reflects soil response to compressive loads, modulating density increase under wheel-induced compaction [19]. Finally, it is important to note that the shear strength parameters of soil are strongly influenced by its water content. Cohesion (c) typically exhibits a non-monotonic trend with increasing moisture—initially rising as water promotes capillary bonding and apparent cohesion, then decreasing beyond the optimum water content due to elevated pore pressures and loss of matric suction [20–22]. Conversely, the internal friction angle (φ) generally decreases with increasing water content, reflecting the reduction in interparticle friction as lubrication and water film thickness increase between soil grains [23,24]. Consequently, at lower moisture levels—such as around 0.4% in the present study, both c and φ are relatively elevated, resulting in greater soil resistance to compaction compared with wetter states, where cohesion and friction are weakened. Similar moisture-dependent behaviors have been widely reported for sandy loam and clayey soils under varying compaction degrees and drying–wetting cycles [22].

Table 1. Agricultural relevance of parameters for sandy loam soil.

Parameter	Description	Agricultural Impact
c	Cohesion	Reduces compaction depth
φ	Friction Angle	Influences the load-bearing capacity
G	Shear Modulus	Affects traction and compaction extent
K	Bulk Modulus	Govern compaction depth under wheels
ε, P	Volumetric-Strain, Pressure	Impacts density increases under compressive loads

These parameters are derived and validated through the two approaches below and applied in a case study (Section 2.4) to simulate soil compaction under wheel loads, ensuring practical value for agricultural engineering.

2.2. *Lab Testing Approach: Parameter Identification*

Triaxial and consolidation tests were conducted on sandy loam soil (73% sand, 27% silt, 0% clay) from the Norfolk region of Virginia, USA, at a moisture content of 0.4%, consistent with the experimental setup described in [6]. The soil was classified as Silty Loam (SL) per the USDA soil classification triangle [14]. The soil's natural moisture content is 0.4%, and its maximum dry density is 1923 kg/m^3.

2.2.1. Triaxial Test

To determine the shear strength parameters of sandy loam soil for use in LS-DYNA simulations with the MAT_005 model, experimental data from unconsolidated-undrained (UU) triaxial tests were utilized [6]. UU triaxial tests are often preferred for studying the shear strength of unsaturated soils due to their cost and time effectiveness [15]. A cylindrical soil specimen was prepared at specified compaction and moisture levels. The specimen was subjected to predetermined confining pressures, followed by a deviator (shearing) load applied vertically at a strain rate of 12.5%/min (Figure 2). The 12.5%/min strain rate was chosen to approximate the soil deformation rate under typical agricultural wheel loading, corresponding to fast but quasi-static compaction during tire passage.

Previous studies have shown that this rate captures soil strength parameters without introducing significant rate-dependent effects, making it representative of machinery-induced loading conditions [15,19,25,26]. The tests were performed per ASTM D2850 [27] using Loadtrac 2 testing equipment (Geocomp Inc., Atlanta, GA, USA). These tests provide critical parameters such as cohesion (c) and friction angle (φ), which are derived by plotting Mohr circles based on the test results [16].

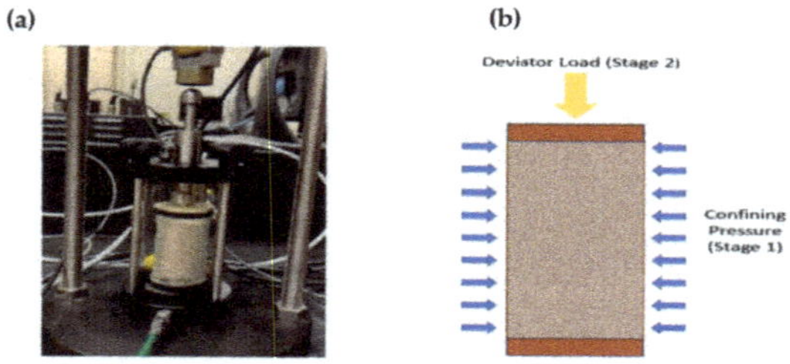

Figure 2. (**a**) Triaxial test apparatus [17]; (**b**) triaxial test schematics [6].

The stress–strain data defined the MAT_005 model's parameters, ensuring consistency with the experimental conditions [6]. These parameters were further verified through numerical simulations of the triaxial test. The numerical simulation of the triaxial test was performed using the Lagrangian method in LS-DYNA software (LSTC, Livermore, CA, USA) [13]. To replicate the experimental triaxial test conditions, a complete three-dimensional cylindrical soil specimen with a radius of 50 mm and a height of 220 mm is modeled (Figure 3). The Lagrangian meshing approach is adopted since the triaxial test involves small to moderate deformations, where large deformation processes are not dominant, making this method suitable for accurately capturing material behavior [7]. Following LS-DYNA best practices, the model is discretized using a structured mesh of 8-node solid elements (hexahedral elements), with sufficient refinement to ensure numerical accuracy and computational efficiency [13]. The bottom boundary of the specimen is fixed in the vertical direction ($u_z = 0$) to prevent rigid body motion, the lateral and bottom surfaces are subjected to a confining pressure of 5 psi to simulate the isotropic consolidation phase, consistent with established soil modeling techniques [8,28].

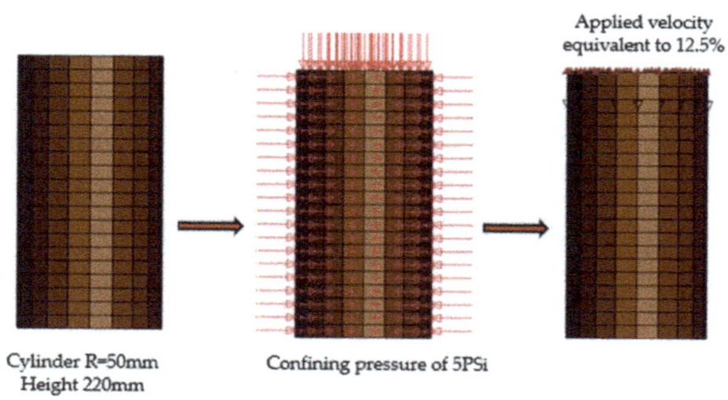

Figure 3. Triaxial simulation set-up using FE mesh.

The simulation is conducted in two stages, consistent with standard triaxial test procedures. In the first stage, the confining pressure of 5 psi is applied uniformly to the bottom and lateral surfaces of the cylindrical specimen. In the second stage, a downward axial velocity ($-u_z$) corresponding to a strain rate of 12.5%/min is imposed on the top surface until an axial strain of 15% is reached. The axial stress–strain response is computed throughout the simulation, and the numerical results are compared with both the experimental data and the results from the Critical State Plasticity (CP) model, as reported in [6]

2.2.2. Consolidation Test

One-dimensional consolidation tests were utilized to study the compressibility of sandy loam soil and obtain pressure–volumetric strain data for the MAT_005 model in LS-DYNA. These tests are essential for defining the hardening behavior of elastic-plastic constitutive material models (Figure 4). Cylindrical soil specimens were prepared and enclosed in a metal ring, then subjected to incremental static loads as per ASTM D2435 [29] using Loadtrac 3 testing equipment (Geocomp Inc., Atlanta, GA, USA) [6].

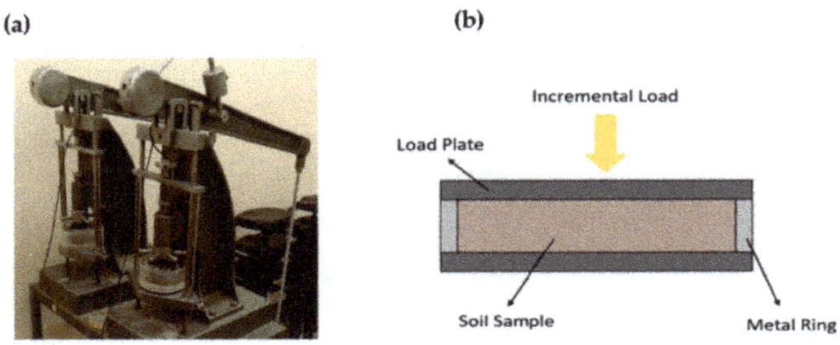

Figure 4. (**a**) Consolidation test apparatus [17]; (**b**) Consolidation test schematics [6].

The parameters for the MAT_005 model in LS-DYNA were derived using a systematic procedure adapted from the Cap Plasticity (CP) model parameters [6], which consists of the following steps (Figure 5):

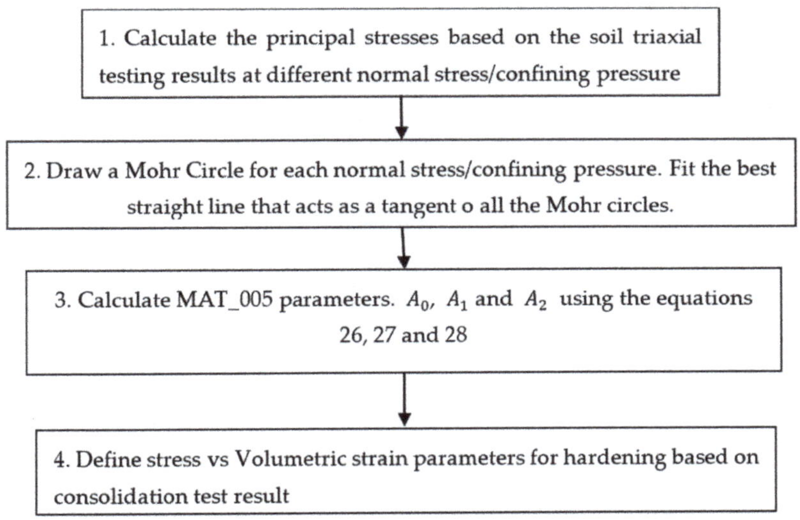

Figure 5. The identification methodology of MAT_005 parameters from experimental data [2].

(1) Obtained Principal Stresses from Triaxial Test Results: Unconsolidated-undrained triaxial test data at confining pressures of 5 psi [6] were used to determine the principal stresses (σ_1, σ_2, σ_3) for the sandy loam soil at 0.4% moisture content.

(2) Draw Mohr Circles and Fit Tangent Line to All Circles: Mohr circles were constructed for each confining pressure using the principal stresses. A best-fit straight line was drawn as the tangent to all Mohr circles, representing the Mohr–Coulomb failure envelope.

(3) Obtain Cohesion (c) as Y-Intercept and Friction Angle (∅) as Slope: From the Mohr–Coulomb failure envelope, the y-intercept provided the cohesion ($c = 0.023$ MPa), and the slope gave the friction angle ($\varnothing = 59.5^{\circ}$) consistent with the values reported by [6] for 0.4% moisture content.

(4) Calculate A_0, A_1, A_2 for MAT_005 via *Yield Function Comparison*: The MAT_005 model in LS-DYNA employs a Drucker–Prager yield function to describe the plastic behavior of soil [14], so on the yield surface:

$$J_2 = A_0 + A_1 P + A_2 P^2 \quad (1)$$

where J_2 is the second invariant of the stress deviator, p is the hydrostatic pressure (positive in compression), and A_0, A_1 and A_2 are yield function constants. These constants were derived by mapping the Mohr–Coulomb failure envelope, obtained from triaxial test data to the Drucker–Prager yield surface using the analytical algorithm by [30]. The derivation leverages the relationship between principal stresses (σ_1, σ_2) from unconsolidated–undrained (UU) triaxial tests and the Mohr–Coulomb shear strength parameters.

Mathematical Derivation:

The Mohr–Coulomb failure criterion defines the shear strength of soil as $\tau = c + \sigma \tan \varphi$ where τ is the shear stress, σ is the normal stress, c is cohesion, and φ is the internal friction angle. In a triaxial compression test, the principal stresses are σ_1 (vertical, major) σ_3 (horizontal, confining, minor), with $\sigma_2 = \sigma_3$. At failure, the shear and normal stresses on the failure plane are related to the principal stresses by [31]:

$$\tau = \frac{\sigma_1 - \sigma_2}{2} \cos \varphi \quad (2)$$

$$\sigma = \left(\frac{\sigma_1 + \sigma_3}{2} - \frac{\sigma_1 - \sigma_3}{2} \right) \sin \varphi \quad (3)$$

The hydrostatic pressure p is defined as:

$$p = \frac{1}{3} (\sigma_1 + 2\sigma_3) \quad (4)$$

The Mohr–Coulomb failure envelope can be expressed in terms of principal stresses using the classic relationship for triaxial compression [1]:

$$\sigma_3 = \sigma_1 \tan\left(45^{\circ} - \frac{\varphi}{2} \right)^2 - 2c\tan\left(45^{\circ} - \frac{\varphi}{2} \right) \quad (5)$$

Rearrange to express the stress difference:

$$\sigma_1 - \sigma_3 = \sigma_1 - \left[\sigma_1 \tan\left(45^{\circ} - \frac{\varphi}{2} \right)^2 - 2c\tan\left(45^{\circ} - \frac{\varphi}{2} \right) \right] \quad (6)$$

$$\sigma_1 - \sigma_3 = \sigma_1 \left[1 - \tan\left(45^{\circ} - \frac{\varphi}{2} \right)^2 \right] + 2c\tan\left(45^{\circ} - \frac{\varphi}{2} \right) \quad (7)$$

Define intermediate variables:

$$C = 1 - \tan\left(45^{\circ} - \frac{\varphi}{2} \right)^2 \quad (8)$$

$$D = 2c\tan\left(45^{\circ} - \frac{\varphi}{2} \right) \quad (9)$$

Thus:

$$\sigma_1 - \sigma_3 = C\sigma_1 + D \quad (10)$$

express p in terms of σ_1:

$$p = \frac{1}{3}\sigma_1 + \frac{2}{3}\sigma_3 \tag{11}$$

Substitute $\sigma_3 = \sigma_1 \tan\left(45^\circ - \frac{\varphi}{2}\right)^2 - 2c\tan\left(45^\circ - \frac{\varphi}{2}\right)$:

$$p = \frac{1}{3}\sigma_1 + \frac{2}{3}\left[\sigma_1\left(45^\circ - \frac{\varphi}{2}\right)^2 - 2c\tan\left(45^\circ - \frac{\varphi}{2}\right)\right] \tag{12}$$

$$p = \sigma_1\left[\frac{1}{3} + \frac{2}{3}\tan\left(45^\circ - \frac{\varphi}{2}\right)^2\right] - \frac{4}{3}c\tan\left(45^\circ - \frac{\varphi}{2}\right) \tag{13}$$

Define:

$$A = \frac{1}{3} + \frac{2}{3}\tan\left(45^\circ - \frac{\varphi}{2}\right)^2 \tag{14}$$

$$B = \frac{4}{3}c\tan\left(45^\circ - \frac{\varphi}{2}\right) \tag{15}$$

$$p = A\sigma_1 - B \tag{16}$$

Solve for σ_1:

$$\sigma_1 = \frac{p + B}{A} \tag{17}$$

Substitute into the stress difference equation:

$$\sigma_1 - \sigma_3 = C\left(\frac{p + B}{A}\right) + D = \frac{C}{A^p} + \frac{CB}{A} + D \tag{18}$$

Define:

$$\alpha = \frac{C}{A} \tag{19}$$

$$\beta = \frac{CB}{A} + D \tag{20}$$

$$\sigma_1 - \sigma_3 = \alpha p + \beta \tag{21}$$

In triaxial compression, the second invariant J_2 is related to the stress difference [32]:

$$J_2 = \frac{1}{3}(\sigma_1 - \sigma_3)^2 \tag{22}$$

$$J_2 = \frac{1}{3}(\alpha p + \beta)^2 \tag{23}$$

Expand:

$$(\alpha p + \beta)^2 = \alpha^2 p^2 + 2\alpha\beta p + \beta^2 \tag{24}$$

$$J_2 = \frac{1}{3}\alpha^2 p^2 + \frac{2}{3}\alpha\beta p + \frac{1}{3}\beta^2 \tag{25}$$

Comparing Equations (1) and (25) we get:

$$A_0 = \frac{1}{3}\beta^2 \tag{26}$$

$$A_1 = \frac{2}{3}\alpha\beta \tag{27}$$

$$A_2 = \frac{1}{3}\alpha^2 \tag{28}$$

(5) Define Stress vs. *Volumetric Strain for Hardening from Consolidation Test Results*: The consolidation test results provided the stress vs. volumetric plastic strain data, which

defines the hardening behavior of MAT_005. This data was adapted from the cap hardening curve of the CP model [6].

2.2.3. Cone Penetrometer Test (CPT)

Cone penetrometer tests (CPTs) were employed (1) to characterize the in-situ behavior of sandy loam soil, providing data to validate the LS-DYNA simulations using the MAT_005 model (parameter identification: lab testing approach), and (2) to be used in the parameter identification using FE simulation-based optimization (approach 2). The CPTs [6] were conducted using an HS-4210 digital static cone penetrometer instrument (Humboldt, Elgin, IL, USA), a 30-degree cone with a base diameter of 0.5 in. (Figure 6). The device recorded the cone resistance force required to penetrate the ground. Soil samples were prepared and compacted to match the conditions of the triaxial and consolidation tests (Figure 6). The penetration was performed at a rate of 20 mm/s to a depth of 150 mm, and the maximum resistance experienced by the cone was recorded.

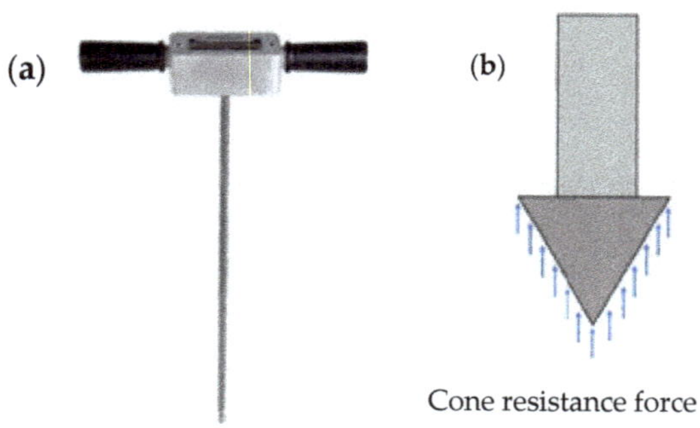

Figure 6. (**a**) CPT apparatus [17] (**b**) CPT schematics [6].

2.2.4. Numerical Setup of CPT Simulations

CPT simulations were performed using three numerical methods in LS-DYNA (version 16.0 solver ls-dyna_mpp_s_R16_0_0_x64): Multi-Material Arbitrary Lagrangian–Eulerian (MM-ALE), Smoothed Particle Hydrodynamics (SPH), and Hybrid-SPH. The cone penetrometer geometry was modeled as a rigid body with dimensions consistent with [6] and a penetration rate of 20 mm/s was applied to a depth of 150 mm. The soil was modeled using the MAT_005 material model with identified parameters. The setups for each method are detailed below, with key parameters summarized in Tables 2–4. The mesh characteristics used in parameter identification were derived from mesh convergence studies (Appendix A).

The Multi-Material Arbitrary Lagrangian–Eulerian (MM-ALE) Setup: The MM-ALE method modeled the soil as a multi-material Eulerian domain to capture large deformations during cone penetration [33,34]. The framework used a Lagrangian mesh for the cone penetrometer and an Eulerian mesh for the soil and air (defined with MAT_140_VACUUM) [35]. Soil and air domains used ELFORM = 11 (1-point ALE multi-material element) for multi-phase interactions [33]. A fine mesh (Figure 7a) ensured accurate soil-cone interaction [36]. Contact was modeled with CONSTRAIN_LAGRANGE_IN _SOLID (CTYPE = 4, DIREC = 2), active in compression, with optimized penalty stiffness (negative PFAC) and coupling forces extracted from DBFSI files, aligning with [37]. Fixed lateral and bottom boundaries (Figure 7b) prevented spurious deformations, per [38]. The setup parameters are in Table 2.

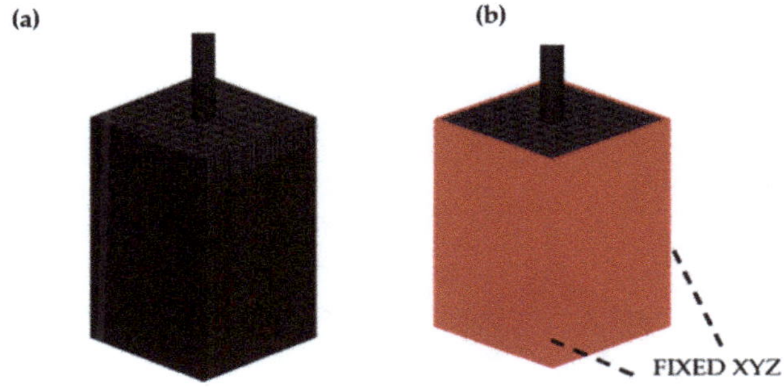

Figure 7. (**a**) MM-ALE model discretization, (**b**) MM-ALE model with boundary conditions.

Table 2. Setup parameters for MM-ALE CPT simulation.

Parameter	Value	Source
Soil Domain Size (mm)	100 × 100 × 220	Domain study (Appendix A)
Vacuum Domain Size (mm)	100 × 100 × 30	Domain study (Appendix A)
Number of Elements	300,000	Convergency study (Appendix A)
Coefficient of Friction	0.3	[19]
Cone Penetration Velocity (mm/s)	20	[39]

Table 3. Setup parameters for SPH CPT simulation.

Parameter	Value	Source
Soil Domain Size (mm)	140 × 140 × 220	Domain study (Appendix A)
Encasement Material	Steel	Domain study (Appendix A)
Encasement Thickness (mm)	5	Domain study (Appendix A)
Number of SPH Particles	350,000	Convergency study (Appendix A)
Coefficient of Friction	0.3	[19]
Cone Penetration Velocity (mm/s)	20	[39]

Table 4. Setup parameters for Hybrid-SPH CPT simulation.

Parameter	Value	Source
SPH Soil Domain Size (mm)	70 × 70 × 220	Domain study (Appendix A)
Lagrangian Domain Size (mm)	140 × 140 × 240	Domain study (Appendix A)
Number of SPH Particles	350,000	Convergency study (Appendix A)
Solid Elements	33,000	Convergency study (Appendix A)
Coefficient of Friction	0.3	[19]
Cone Penetration Velocity (mm/s)	20	[39]

SPH Setup: The Smoothed Particle Hydrodynamics (SPH) method was used to model soil granular behavior and large deformations during cone penetration, as validated by [33]. The soil domain, encased in a rigid steel modeled with solid elements, used CONTACT_AUTOMATIC_NODE_TO_SURFACE in LS-DYNA for robust soil-encasement and soil-cone interactions, following [40]. The cone penetrometer was a rigid body with solid elements. The soil was discretized with 350,000 SPH particles (uniform radius, Figure 8a) for high-resolution deformation capture. For stability, fixed boundary conditions, adapted from [41], were applied to the encasement (Figure 8b). Table 3 summarizes the setup parameters (domain size, encasement thickness, friction, penetration velocity).

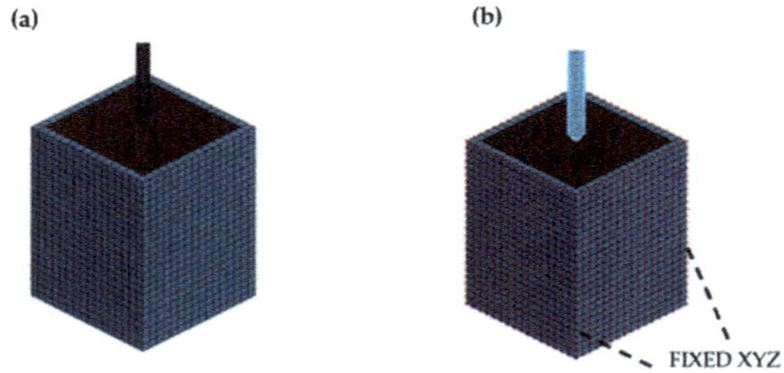

Figure 8. (**a**) SPH model discretization, (**b**) SPH model with boundary conditions.

Hybrid-SPH Setup: The Hybrid-SPH method integrates SPH particles with Lagrangian elements to balance computational efficiency and accuracy in modeling soil behavior during cone penetration, as inspired by [33]. The outer soil domain was modeled with Lagrangian solid elements (geometry per Table 3), while the inner region (60 mm $\times$ 60 mm $\times$ 220 mm) was replaced with 350,000 SPH particles (Figure 9a). The SPH and Lagrangian regions were coupled using CONTACT_AUTOMATIC_NODE_TO_SURFACE_TIE in LS-DYNA for robust interaction, following [40]. The combined soil domain interacted with a rigid cone penetrometer (solid elements) via CONTACT_AUTOMATIC_NODE_TO_SURFACE. Boundary conditions fixed the lateral and bottom surfaces of the domain (Figure 9b), aligning with [41] for stability. Table 4 summarizes setup parameters, including domain sizes, SPH particle count, friction coefficient, and penetration velocity.

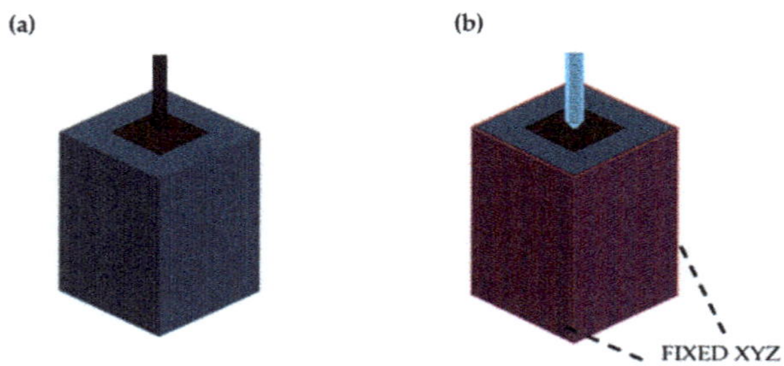

Figure 9. (**a**) Hybrid-SPH model discretization, (**b**) Hybrid-SPH model with boundary conditions.

2.3. Optimization Approach: Parameter Identification

An optimization study was conducted using LS-OPT (Ansys, Canonsburg, PA, USA) to calibrate soil model parameters against in situ CPT data [18]. The goal was to minimize the deviation of the simulated vertical force on the cone penetrometer from the experimental cone resistance force vs. penetration depth corridor, ensuring the simulated force remains within the experimental bounds. The soil was modeled using a Hybrid-Lagrangian-SPH approach, which effectively managed large deformations [6]. The cone penetrometer was modeled with a Lagrangian approach and assigned an initial velocity corresponding to a drop height of 150 mm.

2.3.1. Design Variables and Fixed Parameters

The design variables in this study include the cohesion (c), friction angle (φ), shear modulus (G), and bulk modulus for loading (K), with variable ranges derived from laboratory test data and literature for sandy loam, consistent with cohesive soil parameters from [32] (Table 5). The yield function constants A_0, A_1, *and* A_2 are then computed as dependent variables using expressions derived using the values of c and φ (Equations (26)–(28)). These ranges were estimated by varying c and φ within their defined limits, and the dependent variables are dynamically updated in the LS-DYNA simulation as ϕ and c change during optimization.

Table 5. Design variables used for sensitivity study (units: mm-s-tone).

Design Variable	Description	Initial Value	Lower Bound	Upper Bound	Source
G	Shear modulus of the soil	3.23	3.0	6.00	[32]
K	Bulk modulus of the soil	8.17	7.00	15.00	[42]
c	Cohesion	0.0213	0.018	0.03	[43]
φ	Friction angle	42.5°	42.0°	71.4°	[44]

2.3.2. Optimization Methodology

A Latin Hypercube Sampling (LHS) Design of Experiments (DOE) scheme was employed to generate 43 sampling points, selected based on a LS-Opt guideline of approximately 10–11 samples per design variable (here, four variables: cohesion, friction angle, shear modulus, bulk modulus) to ensure sufficient coverage for an accurate elliptical metamodel while minimizing computational expense [32,42–44]. LS-DYNA simulations were conducted corresponding to a maximum penetration depth of 150 mm

The composite objective function quantifies the deviation of the simulated vertical force from a target corridor of 150–200 N, derived from the mean range (± standard deviation) of multiple in situ CPT tests on sandy loam soil (0.4% moisture) as reported in [6], starting from a penetration depth of 50 mm to 150 mm in 5 mm increments, and is expressed as:

$$err = \sum(CTP\_fotrce(d) - 175)^2$$

No additional constraints are imposed, providing flexibility to align the simulated force within the specified corridor. A final Cone Penetration Test (CPT) simulation was performed in LS-DYNA using the optimized soil model parameters (variables c, φ, G, K and dependent variables A_0, A_1, A_2) to check the calibrated model against the experimental force corridor.

2.4. Case Study: Soil Compaction Simulation Using Hybrid FEM-SPH

This section describes the rigid tire geometry discretization, model preparation, and numerical setup used to simulate agricultural soil compaction under wheel loading with the identified MAT_005 parameters. The simulations focus on a 3 kN normal load with slip ratios ranging from −10% to 30% to enable a comprehensive comparison with the experimental results for soil-tire interaction.

2.4.1. Tire Geometry and FE Discretization

The geometry and FE model of the non-pneumatic tire used in this study were prepared in three steps. First, the tire was scanned with an 3D Einstar handheld scanner to capture tread lugs and shoulder details. The reconstructed geometry was cleaned (hole filling, smoothing only on non-critical surfaces) and rebuilt into a watertight solid in Altair HyperMesh. Finally, the tire model was discretized (Table 6) with hexahedral elements and exported for use in LS-DYNA where it was modeled as rigid to reduce computational time and focus on soil response.

Table 6. Tire geometry and FE discretization.

Parameter	Dimensions (mm)
Tire diameter	720
Tire width	254
Lug height	37
Lug width	45
Number of Lugs	22
Hex mesh size	10
Total elements	50,632

2.4.2. Numerical Simulation Setup

To investigate the soil compaction under the tire of agricultural machinery, the soil was modeled using a Hybrid FEM–SPH soil approach, which captures large deformation beneath the tire while keeping the far field efficient. The tire acts as a rigid body with prescribed translation and rotation to realize the target slip ratio [45]. For both soil material parameter sets (lab-based and optimized), each simulation follows the same three-stage procedure for the normal load and various slip ratios. First, gravity is applied to the soil domain to establish in-situ stresses and allow settlement. Second, the prescribed normal load is ramped onto the rigid tire, and contact is stabilized. Third, longitudinal translation V_y and rotation ω are imposed simultaneously to realize the target driving slip s using $\omega = \frac{v_y}{R(1-s)}$.

All model settings used in these steps—including soil domain size, contact friction, V_y, s, applied normal load, and the resulting ω are summarized in Table 7. The boundary condition and domain discretization are also shown (Figure 10). The contact area, normal and shear contact forces, net traction, and tire sinkage history plots are extracted from the numerical simulations.

Table 7. Setup parameters for Soil-Rigid Tire Interaction Simulation.

Parameter	Value	Source
SPH Soil Domain Size (mm)	2220 × 320 × 200	Domain study (Appendix A)
Lagrangian Domain Size (mm)	2200 × 750 × 240	Domain study (Appendix A)
Number of SPH Particles	500,000	Convergency study (Appendix A)
Solid Elements	300,000	Convergency study (Appendix A)
Applied Longitudinal Velocity, V_y	138 mm/s	[45]
Friction coefficient soil-tire	0.23	[45]
Target slip ratios (driving)	−10% to 30%	[45]
Applied normal load on the tire	3 kN	[45]

The simulation evaluates key performance indicators in tire–soil interaction, including net traction, tire sinkage, stress distribution, and plastic strain, with a focus on comparing the predictive accuracy of lab-tested soil parameters (Soil-A) and optimized parameters (Soil-B) under a 3 kN normal load across slip ratios from −10% to 30%. Traction represents the contribution of the motion resistance force governed by soil shear strength and the friction force at the tire–soil interface. For the driving mode (positive slip ratio), friction contributes positively to net traction, while its sensitivity to slip ratio remains moderate.

Motion resistance increases with shear strain until reaching peak soil strength, after which it stabilizes, influencing the net traction trend.

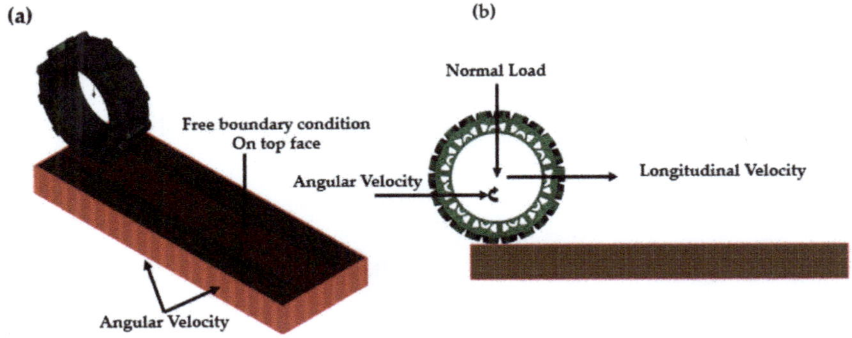

Figure 10. (**a**) Hybrid-SPH model discretization for Tire-soil Interaction, (**b**) Hybrid-SPH model for Tire-soil Interaction with boundary conditions.

3. Results and Discussion

3.1. Lab Testing Approach: Identified Parameters and Simulation Results

This section presents the results of the laboratory-based approach for identifying and verifying the parameters of the MAT_005 model in LS-DYNA, tailored for sandy loam soil at 0.4% moisture content. Subsequent numerical simulations of triaxial and cone CPT tests verify these parameters [6].

3.1.1. Identified Parameters of Soil Constitutive Material Model

Following the methodology outlined previously (Figure 5), the parameters for the MAT_005 model in LS-DYNA were derived from triaxial and consolidation tests (the laboratory testing approach) (Table 8). The hardening behavior, defined by volumetric strain (ε1, ε2, ...) and corresponding pressures (P1, P2, ...), were obtained from the consolidation test data (Figure 10). These parameters provide a robust foundation for the MAT_005 model to simulate the elastic-plastic response of sandy loam soil in triaxial and cone penetrometer test (CPT) simulations, as discussed in subsequent sections. The parameters effectively capture the soil's mechanical behavior, although the model's limitations, such as the lack of strain softening, may influence its performance in specific failure scenarios.

Table 8. Identified parameters for MAT_005 at 0.4% moisture content (units mm-s-tone).

Parameter	Description	Value	Source
ρ	Mass density	1845.3×10^{-9}	Triaxial Test
G	Shear Modulus	4.23	Calculated from E and μ
K	Bulk modulus	9.17	Calculated from E and μ
A_0	Yield function constant	0.0003566	Triaxial Test
A_1	Yield function constant	0.05279	Triaxial Test
A_2	Yield function constant	1.905	Triaxial Test
P_C	Pressure cutoff for tensile fracture (<0)	−0.0328	Adapted from Cap Hardening
VCR	Volumetric crushing option	0.0 (on)	
REF	Use reference geometry to initialize pressure	0.0 (off)	Triaxial Test
$\varepsilon_1 \ldots$	Volumetric strain values	See Figure 11	Triaxial Test
$P1 \ldots$	Pressure corresponding to volumetric strain	See Figure 11	Triaxial Test

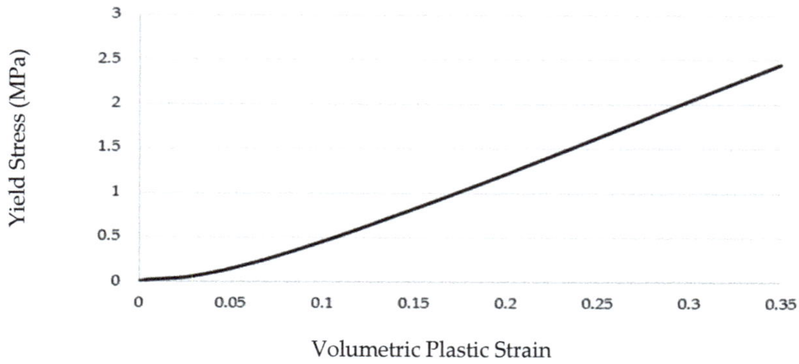

Figure 11. Consolidation compression test data for Sandy Loam [6].

3.1.2. Triaxial Test Simulation Results

The triaxial test simulations were performed to verify the identified parameters of the MAT_005 model at a confining pressure of 5 psi, matching the experimental conditions reported [6]. The deviator stress versus axial strain plot (Figure 12) indicates that the MAT_005 model captures the brittle failure behavior observed in the experimental results, consistent with the CP model's performance in our previous study in Abaqus [6]. However, the MAT_005 model underpredicts the peak deviator stress by approximately 6% compared to the experimental mean, with experimental data showing a standard deviation of 0.038, Cap-Plasticity [6] at 0.050680432, and MAT_005 at 0.052323652, and overall simulation errors ranging from 0–6%. This underprediction is attributed to the simplified yield function in MAT_005, which lacks strain-hardening and softening mechanisms, unlike the CP model that better captures complex behavior in the cap region [5,6]. Despite this, the overall stress–strain trend aligns well with the experimental data, confirming the reliability of the identified parameters for simulating triaxial test conditions in LS-Dyna.

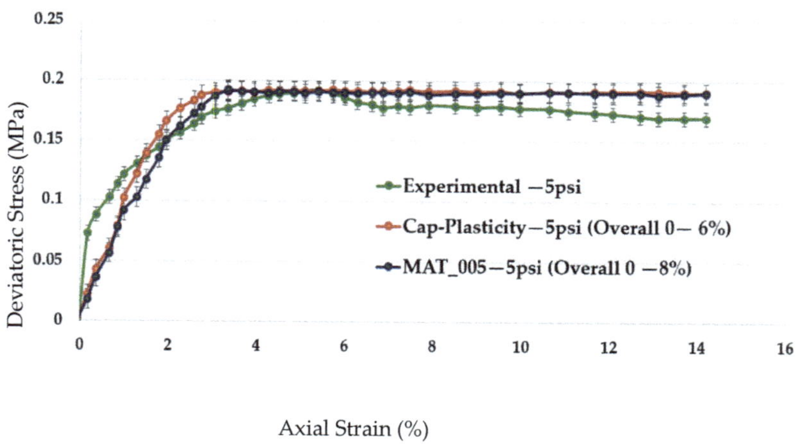

Figure 12. Numerical stress-strain results of MAT_005 vs. experimental and CP model [6] results at 5 psi confining pressure (0.4% moisture content).

3.1.3. Initial CPT Simulation Results (with Experimental Data)

The cone resistance force as a function of penetration depth, obtained from CPT simulations using the MM—ALE, SPH, and Hybrid—SPH numerical methods with the MAT_005 model, was compared against experimental data and CP model results from a previous study [6] results. The results reveal distinct performance differences among the methods. The MM—ALE results of the MAT_005 model are compared with the results of another meshed method (CEL) [6], which used the CP model (Figure 13a). The slightly better performance of MM—ALE (15% error) over CEL (20% error) may be attributed to differences in contact formulations and mesh handling between LS—DYNA and Abaqus. The Hybrid-SPH method provided the best balance of accuracy and computational efficiency. The cone resistance force versus penetration depth obtained with the SPH method and the MAT_005 model (this study) was compared with the SPH results with the CP model [6] (Figure 13b). Both meshless models predict cone resistance forces within the experimental corridor, with the MAT_005 model achieving a 5% error and the CP model a 2% error, as reported by [6]. The close agreement suggests that the MAT_005 model, when paired with the SPH method, effectively captures the granular behavior of sandy loam soil during penetration.

Finally, the cone resistance force across all three numerical methods (MM—ALE, SPH, and Hybrid—SPH) in LS-Dyna was compared with the experimental force corridor (Figure 13c). The meshless methods (SPH and Hybrid—SPH) predict forces within the experimental corridor, with 5% and 4% errors, respectively. In contrast, the meshed techniques, such as the MM—ALE method, predict the resistance force with a 15% error. This discrepancy is likely due to MM—ALE's mesh-based nature, which struggles to capture localized granular behavior compared to the meshless SPH methods.

The stress and strain distributions during the CPT simulations provide insight into the soil's behavior under penetration. The contour plot of the equivalent elastic-plastic strain for the MM—ALE (Figure 14a), SPH (Figure 14b), and Hybrid—SPH (Figure 14c) methods is presented (Figure 14). The SPH and Hybrid-SPH methods exhibit more localized plastic strain around the cone tip, consistent with the granular nature of sandy loam soil, as noted by [6]. In contrast, the MM—ALE method shows a more diffuse distribution of strain, which aligns with its tendency to overpredict cone resistance force due to less accurate modeling of localized deformation. The contour plot of the stress in the Z-direction (S33) for the three methods was also presented (Figure 15). The SPH (Figure 15b) and Hybrid-SPH (Figure 15c) methods predict a more concentrated stress bulb beneath the cone tip than the MM-ALE (Figure 15a) method, indicating better soil compaction and contact area representation at the cone–soil interface.

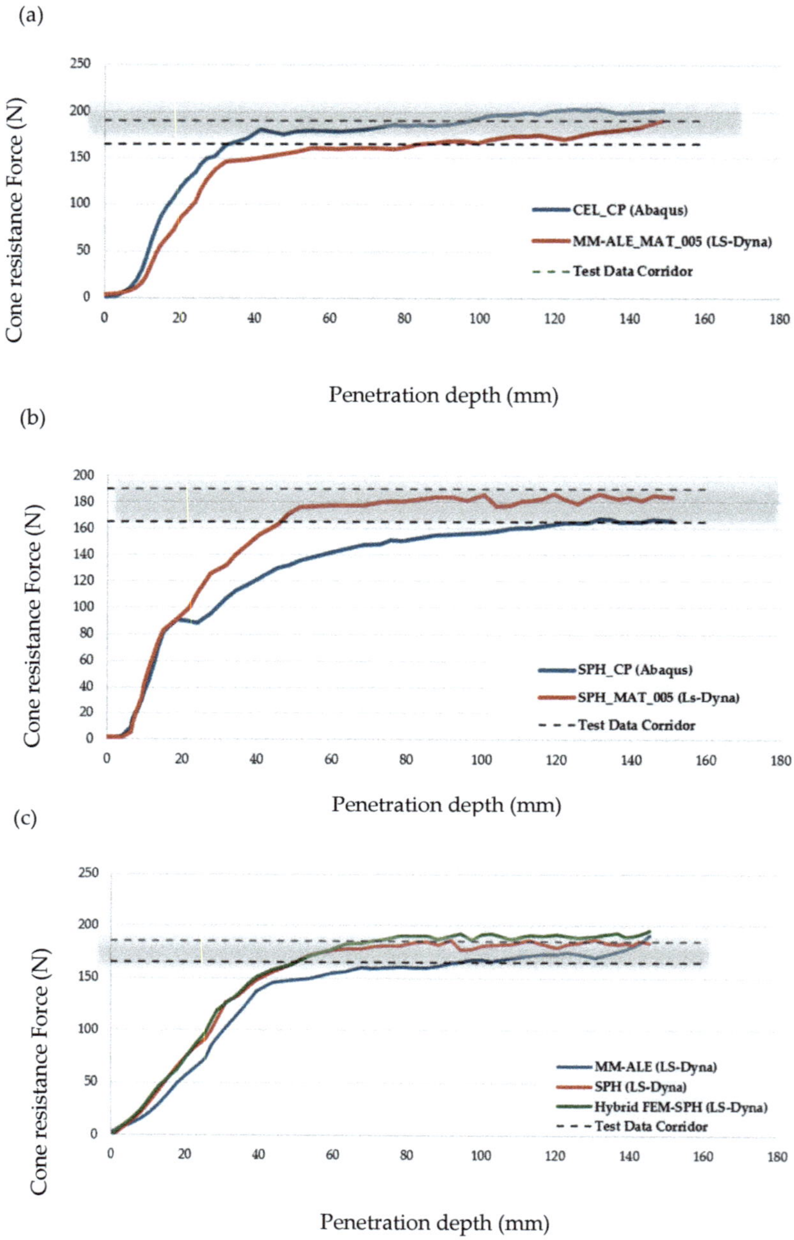

Figure 13. Initial numerical results of MAT_005_SOIL vs. experimental and CP [6] results for CPT simulation at 0.4% moisture content: (**a**) MM—ALE (MAT_005) vs. CEL (CP model). (**b**) SPH method comparison with the CP model. (**c**) All methods (MM—ALE, SPH, Hybrid—SPH) vs. experimental corridor.

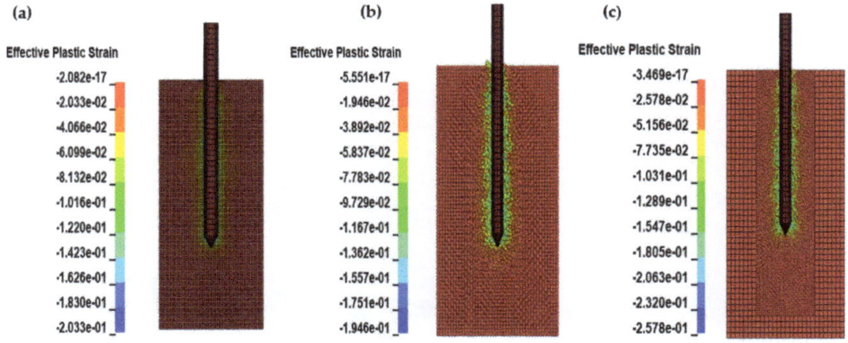

Figure 14. Contour plot of equivalent elastic-plastic strain for CPT simulation (MAT_005 model, 0.4% moisture content) using (**a**) MM—ALE (**b**) SPH, and (**c**) Hybrid—SPH methods.

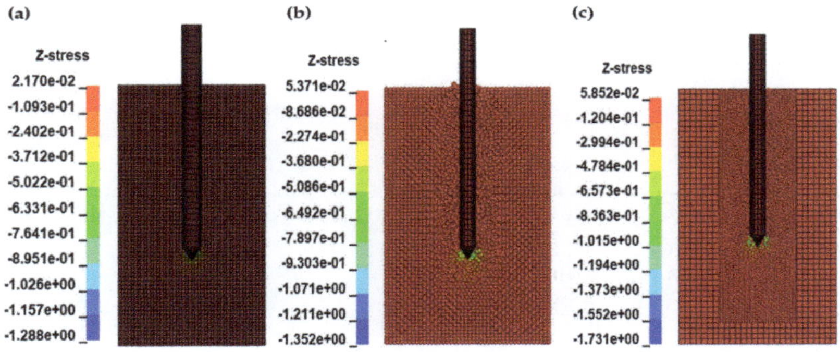

Figure 15. Contour plot of stress in the Z—direction (S33) for CPT simulation (MAT_005 model, 0.4% moisture content) using (**a**) MM—ALE (**b**) SPH, and (**c**) Hybrid—SPH methods.

3.2. Optimization Approach: Identified Parameters and Sensitivity Analysis

3.2.1. Optimization Results

The sensitivity analysis (Figure 16) revealed that the shear modulus (G) was the most influential parameter on the composite output (force—penetration response described in Section 2.2.2), contributing 90.6% to the variation in the cone resistance force. This was followed by bulk modulus (K) with 6.4% influence, the angle of friction parameter (φ) with 2.4%, and cohesion (c) with a minimal impact of 0.6%.

The optimized parameters (Table 9) included a 0.057% increment in A_1, a 0.168% reduction in A_0, a 1.229% increase in G, and a 0.065% reduction in K. Additionally, φ was reduced by 8.017%, c by 12.8%, and A_2 by 11.916%. These parameter adjustments significantly improved the model's ability to capture the peak cone resistance force, enhancing the agreement between the finite element simulation and the experimental force corridor across the 150 mm penetration depth. As a result, the FEM—SPH simulation error decreased from 5% to 2%, as detailed in Section 3.2. The optimization results demonstrated a significantly improved fit to the experimental force corridor, with the optimized cone resistance force consistently lying within the corridor across the 150 mm penetration depth. These results validated the sensitivity analysis for sandy loam soil at 0.4% moisture content, accurately reflecting in situ behavior, particularly for granular soils where localized effects drive penetration resistance (Section 3.1.3) The optimized soil parameters, also listed in the table below, will be used for the final part of the paper's objective, involving a comprehensive CPT simulation employing MM—ALE, SPH, and Hybrid FEM—SPH methods to compare their performance and ensure high-fidelity representation of the soil behavior.

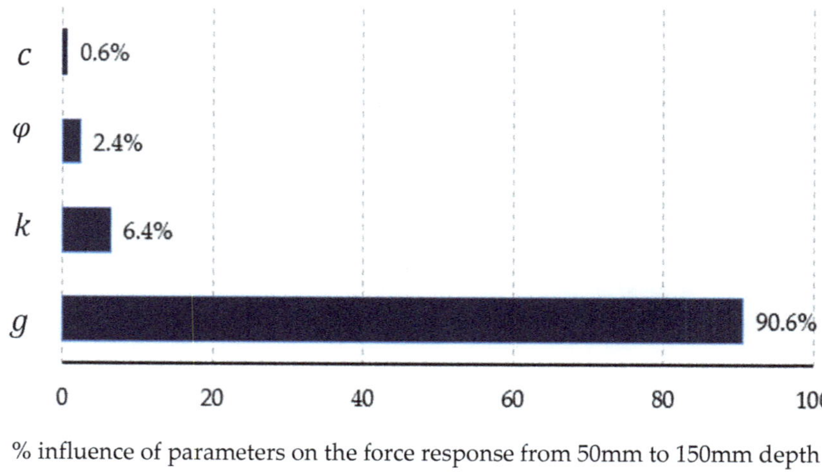

Figure 16. Sensitivity analysis and optimization results: Influence of parameters on cone resistance force corridor.

Table 9. Optimized parameters vs. experimental parameters for MAT_005 at 0.4% moisture.

Parameter	Description	Approach 2	Approach 1	Diff. (%)
G	Shear Modulus	4.283	4.231	+1.2
K	Bulk modulus	9.167	9.173	−0.065
φ	Friction angle	54.73°	59.5°	−8
c	Cohesion	0.02006	0.023	−12.8
A_0	Yield function constant	0.000356	0.0003566	−0.2
A_1	Yield function constant	0.05276	0.05279	−0.1
A_2	Yield function constant	1.678	1.905	−11.916

3.2.2. Final CPT Simulation Results with Optimized Parameters

Final CPT simulations were conducted with the optimized MAT_005 parameters using MM—ALE, SPH, and Hybrid FEM—SPH methods. Force penetration curves for all three methods using optimized parameters align with the experimental corridor across the 150 mm penetration depth (Figure 17). The Hybrid FEM—SPH method achieved the highest accuracy, with an error of only 2% compared to the experimental corridor, significantly improving from the initial 4% error as shown in (Figure 18). This result was expected because this method was used in the optimization process. The SPH method followed with a 3% error, improved from the initial 5% error, aligning closely with the CP model [6], which achieved a 2–3% error range [3]. The MM—ALE method showed an error of 8%, a notable improvement from the initial 15% error (Section 3.1.3), yet it still underperformed compared to the CEL method (20% error) and lagged the SPH-based approaches due to their mesh-based limitations in modeling localized deformations, as discussed in Section 3.1.3.

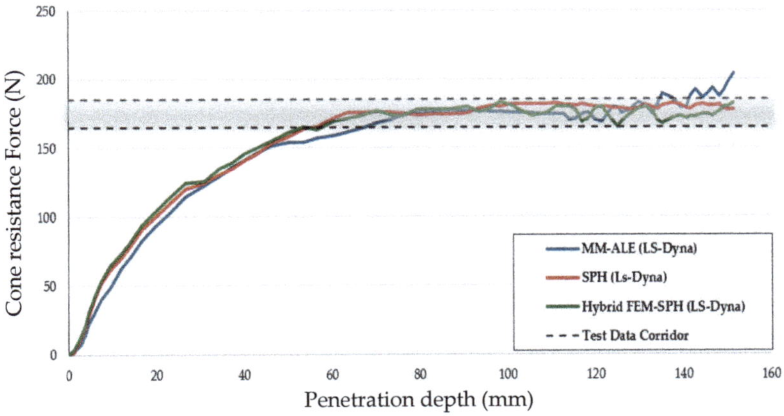

Figure 17. Final force penetration curves for CPT simulations with optimized parameters using MM-ALE, SPH, and Hybrid FEM-SPH methods, compared to the experimental corridor and CP model results [6] at 0.4% moisture content.

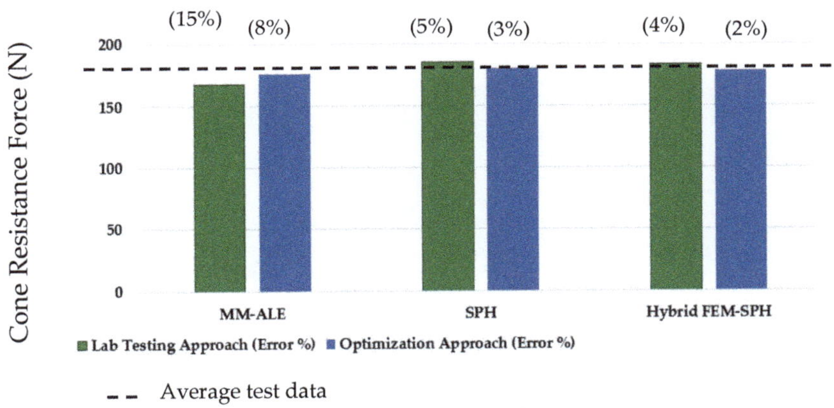

Figure 18. Error comparison of initial and optimized CPT simulations at 0.4% moisture content using MM—ALE, SPH, and Hybrid FEM—SPH methods, compared to the experimental average error [6].

To further highlight the impact of parameter optimization, the percentage errors of the cone resistance force for the initial and optimized simulations across all three methods were compared (Figure 18) alongside the experimental average error range. Comparing the initial and optimized parameter simulations, the optimization process significantly enhanced accuracy across all methods. The MM—ALE method's error decreased from 15% to 8%, the SPH method's error reduced from 5% to 3%, and the Hybrid FEM—SPH method's error improved from 4% to 2%. The optimized curves for SPH and Hybrid FEM—SPH now lie almost entirely within the experimental corridor (Figure 18), showing a tighter fit than the initial simulations, where the curves occasionally deviated at higher penetration depths (beyond 100 mm). The MM—ALE method, while improved, still tends to overpredict the force, especially at deeper penetration depths, consistent with its initial behavior but with a reduced magnitude of error. The experimental average error, ranging between 2% and 3%, serves as a benchmark, confirming that the optimized Hybrid FEM—SPH and SPH methods now match or exceed this accuracy level. This visual comparison underscores the effectiveness of the optimization process in refining the MAT_005 model parameters, particularly for the meshless methods, which now exhibit errors within or below the experimental benchmark.

These results validate the proposed methodology and highlight the potential of LS-DYNA for high-fidelity soil–tire/tool interaction simulations in agricultural applications. The findings align with previous observations. on the advantages of meshless methods, especially the Hybrid FEM-SPH approach, which showed a good balance of precision and computational efficiency, making it a promising method for future studies.

3.2.3. Soil-Tire Interaction Simulation Results

The traction response of the tire–soil system across slip ratios from −10% to 30% is shown in Figure 19. The experimental data demonstrate traction increasing from −600 N at −10% slip to approximately 800 N at 30% slip. Soil-A consistently underpredicts traction, reaching 800 N at 30% slip, while Soil-B closely follows the experimental curve, predicting ~724 N at the same slip level.

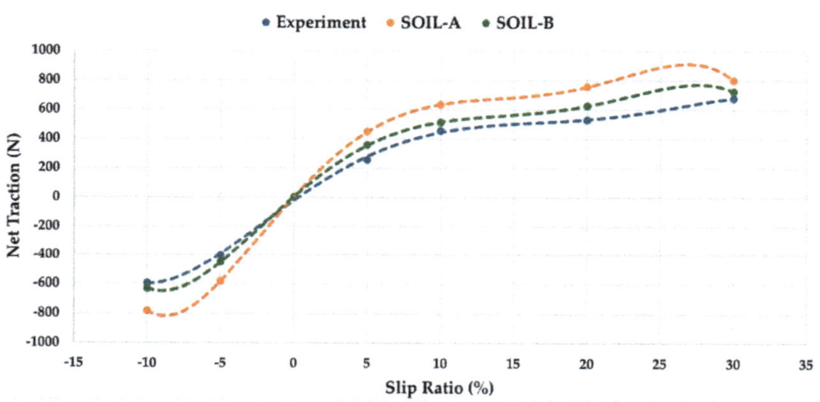

Figure 19. Plot of Net traction against Slip Ratio.

Quantitative error analysis highlights the advantage of Soil-B over Soil-A. Across the full slip range, Soil-A exhibited a root-mean-square deviation (RMSE) of 60 N and a MAPE of 10%, while Soil-B achieved significantly lower errors, with an RMSE of 20 N and a MAPE of 4.2% relative to experimental values.

This reduction in error demonstrates that the optimized Soil-B parameters better capture the nonlinear traction–slip behavior and prevent underestimation of drawbar pull, particularly at higher slip ratios. Accurate traction prediction is critical for agricultural applications, as it directly affects fuel efficiency, soil disturbance, and the development of optimized tire designs for sustainable field operations [46].

The tire sinkage response across slip ratios from −10% to 30% is shown in Figure 20. All three datasets Experiment, Soil-A, and Soil-B—capture the initial compaction phase followed by progressive rut deepening as slip increases. The experimental results indicate sinkage increasing from −7.5 mm at −10% slip to −12.5 mm at 30% slip. Soil-A consistently underestimates sinkage, predicting a final depth of −13.5 mm, while Soil-B provides a closer match with −12.5 mm at 30% slip.

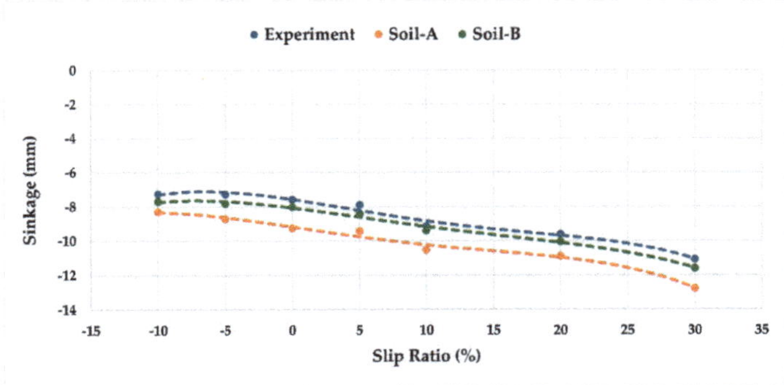

Figure 20. Plot of Tire Sinkage (mm) against slip ratio.

Error analysis further confirms the improved accuracy of the optimized Soil-B parameters. Across the slip ratio range, Soil-A exhibited an RMSE of 0.9 mm and a MAPE of 10%, whereas Soil-B achieved substantially lower errors, with an RMSE of 0.3 mm and a MAPE of 3.5% relative to experimental data.

The improved predictive performance of Soil-B demonstrates its capability to capture soil deformation trends more reliably under operational loading. This is critical in agricultural applications, as excessive rut depth directly correlates with soil compaction, reduced porosity, impaired water infiltration, and potential yield losses of 20–50% depending on soil type and severity [17,38]. By more accurately representing soil deformation, Soil-B supports better evaluation of field traffic impacts and enables tire design and load management strategies that protect long-term soil productivity.

The comparison of normal stress (Figure 21), shear stress (Figure 22), effective plastic strain (Figure 23), and rut profiles (Figure 24) for the lab-tested soil parameters (Soil-A) and optimized parameters (Soil-B) under a 3 kN normal load and 0% slip ratio, benchmarked against experimental 3D-scanned rut profiles [45]. In the standard stress contour plots (Figure 21), both Soil-A and Soil-B captured the stress bulb formation beneath the tire—a key indicator of soil compaction and load transfer in off-road agricultural applications [46,47] but Soil-B produced a distribution and magnitude that more closely matched experimental observations, particularly in the uniformity and depth of the stress bulb. The shear stress plots (Figure 22) further revealed that Soil-B's contours aligned more closely with the experimentally observed slip plane, with a more realistic distribution of tensile stresses in front of the tire and compressive stress behind it, improving the prediction of traction efficiency and energy loss during field operations [48]. In the effective plastic strain plots (Figure 23), Soil-B better reproduced the depth and lateral spread of permanent deformation zones, closely matching the experimental rut shapes, which is vital for assessing cumulative soil compaction effects that can reduce porosity and crop yield [49,50]. Finally, the rut profile comparison (Figure 24) confirmed that Soil-B more accurately replicated rut depth, width, and lug imprint geometry, while Soil-A underestimated both deformation depth and the lateral soil displacement around the rut. Overall, across all four figure sets, the improved agreement of Soil-B with experimental data highlights the advantage of optimized parameters in predicting soil response, making them more reliable for evaluating tire performance and developing strategies to mitigate harmful compaction in agricultural operations.

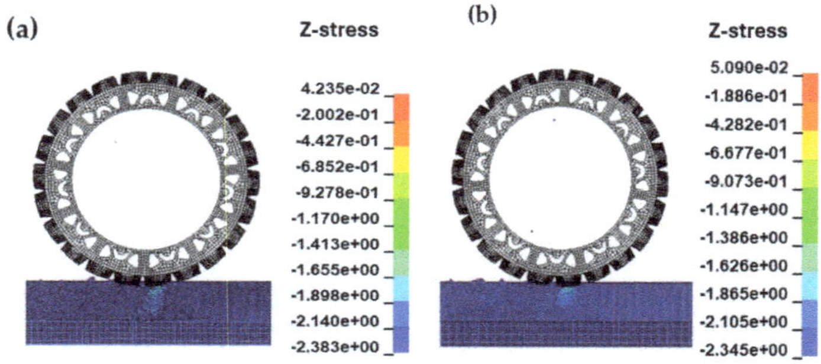

Figure 21. (**a**) Contour plot for normal stress distribution for soil from Lab—Testing Method. (**b**) Contour plot for normal stress distribution for soil from Optimized parameter.

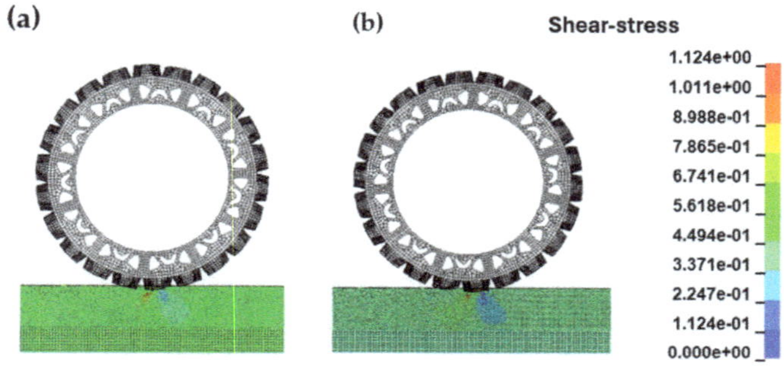

Figure 22. (**a**) Contour plot of shear stress for soil from Lab—Testing Method. (**b**) Contour plot of shear stress for soil from Optimized parameter.

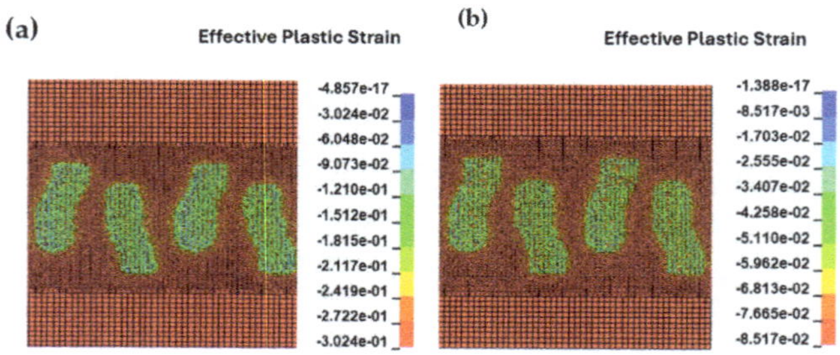

Figure 23. (**a**) Contour plot for effective plastic strain for soil from Lab—Testing Method. (**b**) Contour plot for effective plastic strain for soil from Optimized parameter.

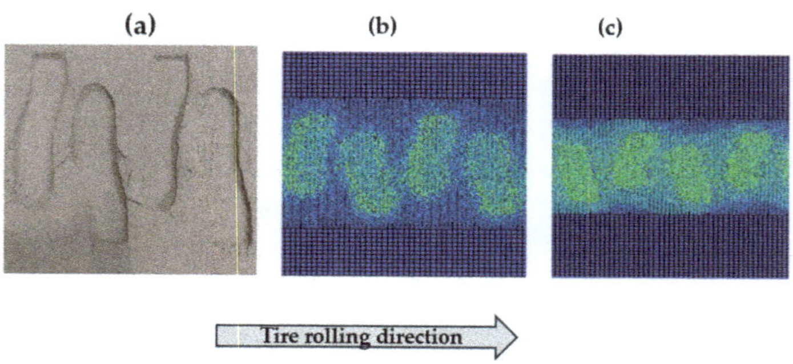

Figure 24. (**a**) Tire rut for 3 kN load at 0% slip. (**b**) Contour plot tire rut for 3 kN load at 0% SR (Lab-Testing Method). (**c**) Contour plot for Tire rut for 3 kN load at 0% slip (Optimized parameter).

3.2.4. Comparative Analysis and Limitations

This study adopts two complementary methodologies for identifying the material parameters of sandy loam soil at 0.4% moisture content in the MAT_005 model for LS-DYNA: a conventional laboratory testing approach and a computational optimization-based approach. As Sections 2.2 and 3.1 outlined, the laboratory approach relied on triaxial and consolidation tests to determine key parameters such as cohesion, friction angle, shear modulus, bulk modulus, and volumetric strain–pressure relationships. These experimentally derived values provided a solid baseline for initial simulations, capturing fundamental soil mechanical behavior under controlled conditions [6]. However, despite their physical reliability, laboratory-derived parameters exhibited limitations when applied directly to complex in situ conditions. The initial Cone Penetrometer Test (CPT) simulations (Section 3.1.3) showed higher prediction errors 15% for MM-ALE, 5% for SPH, and 4% for Hybrid FEM–SPH, indicating that laboratory data alone may not fully capture site-specific variability, boundary effects, and the natural heterogeneity of agricultural soils, as similarly reported in agriculture studies on soil compaction variability [1].

The optimization-based approach using LS-OPT (Sections 2.3 and 3.2) was introduced to address these shortcomings by calibrating the MAT_005 parameters directly against in situ CPT data. This inverse analysis method allowed targeted adjustment of model parameters to match field-measured penetration resistance within a defined corridor, significantly improving predictive accuracy for real-world conditions. Sensitivity analysis identified the shear modulus as the dominant contributor to the force–penetration response, followed by bulk modulus, friction angle, and cohesion. The optimization process produced measurable improvements in CPT simulation accuracy: Hybrid FEM–SPH achieved the lowest error at 2%, SPH followed closely at 3%, and MM-ALE improved to 8%—a notable gain over its pre-optimization performance but still limited by mesh-based representation of granular flow. These findings reinforce earlier conclusions in meshless soil modeling research [1,46] that SPH-based methods excel in replicating localized deformation and stress transmission in granular media.

A closer examination of stress and strain distributions (Figures 13 and 14) highlighted key differences in the numerical approaches. SPH and Hybrid FEM–SPH methods accurately localized plastic strain and stress concentration beneath the cone tip, consistent with observed granular soil mechanics and in line with results reported by and agriculture-focused soil compaction studies [26,47]. In contrast, MM-ALE produced more diffuse stress and strain patterns, which may explain its tendency to overpredict cone resistance at deeper penetrations. These results underscore the importance of selecting numerical methods that align with the physical characteristics of the target soil and the scale of deformation being modeled.

For large soil deformations, the Coupled Eulerian–Lagrangian (CEL) method [6] typically offers higher computational efficiency (e.g., shorter runtimes due to reduced computational volume) but struggles with granular behavior, yielding 20% errors in CPT simulations. In contrast, SPH in LS-Dyna provides superior accuracy (2–5% error) for granular soils but is computationally intensive (e.g., 22 h and high memory for 350,000 particles in our CPT simulation). The Multi-Material Arbitrary Lagrangian–Eulerian (MM-ALE) method takes less computational time (e.g., 16 h with 300,000 elements), while Hybrid-SPH balances accuracy and efficiency, completing simulations by combining SPH and Lagrangian elements [9] (e.g., in 14 h in our CPT simulation).

While validated for sandy loam, the methodology's applicability to other soil types, such as clay (higher cohesion, lower permeability) or silt (intermediate granularity), warrants consideration. In clays, parameter identification may be more sensitive to moisture due to pronounced pore pressure effects, potentially increasing optimization iterations and errors (e.g., 5–10% higher than sandy loam) without strain softening. The Hybrid-SPH method could excel in capturing mixed granular-cohesive behavior for silts, but field CPT data calibration would need adjustment for texture-specific yield surfaces [3,5]. Future studies should test these adaptations to broaden the model's utility across soil textures.

The benefits of optimized parameters extended beyond penetration simulations to the soil–tire interaction case study (Section 3.2.3). Under identical loading and slip conditions, optimized parameters (Soil-B) consistently produced results closer to experimental benchmarks than lab-based parameters (Soil-A). Notably, Soil-B predicted tire sinkage at 13.8 mm—within 0.6 mm of measured 13.2 mm, while Soil-A underestimated at 12 mm. Similar improvements were seen in net traction prediction, stress bulb formation, shear band development, and rut geometry reproduction, reflecting a better alignment with experimental 3D rut profiles. These enhancements are consistent with findings from [46,48] that accurate soil–tire interaction models are critical for evaluating field traffic impacts, optimizing wheel load management, and mitigating soil compaction in precision agriculture.

Despite these improvements, the MAT_005 model retains inherent limitations. Its simplified elastic–perfectly plastic yield surface lacks strain hardening and softening mechanisms, limiting its ability to model progressive failure and post-peak softening in real soils under cyclic or repeated loading. This is evident in the triaxial simulations (Section 3.1.2), where peak deviator stress was slightly underpredicted compared to both experimental data and the Cap Plasticity (CP) model results [26]. The study's focus on a single moisture content (0.4%) restricts its applicability across the full range of moisture conditions encountered in agricultural soils. As demonstrated in [5,6], soil compaction response can vary significantly with moisture content, texture, and structure, necessitating broader experimental coverage. Furthermore, the available CPT dataset was limited to peak cone resistance values without continuous force–depth profiles, constraining detailed validation of simulation curves.

Overall, the optimized MAT_005 model combined with SPH and Hybrid FEM–SPH methods demonstrated predictive capabilities comparable to the CP model for granular soils while maintaining compatibility with LS-DYNA's advanced numerical frameworks. These results validate the integration between field-calibrated parameter identification and meshless methods for accurate, computationally efficient modeling of soil compaction under agricultural traffic. Future research should aim to integrate strain softening into the MAT_005 formulation, extend the methodology to other soil types (including cohesive clays and partially frozen soils), explore moisture-dependent parameterization, and employ high-resolution CPT equipment for capturing continuous penetration resistance profiles. Such developments would expand the applicability of this workflow in precision agriculture, enabling improved design of low-compaction machinery and more sustainable soil management practices [46,47].

4. Conclusions

In this study two methodologies for identifying parameters of the MAT_005 soil model in LS-Dyna, focusing on soil-tire interaction simulations for agricultural applications, were developed and compared. The soil was represented by sandy loam soil (0.4% moisture content), but these methodologies could be applied to other soils as well. The laboratory-based approach, using triaxial and consolidation tests from [6], provided a robust baseline but showed limitations in capturing in-situ soil variability, with initial CPT simulation errors of 15% (MM-ALE), 5% (SPH), and 4% (Hybrid-SPH). The optimization-based approach, leveraging in-situ CPT data with optimization algorithms, significantly improved accuracy, reducing errors to 2% for Hybrid FEM-SPH and 3% for SPH, outperforming lab-based results and matching or exceeding the Cap Plasticity model's performance (2–3% error) [6].

In addition, optimization approach avoids expensive and time-consuming soil lab tests. Sensitivity analysis identified shear modulus as the dominant parameter (90.6% influence), guiding precise calibration.

In soil-tire interaction simulations, optimized parameters (Soil-B) predicted net traction (~724 N vs. approximately 678 N experimental at 30% slip) and tire sinkage (12.5 mm vs. 11.1 mm experimental at 30% slip) with higher fidelity, closely matching experimental rut profiles and stress distributions. The Hybrid FEM-SPH method balanced accuracy and computational efficiency (14 h runtime vs. 22 h for SPH), making it ideal for high-fidelity soil modeling. Quantitative error analysis across the full slip range from −10% to 30% showed Soil-A with an RMSE of 60 N and 0.9 mm, and a MAPE of 9.1% and 9.0% for traction and sinkage, respectively, while Soil-B achieved significantly lower errors with an RMSE of 20 N and 0.33 mm, and a MAPE of 4.2% and 3.5% for traction and sinkage, respectively.

For practical implementation in agricultural machinery design, follow these steps: (1) conduct in situ CPT tests to capture seasonal moisture variations and establish force corridors; (2) calibrate MAT_005 parameters quarterly using the optimization approach, prioritizing shear modulus and validating against CPT data; (3) simulate wheel loads with Hybrid-SPH to evaluate compaction, iterating tire parameters (e.g., pressure, tread) for low-impact designs, reducing field trial needs. This workflow promotes sustainable practices. Despite limitations (e.g., lack of strain softening) in soil material model (MAT_005) and focus on a single moisture content, the approach demonstrates practical value for sustainable soil management. Future work should incorporate strain softening, explore varied soil types and moisture levels, and use high-resolution CPT data for broader applicability in precision agriculture.

Author Contributions: Conceptualization, A.S. and C.U.; methodology, A.S. and C.U.; software, A.S.; validation, A.S., D.J. and C.U.; formal analysis, A.S.; investigation, A.S.; data curation, A.S. and D.J.; writing—original draft preparation, A.S. and D.J.; writing—review and editing, C.U. and A.S.; visualization, A.S. and D.J.; supervision, C.U.; project administration, C.U.; funding acquisition, C.U. and A.S. All authors have read and agreed to the published version of the manuscript.

Funding: This research received no external funding.

Institutional Review Board Statement: Not applicable.

Data Availability Statement: All experimental data are provided in our previous paper [6].

Conflicts of Interest: The authors declare that they have no conflicts of interest.

Abbreviations

The following abbreviations are used in this manuscript:

ASTM	American Society for Testing and Materials
CP	Cap Plasticity model
CPT	cone penetrometer tests
DOE	Design of Experiments
MAPE	Mean absolute percentage error
MM-ALE	Multi-Material Arbitrary Lagrangian–Eulerian
RMSE	Root-mean-square deviation
SPH	Smoothed Particle Hydrodynamics
FEM	Finite Element Methods
CEL	Combine Eulerian Lagrangian

Nomenclature

The following symbols are used in this manuscript:

ρ	Mass density
E	Young's modulus
G	Shear modulus
K	Bulk modulus
c	Material Cohesion
φ	Material Friction Angle
A_0	Yield function constant (cohesion parameter)
A_1	Yield function constant (friction parameter)
A_2	Yield function constant
P	Hydrostatic stress (pressure)

Appendix A Convergence Studies

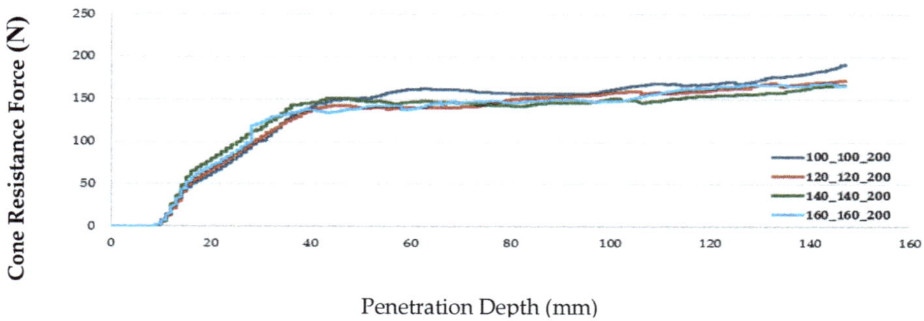

Figure A1. CPT simulation soil domain study with MM-ALE domain shape.

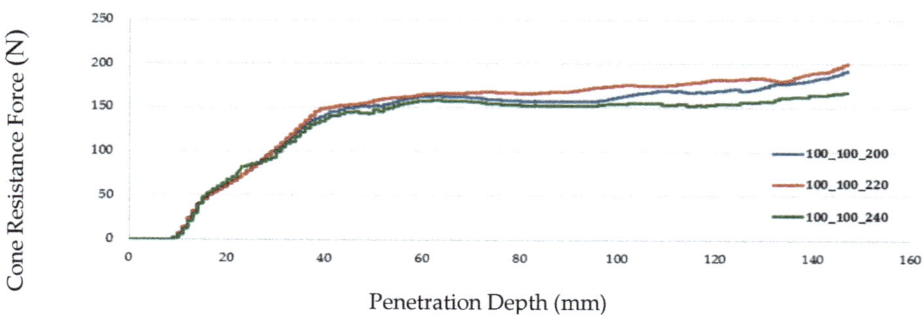

Figure A2. CPT simulation soil domain study with MM-ALE domain height.

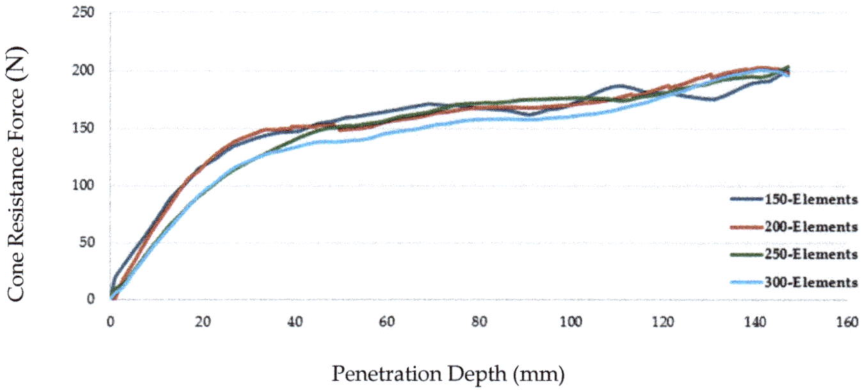

Figure A3. CPT simulation, convergence study with MM-ALE meshing method.

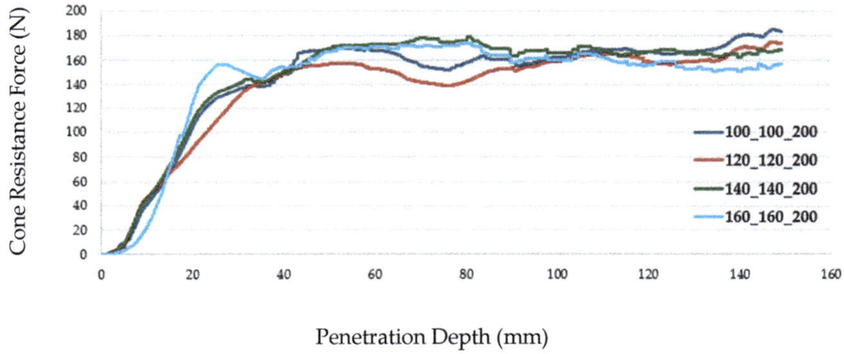

Figure A4. CPT simulation soil domain study with SPH domains cross-section.

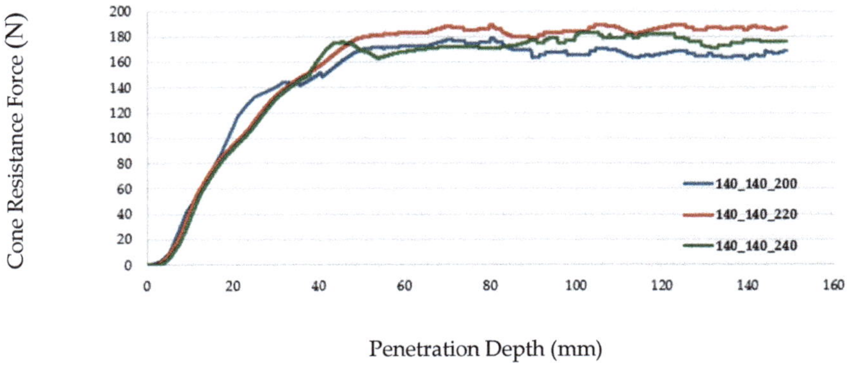

Figure A5. CPT simulation soil domain study with SPH domain height.

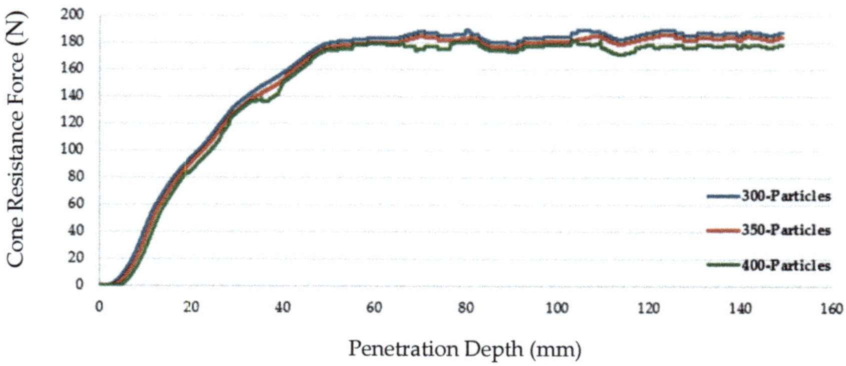

Figure A6. CPT simulation CONVERGENCE study with SPH meshing method.

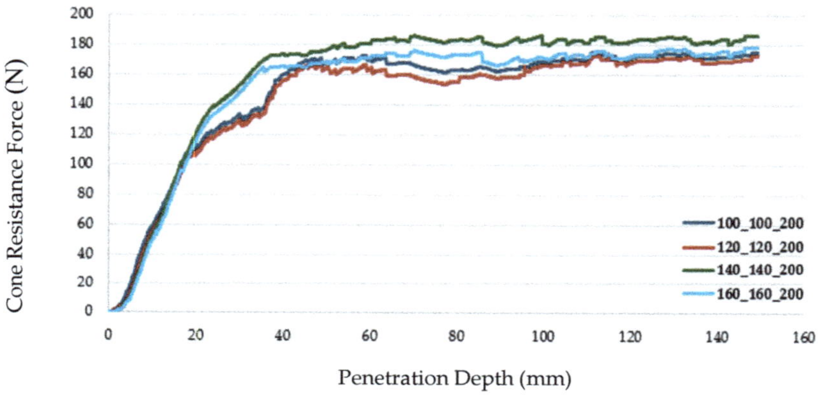

Figure A7. CPT simulation soil domain study with Hybrid-FEM-SPH domains shape.

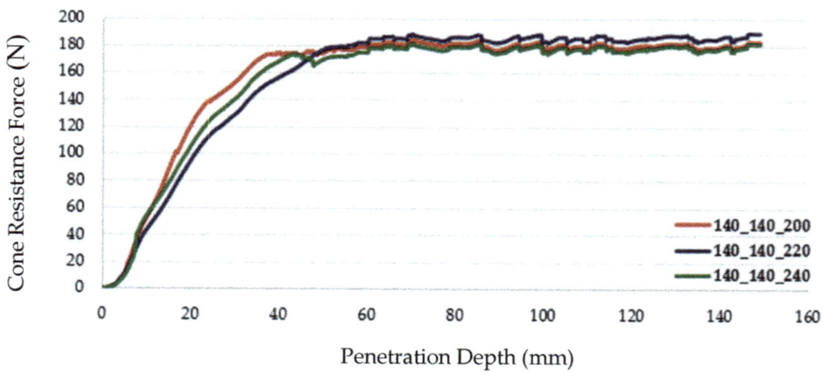

Figure A8. CPT simulation soil domain study with Hybrid-FEM-SPH domain height.

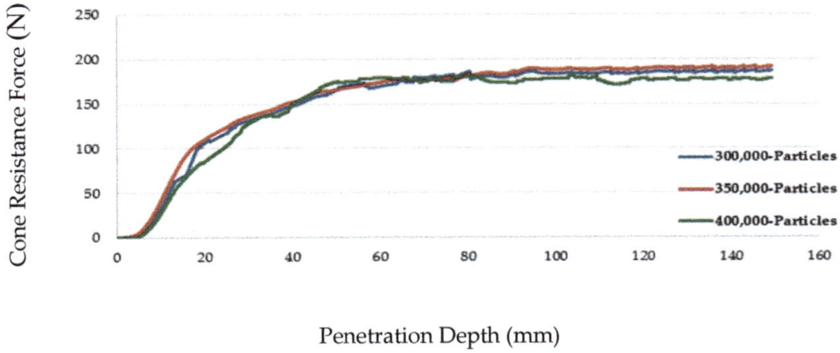

Figure A9. CPT simulation CONVERGENCE study with Hybrid-FEM-SPH meshing method.

References

1. Zeng, H.; Zhao, C.; Chen, S.; Xu, W.; Zang, M. Numerical Simulations of Tire-Soil Interactions: A Comprehensive Review. *Arch. Comput. Methods Eng.* **2023**, *30*, 4801. [CrossRef]
2. Kalantari, D.; Askari, M.; Mousavi-Seyedi, S.R.; Asadi, N. Investigating the Compressibility of Sandy-Clay Soil in the Laboratory Soil Bin Caused by Agricultural Tire Traffic and Analyzed by the Adaptive Neuro–Fuzzy Inference System. *Appl. Environ. Soil Sci.* **2025**, *2025*, 4471543. [CrossRef]
3. Jasoliya, D.; Untaroiu, A.; Untaroiu, C. A review of soil modeling for numerical simulations of soil-tire/agricultural tools interaction. *J. Terramech.* **2024**, *111*, 41–64. [CrossRef]
4. Helwany, S. *Applied Soil Mechanics with ABAQUS Applications*; John Wiley & Sons: Hoboken, NJ, USA, 2007.
5. DiMaggio, F.L.; Sandler, I.S. Material model for granular soils. *J. Eng. Mech. Div.* **1971**, *97*, 935–950. [CrossRef]
6. Jasoliya, D.; Untaroiu, C. Identification, verification and validation of soil constitutive material model parameters for soil-tire/tool interaction numerical simulations. *Comput. Electron. Agric.* **2025**, *229*, 109787. [CrossRef]
7. Fasanella, E.L.; Lyle, K.H.; Jackson, K.E. Developing soil models for dynamic impact simulations. In Proceedings of the AHS International 65th Forum and Technology Display, Grapevine, TX, USA, 27–29 May 2009.
8. Saez Barrios, D.O. Determination of Soil Properties for Sandy Soils and Road Base at Riverside Campus Using Laboratory Testing and Numerical Simulation. Master's Thesis, Texas A&M University, College Station, TX, USA, 2010.
9. Untaroiu, C.D.; Meissner, M.U.; Crandall, J.R.; Takahashi, Y.; Okamoto, M.; Ito, O. Crash reconstruction of pedestrian accidents using optimization techniques. *Int. J. Impact Eng.* **2009**, *36*, 210–219. [CrossRef]
10. Untaroiu, C.D.; Crandall, J.R.; Takahashi, Y.; Okamoto, M.; Ito, O.; Fredriksson, R. Analysis of running child pedestrians impacted by a vehicle using rigid-body models and optimization techniques. *Saf. Sci.* **2010**, *48*, 259–267. [CrossRef]
11. Untaroiu, C.D.; Shin, J.; Crandall, J.R. A design optimization approach of vehicle hood for pedestrian protection. *Int. J. Crashworthiness* **2007**, *12*, 581–589. [CrossRef]
12. Untaroiu, C.D.; Untaroiu, A. Constrained design optimization of rotor-tilting pad bearing systems. *J. Eng. Gas Turbines Power* **2010**, *132*, 122502. [CrossRef]
13. Boldyrev, G.; Muyzemnek, A. The modeling of deformation process in soils with use of ansys and Ls-Dyna programs. In Proceedings of the International Conference on Case Histories in Geotechnical Engineering, Arlington, VA, USA, 13 August 2008.
14. Garc, A.; Frankenstein, S. *USCS and the USDA Soil Classification System: Development of A Mapping Scheme*; U.S. Army Engineer Research and Development Center, Cold Regions Research and Engineering Laboratory: Hanover, NH, USA, 2015.
15. Lade, P.V. *Triaxial Testing of Soils*; John Wiley & Sons: Hoboken, NJ, USA, 2016.
16. Swamy, V.S.; Mason, D.; Yerro, A.; Sandu, C.; Sebeck, K.; Gorsich, D. Numerical Evaluation of the Influence of Terrain Properties in Clay-Tire Interactions. In Proceedings of the International Congress of Automotive and Transport Engineering, Brasov, Romania, 6–8 November 2024; pp. 353–363.
17. Pandit, R. Characterization of Soft Clay and Clay-tire Interaction for the Prediction of Ground Mobility. Master's Thesis, Virginia Polytechnic Institute and State University, Blacksburg, VA, USA, 2023.
18. Stander, N.; Goel, T. An Overview of LS-OPT® Version 4.1. In Proceedings of the 11th International LS-DYNA Users Conference, Dearborn, MI, USA, 6–8 June 2010.
19. Shoop, S.; Affleck, R.; Haehnel, R.; Janoo, V. Mechanical behavior modeling of thaw-weakened soil. *Cold Reg. Sci. Technol.* **2008**, *52*, 191–206. [CrossRef]
20. Wang, X.; Qin, X.; Tan, J.; Yang, L.; Ou, L.; Duan, X.; Deng, Y. Effect of the moisture content and dry density on the shear strength parameters of collapsing wall in hilly granite areas of South China. *Int. Soil Water Conserv. Res.* **2024**, *12*, 697–713. [CrossRef]

21. Feng, X.; Teng, J.; Wang, H. Influence mechanism of water content and compaction degree on shear strength of red clay with high liquid limit. *Materials* **2023**, *17*, 162. [CrossRef] [PubMed]
22. Zhang, Y.; Lu, J.; Han, W.; Xiong, Y.; Qian, J. Effects of moisture and stone content on the shear strength characteristics of soil-rock mixture. *Materials* **2023**, *16*, 567. [CrossRef] [PubMed]
23. Keskin, İ.; Atayşen, E.Ö.; Kahraman, S.; Vakili, A.H.; Salimi, M. An evaluation on the effect of stress and wetting conditions on engineering properties of clay–sand mixtures. *Q. J. Eng. Geol. Hydrogeol.* **2025**, *58*, qjegh2025-2008. [CrossRef]
24. Chen, M.; Zhang, J.; Qiu, P.; Liu, F. The effects of dry density and moisture content on the shear characteristics of clay-type muddy interlayer. *Sci. Rep.* **2025**, *15*, 5732. [CrossRef]
25. Kornfield, T.; Zubeck, H. Triaxial Testing of Frozen Soils—State of the Art. In *Symposium on Mechanical Properties of Frozen Soil*; ASTM International: West Conshohocken, PA, USA, 2013; pp. 76–85.
26. Hamza, M.; Anderson, W.K. Soil compaction in cropping systems: A review of the nature, causes and possible solutions. *Soil Tillage Res.* **2005**, *82*, 121–145. [CrossRef]
27. *ASTM D2850*; Standard Test Method for Unconsolidated-Undrained Triaxial Compression Test on Cohesive Soils. ASTM International: West Conshohocken, PA, USA, 2015; Volume D2850-15, pp. 1–6.
28. Lascarro, C.; Tafili, M.; Fuentes, W.; Duque, J. Comparative analysis of two intergranular strain-based hypoplastic models through elemental and centrifuge testing. *Soil Dyn. Earthq. Eng.* **2024**, *180*, 108572. [CrossRef]
29. *ASTM D2435*; Standard Test Methods for One-Dimensional Consolidation Properties of Soils Using Incremental Loading. ASTM International: West Conshohocken, PA, USA, 2004.
30. Klasztorny, M.; Nycz, D.B.; Dziewulski, P.; Gieleta, R.; Stankiewicz, M.; Zielonka, K. Numerical modelling of post-ground subsystem in road safety barrier crash tests. *Eng. Trans.* **2019**, *67*, 513–534.
31. Wu, W.; Thomson, R. A study of the interaction between a guardrail post and soil during quasi-static and dynamic loading. *Int. J. Impact Eng.* **2007**, *34*, 883–898. [CrossRef]
32. Holtz, R.D.; Kovacs, W.D.; Sheahan, T.C. *An Introduction to Geotechnical Engineering*; Prentice-hall: Englewood Cliffs, NJ, USA, 1981; Volume 733.
33. Kulak, R.F.; Bojanowski, C. Modeling of cone penetration test using SPH and MM-ALE approaches. In Proceedings of the 8th European LS-DYNA Users Conference, Strasbourg, France, 23–24 May 2011; pp. 1–10.
34. Aubram, D. Development and experimental validation of an arbitrary Lagrangian-Eulerian (ALE) method for soil mechanics. *Geotechnik* **2015**, *38*, 193–204. [CrossRef]
35. Aubram, D. *An Arbitrary Lagrangian-Eulerian Method for Penetration into Sand at Finite Deformation*; Technische Universität Berlin: Berlin, Germany, 2013.
36. Tian, Y.; Wang, D.; Cassidy, M. Large deformation finite element analysis of offshore penetration tests. In Proceedings of the Second International Symposium on Computational Geomechanics (ComGeo II), Cavtat-Dubrovnik, Croatia, 27–29 April 2011.
37. Nazem, M.; Carter, J.P.; Airey, D.W.; Chow, S. Dynamic analysis of a smooth penetrometer free-falling into uniform clay. *Géotechnique* **2012**, *62*, 893–905. [CrossRef]
38. Carter, J.P.; Nazem, M.; Airey, D.W.; Chow, S. Dynamic analysis of free-falling penetrometers in soil deposits. In *GeoFlorida 2010: Advances in Analysis, Modeling & Design*; American Society of Civil Engineers: Reston, VA, USA, 2010; pp. 53–68.
39. He, R. Systematic Tire Testing and Model Parameterization for Tire Traction on Soft Soil. Ph.D. Thesis, Virginia Polytechnic Institute and State University, Blacksburg, VA, USA, 2020.
40. Bui, H.H.; Fukagawa, R. An improved SPH method for saturated soils and its application to investigate the mechanisms of embankment failure: Case of hydrostatic pore-water pressure. *Int. J. Numer. Anal. Methods Geomech.* **2013**, *37*, 31–50. [CrossRef]
41. Zhao, S.; Bui, H.H.; Lemiale, V.; Nguyen, G.D.; Darve, F. A generic approach to modelling flexible confined boundary conditions in SPH and its application. *Int. J. Numer. Anal. Methods Geomech.* **2019**, *43*, 1005–1031. [CrossRef]
42. Lambe, T.W.; Whitman, R.V. *Soil Mechanics SI Version*; John Wiley & Sons: Hoboken, NJ, USA, 2008.
43. Wang, Y.; Akeju, O.V. Quantifying the cross-correlation between effective cohesion and friction angle of soil from limited site-specific data. *Soils Found.* **2016**, *56*, 1055–1070. [CrossRef]
44. Watkins, R.; Keil, B.; Mielke, R.; Rahman, S. Pipe zone bedding and backfill: A flexible pipe perspective. In *Pipelines 2010: Climbing New Peaks to Infrastructure Reliability: Renew, Rehab, and Reinvest*; American Society of Civil Engineers (ASCE): Reston, VA, USA, 2010; pp. 426–438.
45. Jasoliya, D.R. Characterization and Modeling of Deformable Soils for Tire Performance Simulations. Ph.D. Thesis, Virginia Polytechnic Institute and State University, Blacksburg, VA, USA, 2025.
46. Liu, K.; Benetti, M.; Sozzi, M.; Gasparini, F.; Sartori, L. Soil compaction under different traction resistance conditions—A case study in North Italy. *Agriculture* **2022**, *12*, 1954. [CrossRef]
47. Shaheb, M.R.; Venkatesh, R.; Shearer, S.A. A review on the effect of soil compaction and its management for sustainable crop production. *J. Biosyst. Eng.* **2021**, *46*, 417–439. [CrossRef]

48. Keller, T.; Sandin, M.; Colombi, T.; Horn, R.; Or, D. Historical increase in agricultural machinery weights enhanced soil stress levels and adversely affected soil functioning. *Soil Tillage Res.* **2019**, *194*, 104293. [CrossRef]
49. Håkansson, I.; Reeder, R.C. Subsoil compaction by vehicles with high axle load—Extent, persistence and crop response. *Soil Tillage Res.* **1994**, *29*, 277–304. [CrossRef]
50. Batey, T. Soil compaction and soil management—A review. *Soil Use Manag.* **2009**, *25*, 335–345. [CrossRef]

agriculture

MDPI

Article

Structural Improvement of Sugarcane Harvester for Reducing Field Loss When Harvesting Lodged Canes

Jiaoli Jiang [1], Xueting Han [2], Qingting Liu [2,*], Hai Xu [2], Tao Wu [2], Jiamo Feng [2], Xiaoping Zou [2] and Yuejin Li [2]

[1] Guangdong Province Agricultural Technology Extension Center, Guangzhou 510515, China; jjl0626@126.com
[2] School of Engineering, South China Agricultural University, Guangzhou 510642, China; 811992hxt@stu.scau.edu.cn (X.H.); haixu@polyu.edu.hk (H.X.); wt55pub@scau.edu.cn (T.W.); fengjiamu_@126.com (J.F.); zxp9v742@126.com (X.Z.); 18881939600@126.com (Y.L.)
* Correspondence: qingting@scau.edu.cn

Abstract: Sugarcane, a key sugar crop in China, is predominantly manually harvested. In the main sugarcane-producing areas of China, typhoons cause canes to become lodged, resulting in high field losses and low harvesting efficiency. This study aimed to reduce these losses by analyzing the causes: ineffective stalk pickup, transfer, and conveyance. The tests showed the stalk–steel static friction coefficient (SFC) was lower than the stalk–soil SFC. Conventional basecutters use raised patterns to enhance friction, but soil adhesion makes them ineffective, hindering lodged stalk pickup. Bent stalks also struggle to enter butt lift rollers or pass through roller trains, increasing losses. The proposed improvements included adding toothed plates on the cutter discs, optimized disc–roller positioning, and using fewer rollers (one butt lift and one feed roller pair). Theoretical analysis confirmed the toothed plates improved pickup via grabbing force, while using fewer rollers stopped the stalks detaching from and blocking the roller train. A prototype was tested via orthogonal experiments, showing a field loss ratio of 1.21%, a feed rate of 13.09 kg/s, and a billet qualification rate of 95.82% with optimal settings (chopper speed: 390 rpm; 10 stalks/group; roller speed: 230 rpm; ground speed: 1.41 m/s). Field tests achieved 2.0% loss, demonstrating effectiveness for severely lodged cane, a significant improvement over the conventional harvesters (15–20% loss). These findings aid low-loss-level harvester development.

Keywords: lodged sugarcane; chopper harvester; field loss ratio of harvesting; structural improvements

1. Introduction

Sugarcane is one of the most important global crops and the primary sugar-producing crop in China [1,2]. In major production regions, such as Guangxi, Guangdong, and Hainan, sugarcane plants usually become lodged due to annual typhoons and monsoons in July and August [2,3]. During subsequent growth, the tops of the lodged canes curve upward, often resulting in stalks with one or multiple bends [4]. Based on the lodging angle (α), sugarcane lodging is classified into three levels: slight (α = 0–30°), moderate (α = 30–60°), and severe (α = 60–90°) [5]. Severely lodged canes often lay flat on the ground, with adjacent rows overlapping, as illustrated in Figure 1 [4].

Figure 1. Canes exhibiting severe lodging in field conditions.

Lodging complicates mechanized harvesting [6,7]. When chopper harvesters were used on severely lodged canes, the field loss ratio reached approximately 15–20% [8,9], which is from two to three times of that of upright canes. This elevated loss ratio is a key deterrent to farmers adopting mechanized harvesting [8]. Consequently, developing technologies to minimize field losses during lodged cane harvesting has become critical.

A typical chopper harvester comprised a topper, two crop dividers, a knockdown roller, a basecutter, a roller train, a chopper, an elevator, and one or two extractors [10–12]. During operation, the dividers separate and lift lodged canes from adjacent rows. Research has focused on increasing the divider lifting height [3]. Adopted a variable pitch could increase the lifting angle [5]. Some studies suggested that chopper harvesters can improve cutting quality and reduce losses by providing support to the basecutter when the stalk is being cut [13]. However, the basecutters of existing chopper harvesters cut stalks after they have already disengaged from the divider.

The basecutter's primary function is to cut stalks at the ground level and transfer them. An extensive number of studies examined cutting forces, energy consumption [14], and stubble damage mechanisms [15]. Minimizing the stubble height reduced field losses and stubble damage, prompting research into basecutter height control [16,17]. Additionally, the basecutter aids in conveying stalks to the feeding rollers [7], with prior work analyzing stalk transfer from the basecutter to the lift rollers [18]. However, few studies addressed the pickup and transfer of lodged canes by the basecutter.

The roller train conveys stalks to the chopper, with the existing literature exploring roller design and the movement of straight stalks [10,11]. These studies primarily assessed the roller train's ability to handle straight stalks and the influence of feed/chopper roller speed synchronization on billet length. Research on losses and breakage of curved stalks in the roller train remains limited.

Overall, few studies addressed field loss reduction during lodged cane harvesting. Based on observation and experimentation, the authors found the main causes of the high field loss rate and proposed relevant improvement measures. A prototype was developed, and experiments were conducted for verification. This article mainly presents this research. Compared with the existing field loss rate of 15–20% for sugarcane chopper harvesters, the 2% loss rate in this study is beneficial for enhancing the enthusiasm of farmers and sugar mills to adopt mechanical harvesting.

2. Materials and Methods

In this study, the authors observed that losses occur primarily during three key operational stages: (1) stalk pickup by the basecutter, (2) stalk transfer to the butt roller train, and (3) stalk conveyance to the chopper. They analyzed the underlying causes of these losses and proposed targeted solutions. Figure 2 illustrates a schematic flow diagram of the developmental framework for the field loss reduction technology investigated in this study.

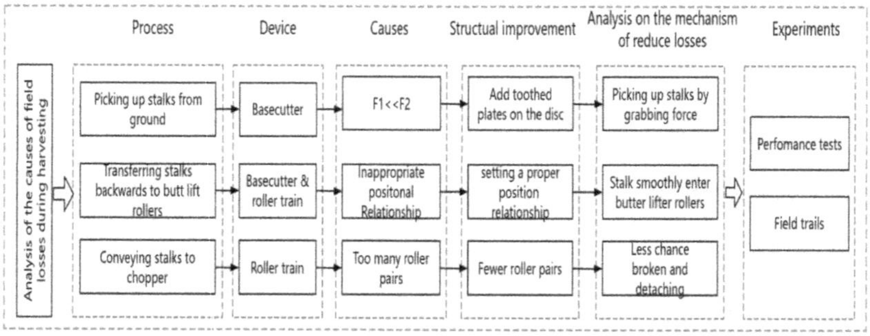

Figure 2. Methodology followed for the development of technology for reducing field losses during lodged sugarcane harvesting.

2.1. Experiment on the Friction Coefficient Between Sugarcane and Soil as Well as Between Sugarcane and Steel Plates

In analyzing the causes of field loss during harvesting, the static friction coefficient (SFC) between sugarcane and soil, as well as between sugarcane and steel plates, was considered. While some studies reported SFC values for sugarcane and steel plates [19], no prior studies were found regarding the SFC between sugarcane and soil. To ensure data comparability, we measured the SFC for both sugarcane–soil and sugarcane–steel interactions in this study.

The sugarcane variety used was Yuetang 55, and the soil type was red soil. Both were obtained from the Suixi Test Station in China on 16 July 2025. The field area measured 10 hectares, with row spacing of 1.2 m. The steel plates used in the experiment were made of Q235A grade steel.

Sugarcane stem samples were prepared by cutting them into 80 mm billets. Two sections were bonded together to form a single sample. A total of six samples was prepared: three with leaves attached (WL) and three de-leafed (DL). The leaves were affixed to the stems using adhesive.

Based on the relevant literature [20], soil samples with three moisture contents (*MCs*)—5%, 10%, and 15%—were prepared to test the sugarcane–soil SFC. Different moisture levels were achieved by adding water to dry soil. The drying process involved placing soil cakes in a DHG-9073A electric forced air drying oven at 50 °C. The samples were weighed hourly using an RS232 electronic scale (resolution: 0.1 g). Drying was considered complete when the weight difference between two consecutive measurements fell below 1%.

The required water mass for each target *MC* was calculated based on the dry soil weight. After adding an appropriate amount of water, the soil was thoroughly mixed and finely broken up to ensure a flat surface for testing.

The static friction coefficient (SFC) was determined using the inclined plate method. The samples were placed in both horizontal (H) and vertical (V) orientations on the inclined plate as illustrated in Figure 3.

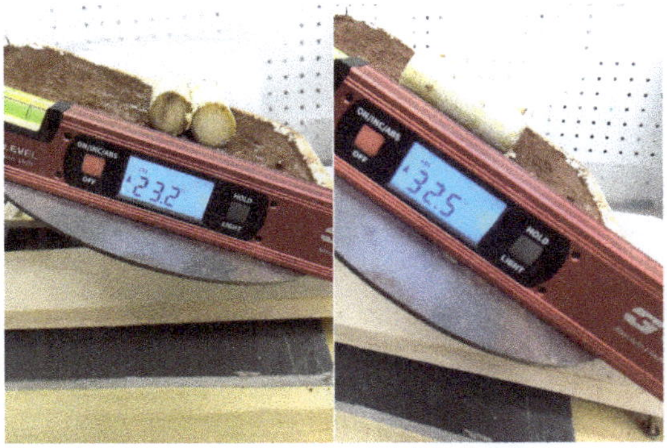

(**a**) (**b**)

Figure 3. Experimental device and samples: (**a**) samples placed horizontally; (**b**) samples placed vertically.

Soil was filled into a plastic tray, and its surface was leveled into a flat plane. The tray was then fixed onto an inclined plate. The inclination angle of the plate relative to the horizontal plane was gradually increased, and the critical angle (ψ) at which the sample began to slide downward was recorded. The static friction coefficient (SFC) was calculated as SFC = tan(ψ).

The following definitions were established:

f_1: SFC between the stalk and steel.

f_2: SFC between the stalk and soil.

f_{1vdl}: SFC between the stalk (de-leafed) and steel in the vertical orientation.

f_{1hdl}: SFC between the stalk (de-leafed) and steel in the horizontal orientation.

f_{2vwl}: SFC between the stalk (with leaf) and soil in the vertical orientation.

f_{2hwl}: SFC between the stalk (with leaf) and soil in the horizontal orientation.

2.2. Measurement of Lodged Sugarcane and Definition of Lodging Angle

2.2.1. Measurement of the Spatial Posture of Lodged Sugarcane

Using the method for measuring the spatial posture of lodged canes developed by an author of this article in his previous research [4], the spatial posture of lodged sugarcane in the experimental field was measured.

(1) Testing equipment

The testing equipment included a T-square with a measurement length of 1 m; a tape measure with a length of 3 m; and a flexible ruler with a length of 3 m.

(2) Coordinate system of sugarcane

The intersection point of the sugarcane base and the ground was designated as the coordinate origin *O*. The *X*-axis was aligned with the sugarcane ridge, while the *Y*-axis was set perpendicular to the ridge along the ground plane. The *Z*-axis was defined as the direction orthogonal to the XOY plane, oriented upward. The positive *X*- and *Y*-axes were oriented such that the sugarcane lay within the first quadrant of the *XOY* coordinate system. This configuration facilitated the measurement of coordinate values for points along the sugarcane.

(3) Selection and measurement of key points for lodged canes

Key points were selected to describe the spatial posture of the sugarcane. These points included point O, the two endpoints of a straight section and the two endpoints and three middle points of a curved section. For sugarcane stalks with multiple curved sections, the key points were chosen separately for each curve.

2.2.2. Definition of Lodging Angle (α)

The literature provides definitions of the lodging angle and the lateral deflection angle of the straight lodged cane. There is no definition for the lodging angle or lateral deflection angle of curved stalks. The base of a lodged cane is generally straight. In this study, the angle formed by this straight line segment and the z-axis was defined as the lodging angle, while the angle between this segment and the x-axis was defined as lateral deflection angle, as shown in Figure 4.

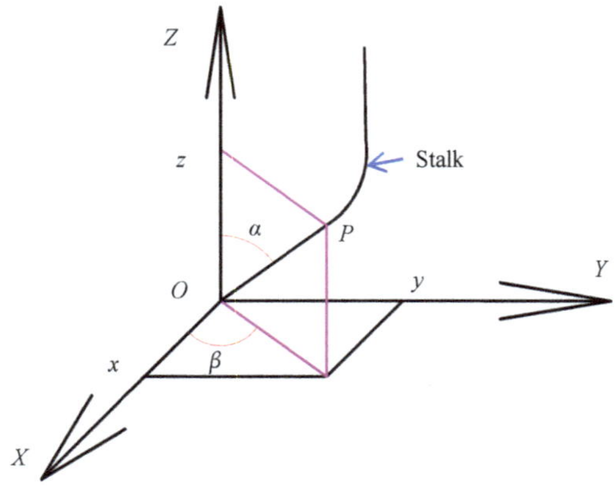

Figure 4. The definitions of the lodging angle and lateral deviation angle of sugarcane.

2.3. Analysis of the Causes of High Field Loss Ratio During Lodged Sugarcane Harvesting

2.3.1. Understanding Sugarcane Chopper Harvester and Its Field Losses During Harvesting

Understanding the harvester's operational workflow is essential for analyzing the causes of field losses during harvesting. Typically, a sugarcane chopper harvester's process follows the sequence illustrated in Figure 5 [10].

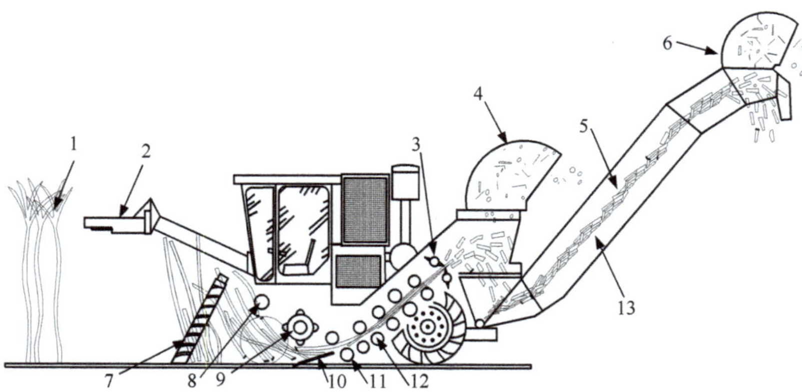

Figure 5. Operational workflow of CASE A8000 sugarcane harvester. 1—sugarcane. 2—topper. 3—chopper rollers. 4—primary extractor. 5—billets. 6—extractor. 7—crop divider. 8—knock down roller. 9—power feed roller. 10—basecutter. 11—butt lifter roller. 12—roller train. 13—elevator.

As shown in Figure 5, the harvesting process begins with the topper cutting off the top of the sugarcane, the dividers separating the adjacent rows of sugarcane, and the knockdown roller pushing over the sugarcane stalks, after which the basecutter cuts them near ground level. Subsequently, the basecutter transfers the stalks backward to the butt lifter roller, from which they are fed into the chopper via a series of rollers (roller train). Finally, the chopped billets are ejected into the extractor for trash removal, and the cleaned billets are transported to the haul vehicle by the elevator.

According to the JB/T 6275-2019 Sugarcane Combine Harvester standard [21], field losses in sugarcane harvesting comprise two primary types: lost losses and stubble losses. Lost losses refer to sugarcane stems and billets that remain in the field after harvest, while stubble losses are defined as stubble segments exceeding 30 mm in height. Additionally, juice losses occur during the chopping of stalks into billets, typically ranging from 2 to 5 tonnes per hectare [10]. Theoretically, if the basecutter blades were able to cut stalks below ground level, stubble losses could be eliminated entirely. Consequently, this study focused on strategies to minimize lost losses.

The elevated field loss ratio observed during the harvesting of lodged sugarcane is attributed to three key factors: (1) severely lodged stalks are challenging for the basecutter to retrieve, (2) some stalks are not effectively conveyed to the butt lifter roller, and (3) bent stalks tend to break and detach from the roller train.

2.3.2. Harvest Losses During Sugarcane Pickup by Basecutter

(1) Mechanism of sugarcane pickup from the ground by basecutter

The basecutter is responsible for cutting sugarcane stalks, collecting them from the ground, and transferring the cut canes to the butt lifter roller [10]. Typically, it consisted of a double-disc cutting system, with each disc equipped with from three to six blades [22,23] (Figure 6). Figure 6b displays a basecutter of an AC 60 sugarcane harvester photographed by the authors of this study.

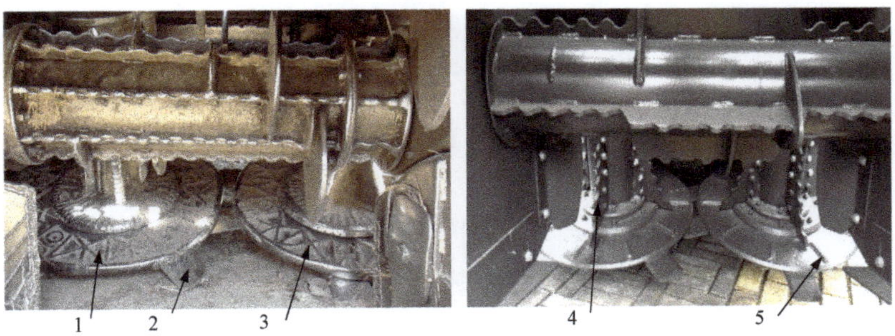

(a) (b)

Figure 6. Structure of basecutter: (**a**) raised patterns on upper surfaces; (**b**) deformed steel bars on upper surfaces. 1—disc. 2—blade. 3—raised patterns. 4—toothed paddle. 5—deformed steel bar.

During harvesting, the blades cut the canes close to or below the ground surface, as illustrated in Figure 7a. Subsequently, the stalks are picked up and transferred backward by the upper surface of the disc and the paddles mounted on its spindle. The disc picks up and transfers the sugarcane through friction between its surface and the stalks. Figure 7b,c presents the force analysis of sugarcane being picked up and transferred by the disc. As shown in Figure 7a, the sugarcane stalk was cut at its base. Once severed, the stalk fell downward. During its descent, the harvester continued moving forward, causing the basecutter's disc to come into contact with the stalk's base, as illustrated in Figure 7b,c.

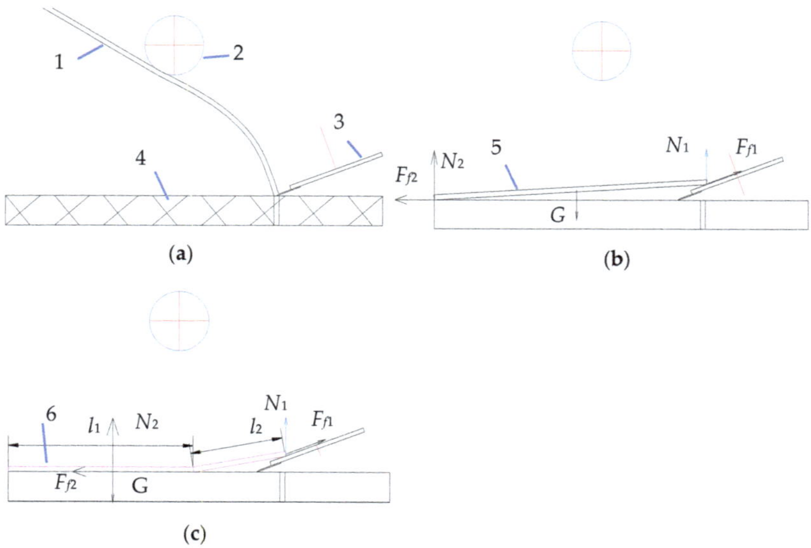

Figure 7. Force analysis of sugarcane being picked up and transferred by disc of basecutter: (**a**) stalk was cut at its base; (**b**) force analysis of straight stalk; (**c**) force analysis of curved stalk. 1—sugarcane. 2—knock down roller. 3—disc of basecutter. 4—ground of field. 5—straight stalk. 6—curved stalk.

Figure 7b,c depicts a straight stalk and a curved stalk, respectively. G represents the gravitational force acting on the sugarcane, while N_1 and N_2 denote the reaction forces exerted by the disc and the ground, respectively. Additionally, F_{f1} corresponds to the frictional force between the sugarcane and the disc, and F_{f2} represents the friction between the sugarcane and the ground. The relationship between these forces is as follows:

$$G = N_1 + N_2, \tag{1}$$

$$F_{f1} = N_1 \times f_1, \tag{2}$$

$$F_{f2} = N_2 \times f_2, \tag{3}$$

f_1 and f_2 are the SFCs of stalk–steel and stalk–soil interactions, respectively. For the disc to successfully pick up and transfer sugarcane, the condition $F_{f1} > F_{f2}$ has to be satisfied.

As shown in Figure 7b, the base end of the sugarcane rested on the disc, while the top end remained on the ground, with $N_1 = N_2$. However, as the SFC test results indicate (showed in the Section Results and Analysis, $f_1 < f_2$), if the top surface of disc was smooth, $F_{f1} < F_{f2}$, according to Equations (1)–(3).

To increase F_{f1}, most conventional sugarcane harvesters incorporate raised patterns or deformed steel bars on the disc's flat surface (Figure 7). Toothed paddles were also found to enhance backward transferring efficiency.

(2) Reason for losses during sugarcane pickup

Lodged sugarcane stalks are typically curved, with severely lodged stalks lying almost flat on the ground. When the disc makes contact with the base of the stalk, a significant portion of the stalk remains in contact with the soil, as illustrated in Figure 7c. Consequently, N_1 is much smaller than N_2. Meanwhile, $f_1 < f_2$. Based on Equations (1)–(3), F_{f1} is much smaller than F_{f2}.

If the raised patterns or deformed steel bars on the disc lack sufficient roughness to ensure $F_{f1} > F_{f2}$, the disc struggles to lift and convey severely lodged stalks. This issue is particularly pronounced in sticky soil conditions, where soil accumulation on the disc is a problem. When the raised patterns and the steel bars are clogged with sticky soil (Figure 8), their effectiveness is diminished, resulting in $f_1 \leq f_2$ and $F_{f1} << F_{f2}$ for severely lodged stalks. This is the primary cause of the high field loss ratio during the harvesting of severely lodged sugarcane. Figure 8 shows the basecutter of a CASE 7000 sugarcane harvester photographed by the authors of this study.

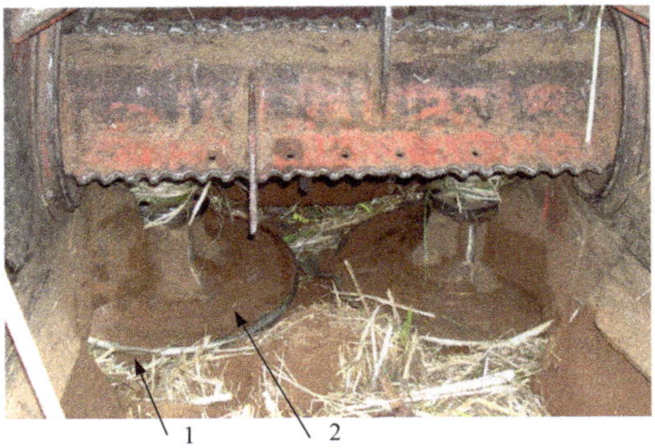

Figure 8. Upper surface of disc with sticky soil: 1—blade; 2—sticky soil.

2.3.3. Harvest Loss During the Transfer of Canes Backwards to Butt Lift Roller by Basecutter

(1) Mechanism of transferring canes backwards to butt lift roller by basecutter

The cut canes are supposed to enter the gap (channel) between the upper and lower butt lift rollers. If they fail to do so, the canes remain in the field, resulting in harvest losses. Due to their naturally bent or curved shape, some stalks have difficulty entering the roller gap, as illustrated in Figure 9. Additionally, an improper positional relationship between the basecutter disc and the butt lift rollers occasionally prevents even straight canes from entering the channel between the rollers. Therefore, establishing the correct geometric alignment between the basecutter and the butt lift rollers is crucial to ensure smooth cane entry into the channel.

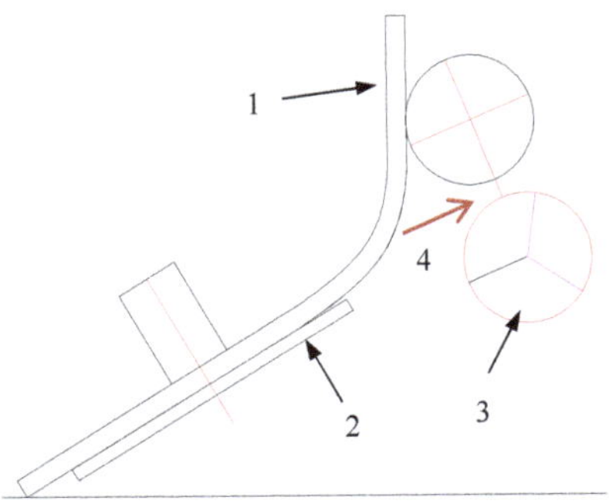

Figure 9. Cane that failed to enter channel between upper and lower rollers. 1—curved cane. 2—basecutter. 3—butt lift roller. 4—channel.

(2) Analysis of position relationship between basecutter and butt lift rollers

To analyze the positional relationship, several key definitions were established, as illustrated in Figure 10.

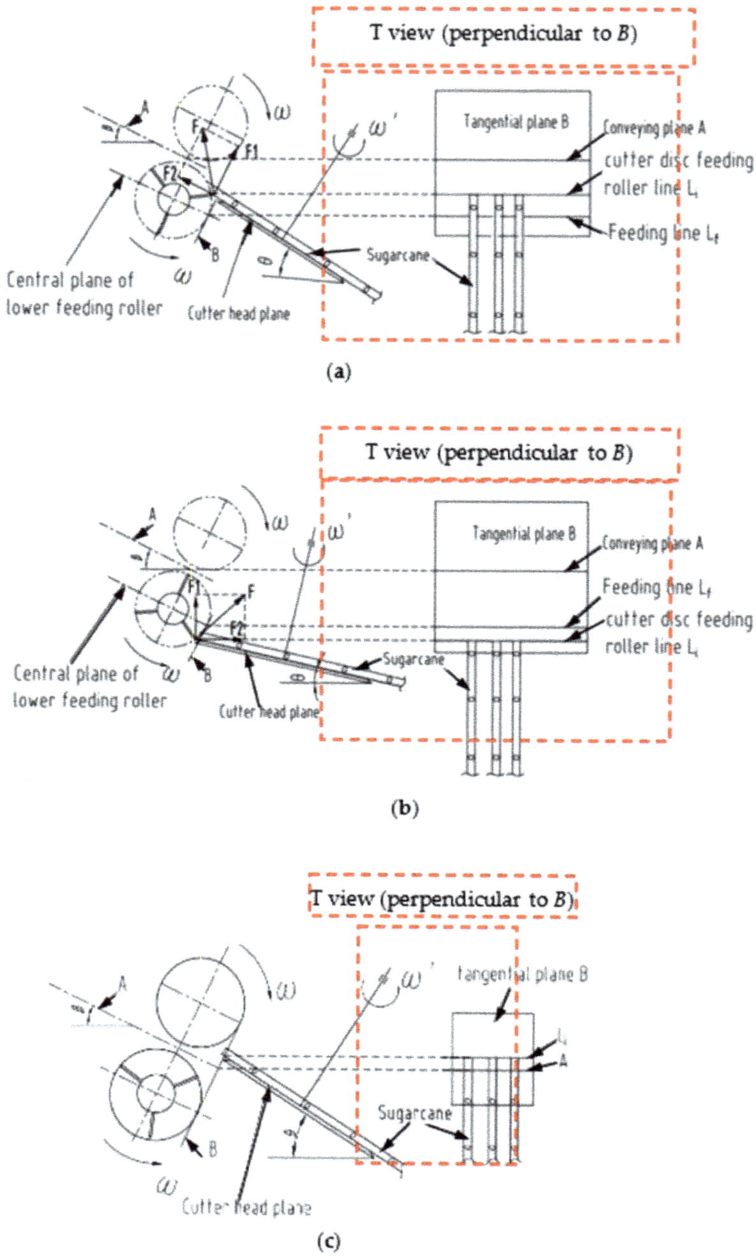

Figure 10. Analysis of the positional relationship between the basecutter and butt lift rollers. (**a**) L_i was above L_f. (**b**) L_i was below L_f. (**c**) position relationship between L_i and A. A, conveying plane; B, tangential plane; L_f, feeding line; L_i, cutter disc feeding roller line.

Conveying Plane (A): The symmetrical plane between the upper and lower rollers was designated as the conveying plane and labeled A.

Outer Cylindrical Surface: When the lower butt lift roller rotated, the outer edges of its blades formed a cylindrical surface, referred to as the outer cylindrical surface.

Tangential Plane (B): The plane perpendicular to the conveying plane (A) and tangent to the outer cylindrical surface was defined as the tangential plane and labeled B.

Feeding Line (L_f): The line of intersection between the central plane of the lower butt lift roller and the tangential plane (B) was termed the feeding line and denoted L_f.

Cutter Disc Feeding Roller Line (L_i): The intersection line between the top surface of the cutter disc and the tangential plane (B) was named the cutter disc feeding roller line and labeled L_i.

Position relationship between L_i and L_f. In Figure 10a,b, the force F applied to the sugarcane stalk by the blade of the lower butt lift roller is resolved into two components: F_1 and F_2. The vertical component F_1 lifts the stalks upward.

When L_i was positioned above L_f (Figure 10a), the horizontal component F_2 conveys the stalks toward the butt lift rollers. Conversely, when L_i is below L_f (Figure 10b), F_2 acts to push the stalks away from the rollers. Consequently, successful feeding of sugarcane into the rollers only occurs when L_i is maintained above L_f.

Position relationship between L_i and A. When L_i is positioned above plane A (Figure 10c), the blades of the lower butt lift roller fail to engage properly with the sugarcane stalks. This ineffective contact prevents successful feeding of the stalks into the roller mechanism. The analysis demonstrated that proper feeding requires L_i to be maintained below plane A.

2.3.4. Harvest Losses During Sugarcane Conveyance in the Roller Train

(1) Structure of roller train

Modern sugarcane chopper harvesters typically feature multiple pairs of rollers positioned between the basecutter and the chopper, collectively referred to as the roller train. For instance, the CASE 8000 harvester is equipped with 5.5 pairs of rollers in its roller train, as illustrated in Figure 5.

(2) Structure of butt lift rollers and their sugarcane conveyance mechanism

Structure of butt lift rollers. The butt lift rollers are responsible for receiving sugarcane stalks from the basecutter and transporting them to the feed rollers. The system consists of an upper and a lower roller (Figure 11a). The upper butt lift roller was designed as a floating roller, allowing for it to adjust vertically based on the volume of sugarcane passing through. To enhance stalk-gripping efficiency, the lower butt lift roller has rebar rods welded along the outer edges of each blade.

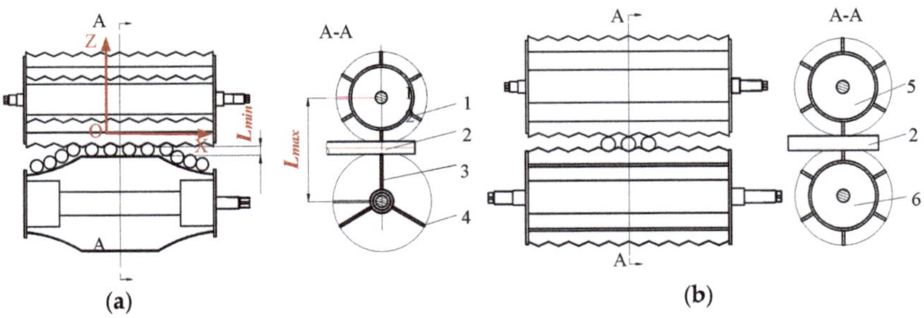

Figure 11. Schematic diagram of butt lift rollers and feed rollers: (**a**) butt lift rollers; (**b**) feed rollers. 1—blade of upper roller. 2—sugarcane. 3—blade of lower butt lift roller. 4—rebar rod along edge. 5—upper roller. 6—lower roller.

Conveyance Mechanism of the Butt Lift Rollers. As illustrated in Figure 11a, the upper roller consists of six blades, while the lower roller has three. Both the rollers rotate at the same angular velocity, causing the clearance between their blade tips to vary cyclically from the minimum (L_{min}) to the maximum (L_{max}).

The clearance reaches its maximum when the blades of both rollers are perpendicular to the XOZ plane. At this position, the lower roller blade lifts the cane butt and feeds it into the gap between the upper and lower rollers. Conversely, the clearance minimizes when the blades align parallel to the XOZ plane, causing the upper and lower blades to push the cane backward.

As the clearance transitions from L_{max} to L_{min} the rollers clamp the stalks and propel them forward toward the subsequent roller pairs.

(3) Structure of feed rollers and their sugarcane conveyance mechanism

The feed rollers are responsible for transporting sugarcane stalks from the butt lift rollers to the chopper. Additionally, they help stabilize the stalks' moving speed and provided support, ensuring uniform billet length from the chopper.

As shown in Figure 11b, the feed rollers consist of an upper and a lower roller, each equipped with six uniformly spaced wavy blades. The minimum clearance between the crest of one blade and the trough of the opposing blade match the average diameter of sugarcane stalks. The upper roller was designed as a floating roller, using its own weight to press the stalks firmly against the lower roller.

(4) Analysis of curved cane detaching from the roller train

Figure 12a illustrates the common roller train layout of a chopper harvester (i.e., the CASE 7000 chopper harvester). In this figure, o_m represents the center of the m-th roller, referred to as the o_m roller. D_{om} denotes the diameter of the o_m roller, while L_{omn} indicates the distance between o_m and o_n. The O_5 roller serves as a transitional roller.

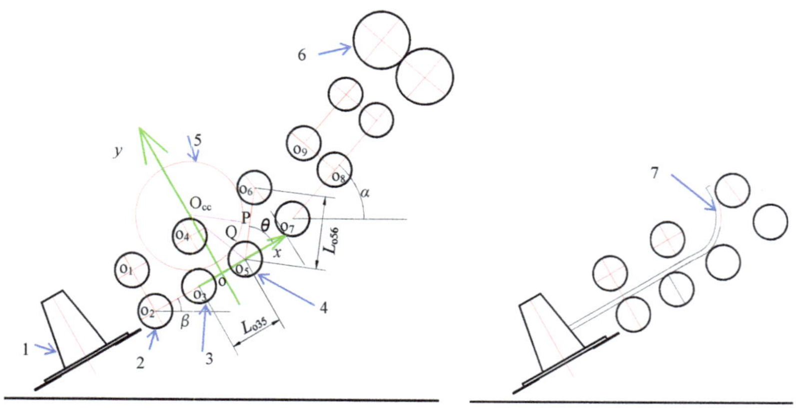

(a) (b)

Figure 12. Schematic of curved sugarcane stalk detachment from roller train: (**a**) roller arrangement and positional relationships; (**b**) detachment process of curved stalk. 1—basecutter; 2—butt lift roller; 3—feed roller; 4—transitional roller; 5—critical circle; 6—chopper; 7—sugarcane stalk.

The angle between the line o_2o_3 and the horizontal is labeled α, whereas the angle between o_7o_8 and the horizontal is denoted as β. Additionally, θ represents the angle between o_2o_3 and line o_5o_6.

In Figure 12a, since $\beta > \alpha$, the distance L_{o46} is greater than any other L_{omn} between the adjacent rollers (m and n). This creates an opportunity for bent stalks to detach from the roller train.

Figure 12a also depicts a circle O_{cc}, which is simultaneously tangent to rollers o_3, o_5, and o_6. This circle is termed the critical circle. If the radius of a stalk's curved arc exceeds that of the critical circle (R_{cc}), the stalk could detach from the gap between the o_4 and o_6 rollers, as demonstrated in Figure 12b.

In Figure 12a, a Cartesian coordinate system xoy is established. The y-axis is perpendicular to and bisected the line o_3o_5. The X-axis is o_3o_5. $O_{cc}P$ is perpendicular to and bisects o_5o_6, and point P is the intersection of o_5o_6 and line $O_{cc}P$. The straight line $O_{cc}P$ is denoted by Equation (4). The distance between points O_{cc} and o_5 is expressed by Equation (5). The radius of circle O_{cc} (R_{Occ}) was defined by Equation (6).

$$y = -tan(90^\circ - \theta) \times (x - L_{o35}/2 - L_{o56}/(2cos(\theta)) \quad (4)$$

$$L_{Occo5} = Squre({x_{o5}}^2 + {y_{Occ}}^2) \quad (5)$$

$$R_{Occ} = L_{Occo5} - R_{o5} \quad (6)$$

In Figure 12a, x_{Occ}, y_{Occ}, and x_{o5} represent the coordinate values of the points O_{cc} and o_5.

The geometric parameters of the harvester in Figure 12a are as follows: L_{o35} = 330 mm, L_{o56} = 439.5 mm, L_{o14} = 410 mm, L_{o46} = 500 mm, L_{o12} = 288 mm, α = 30°, β = 50°, and θ = 53°.

Using Equations (1)–(4), the calculated value of R_{Occ} is 327.2 mm.

(5) Analysis of curved cane clogging in the roller train

Stalk conveyance is driven by friction between the stalks and the rollers (Figure 13). In this figure, the normal force (N) acting on the cane is generated by the weight of the upper rollers, while F represents the frictional force between the blade of roller and cane. Figure 13b illustrates the forces acting on a straight cane stalk, while Figure 13c shows the forces on a curved cane. When one end of the bent stalk is pressed against the side wall of the channel, the side wall exerts a force (F_c, in Figure 13c) that hinders the movement of the stalk.

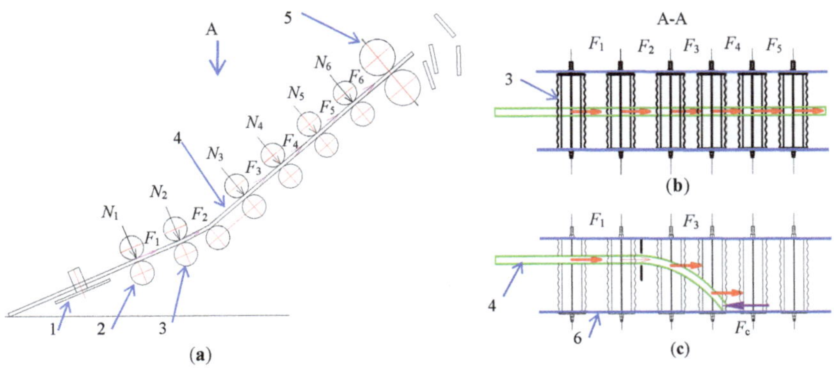

Figure 13. Schematic diagram of force analysis of cane in roller train: (**a**) forces on cane from roller train; (**b**) forces on straight cane; (**c**) forces on bending cane. 1—basecutter. 2—butter lift roller. 3—feed roller. 4—cane. 5—chopper. 6—side wall of roller train.

2.4. Performance Enhancement of Chopper Harvesters: Structural Improvements and Field Loss Mitigation Mechanisms

2.4.1. Basecutter Structural Modifications and Stalk Pickup Mechanism

As analyzed in Section 2.2.2, conventional basecutter disc structures struggle to effectively pick up and convey severely lodged sugarcane stalks, particularly in sticky soil conditions. To address this limitation, this study proposed adding small-toothed plates on the upper surface of the disc to enhance lodged stalk recovery. The design and operating mechanism of these toothed plates are illustrated in Figure 14.

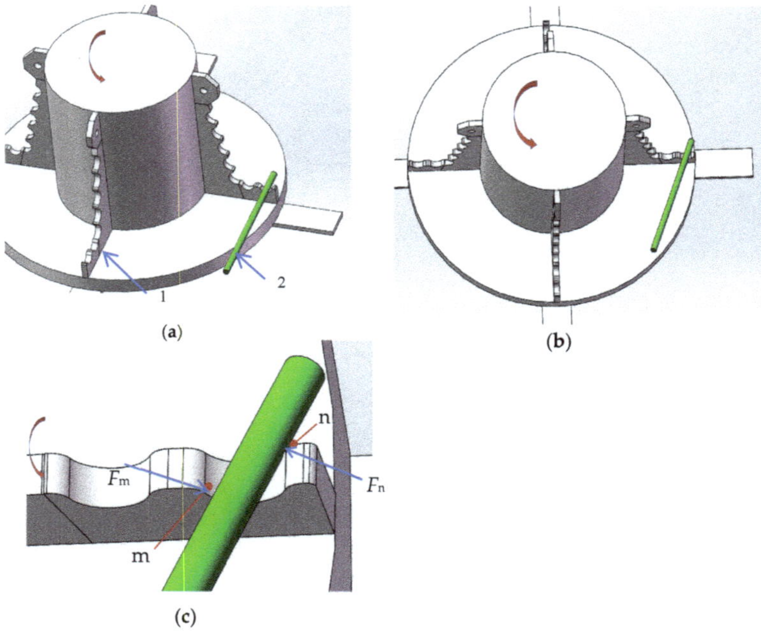

Figure 14. Mechanism of lodged sugarcane stalk engagement using toothed plates: (**a**) stalk is perpendicular to toothed plate; (**b**) stalk is not perpendicular to toothed plate; (**c**) contact points between stalk and toothed plate. 1—small-toothed plate. 2—sugarcane.

During operation, the stalk butt makes contact with a toothed plate after being cut. When the stalk is perpendicular to the plate (Figure 14a), it settles into the groove between the teeth. With the disc rotating, the angular relationship between the stalk and plate changes (Figure 14b), causing the stalk to be gripped by both sides of the groove (Figure 14c). Forces F_m and F_n act backward on the stalk to maintain engagement.

Given typical sugarcane stalk diameters of 20–40 mm [24,25], the groove radius was designed at 22.5 mm to ensure proper stalk containment (exceeding the minimum 20 mm requirement). Relative to the raised pattern and steel bar configurations shown in Figure 5, the tooth plate's peak-and-trough design exhibited a superior performance in sticky soil environments.

2.4.2. Position Relationship Between the Disc and Butt Lift Rollers

To maintain the fixed positional relationship among L_i, L_f, and the conveying plane A, the basecutter, butt lift roller, feed roller, and chopper were mounted on the same frame, as illustrated in Figure 15.

The angles θ and β represent the inclination of the disc and conveying plane A relative to the horizontal plane, respectively. In this study, θ was designed to be 2° greater than β (i.e., $\theta = \beta + 2°$).

During harvesting, sugarcane was cut at its base, 0–50 mm below the ground level. Previous studies indicated that the disc inclination angle should be maintained within 5–15° [26]. However, to accommodate varying sugarcane row ridge heights across different fields, θ could be adjusted accordingly. A hydraulic cylinder was employed as the adjustment mechanism, as depicted in Figure 15. The hydraulic cylinder enabled real-time angle adjustments during operation.

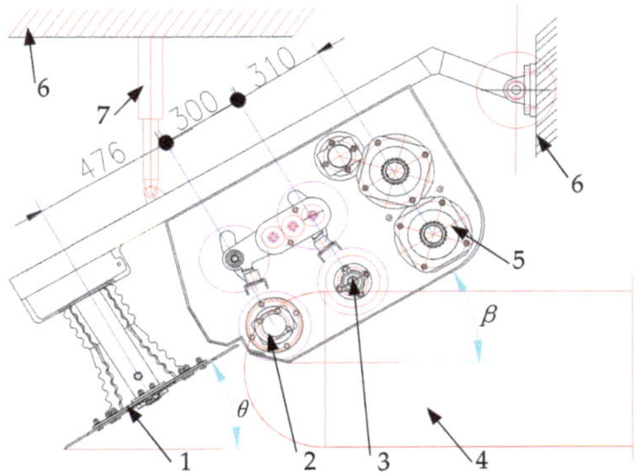

Figure 15. Geometric relationship between basecutter orientation and roller train configuration. 1—disc of basecutter. 2—butter lift roller. 3—feed roller. 4—crawler wheel. 5—chopper roller. 6—fixed frame. 7—hydraulic cylinder.

2.4.3. Structural of Roller Train

Analysis of the Structure Without Roller Train Between Basecutter and Chopper

The original CLASS 1400 design featured a sequential arrangement of (1) a basecutter, (2) a chopper, (3) a billet conveying device, and (4) an extractor, without intermediate roller train integration (Figure 16 [11]).

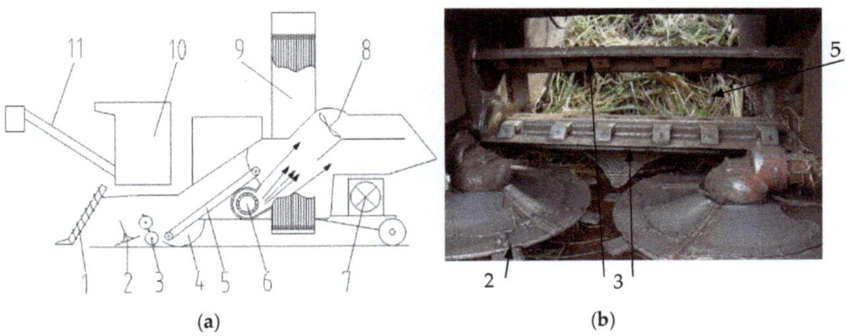

Figure 16. Schematic diagram of CLASS 1400 sugarcane harvester's structural layout: (**a**) structure diagram; (**b**) structure image. 1—divider. 2—basecutter. 3—chopper. 4—walking wheel. 5—billet conveyor. 6, 8—extractor fan. 7—engine. 9—billet elevator. 10—cab. 11—topper.

During the harvesting of curved sugarcane stalks, the stalks were directly processed into billets by the chopper without passing through any rollers (Figures 15 and 16). This configuration eliminated the possibility of stalk breakage by the roller train components. However, the chopping process introduced another issue: some billets were frequently recaptured by the blades of the upper chopper roller, leading to significant field losses (Figure 17). The authors observed this phenomenon during field trails at Maui in Hawaii (April 2013).

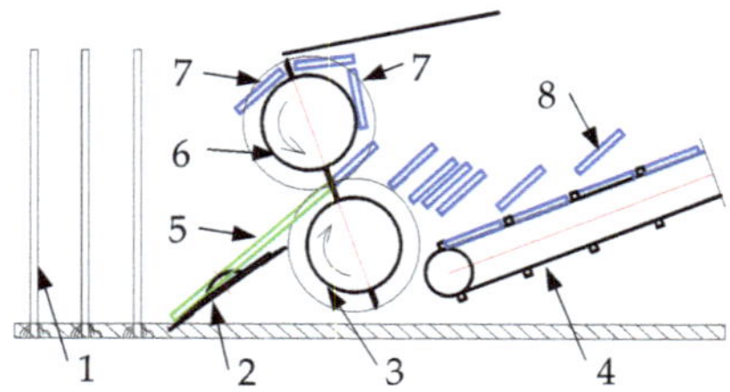

Figure 17. Schematic diagram of cane base cutting and chopping. 1,5—sugarcane. 2—basecutter. 3—lower chopper roller. 4—billet conveyor. 6—upper chopper roller. 7—billets brought back by blade. 8—billets fallen onto the conveyor.

Structural Improvements of Roller Train

The preceding analysis revealed that an excessive number of rollers between the basecutter and the chopper (Figures 11 and 12) increased the likelihood of curved sugarcane stalks detaching from the roller train or blocking the channel. Conversely, the absence of rollers in this region (Figures 16 and 17) led to a higher probability of billets being recaptured by the blades of the upper chopper roller.

To optimize performance, this study developed an improved roller train configuration consisting of a single pair of butt lift rollers and a single pair of feeding rollers positioned between the basecutter and the chopper (Figure 15).

Compared to the conventional multiple-roller train design (Figure 5), the reduced number of rollers minimized stalk blockage and detachment. Additionally, unlike the roller-free structure (Figures 16 and 17), the new design prevented billet recapture by the chopper blades. These findings suggest that a reduced roller count enhances the chopper harvester performance, a conclusion subsequently validated through further experimentation.

2.5. Development of Improved Sugarcane Harvester

2.5.1. Design of Improved Harvester

Based on the preceding analysis, the research team developed an enhanced sugarcane chopper harvester incorporating the optimized designs of the basecutter, roller train, and billet conveyor systems. The key structural parameters of these components are given as follows.

Basecutter Structural Configuration

The modified basecutter's dimensional specifications are illustrated in Figure 18.

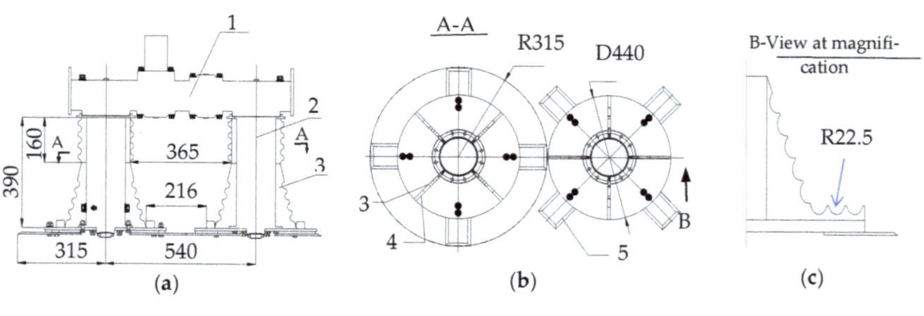

Figure 18. This figure illustrates key dimensional parameters of basecutter assembly: (**a**) structural dimensions of basecutter; (**b**) A-A view; (**c**) small-toothed plate. 1—gear box. 2—axis of disc. 3—toothed paddle. 4—small-toothed plate. 5—blade.

The disc diameter measured 440 mm, while the blades operated at a maximum rotational radius of 315 mm. Axial spacing between the components was maintained at 540 mm. A vertical clearance of 390 mm was specified between the gearbox's lower surface and the disc's upper surface. The small-tooth plate featured a groove radius of 22.5 mm and an overall height of 25 mm.

Structural Dimension of the Roller Train

The positional relationship between the basecutter and roller train is illustrated in Figure 15. Figure 19 presents the dimensional specifications of both the roller train and chopper rollers. The key measurements in Figure 19 are as follows: length (L) = 1017 mm, width (W) = 776 mm, total height (H) = 629 mm, H_1 = 264.5 mm, L_1 = 300 mm, L_2 = 310 mm, L_3 = 490 mm, L_4 = 495 mm, D_1 = D_3 = 232 mm, D_2 = 252 mm, and D_4 = 266 mm.

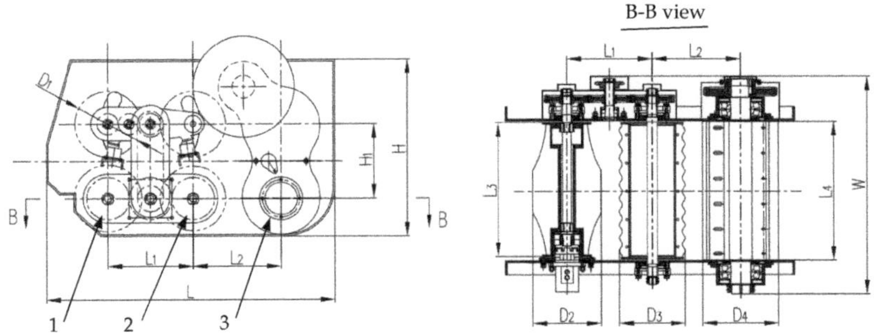

Figure 19. Dimensions schematic of roller train and chopper rollers. 1—butter lift rollers. 2—feed rollers. 3—chopper rollers.

Dimensional Configuration of the Chopper Assembly

The chopper roller was composed of an upper roller and a lower roller (Figures 19 and 20). The pairing relationship of the upper and lower rollers' blades is shown in Figure 20 [27].

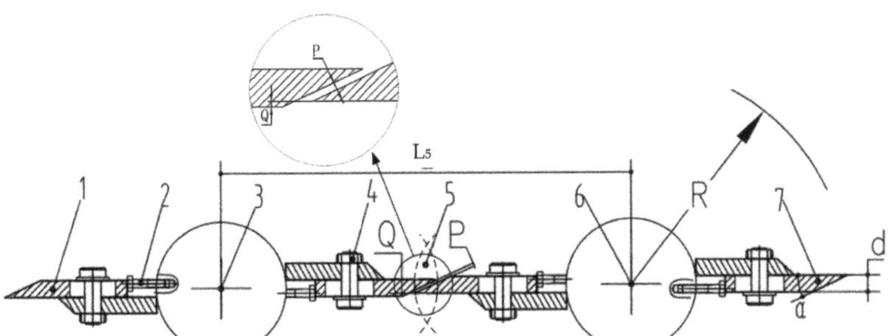

Figure 20. Dimensional drawing of chopper assembly. 1—blade. 2—adjusting bolt. 3—upper chopper roller. 4—fastening bolt. 5—overlap area. 6—lower chopper roller. 7—blade.

In this research, the center-to-center distance (L_5) between the upper and lower chopping rollers measured 252 mm. Each roller incorporated two symmetrically arranged blades secured using bolt fasteners. The blade geometry was characterized by three critical parameters: gyration radius (R = 133 mm), thickness (d = 10 mm), and edge angle (α = 25°). When the paired blades of the upper and lower rollers were in alignment, inter-blade clearance (Q) was maintained at 6 mm. Additionally, the chopping edges featured a bevel gap (P) of 2 mm.

Structural Dimension of Billet Conveyor System

The billet conveyor was positioned downstream of the chopper rollers, with its complete dimensional specifications detailed in Figure 21. This conveying system operated by transporting billets between uniformly spaced scrapers via continuous chains toward the extractor. The design incorporated several key features: a perforated plate that facilitated soil separation during operation, an effective internal clearance width (W_1) of 566 mm, scraper blades with a vertical height (H_2) of 160 mm, and consistent inter-scraper spacing (L_6) of 450 mm.

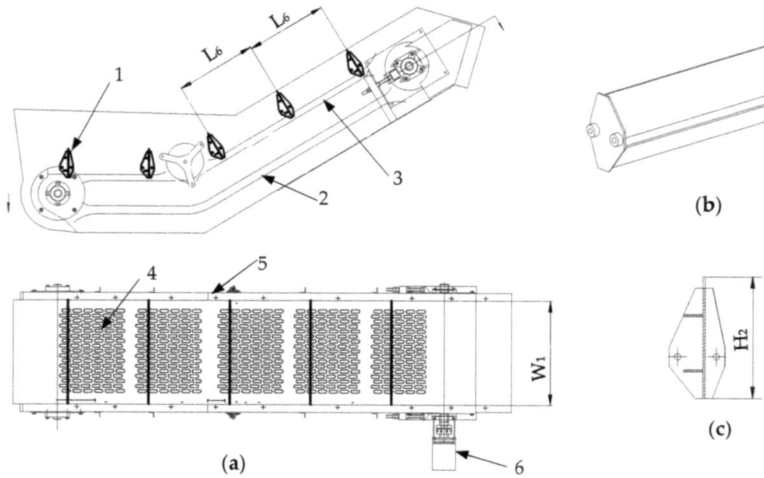

Figure 21. Structure dimension of billet conveyor: (**a**) structure of billet conveyor; (**b**) isometric views of scraper; (**c**) structural dimension of scraper. 1—scraper. 2—lower chain track. 3—upper chain track. 4—porous plate. 5—bed of conveyor. 6—hydraulic motor.

2.5.2. Structural Dimensions and Parameters of the Improved Harvester Prototype

A prototype of the improved harvester was developed as a single-row harvester. Designated as the HN4GDL-91 sugarcane harvester, its schematic diagram is presented in Figure 22. The key structural dimensions and design parameters are summarized in Table 1.

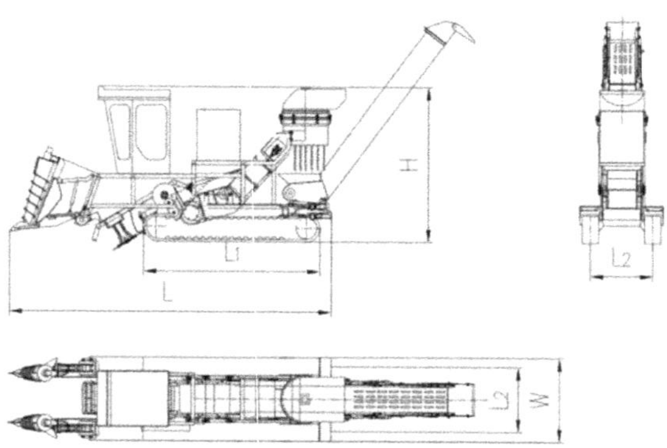

Figure 22. Design diagram of HN4GDL-91 sugarcane harvester.

Table 1. Parameters of HN4GDL-91 sugarcane harvester.

Parameters	Values
Overall size L × W × H (mm)	6355 × 1780 × 2940
Track length L_1/track gauge L_2 (mm)	3480/1230
Engine rated power (kW)	91
Displacement of the motor of knockdown roller (mL/r)	160
Displacement of the motor of basecutter (mL/r)	80
Displacement of the motor of butt lift/feed roller (mL/r)	200
Displacement of the motor of chopper (mL/r)	200
Displacement of the motor of billet conveying device (mL/r)	125

Its total weight was 6000 kg, and it was powered by a 97 kW engine. During operation, the maximum ground speed was designed at 5 km/h (1.39 m/s), with a maximum feed capacity of 20 kg/s.

2.6. *Performance and Field Test of the Harvester*

2.6.1. Experimental Instruments and Equipment

Field performance evaluations were conducted using the HN4GDL-91 sugarcane chopper harvester (Figure 22). The following instruments and tools were employed for data collection and measurements.

(1) Measurement devices:

Electronic balance (precision: ±0.01 kg).

Photoelectric tachometer (DT2234C, measurement range: 2.5–999,999 RPM).

Vernier caliper (resolution: 0.02 mm).

Measuring tape (length: 50 m).

Protractor (angle measurement).

Soil penetrometer (TJSD-750, measurement range: 0–7000 kPa).

Tachometer (EMT260D, measurement range: 1–99,999 rpm).

(2) Supporting equipment:

Digital camera (for photographic documentation).

Agricultural tools (sickle and shovel for field preparation).

A large plastic sheet was placed on the ground to collect the discharged billets, as illustrated in Figure 23.

Figure 23. HN4GDL-91 harvester field test configuration illustrating standardized billet collection method.

2.6.2. Field Testing Conditions and Experimental Setup

The field experiments were conducted in January 2018 at the Zhanjiang Test Station of the China Sugarcane Research System. The test site comprised a 100 hectare sugarcane field with the following characteristics.

(1) Soil Properties

Soil type: Red soil. Moisture content: 13–15% at 0–50 mm depth and 18–20% at 50–100 mm depth. Mean soil hardness: 2837 kPa (0–50 mm depth) and 3355 kPa (50–100 mm).

(2) Field Configuration

Dimensions: 330 m (length) × 300 m (width). Row spacing: 1.2 m. Total rows: 250.

(3) Crop Status

Severely lodged due to typhoon impact. Sugarcane variety: Yuetang 55. Planting density: 7–9 stalks/m.

2.6.3. Experiment Methods

Field Sampling Methodology

The experimental field contained 250 rows of sugarcane. For testing purposes, 10 rows on one side were reserved for the field test of the harvester. The remaining 240 rows were systematically numbered from 1 to 240. Each row was then subdivided into 33 equal block to establish a sampling grid.

The sampling protocol involved the following steps:

(1) Random selection of both row numbers (1–240) and segment numbers (1–33) for each sample group.

(2) Manual cutting of selected stalks at ground level.

(3) Random allocation of sample groups to experimental treatments.

Performance Evaluation Methodology

To assess the structural improvements' effectiveness in reducing field losses, key performance indicators were selected in accordance with JB/T 6275-2019 Sugarcane Combine Harvester standard [21], including the following:

(1) Field loss ratio (S_z)

The field loss ratio (S_z, %) was defined as the percentage of total sugarcane stem mass remaining in the field (W_{loss}) after harvest relative to the harvestable stem mass (W_q) in each test plot. The calculation is given as follows:

$$S_z = \frac{W_{loss}}{W_q} \times 100 \tag{7}$$

(2) Feed quantity (Q_w)

The feed quantity (Q_w, kg/s) was defined as the total mass of sugarcane processed by the harvester per unit time. This parameter was calculated as follows:

$$Q_w = \frac{W_s + W_{pa}}{t} \tag{8}$$

where W_s is the total mass of harvested material collected from the harvester's elevator output in the test plot, kg; W_{pa} is the processed material obtained from the extractor discharge in the test plot, kg; and t is the total harvesting time recorded for the plot.

Billet Qualification Ratio (C_{dh})

The billet qualification ratio (C_{dh}, %) was defined as the percentage of qualified billet mass (C_{dh}, kg) relative to the total billet mass (W_{dz}, kg) collected from the harvester during each test. The qualification criterion followed the JB/T 6275-2019 Sugarcane Combine Harvester standard [21]. The ratio was calculated as follows:

$$C_{dh} = \frac{W_{dh}}{W_{dz}} \times 100 \tag{9}$$

Performance Test Methods

(1) Test methods

In the field, sugarcane often grows in clusters due to tillering. Consequently, the harvester needed to be capable of harvesting clustered sugarcane. To simulate severely lodged sugarcane clumps, multiple stalks were placed closely together, with their bases oriented toward the harvester. These stalks were treated as a single group. Three such groups were sequentially arranged in the field as illustrated in Figure 24. Since the stalks entered the roller train base first, all the groups were harvested in the direction from their bases to their tops during the tests.

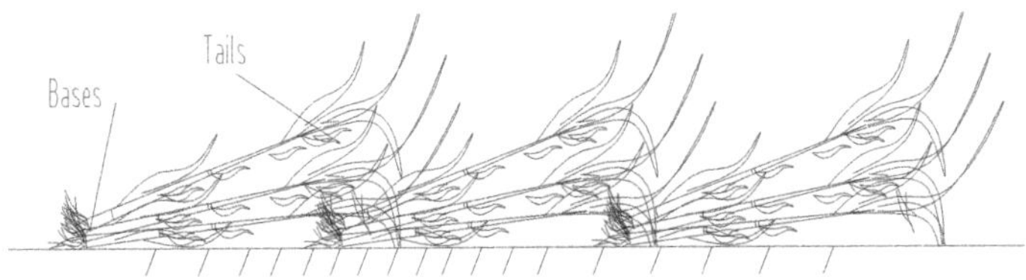

Figure 24. Layout schematic of sugarcane laying in field.

Test factors and experimental design. The rotational speed of the lift/feed rollers was positively correlated with the moving speed of stalks in the roller train. The moving speed affected the feed quantity (Q_w). The ground speed also affected the feed quantity (Q_w). The rotational speed of the chopper determined its cutting speed and affected the quality of cutting billets (C_{dh}). Therefore, the test factors included the chopper's rotational speed, the number of sugarcane stalks per group, the rotational speed of the butt lift/feed rollers, and the harvester's ground speed. Prior to the experiment, preliminary field tests were performed to determine the operational ranges of all the conditions under which the harvester functioned smoothly. Based on those results, the specific level values for each factor were established. Table 2 presents the factors and their respective levels. An $L_{27}(3^{13})$ orthogonal array was selected to conduct the tests. Each treatment was replicated three times.

Table 2. Factors and level assignments for $L_{27}(3^{13})$ orthogonal test.

	A: Rotational Speed of Chopper (r/min)	B: Numbers of Stalks in a Group	C: Rotational Speed of Butt Lift/Feed Rollers (r/min)	D: Ground Speed (m/s)
1	230	5	150	slow (0.65)
2	310	10	190	medium (1.09)
3	390	15	230	fast (1.41)

(2) Method for determination of Optimal Factor Combination

The comprehensive balance method was employed to identify the optimal factor combination for improving the harvester performance. The optimization process followed these methodological steps:

Performance Priority Establishment. Based on the study objectives, the performance indicators were prioritized as follows: primary, field loss ratio (S_z); secondary, feed quantity (Q_w); and tertiary, billet qualification ratio (C_{dh}).

Factor Selection Criteria. Statistical analysis revealed that factors showing significant effects ($p < 0.05$) on S_z and Q_w received priority consideration. Interaction effects between factors were incorporated into the optimization model.

Optimization Procedure. The comprehensive balance method integrated the range analysis results from orthogonal testing, the variance decomposition outcomes, and practical operational constraints.

Field Test Methods of the Harvester

During the field test, the harvester reaped lodged sugarcane in the field. In accordance with the *sugarcane harvester test standard* [21], the test area was divided into a stabilization zone, a measurement zone, and a parking zone, with lengths of 20 m, 10 m, and 10 m, respectively. The harvester traveling in the direction the lodged stalks fell over.

Billets collected from the plastic sheet corresponding to the test area were used as samples (Figure 24). The harvester's ground speed, chopper rotational speed, and butt lift/feed roller rotational speed were set according to the optimized parameter combination obtained from prior performance tests. Each treatment was replicated three times. The field trials were conducted by Guangdong Provincial Agricultural Machinery Appraisal Station staff.

3. Results and Analysis

3.1. The Results and Analysis of the SFC Tests

The results of the static friction coefficient (SFC) tests are presented in Table 3. The SFC values between the sugarcane stalks (de-leafed) and soil ranged from 0.59 to 0.84 in the vertical direction and 0.5 to 0.8 in the horizontal direction. For the stalks with leaves, the SFC values in relation to soil were 0.50–0.83 (vertical) and 0.64–0.9 (horizontal). Additionally, the SFC between the de-leafed stalks and steel was 0.48, while for the leafed stalks, it ranged from 0.4 to 0.48.

Table 3. Test results of the static friction coefficient (SFC) between sugarcane and soil/sugarcane and steel.

	Orientation	Sample	MC	1	2	3	Mean	Standard Error
Soil	Vertical	DL	5%	0.7	0.79	0.71	0.73	0.03
			10%	0.8	0.8	0.91	0.84	0.04
			15%	0.59	0.62	0.56	0.59	0.02
		WL	5%	0.74	0.73	0.85	0.77	0.04
			10%	0.79	0.88	0.82	0.83	0.03
			15%	0.41	0.51	0.56	0.5	0.04
	Horizontal	DL	5%	0.72	0.76	0.8	0.76	0.02
			10%	0.69	0.79	0.93	0.8	0.07
			15%	0.4	0.56	0.53	0.5	0.05
		WL	5%	0.92	0.93	0.84	0.9	0.03
			10%	0.62	0.71	0.59	0.64	0.03
			15%	0.75	0.73	0.72	0.73	0.01
Steel	Vertical	DL		0.48	0.46	0.49	0.48	0.01
		WL		0.4	0.39	0.42	0.4	0.01
	Horizontal	DL		0.51	0.44	0.49	0.48	0.02
		WL		0.47	0.49	0.47	0.48	0.01

Note: DL was sample de-leafed, WL was sample with leaves attached.

This study investigated the ability of the basecutter to pick up sugarcane. What is needed to be known are the SFC between the sugarcane stalks and soil and the SFC between the stalks and the steel plates. The base of sugarcane stalks in fields sometimes have leaves, and sometimes they do not. Therefore, the results of the experiments in Table 3 could be summarized as follows: the SFC between the stalks and steel was 0.4–0.48, and it was 0.5–0.9 between the stalks and soil.

In another study [19], the SFC between sugarcane stalks (de-leafed) and steel was 0.377–0.515. Overall, the SFC between the sugarcane stalks and steel is lower than that between the stalks and soil, that is, $f_1 < f_2$.

3.2. Sugarcane Field Lodging Severity

(1) Dimensions of sugarcane stalk

Sample size: 125.

Diameter (Mean, standard deviation): 23.45, 2.54 mm (at base); 24.48, 3.08 mm (at middle); and 20.94, 2.91 mm (at top).

Length: (Mean, standard deviation): 2.26, 0.276 m.

(2) Sugarcane Field Lodging Severity

Sample size: 2500. Sampling method: Randomly select one 2 m long block per row, with 10 sugarcane stalks in each block.

Lodging percentage in field: slight lodging (0–30°), 15.4%; moderate lodging (30–60°), 47.2%; and severe lodging (60–90°), 37.4%.

Bending status of lodged canes: 73.6% lodged canes had one curved section (Figure 25a), and 26.4% lodged canes had two curved sections (Figure 25b).

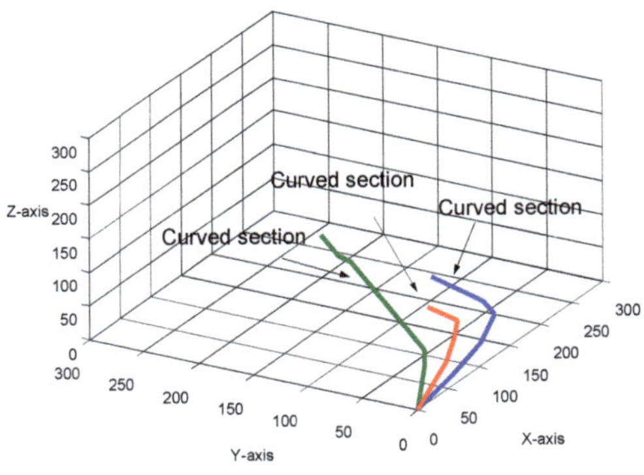

(a)

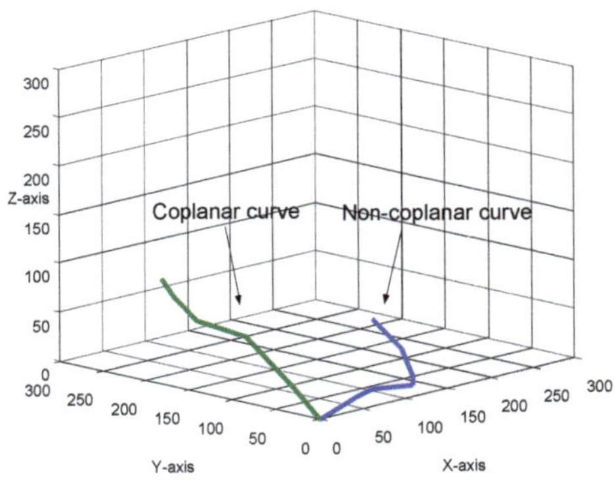

(b)

Figure 25. Lodged stalks with curved sections: (**a**) one curved section; (**b**) two curved sections.

3.3. Results and Analysis of the Performance Test

The test results are presented in Table 4, where the values represent the averages of three repeated treatments. Variance analysis of the test results is provided in Table 5. The statistical analysis revealed the following key findings: Factor A had an extremely significant effect ($p < 0.01$) on indicator C_{dh}. Factor B showed significant effects ($p < 0.05$) on indicators S_z and Q_w. The interaction effects A×D and B×D exhibited extremely significant influences ($p < 0.01$) on all the three indicators (S_z, Q_w, and C_{dh}).

Table 4. Results of orthogonal test of performance.

No.	A	B	A×B	C	A×C	B×C	D	A×D	B×D	C×D	S_z/%	Q_w/kg/s	C_{dh}/%
1	1	1	1	1	1	1	1	1	1	1	9.00	2.30	89.90
2	1	1	1	1	2	2	2	2	2	2	1.76	6.02	87.21
3	1	1	1	1	3	3	3	3	3	3	6.30	5.65	90.66
4	1	2	2	2	1	1	2	2	2	3	2.55	9.50	86.29
5	1	2	2	2	2	2	3	3	3	1	2.95	9.17	86.48
6	1	2	2	2	3	3	1	1	1	2	3.30	7.61	87.57
7	1	3	3	3	1	1	3	3	3	2	2.18	18.77	78.71
8	1	3	3	3	2	2	1	1	1	3	1.84	9.54	92.78
9	1	3	3	3	3	3	2	2	2	1	2.57	9.40	95.39
10	2	1	2	3	1	2	1	3	3	1	3.44	3.12	95.24
11	2	1	2	3	2	3	2	1	1	2	6.11	5.41	92.20
12	2	1	2	3	3	1	3	2	2	3	8.32	6.00	90.24
13	2	2	3	1	1	2	2	1	1	3	3.81	9.41	96.10
14	2	2	3	1	2	3	3	2	2	1	2.12	9.85	95.99
15	2	2	3	1	3	1	1	3	3	2	1.92	6.69	96.45
16	2	3	1	2	1	2	3	2	2	2	2.00	2.79	95.53
17	2	3	1	2	2	3	1	3	3	3	1.80	9.44	94.76
18	2	3	1	2	3	1	2	1	1	1	1.51	13.67	96.91
19	3	1	3	2	1	3	1	2	2	1	5.57	3.00	91.01
20	3	1	3	2	2	1	2	3	3	2	1.64	5.24	94.30
21	3	1	3	2	3	2	3	1	1	3	4.34	7.42	94.95
22	3	2	1	3	1	3	2	3	3	3	2.01	12.04	96.02
23	3	2	1	3	2	1	3	1	1	1	1.21	13.09	97.65
24	3	2	1	3	3	2	1	2	2	2	2.19	6.76	97.26
25	3	3	2	1	1	3	3	1	1	2	1.37	13.34	96.43
26	3	3	2	1	2	1	1	2	2	3	2.41	7.02	97.83
27	3	3	2	1	3	2	2	3	3	1	2.07	13.52	97.36

Note: The factors and their levels are as shown in Table 3.

Table 5. Analysis of variance (ANOVA) for orthogonal test results.

Performance Indicators	S_z, %	Q_w, kg/s	C_{dh},%
Source	*sig*	*sig*	*sig*
A	0.421	0.540	0.004 **
B	0.010 *	0.011 *	0.435
A×B	0.679	0.858	0.575
C	0.774	0.512	0.442
A×C	0.328	0.982	0.394
B×C	0.618	0.588	0.610
D	0.57	0.098	0.516
A×D	0.0 **	0.0 **	0.0 **
B×D	0.0 **	0.0 **	0.0 **
C×D	0.367	0.943	0.428

Note: * represents a significant impact with a 95% confidence interval; ** represents a significant impact with a 99% confidence interval.

The intuitive analysis of the test results is illustrated in Figure 26, while the interaction effects between D and A, as well as D and B, are graphically represented in Figure 27.

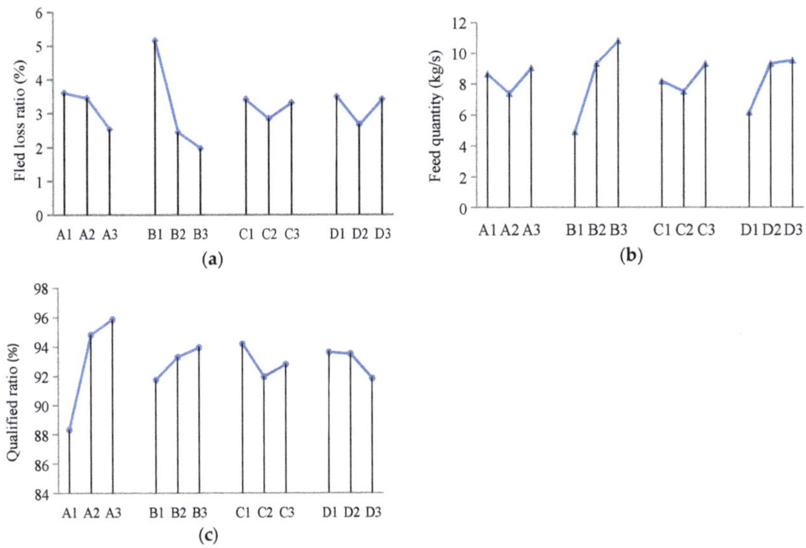

Figure 26. Influence trends and range analysis of factors on indicators: (**a**) factor effects on field loss ratio; (**b**) factor effects on feed quantity; (**c**) factor effects on qualification rate of billets.

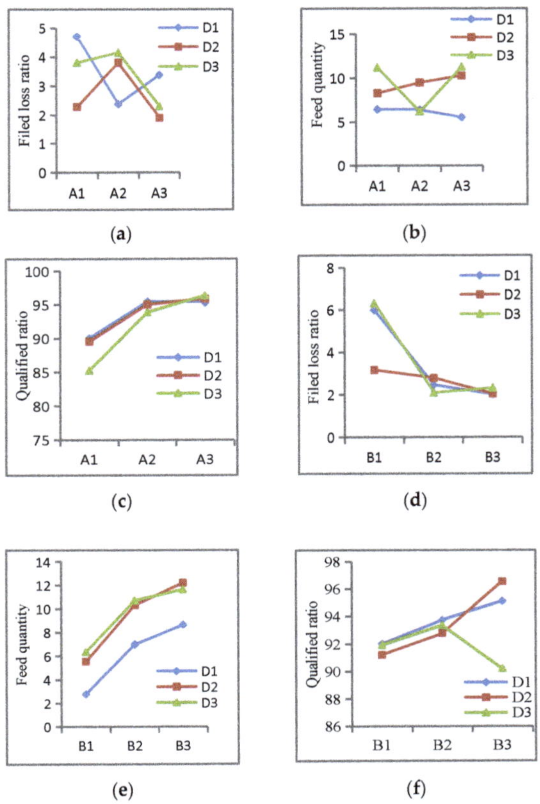

Figure 27. Effects of A×D and B×D interactions on performance indicators: (**a**) effects of A×D on S_z; (**b**) effects of A×D on Q_w; (**c**) effects of A×D on C_{dh}; (**d**) effects of B×D on S_z; (**e**) effects of B×D on Q_w; (**f**) effects of B×D on C_{dh}.

(1) Effects of operational factors on field loss ratio (S_Z)

Figure 26a demonstrates that S_z exhibited a decreasing trend with increasing levels of both Factor A and Factor B. When Factor A's level was elevated, the cutting speed of the chopping blades correspondingly increased. This enhanced cutting speed proved effective in improving the quality of the sugarcane cutting billets [10].

Factor B (number of sugarcane stalks per feeding group) exerted a more pronounced influence on the field loss ratio compared to Factor A. This phenomenon occurred because the basecutter successfully captured nearly all the stalks that came into contact with its disc, subsequently transporting them to the butt lift rollers. Consequently, only minimal billet quantities remained uncollected in the field, resulting in negligible harvest losses.

(2) Effects of factors on the feed quantity (Q_w)

Figure 26b illustrates that the feed quantity increased progressively with higher levels of Factors B and D. This observation aligns with the fundamental definition of feed quantity in harvesting operations.

The experimental results demonstrated that the harvester achieved an optimal feeding performance at the tested levels of Factors B and *D*.

Specifically, Factor *B* (number of stalks per feeding group) directly influenced the material intake capacity. Factor D (ground speed) affected the continuous flow of feed stock.

The combined effect of these factors ensured efficient material handling throughout the harvesting process.

(3) Effects of factor levels on billet qualification ratio (C_{dh})

As shown in Figure 26c, the qualification ratio of billets increased with the increase in levels of Factors A and B. With the level of Factor *A* increased, the cutting speed of the chopping blades increased. A higher cutting speed was helpful to cutting quality of the billets. As Factor *B* increases, there were more stalks in the roller train. More stalks was helpful to increase the force of which exerted on the stalks by the upper and lower rollers. This force acted as a support for the chopping blades cutting the stalks. The supported cutting quality usually was better than that of unsupported cutting.

(4) Effects of A×D

When Factors B and C were held constant, the following relationship was observed: increasing factor D (ground speed) resulted in a higher feed quantity (Q_w). Higher levels of Factor A (cutting speed) produced the improved cutting quality and reduced the billet length.

As shown in Figure 27c, when Factor A increased from level A1 to A2, the billet qualification rate (C_{dh}) increased significantly from 85% to 95%. This improvement plateaued near 95% due to random cutting positions along the stalks (including nodal regions and stalk ends).

The mechanisms underlying the results in Figure 25a,b remain unclear and warrant further investigation to better understand the interaction dynamics.

(5) Effects of B×D

When Factor B (stalk grouping number) was maintained at constant levels, the feed quantity (Q_w) showed a positive correlation with increasing factor D (ground speed). When the D levels were fixed, Q_w increased proportionally with higher B values, as shown in Figure 27e.

This synergistic relationship was attributed to two key mechanical advantages of the harvesting system: the basecutter's toothed disc assembly exhibited an exceptional lodged cane recovery performance, and the optimized roller configuration demonstrated high conveying efficiency, particularly for densely grouped stalks. Consequently, the combined losses throughout these processes remained consistently low, confirming the strong positive interaction between these operational parameters (Figure 27d).

(6) Determination of Optimal Factor Combination

The optimal treatments for test indicators S_z, Q_w, and C_{dh} shown in Table 4 were A3B2C3D3, A1B3C3D3, and A3B3C1D3, respectively.

The optimal treatments for test indicators S_z, Q_w, and C_{dh} shown in Figure 26a–c were A3B3C2D2, A3B3C3D3, and A3B3C1D1, respectively. Therefore, A3B3 represents the optimal level.

For A×D interaction (Figure 27a–c), A3D2 and A3D3 were better than A_3D_1 in terms of both S_z and Q_w. All the three configurations (A3D1, A3D2, and A3D3) showed an equivalent performance for C_{dh}.

For B×D interaction (Figure 27d–f), all the three configurations (B3D1, B3D2, and B3D3) showed an equivalent performance for S_z. Both B3D2 and B3D3 were better than B3D1 for Q_w. Both B3D1 and B3D2 were better than B3D3 for C_{dh}. Therefore, D2 is the optimal choice in the case of considering interaction between A×D and B×D.

In Table 5, C, A×C, B×C, and C×D had no significant effect on the three test indicators. In Figure 26a, C2 was best for S_z. In Figure 26b, C3 was best for Q_w In Figure 26c, C1was best for C_{dh}. In this research, S_z had a highest priority than Q_w and C_{dh}, so C2 was selected as the optimal level.

Based on the comprehensive analysis of factor effects, the combination A3B3C2D2 was identified as the theoretically optimal treatment. However, this specific combination was not included in the original orthogonal array (Table 4).

To validate these findings, additional verification tests were conducted following the experimental methodology described in Section 2.6.3.

The experimental results for treatments A3B3C2D2 and A3B2C3D3 (the latter is included in Table 4) are presented in Table 6. Comparative statistical analysis yielded the following findings:

Table 6. Test results of treatments A3B3C2D2 and A3B2C3D3.

		S_z, %	Q_w, kg/s	C_{dh},%
A3B3C2D2	1	1.81,	13.41	97.81
	2	1.74	14.5	98.28
	3	1.58	12.92	96.85
	Mean	1.71	13.61	97.65
	Standard error	0.065	0.467	0.421
A3B2C3D3	1	1.26	13.29	95.62
	2	1.15	13.65	97.41
	3	1.22	12.34	94.42
	Mean	1.21	13.09	95.82
	Standard error	0.032	0.391	0.869
t-test value		6.6421	0.8486	−1.8958

S_z: $t = 6.6421 > t(4)_{0.01} = 3.7496$ ($p < 0.01$).

Q_w: $t = 0.8486 < t(4)_{0.05} = 2.138$ ($p > 0.05$).

C_{dh}: $t = 1.895 < t(4)_{0.05} = 2.1318$ ($p > 0.05$).

S_z showed highly significant differences between treatments. Neither Q_w nor C_{dh} demonstrated statistically significant differences.

Therefore, A3B2C3D3 is the optimal treatment in this research.

3.4. Results and Analysis of the Field Test

Field testing was conducted under severe sugarcane lodging conditions. The operational parameters of ground speed, rotational speed of chopper, and rotational speed of butt lift/feed rollers were set as 1.1 m/s, 390 r/min, and 230 r/min, respectively. Each treatment was replicated three times. The results are as shown in Table 7.

Table 7. Results of field test.

Parameters	1	2	3	Mean	Standard Error
Feed quantity, kg/s	10.1	11.2	10.2	10.5	0.287
Ratio of stubble broken, %	12.9	13.2	12.3	12.8	0.215
Ratio of trash content, %	5.9	4.2	4	4.7	0.492
Field loss ratio, %	2.1	1.7	2.2	2.0	0.125
Qualification rate of the billets, %	92.9	95.1	94.6	94.2	0.544
Unit diesel consumption ratio, kg/ha	125.8	127.2	127.4	126.8	0.411

The field test results demonstrated that the developed harvester operated effectively in severely lodged sugarcane fields.

Under typical conditions in the Zhanjiang region, the lodged field yield average is 45,000–60,000 kg/ha, and the conventional field losses during harvesting by a chopper harvester range from 7500 to 12,000 kg/ha. The field loss ratio is about 16–20%.

Comparative analysis revealed that the prototype harvester achieved a significantly lower field loss ratio than the commercial harvesters under identical conditions.

4. Discussion and Conclusions

4.1. Discussion

This study categorized sugarcane harvesting losses into lost losses (unrecovered stalks) and stubble losses (remaining stalk portions) in accordance with the JB/T 6275-2019 Sugarcane Combine Harvester Standard [21]. However, an additional loss mechanism—juice loss during billet chopping—was identified, with the reported losses ranging from 2 to 5 tonnes per hectare [10].

The row profile significantly influenced the field loss ratios during harvesting. Based on prior research [28], the optimal configuration was a slightly rounded mound (100–150 mm elevation above flat inter-rows). In contrast, crater-shaped rows increased the cane losses due to uneven cutting dynamics.

In this study, the test field featured a 150 mm mound profile. This design enabled subsurface cutting (blades operating below ground level), zero stubble loss (complete stalk recovery), and isolation of lost losses (primary focus of optimization). Consequently, the measured field loss ratio exclusively reflected lost losses, aligning with the research objective of minimizing the amount of unrecovered cane.

The experimental results demonstrated that cane losses were primarily governed by two operational parameters of extractor speed (air velocity) and harvester pour rates. In this research, the harvester maintained an average extractor air velocity range of 14–18 m/s during testing, which consistently achieved the target trash content of about 5%.

The current standard chopper harvester configuration utilized an axial flow exhaust fan extractor system. This design required a specific vertical elevation (about 1–2 m) between the billet intake and the elevator billet collection box. As illustrated in Figures 4 and 12, conventional multi-roller systems serve two primary functions of pre-chopping stalk elevation and soil separation prior to extraction.

In this study, the prototype incorporated only two roller pairs, relying instead on an extended billet conveyor for material transport to the extractor. This configuration demonstrated two key characteristics of being beneficial for reducing the harvester length and increasing complexity in conveyor routing. How to reduce the conveyor length, while meeting the impurity removal requirements is the key to realizing the miniaturization of harvesters.

When blades cut stalks beneath the surface of the ground, some soil is carried into the roller train. Multiple rollers were beneficial for discharging the soil before it entered the chopper rollers. Using fewer rollers was not conducive to the removal of soil. This requires the blade to cut stalks at the surface of the ground rather than beneath it.

In China, sugarcane is mainly harvested by hand. The high cost of manual cutting leads to a higher production cost. It further leads to a decrease in planting areas. As a result, sugar mills have not been able to obtain enough sugarcane in recent years. Mechanized harvesting is one of the important measures used to reduce the production cost of sugarcane. However, field losses during mechanized harvesting reduce the available yield. This is a key drawback for farmers and sugar mills when deciding whether to adopt mechanized harvesting. Compared with the existing field loss rate of 15–20% for sugarcane chopper harvesters, the 2% loss rate in this study was beneficial for encouraging farmers and sugar mills to adopt mechanical harvesting. Since 2022, sugarcane chopper harvesters have been difficult to sell in China. The main reason was also the high field loss rate. The technology used in this research is expected to solve this problem.

These sugarcane harvesters have a structural complexity that is basically the same as that of traditional harvesters. Therefore, their durability and manufacturing cost are also basically the same.

4.2. Conclusions

(1) Analysis of high field losses revealed they are caused by three factors: picking up stalks from the ground and conveying them backwards to the butt lift rollers by the basecutter, and conveying them by the roller train.

When the raised patterns and the steel bars on the basecutter disc were covered by sticky soil, they no longer functioned effectively. As a result, the disc failed to lift severely the lodged sugarcane. Additionally, due to their bent shape, curved stalks exiting the basecutter sometimes struggled to enter the gap between the upper and lower butt lift rollers. When these bent stalks passed through multiple roller pairs, they were more likely to contact the side walls of the roller train and detach from it. These issues led to channel blockages, stalk breakage, and stalks being left in the field.

(2) Structural improvements were proposed for the chopper harvester, including installing toothed plates on the upper surface of the cutter disc, establishing an optimal positional relationship between the disc and the butt lift rollers, and reducing the number of roller pairs between the basecutter and the chopper to just one pair of butt lift rollers and one pair of feed rollers.

The mechanisms for reducing field loss were analyzed. The theoretical analysis results indicated the following: The conventional basecutter disc relied on friction force to pick up stalks, whereas the small-toothed plates utilized grabbing force, enabling smoother pickup of severely lodged cane. The disc's inclination angle needed to be 2° greater than that of the roller train. When L_i (the intersection line between the top surface of the cutter disc and the tangential plane) was above L_f (the line of intersection between the central plane of the lower butt lift roller and the tangential plane), but below A (The symmetrical plane between the upper and lower rollers), the stalks could smoothly enter the butt lifter rollers. Using fewer rollers reduced the likelihood of curved canes blockage or detaching from the roller train.

(3) After implementing the above improvement measures, a prototype sugarcane harvester was designed and developed. Four-factor, three-level orthogonal tests were conducted to evaluate its performance in harvesting severely lodged sugarcane. Severely lodged sugarcane was simulated by stalks laid on the ground in the field. Field tests were carried out to assess the harvester's effectiveness in harvesting growing lodged cane.

The performance test results showed that under the optimal parameter combination—a chopper rotational speed of 390 r/min, 10 sugarcane stalks per group, a butt lift/feed roller rotational speed of 230 r/min, and a ground speed of 1.41 m/s—the field loss ratio, feed quantity, and the billet qualification ratio were 1.21%, 13.09 kg/s, and 95.82%, respectively.

The field tests demonstrated that the harvester could smoothly harvest severely lodged sugarcane. In these tests, the field loss ratio, feed quantity, and the billet qualification ratio were 2.0%, 10.05 kg/s, and 94.2%, respectively. Compared to the 15–20% field loss ratio demonstrated by existing sugarcane chopper harvesters when harvesting lodged cane, the proposed structural improvements significantly reduced the field losses.

Author Contributions: Conceptualization, J.J. and Q.L.; validation, X.H., H.X. and T.W.; formal analysis, X.H.; investigation, J.J., J.F., Y.L. and X.Z.; resources, X.H.; data curation, J.F. and Y.L.; writing—original draft preparation, J.J.; writing—and editing, J.J., H.X. and Q.L.; visualization, X.H.; supervision, Q.L.; project administration, Q.L.; funding acquisition, T.W. All authors have read and agreed to the published version of the manuscript.

Funding: This research was funded by the "China National Major Agricultural Research Project" and the "China National Modern Agricultural Industry Technology System", grant number 170402.

Data Availability Statement: This experiment was conducted in January 2018. The data presented in this study are available on demand from the communication author at qingting@scau.edu.cn.

Conflicts of Interest: The authors declare no conflicts of interest.

References

1. Younos, M.A.; Embaby, E.M. Natural occurrence of fungi and aflatoxins associated sugarcane plant and sugarcane juice and their Control. *Indian J. Microbiol.* **2024**, *64*, 500–510. [CrossRef] [PubMed]
2. Wen, C.; Yan, Y.; Wu, W.; Li, J.; Hou, B.; Cui, W.; Huang, Y.; Li, K.; Long, X.; Nong, H. Height detection of crop divider toes of sugarcane harvester based on Kalman adaptive adjustment. *Sci. Rep.* **2023**, *13*, 16817. [CrossRef] [PubMed]
3. Bai, J.; Ma, S.; Wang, F.; Xing, H.; Ma, J.; Wang, M. Performance of crop dividers with reference to harvesting lodged sugarcane. *Sugar Tech* **2020**, *22*, 812–819. [CrossRef]
4. Liu, Q.; Chen, J.; Ou, Y. Spatial attitude of lodged sugarcane stalk in field. In Proceedings of the ASABE's Annual International Meeting 2012, Dallas, TX, USA, 29 July–1 August 2012.
5. Bai, J.; Ma, S.; Wang, F.; Xing, H.; Ding, Z.; Hu, J. Lifting performance of crop dividers with variable pitch spiral scrolls. *Sugar Tech.* **2022**, *24*, 1877–1886. [CrossRef]
6. Wu, T.; Ren, J.; Liu, Q.; Zou, X.; Zhang, Z.; Zhou, S.; Feng, J.; He, L. Development of the 4GDZ-132 sugarcane chopper combine harvester. *Trans. Chin. Soc. Agric. Eng.* **2022**, *38*, 30–38. (In Chinese)
7. Ma, S.; Scharf, P.A.; Karkee, M.; Zhang, Q. Performance evaluation of a chopper harvester in Hawaii sugarcane fields. In Proceedings of the 2014 ASABE International Annual Meeting, Montreal, Canada, 13–16 July 2014; American Society of Agricultural and Biological Engineers: St. Joseph, MI, USA, 2014; p. 1.
8. Lu, H.; Wu, T. Reflection on development status of mechanized sugarcane harvesting in Guangxi. *Sugarcane Canesugar* **2023**, *52*, 53–59.
9. Jiang, J.; Yang, Y.; Wu, C.Y.; Liu, Q.T.; Ou, Y.G.; Mo, Q.G.; Xia, M.; Wu, T.; Liang, S. Analysis on reasons for the slow advance of sugarcane harvesting mechanization in China. *Sugarcane Canesugar* **2022**, *51*, 1–5.
10. Gary, S.; John, A. *The Harvesting Best Practice Manual for Chopper-Extractor Harvester*; Bureau of Sugar Experiment Station: Brisbane, Australia, 2002.
11. Kerr, B.; Blyth, K. *100 Years of Mechanical Cane Harvesting*; Canegrowers: Ingham, Australia, 1993.
12. Dias Neto, A.F.; Albiero, D.; Rossetto, R.; Biagi, J.D. Modeling of Mechanized Sugarcane Harvesting to Support Decision-Making on Asset Management. *Sugar Tech. Int. J. Sugar Crops Relat. Ind.* **2022**, *24*, 798–812. [CrossRef]

13. Wang, Q.; Zhou, G.; Huang, X.; Song, J.; Xie, D.; Chen, L. Experimental research on the effect of sugarcane stalk lifting height on the cutting breakage mechanism based on the sugarcane lifting-cutting system (SLS). *Agriculture* **2022**, *12*, 2078. [CrossRef]
14. Qiu, M.; Meng, Y.; Li, Y.; Shen, X. Sugarcane stem cut quality investigated by finite element simulation and experiment. *Biosyst. Eng.* **2021**, *206*, 135–149. [CrossRef]
15. Qian, J.; Ma, S.; Xu, Y.; Liang, W.; Zhou, B.; Li, W. Analysis of crack stress intensity factor of sugarcane cutting based on fracture mechanics. *Trans. Chin. Soc. Agric. Mach.* **2023**, *54*, 101–109. (In Chinese)
16. Ding, Z.; Ma, S.; Zhang, X.; Liang, W.; Li, L.; Su, C. Ultrasonic sensor-based basecutter height control system of sugarcane harvester. *Sugar Tech* **2023**, *25*, 453–459. [CrossRef]
17. Li, H.; Yang, J.; Song, J.; Sun, M.; Wei, W.; Zhang, S.; Wu, T.; Li, Y. Stem base detection in sugarcane plants using improved YOLOv5m model. *Sugar Tech* **2025**, 1–11. [CrossRef]
18. Davis, R.J.; Norris, C.P. Improving the feeding ability of sugarcane harvesters. In Proceedings of the 2002 Conference of the Australian Society of Sugar Cane Technologists, Cairns, Australia, 29 April–2 May 2002; pp. 190–198.
19. Wang, M.; Liu, Q.; Ou, Y.; Zou, X. Determination of physical and mechanical properties of sugarcane single-bud billets. *J. ASABE* **2022**, *65*, 1029–1038. [CrossRef]
20. Ma, L.; Huang, W.; Li, S.; Huang, Z.; Teng, X. Parameters calibration of discrete element model for latosolicred soil of sugarcane in Guangxi Autonomous Region. *J. Agric. Mech. Res.* **2023**, *45*, 18–26.
21. *JB/T 6275-2019*; Sugarcane Combine Harvester. Ministry of Industry and Information Technology of the People's Republic of China: Beijing, China, 2019.
22. *CASE A8000 and A8800 Cane Harvesters Operator's Manual, 84158982-01*; Case IH: Sydney, Australia, 2009.
23. Available online: https://www.deere.com/en/search/?term=Sugar+Cane+Harvester (accessed on 10 April 2025).
24. Wang, F.; Ma, S.; Xing, H.; Bai, J.; Ma, J.; Hu, J.; Yang, Y. High-speed photography analysis on sugarcane base cutting process of contra-rotating basecutters. *Sugar Tech* **2021**, *23*, 1118–1125. [CrossRef]
25. Qiu, C.; Mo, H.; Li, S.; He, G.; Zeng, B.; Yang, D. Experimental research on influence factors of the sugarcane ratoon cutting quality undervibration conditions. In *Proceeding of the IncoMe-VI and Tepen 2021, Tianjin, China, 20–23 October 2021*; Springer: Cham, Switzerland, 2023. [CrossRef]
26. Wang, F.L.; Ma, S.C.; Xing, H.N.; Bai, J.; Ma, J.-Z.; Wang, M.-L. Effect of contra-rotating basecutter parameters on basecutting losses. *Sugar Tech Int. J. Sugar Crops Relat. Ind.* **2021**, *23*, 278–285. [CrossRef]
27. Xie, L.; Wang, J.; Cheng, S.; Zeng, B.; Yang, Z. Performance evaluation of a chopper system for sugarcane harvester. *Sugar Tech.* **2019**, *21*, 825–837. [CrossRef]
28. Hogarth, D.M.; Allsopp, P.G. *Manual of Canegrowing*; Sugar Research Australia: Brisbane, Australia, 2000.

agriculture

MDPI

Article

Design and Experiment of Autonomous Shield-Cutting End-Effector for Dual-Zone Maize Field Weeding

Yunxiang Li [1,2], Yinsong Qu [1,2], Yuan Fang [2], Jie Yang [1] and Yanfeng Lu [2,*]

1 Beijing Institute of Technology, Beijing 100081, China; 3120235332@bit.edu.cn (Y.L.); 3120246387@bit.edu.cn (Y.Q.); 3120225214@bit.edu.cn (J.Y.)
2 The State Key Laboratory of Multimodal Artificial Intelligence Systems, Institute of Automation, Chinese Academy of Sciences, Beijing 100190, China; fangyuan2023@ia.ac.cn
* Correspondence: yanfeng.lv@ia.ac.cn

Abstract: This study presented an autonomous shield-cutting end-effector for maize surrounding weeding (SEMSW), addressing the challenges of the low weed removal rate (WRR) and high seedling damage rate (SDR) in northern China's 3–5 leaf stage maize. The SEMSW integrated seedling positioning, robotic arm control, and precision weeding functionalities: a seedling positioning sensor identified maize seedlings and weeds, guiding XYZ translational motions to align the robotic arm. The seedling-shielding anti-cutting mechanism (SAM) enclosed crop stems, while the contour-adaptive weeding mechanism (CWM) activated two-stage retractable blades (TRWBs) for inter/intra-row weeding operations. The following key design parameters were determined: 150 mm inner diameter for the seedling-shielding disc; 30 mm minimum inscribed-circle for retractable clamping units (RCUs);
40 mm ground clearance for SAM; 170 mm shielding height; and 100 mm minimum inscribed-circle diameter for the TRWB. Mathematical optimization defined the shape-following weeding cam (SWC) contour and TRWB dimensional chain. Kinematic/dynamic models were introduced alongside an adaptive sliding mode controller, ensuring lateral translation error convergence. A YOLOv8 model achieved 0.951 precision, 0.95 mAP50, and 0.819 mAP50-95, striking a balance between detection accuracy and localization precision. Field trials of the prototype showed 88.3% WRR and 2.2% SDR, meeting northern China's agronomic standards.

Keywords: maize; weed detection; dual-zone weeding; end-effector; adaptive sliding mode; field validation

1. Introduction

Weeds severely impact agricultural production as they compete with crops for essential resources such as light, water, and nutrients. This competition restricts crop growth, reduces yields, deteriorates quality, and decreases farmers' incomes [1–4]. Maize, one of the world's vital cereal crops, is especially prone to weed interference during its early growth stages, specifically the 3–5 leaf stage. When weeds are not promptly removed, maize may experience growth retardation and even yield decline [5]. Traditional weeding methods demand significant labor input and are inefficient, thus being unable to meet the precision management requirements of modern agriculture. Consequently, the development of maize weeding robots with precise weeding capabilities and low SDR is crucial. These robots should be adaptable to automated operations in narrow fields, reducing labor intensity, enhancing weeding efficiency, and facilitating the intelligent and precise development of agriculture [6,7].

Weed management in maize fields predominantly involves three techniques: mechanical, chemical, and laser weeding. Mechanical weeding, in contrast to chemical weeding, provides a greener alternative by decreasing herbicide application, thus mitigating environmental contamination. When compared with laser weeding, mechanical weeding demonstrates lower reliance on high-precision visual recognition systems, higher adaptability, and reduced energy consumption. Consequently, mechanical weeding was chosen as the operation mode for the maize weeding robot in this research.

The core challenge in designing maize weeding robots resides in the end-effector. Current research is primarily divided into two categories: contact-type and non-contact-type mechanisms. Contact-type weeding devices (contact with the soil surface) act directly on weed roots by penetrating the soil, yielding remarkable weeding results. A representative example is Lemken's EC-weeder [8]. It utilizes the downward pressure of a compression spring to enable finger-clamp weeders to penetrate compacted soil and efficiently remove inter-row weeds. Liu et al. [9] developed a double-disc cultivator weeder that integrates disc-type weeding components with single- (double-) winged shovels, thereby improving operational quality. Chen et al. [10,11] developed an intra-row brush weeding manipulator, where a hydraulic motor rotates an eccentric cutter bar via gear transmission, enabling brush weeders to operate around crop plants. After parametric optimization, this system achieves 83.4% WRR with 2.5% SDR. To adapt to different plant-spacing, row-spacing, and growth stages, Gobor et al. [12] developed a horizontally rotating integrated arm weeding tool along crop rows. Comprising an arm holder and three or more integrated arms, the tool primarily detected the distance between adjacent plants and the angular position of the weeding mechanism.

Non-contact weeding mechanisms typically employ rotating blades or clamping and pulling devices to perform weeding operations while minimizing soil disturbance, reducing operational resistance, and alleviating crop roots damage. Nørremark et al. [13,14] developed a claw-type intra-row weeding robot equipped with four S-shaped vertical ground-penetrating tines for precise weed removal around plants. Pérez-Ruíz et al. [15] designed a pneumatic oscillating weeding device that avoids seedlings while weeding. Tested in precision-transplanted row crops, this design might also suit direct-seeded row crops. Chen et al. [16,17] developed a weeding robot applicable to the maize seedling stage. This robot integrates mechanical and chemical weeding approaches. It first cuts the weeds and then sprays herbicides to impede weed regrowth. Jia et al. [18] designed an intra-row weeding device for the maize-jointing stage. Once the mechanical flexible shaft detects maize seedlings, the weeding component drives the comb teeth to rotate intermittently, thus avoiding the plants. This device attains 94.7% WRR and 5.9% SDR.

Recent advances in machine vision have significantly enhanced precision weeding through innovative weed detection technologies, providing essential technical support for the mechanical weeding devices discussed in preceding sections. Early studies primarily focused on integrating color features with morphological analysis for plant recognition. Slaughter et al. [19,20] pioneered real-time color segmentation using dual-camera systems to monitor crop rows, achieving positioning accuracy ranging from 4.2 mm in weed-free conditions to 12 mm under high weed densities. To address species differentiation challenges, Pantazi et al. [21] introduced an active learning framework utilizing spectral reflectance differences, where Self-Organizing Maps and Mixture of Gaussians classifiers demonstrated 100% crop recognition accuracy with weed identification rates between 53% and 94%. Subsequent research leveraged deep learning for improved robustness, with Sa et al. [22] developing an encoder–decoder CNN architecture that processes multispectral aerial imagery through NDVI-based vegetation segmentation, enabling automated crop–weed distinction in complex field conditions. Simultaneously, Dyrmann et al. [23] overcame occlusion challenges in cereal fields through a fully convolutional neural network trained on 17,000 annotated field images, effectively detecting weeds despite overlapping foliage. In recent years, YOLO (You Only Look Once), a cutting-edge deep-learning object detection framework, has gained prominence in agricultural applications. Its unique architecture enables rapid and accurate real-time detection, making it well-suited for the dynamic and complex scenarios of precision agriculture. By integrating YOLO with existing technologies, such as spectral analysis, aerial imaging, and occlusion-handling algorithms, researchers have established a robust multimodal data fusion paradigm. This approach effectively meets the stringent demands of real-time operation and environmental adaptability in modern farming. The successful implementation of YOLO-based models in previous studies has laid a solid foundation for further optimizing weeding mechanisms, as demonstrated in the subsequent sections of this paper.

Existing studies have primarily focused on single-zone weeding (either inter-row or intra-row) in maize fields, resulting in suboptimal mechanical weeding performance. Additionally, insufficient precision in seedling–weed identification and motion control often leads to crop damage. Furthermore, there is a lack of innovative research on the theory and mechanism design for simultaneous inter-row and intra-row weeding. To address these limitations, this study proposes an SEMSW. First, the seedling positioning sensor localizes the positions of maize seedlings and weeds. The control system drives the weeding robotic arm to translate in the X, Y, and Z directions according to the feedback position coordinates, aligning the SEMSW with the maize seedlings. The SAM then encloses and protects the crop stems, while the CWM drives the TRWB through rotational shaping motions to cut weeds in both inter-row and intra-row areas.

The main research aims of this study are as follows. First, critical parameters for the SEMSW's key components are determined through theoretical modeling. Second, by modeling the structure of the weeding robotic arm, its kinematic and dynamic models are established, and a lateral translation error model of the robotic arm is constructed. An adaptive sliding mode controller for the lateral translation mechanism is designed, and its stability is verified to ensure the convergence of lateral translation errors. Additionally, a high-performance weed detection model based on YOLOv8 is developed to enhance the SEMSW's operational precision. Finally, prototype fabrication and field trials validate its operational efficacy, providing valuable references for maize weeding research in the agricultural systems of northern China.

2. Materials and Methods

2.1. Working Principle and Machine Structure

2.1.1. Working Principle

The SEMSW operational workflow consisted of two sequential phases: seedling-shielding and weeding. During the seedling-shielding phase, the seedling positioning sensor (16) in the SAM (1) first precisely located target maize seedlings. The end-effector subsequently descended vertically, synchronously lowering the anti-cutting rotating shield (9) of SAM (1) to envelop the identified seedlings. Finally, due to the close connection between the RCU (8), when the shielding drive motor (10) rotated counterclockwise, the anti-cutting rotating shield (9) was subjected to a counterclockwise rotational driving force. This force caused the sliding groove at the connection between the anti-cutting rotating shield (9) and the RCU (8) to rotate, pushing the RCU axle (12) to translate toward the center along the sliding groove on the seedling-shielding fixed disc (6). As a result, all RCUs (8) contract toward the center, completing the contraction movement to enclose and protect the maize plants, thus avoiding damage.

In the weeding phase, first, the transmission gearbox (4) converted torque from the weeding drive motor (35) into rotational motion of the drive rotary disk (27) on the CWM (2), energizing the TRWB (29) with cutting forces. Second, the stationary SWC (26) co-operated with the rotating disk (27), enabling the telescopic control axis (30) to move along the contour track set by the weeding blade limiting chute on the SWC (26), driving the TRWB (29) to perform contour-adaptive cutting motions. The high-speed rotating TRWB (29) simultaneously cleared weeds in both inter-row and intra-plant spaces surrounding the maize seedlings.

Upon completing weeding, the shielding drive motor (10) rotated clockwise to retract the RCUs (8), disengaging from the seedlings. The end-effector ascends until achieving full separation from the plants, completing one operational cycle of the SEMSW.

2.1.2. Structure of the SEMSW

The SEMSW was installed below the weeding robotic arm, composed of the SAM (1), the CWM (2), the end-effector bracket (3), the transmission gearbox (4), and the weeding drive mechanism (5). Its overall structure and installation location are shown in Figure 1. The SAM (1) was installed on the inner side of the end-effector bracket (3), while the CWM (2) was mounted on its outer side. The weeding drive mechanism (5) was connected to the CWM (2) through the transmission gearbox (4).

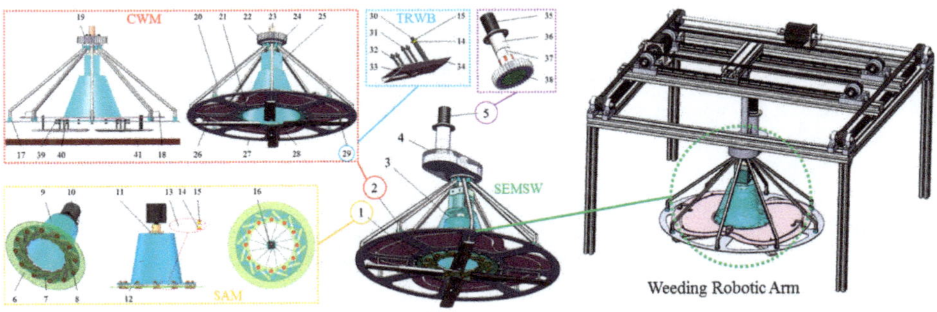

Figure 1. SEMSW with its specific position. Note: 1, SAM; 2, CWM; 3, End-effector bracket; 4, Transmission gearbox; 5, Weeding drive mechanism; 6, Seedling-shielding fixed disc; 7, Anti-cutting rotating disc; 8, RCU; 9, Anti-cutting rotating shield; 10, Shielding drive motor; 11, Rotating shield connector; 12, RCU axle; 13, RCU baffle; 14, Sliding flange bearing; 15, Open pin; 16, Seedling positioning sensor; 17, Support tube connector; 18, Contour cam linkage; 19, Gear fixation bearing; 20, Umbrella-shaped support connecting tube; 21, Rolling flange bearing; 22, Weeding-driven gear; 23, Effector fixation shaft; 24, Drive-rotating coupling; 25, Shielding motor enclosure; 26, SWC; 27, Drive rotary disc; 28, Cover of SAM; 29, TRWB; 30, Telescopic control axis; 31, Weed blade fixing bracket; 32, Fixed blade body; 33, Limit slider; 34, Telescopic blade body; 35, Weeding drive motor; 36, Weeding motor mounting bracket; 37, Coupler; 38, Weeding drive gear shaft; 39, Disc fixation nut; 40, Disc fixation bolt; 41, Soil surface.

2.2. Adaptive Sliding Mode Control of Weeding Robotic Arm

2.2.1. Control Process of Weeding Robotic Arm

The structure and control process of the weeding robotic arm are illustrated in Figure 2. The seedling positioning sensor first detected the location of maize seedlings and fed the position information back to the control system. Based on the sensor data, the control system calculated the motion path of the robotic arm and drove the longitudinal translation mechanism (1), lateral translation mechanism (2), and vertical telescopic mechanism (3) to work in coordination.

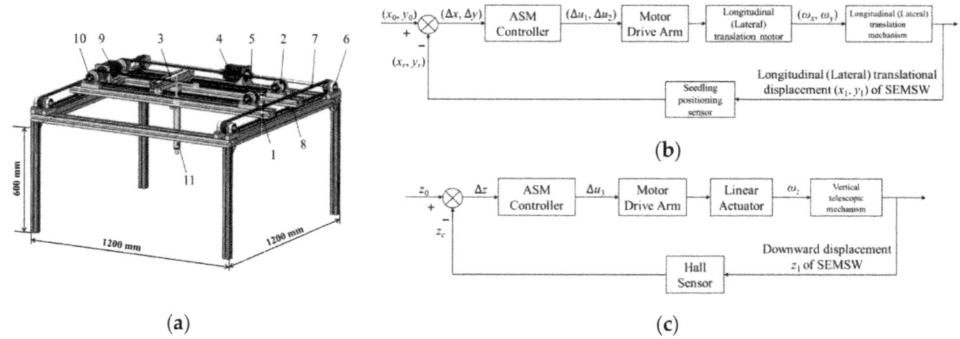

Figure 2. Structure and control block diagram of weeding robotic arm. (**a**) Structure of weeding robotic arm; (**b**) Control block diagram of longitudinal (lateral) translation mechanism; (**c**) Control block diagram of vertical telescopic mechanism. Note: 1, Longitudinal translation mechanism; 2, Lateral translation mechanism; 3, Vertical telescopic mechanism; 4, Longitudinal translation motor; 5, Longitudinal motor-connecting shaft; 6, Synchronous belt pulley; 7, Synchronous belt; 8, Slide slider; 9, Lateral translation motor; 10, Lateral motor-connecting shaft; 11, Linear actuator.

The longitudinal translation motor (4) drove the synchronous belt pulley (6) to rotate through the longitudinal motor-connecting shaft (5), driving the synchronous belt (7) to move. The synchronous belt (7) then drove the slide slider (8) and the longitudinal translation mechanism (1) to achieve longitudinal movement. The lateral translation motor (9) adopted a similar approach, driving the synchronous belt transmission mechanism through the lateral motor-connecting shaft (10) to enable the lateral translation mechanism (2) to complete lateral movement. The linear actuator (11) adjusted the vertical position of the robotic arm through telescopic actions and achieved precise control by integrating feedback signals from Hall sensors. Through the collaborative interaction of sensors, the control system, and mechanical structures, the robotic arm could achieve precise positioning in three degrees of freedom (longitudinal, lateral, and vertical), effectively meeting the requirements of weeding operations.

2.2.2. Derivation of Kinematic and Dynamic Models

From the motion principle of the robotic arm, the linear movement of the lateral and longitudinal translation mechanisms was driven by the rotation of the motor shaft, while the telescopic motion of the vertical telescopic mechanism was realized by the linear actuator motor driving the ball screw to rotate. Neglecting the speed loss caused by friction during motion, according to the kinematic equation $v = \omega r$, the linear velocity of the translation (telescopic) mechanism was equal to the product of the angular velocity of the motor shaft and the distance between them. This converted the differential form of the translation mechanism's displacement into the angular velocity of the motor shaft, thereby constructing the kinematic equation of the weeding robotic arm:

$$\frac{\mathrm{d}}{\mathrm{d}t}\begin{bmatrix} x_1(t) \\ y_1(t) \\ z_1(t) \end{bmatrix} = \begin{bmatrix} \omega_x(t) \cdot r_w \\ \omega_y(t) \cdot r_w \\ i\omega_z(t) \cdot l_s/2\pi \end{bmatrix}, t \in [0, t_f] \quad (1)$$

where x_1, y_1, and z_1 represent the coordinates of the weeding robotic arm in the spatial coordinate system (Figure 2a), m; ω_x, ω_y, and ω_z denote the output angular velocities of the lateral translation motor, longitudinal translation motor, and linear actuator motor, rad/s; r_w is the radius of the synchronous pulley, m; i is the synchronous transmission ratio between the linear actuator motor and the screw; l_s is the lead of the screw, m; and t_f is the time required for the weeding robotic arm to move and align with the target point of the maize seedling (a non-fixed value), s.

As shown in Figure 3, the lateral translation motor provides a driving torque M_{xa} for the lateral translation mechanism. During the process of driving the lateral translation mechanism to achieve lateral movement, the motor transmitted power to the synchronous belt through the synchronous belt pulley, which was a first-level mechanical transmission with a transmission efficiency η_1 of 98%. Additionally, the lateral translation of the lateral unit needed to overcome the frictional resistance f_{xa} generated when the slider slides on the slide rail.

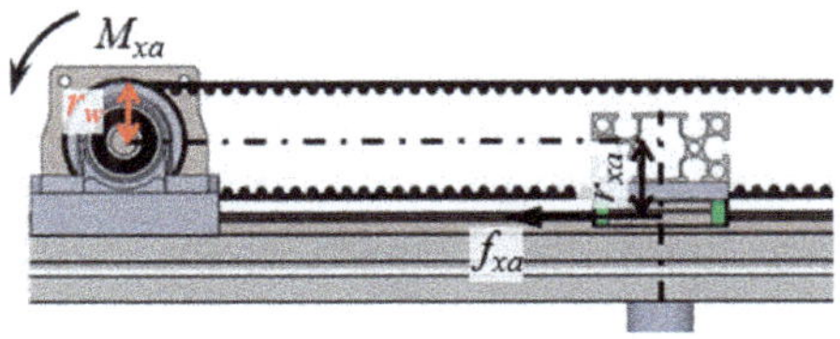

Figure 3. Torque analysis of lateral translation mechanism. Note: M_{xa} is the riving torque provided by the lateral translation motor, N·m; r_w is the radius of the synchronous belt pulley, m; f_{xa} is the sliding friction resistance of lateral translation, N; and r_{xa} is the length of the arm of the sliding friction force f_{xa}, m.

The driving principle of the longitudinal translation mechanism was similar to the lateral translation mechanism and was not repeated here. Since the linear actuator was directly connected to the SEMSW, energy loss can be ignored. Based on the above analysis and combining with the kinematic Equation (1) of the robotic arm, the dynamic equation is constructed as follows:

$$\frac{\mathrm{d}}{\mathrm{d}t}\begin{bmatrix} J_1\omega_x(t) \\ J_2\omega_y(t) \\ J_3\omega_z(t) \end{bmatrix} = \begin{bmatrix} M_{xa}\cdot\eta_1 - f_{xa}\cdot r_{xa} \\ M_{ya}\cdot\eta_1 - f_{ya}\cdot r_{ya} \\ T_{za}/i \end{bmatrix}, t \in [0, t_f] \tag{2}$$

where J_1, J_2, and J_3 represent the moments of inertia of the lateral translation mechanism, longitudinal translation mechanism, and vertical telescopic mechanism during motion, respectively, kg·m^2; M_{xa} is the driving torque provided by the lateral translation motor, N·m; f_{xa} is the sliding friction force experienced by the mechanism during lateral translation, N; r_{xa} is the force arm length of the sliding friction force f_{xa}, m; M_{ya} is the driving torque provided by the longitudinal translation motor, N·m; f_{ya} is the sliding friction force experienced by the mechanism during longitudinal translation, N; r_{ya} is the force arm length of the sliding friction force f_{ya}, m; and T_{za} is the torque of the linear actuator, N·m.

2.2.3. Problem Formulation

Agricultural robotic arm control systems encountered multiple challenges in practical field applications. The structural complexity of agricultural machinery, variability of farmland operational environments, and inherent system uncertainties collectively limited the grasping and alignment performance of robotic manipulators. Due to the strong coupling effects and significant nonlinear uncertainties in Equation (2), direct implementation of adaptive sliding mode control was difficult. Thus, an equivalent model transformation of System (2) was required to simplify the controller design. Through parameter substitution $\theta_1 = J_1$ and the definition of new state variables $x_2 = \omega_x(t)\cdot r_w$, with control input $u_1 = M_{xa}\cdot\eta_1/r_w$ and disturbance term $\Delta_1 = -f_{xa}\cdot r_{xa}/r_w$, System (2) can be reconstructed into the following state-space representation:

$$\begin{cases} \dot{x}_1 = x_2 \\ \theta_1\dot{x}_2 = u_1 + \Delta_1 \end{cases} \quad (3)$$

where J_1 denotes the unknown rotational inertia of the lateral translation system, which constitutes a positive constant. The disturbance term Δ_1 represents the aggregate uncertainty encompassing external disturbances and unmodeled dynamics.

Assumption 1: *The upper and lower bounds of the uncertain parameter θ_1 are defined as:*

$$\theta_1 \in \Omega \triangleq \{\theta_1 : 0 < \theta_{\min} \leq \theta_1 \leq \theta_{\max}\} \quad (4)$$

Assumption 2: *The disturbance term Δ_1 is bounded, expressed as:*

$$|\Delta_1| \leq D \quad (5)$$

2.3. Dataset and Weed Detection

2.3.1. Dataset

The dataset was obtained from https://github.com/zhangchuanyin/weed-datasets (accessed on 1 May 2025), a publicly accessible resource [24] available for download. Comprising 1390 training, 595 validation, and 523 test images of maize and weeds, it closely resembled field images captured during real-world experiments with the weeding device. The dataset included plant images under diverse lighting (day/night) and soil conditions. Representative samples are shown in Figure 4.

Figure 4. The overview of dataset.

To meet the requirements for subsequent image detection model training, this paper annotated the dataset using LabelMe, a widely adopted tool in image detection tasks. The annotation process ensures precise detection of targets within the images.

2.3.2. Weed Detection

Considering its validated performance in similar vegetation detection tasks, compatibility with system computational limits, and field applicability, YOLOv8 [25] was adopted to develop a robust weed recognition model for the SEMSW. As shown in Figure 5, this advanced single-stage detector features an optimized Cross-Stage Partial (CSP) backbone and modified Path Aggregation Network (PANet) for efficient multi-scale feature extraction. Its anchor-free architecture with decoupled classification/regression branches balances detection accuracy and real-time performance, making it suitable for on-board agricultural machinery deployment.

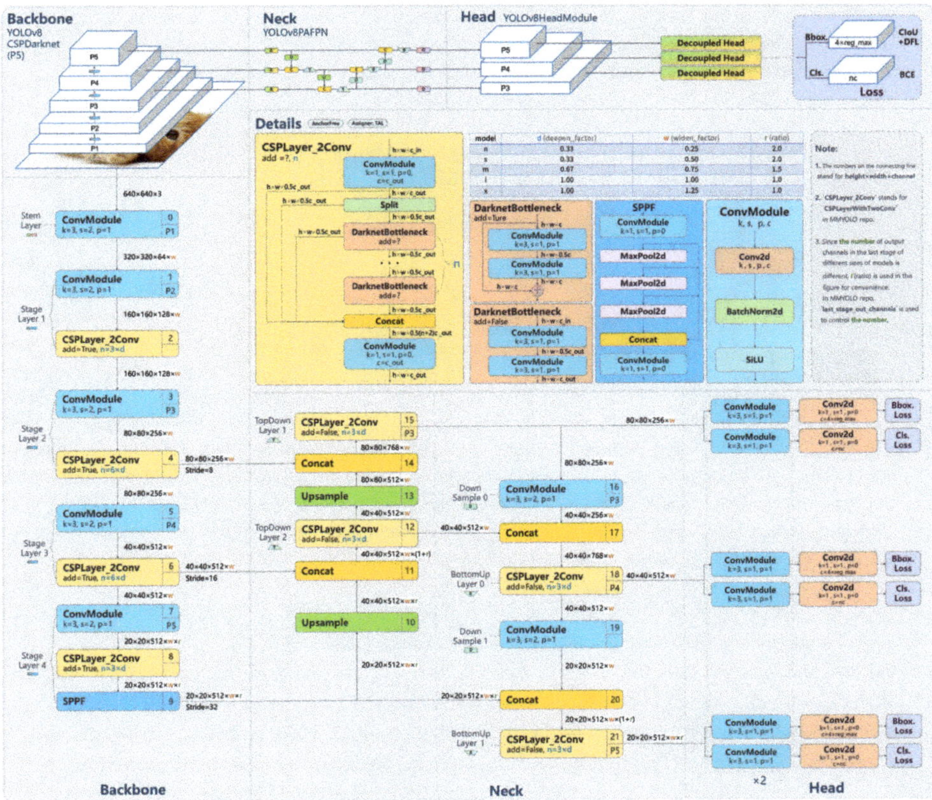

Figure 5. The overview of YOLOv8. Cited from: https://github.com/open-mmlab/mmyolo/blob/dev/configs/yolov8/README.md (accessed on 1 May 2025).

3. Results and Discussion

3.1. Parametric Design of SAM

The SAM effectively protected maize seedlings by preventing collision damage to cotyledons and stalks during operation. Through parameter and structural optimization, the end-effector's stability during seedling-shielding was enhanced, while rigid contact between the shield and seedlings was avoided to reduce breakage risks.

Therefore, this study conducted structural and parametric designs of the SAM based on the actual maize cultivation conditions in northern China (Figures 6 and 7).

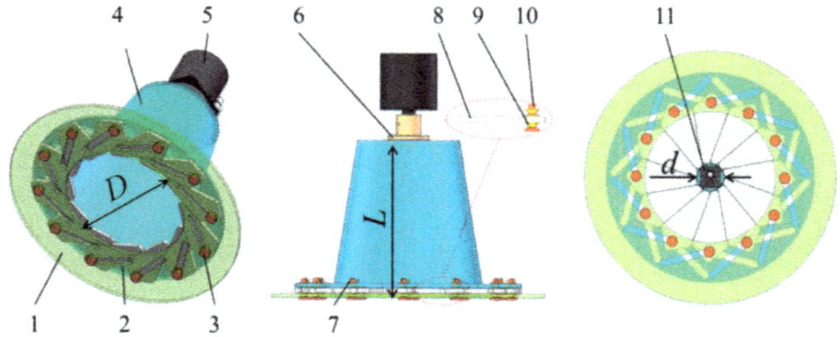

Figure 6. Key parameters of SAM. Note: 1, Seedling-shielding fixed disc; 2, Anti-cutting rotating disc; 3, RCU; 4, Anti-cutting rotating shield; 5, Shielding drive motor; 6, Rotating shield connector; 7, RCU axle; 8, RCU baffle; 9, Sliding flange bearing; 10, Open pin; 11, Seedling positioning sensor. *D* is the inner diameter of the seedling-shielding disc, mm; *L* is the seedling-shielding height, mm; *d* is the minimum inscribed-circle diameter d of the RCUs, mm.

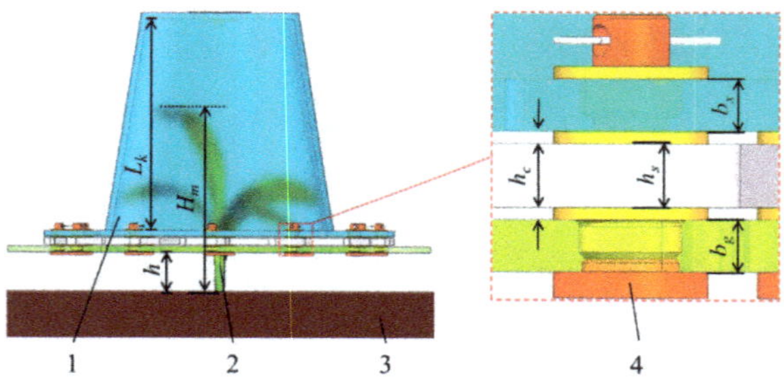

Figure 7. The process of seedling-shielding with SAM. Note: 1, SAM; 2, Maize seedling; 3, Soil surface; 4, Partial schematic diagram. L_k is the height of the anti-cutting rotating shield, mm; H_m is the maximum height of maize at the 3–5 leaf stage, mm; h is the height of the SAM from the ground, mm; h_c is the gap between RCUs and disc, mm; h_s is the height of the RCU, mm; b_x is the thickness of the anti-cutting rotating disc, mm; b_g is the thickness of the seedling-shielding fixed disc, mm.

3.1.1. Inner Diameter D of the Seedling-Shielding Disc

The seedling-shielding fixed disc must maintain sufficient inner aperture. During downward SAM operation for seedling envelopment, rigid contact between maize cotyledons and the disc should be avoided. Cotyledons should either smoothly enter the disc's aperture or transition into the anti-cutting rotating shield without physical interaction, thereby preventing seedling injury, breakage, or soil-surface compression of cotyledons.

The inner diameter *D* requires precise calibration. Excessive *D* increased the structural footprint of the CWM, elevating collision risks with adjacent seedlings. Thus, *D* was optimized to the minimum value, ensuring reliable seedling envelopment. Field measurements of 100 groups of maize seedlings at the 3–5 leaf stages revealed that the maximum cotyledon spans were 8~15 cm. Hence, the finalized *D* was set as 150 mm.

3.1.2. Minimum Inscribed-Circle Diameter d of the RCUs

During counterclockwise rotation of the anti-cutting rotary disc, the RCUs progressively contracts toward the center of the seedling-shielding fixed disc, as shown in Figure 6. At its maximum contraction position, the minimum inscribed-circle diameter formed by the tops of the RCUs components was defined as *d*.

Parametric design requires that d slightly exceeds the maize stem diameters to prevent clamping-induced injury. Through measurement and statistics, the diameter d_{m} of the majority of the maize stalks was found to be 1.5~2.5 cm. Consequently, d was optimized to 30 mm to ensure safe stem clearance while maintaining mechanical reliability.

3.1.3. Ground Clearance h of the SAM

Field-surface irregularities necessitated avoiding direct contact between the end-effector and the ground to minimize soil disturbance, reduce operational resistance, and ensure weeding efficacy. However, excessive ground clearance would prevent the SAM from effectively surrounding and protecting maize plants. As shown in Figure 7, the SAM typically envelops the cotyledon part of the maize plants, and exposing the bottom part of the stalk meets the operation requirements. Based on these constraints, the ground clearance h of the SAM was designed to be 40 mm.

3.1.4. Seedling-Shielding Height L

The seedling-shielding height L pertained to the required height for the SAM to cover maize seedlings from top to bottom through the inner hole of the seedling-shielding fixed disc under the natural growth posture of the maize seedlings. As shown in Figures 6 and 7, with the premise of ensuring a sufficient height h from the ground, it was necessary to ensure that the top inside the anti-cutting rotating shield did not come into contact with the maize seedlings to prevent bruising or breaking. The calculation equation for the seedling-shielding height L was as follows:

$$\begin{cases} L = L_k + b_x + b_g + h_s + 2h_c \\ L > L_k > H_{\mathrm{m}} - h \end{cases} \tag{6}$$

where L is the seedling-shielding height, mm; L_k is the height of the anti-cutting rotating shield, mm; b_x is the thickness of the anti-cutting rotating disc, which is taken as 4 mm; b_g is the thickness of the seedling-shielding fixed disc, which is taken as 4 mm; h_s is the height of the RCU, which is taken as 5 mm; h_c is the gap between RCUs and disc, which is taken as 1 mm; H_{m} is the maximum height of maize at the 3–5 leaf stage, mm; and h is the height of the SAM from the ground, mm.

To ensure that the anti-cutting rotating shield did not come into contact with maize seedlings, its height had to be greater than the difference between the maximum height H_{m} of maize seedlings and the height h of the SAM from the ground. Through field measurements, the height range of maize seedlings in the 3–5 leaf stage was found to be 12–19 cm. The maximum height H_{m} of maize in this stage was set as 190 mm. According to the previous analysis, the height h of the SAM from the ground was 40 mm. Thus, from Equation (6), $L_k > H_{\mathrm{m}} - h$ = 150 mm. To meet this condition, the height L_k of the anti-cutting rotating shield was designed to be 155 mm. Finally, by substituting L_k and the other parameter values into Equation (6) again, the seedling-shielding height L was designed to be 170 mm.

3.2. Parametric Design of CWM

The core design challenges for the CWM involve two critical components: the contour curve design of the SWC and the dimensional chain design of the TRWB. As depicted in Figure 8, the schematic clarifies the CWM's key geometric parameter and its spatial configuration, specifically highlighting the minimum inscribed-circle diameter d_1 of the TRWB.

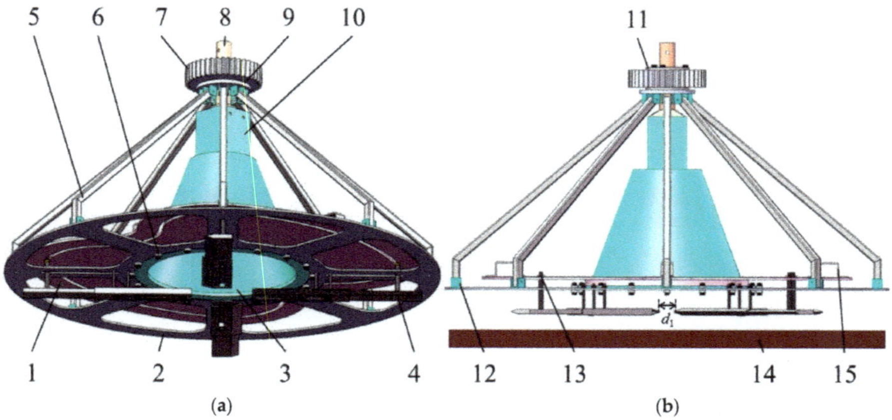

Figure 8. Key parameters of CWM. (**a**) Oblique view; (**b**) Front view. Note: 1, SWC; 2, Drive rotary disc; 3, Cover of SAM; 4, TRWB; 5, Umbrella-shaped support connecting tube; 6, Rolling flange bearing; 7, Weeding-driven gear; 8, Effector fixation shaft; 9, Drive-rotating coupling; 10, Shielding motor enclosure; 11, Gear fixation bearing; 12, Support tube connector; 13, Sliding flange bearing; 14, soil surface; 15, Contour cam linkage; d_1 is the minimum inscribed-circle diameter of the TRWB, mm.

3.2.1. Minimum Inscribed-Circle Diameter d_1 of the TRWB

As shown in Figure 9, the minimum inscribed-circle diameter d_1 of the TRWB refers to the diameter of the non-cutting central zone formed by the TRWB during counterclockwise rotation. This parameter had to balance seedling protection and weeding efficacy. If d_1 was excessively small, the TRWB risked contacting the protected maize seedlings at the inner fixed ring of the fixed weeding blades. Conversely, if d_1 was overly large, the non-cutting areas increased, thus elevating the likelihood of incomplete weed removal near the seedlings.

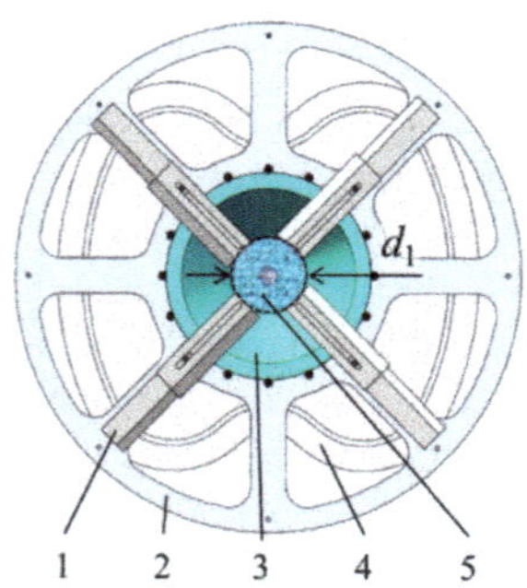

Figure 9. Minimum inscribed-circle diameter d_1 of the TRWB. Note: 1, TRWB; 2, Drive rotary disk; 3, Cover of SAM; 4, SWC; 5, Non-cutting area. d_1 is the minimum inscribed-circle diameter of the TRWB.

Field measurements indicated that the diameter range of the maize seedling stalks d_m was 1.5~2.5 cm. Therefore, the minimum inscribed-circle diameter d_1 of the TRWB should not have been less than 25 mm. Meanwhile, considering the diversity of plant morphologies, seedling-enclosing postures, and the uncertainties of the operating environment such as mechanical vibrations and sensor-positioning errors, a safety factor $K = 4$

was adopted. The minimum inscribed-circle diameter d_1 of the TRWB was designed to be 100 mm to minimize the risk of bruising the seedlings as much as possible.

3.2.2. Contour Curve Design of the SWC

The significant variability in maize row and plant-spacing constituted one of the primary technical challenges for simultaneous inter-row and intra-row weeding [26]. In northern China, conventional agronomic practices dictated maize planting with a row-spacing of 55–60 cm and plant-spacing of 22–27 cm [27–29]. Figure 10a illustrates the theoretical planting model under these parameters. Specifically, when ensuring that the TRWB do not touch the surrounding eight maize plants, the orange-shaded area in Figure 10b represents the minimum cutting area required for weeding through a rotational cutting method.

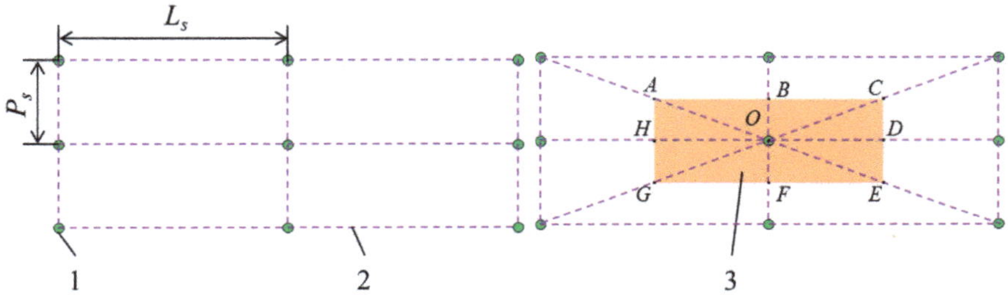

Figure 10. Agronomy conditions and minimum cutting area for maize sowing. (**a**) Agronomy conditions (top view); (**b**) minimum cutting area (top view). Note: 1, Maize plants; 2, Weeding segmentation grid; 3, Minimum theoretical cutting area.

Conventional rotary cutters struggled to match rectangular working areas due to their circular motion trajectories. In this study, a planar disc cam mechanism was innovatively employed. A customized contour curve guided the cutter to generate a quasi-rectangular motion path. The maximum row-spacing value (L_s = 60 cm) and the minimum plant-spacing value (P_s = 22 cm) were selected as reference values for the design of the SWC. As shown in Figure 11b, the boundary points *A-H* of the minimum cutting area are identified as key trajectory points.

As shown in Figure 11a, the cam contour was composed of eight segments of composite curves, and $\odot B'YF'X$ was the base circle of the cam. First, to prevent the weeding blade from touching adjacent maize seedlings between plants, it was essential to ensure that the length of OB' was less than the difference between the plant-spacing P_s and the radius r_m of the maize stalk. The base-circle radius r_b of the cam was designed to be 217.5 mm, and the cam height $B'F'$ was 435 mm. Second, to ensure that the weeding blade could cover in the row-spacing direction, the width $H'D'$ of the cam should not be less than the minimum theoretical cutting width HD. The width $H'D'$ of the cam was designed to be 600 mm. Finally, the diagonal $A'E'$ of the cam should not be less than the length of the diagonal of the rectangle formed by the maize sowing row-spacing and plant-spacing. The length z_b of the diagonal $A'E'$ of the cam was designed to be 660 mm. The above-mentioned calculation methods are shown in Equation (7).

$$\begin{cases} \overline{OB'} = \overline{OF'} = \overline{OX} = \overline{OY} = \frac{1}{2}\overline{B'F'} = \frac{1}{2}\overline{XY} < P_s - r_m \\ \overline{OD'} = \overline{OH'} = \frac{1}{2}\overline{H'D'} \geq \frac{1}{2}L_s \\ \overline{OA'} = \overline{OC'} = \overline{OE'} = \overline{OG'} = \frac{1}{2}\overline{A'E'} = \frac{1}{2}\overline{C'G'} \geq \frac{1}{2}\sqrt{P_s^2 + L_s^2} \end{cases} \tag{7}$$

where P_s is the plant-spacing of maize sowing, cm; r_m is the radius of the maize stalk, cm; L_s is the row-spacing of maize sowing, cm; and the remaining alphabetic symbols denote the specified points and their connecting lines in Figure 11.

As depicted in Figure 11b,c, the cam trajectory efficiently covers the minimum theoretical cutting area while avoiding contact with adjacent maize plants. To mitigate rigid impact, the contour curve was designed with continuous gradient arcs. Simulation results demonstrated that this approach would increase the coverage rate of the SWC from 69.3% to 98.5%. The minor uncut regions caused by the continuity of arcs $A'G'$ and arcs $C'E'$ will be resolved in subsequent research through a telescopic compensation mechanism integrated into the TRWB.

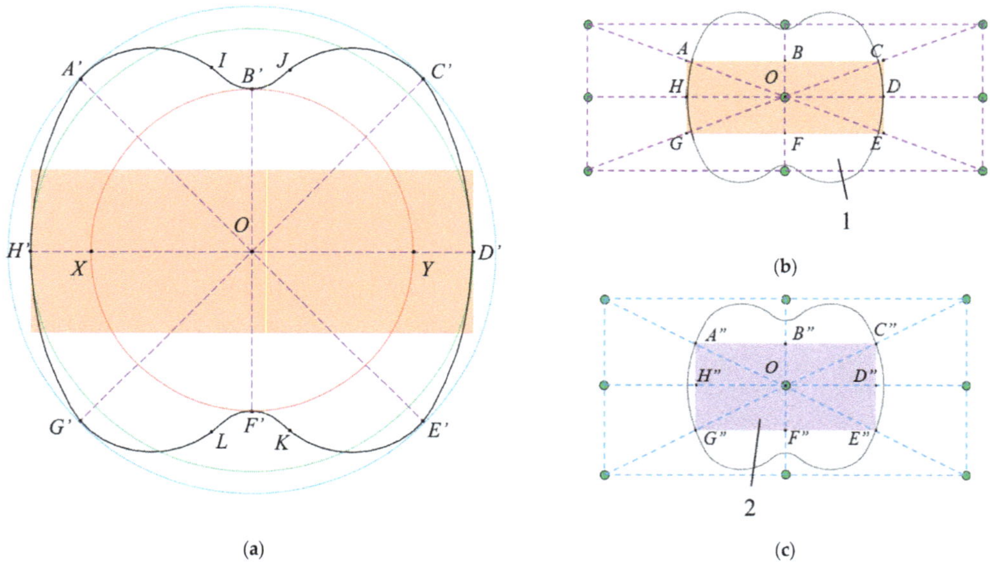

Figure 11. The process of seedling-shielding with SAM. (**a**) Structural parameters of cam contour curve; (**b**) L_s = 60 cm, P_s = 22 cm; (**c**) L_s = 55 cm, P_s = 27 cm. Note: 1, Cam contour curve; 2, Other situations cutting area.

3.2.3. Dimensional Chain Design of the TRWB

To ensure precise retraction and extension of the TRWB at critical positions, it was necessary to design their dimensional chain, as shown in Figure 12. The key parameters included: the minimum retraction distance W_1 of the TRWB, the maximum extension distance W_2 of the TRWB, the length l_1 of the fixed blade body, and the length l_2 of the telescopic blade body.

First, to guarantee structural rigidity and strength, the lengths of the fixed blade body and the telescopic blade body should have been comparable. Referring to the Mechanical Design Handbook [30], both lengths were recommended to be set as half of W_2. To optimize telescopic performance, the telescopic blade body was slightly longer than the fixed blade body, satisfying the relationship:

$$l_2 = l_1 + q_3 \tag{8}$$

where l_1 is the length of the fixed blade body, mm; l_2 is the length of the telescopic blade body, mm; and q_3 represent the unknown lengths in Figure 12b, mm.

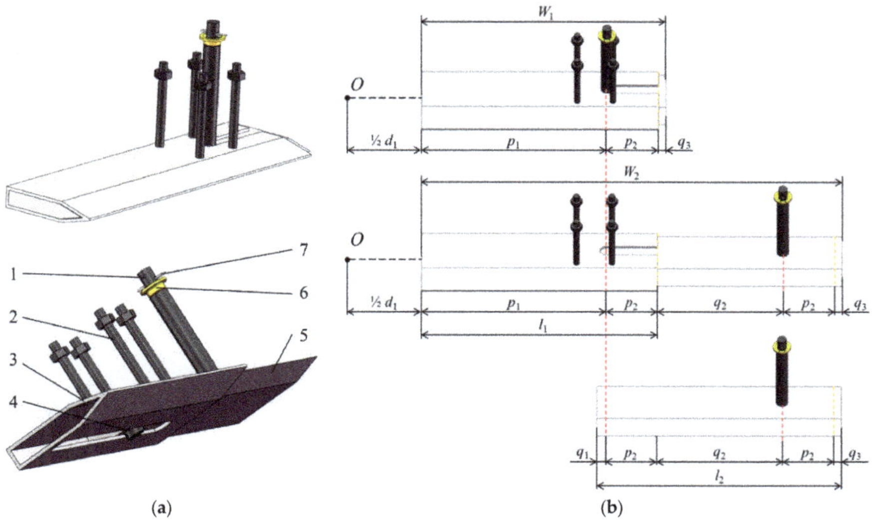

Figure 12. Design of TRWB structure and size chain. (**a**) Schematic diagram of TRWB; (**b**) dimensional chain design of TRWB. Note: 1, Telescopic control axis; 2, Weed blade fixing bracket; 3, Fixed blade body; 4, Limit slider; 5, Telescopic blade body; 6, Sliding flange bearing; 7, Open pin.

Second, when designing the dimensional chain of the TRWB, it was also necessary to consider its motion states at different positions. When the weeding blade rotated to the OB' or OF' position, it had to be ensured that it could contract to the minimum length W_1. At this time, the tip of the telescopic blade body should be within the base circle $B'YF'X$ (as shown in Figure 11a). When it rotated to the OA, OC, OE, or OG position, it had to be ensured that it could stretch to the maximum length W_2. At this moment, the tip of the telescopic blade body should be outside the cam contour curve (as shown in Figure 11b). This was to ensure that the TRWB had a sufficient telescopic range to prevent missed cutting. As can be seen from Figure 12b, the dimensional chain equations are expressed as follows:

$$\begin{cases} \overline{OB'} = 1/2d_1 + p_1 + p_2 + q_3 < P_s - r_m \\ \overline{OA'} = 1/2d_1 + p_1 + 2p_2 + q_2 + q_3 \geq 1/2\sqrt{P_s^2 + L_s^2} \\ W_1 = p_1 + p_2 + q_3 < r_b - 1/2d_1 \\ W_2 = p_1 + 2p_2 + q_2 + q_3 > z_b - 1/2d_1 \\ l_1 = p_1 + p_2 \\ l_2 = 2p_2 + q_1 + q_2 + q_3 \end{cases} \tag{9}$$

where l_1 is the length of the fixed blade body, mm; l_2 is the length of the telescopic blade body, mm; W_1 is the minimum contraction distance of the TRWB (W_1 = l_2) in mm; W_2 is the maximum stretching distance of the TRWB, mm; d_1 is the minimum inscribed-circle diameter of the TRWB, mm; p_1, p_2, q_1, q_2, q_3 represent the unknown lengths in Figure 12b; and the design q_3 = q_1 = 5 mm. *Ps* is the plant-spacing of maize sowing, cm; r_m is the radius of the maize stalk, cm; and L_s is the row-spacing of maize sowing, cm.

During the design process of the dimensional chain for the TRWB, the parameter values of the cam contour curve also needed to be considered. Incorporating the aforesaid design content, by substituting the parameters into Equation (9), it could be deduced that the minimum contraction distance W_1 of the TRWB should be less than 167.5 mm. Consequently, W_1 was designed as 165 mm. Similarly, it could be inferred that the maximum stretching distance W_2 of the TRWB should be greater than 280 mm. Therefore, W_2 was designed as 285 mm.

Upon substituting W_1 = 165 mm and W_2 = 285 mm into Equation (9) and performing the calculations, p_1 was calculated as 125 mm, p_2 as 35 mm, and q_2 as 85 mm. Eventually, the length l_1 of the fixed blade body was determined to be 160 mm, and the length l_2 of the telescopic blade body was 165 mm. Such parameter design satisfied the requirements of the dimensional chain design and also guaranteed the mechanical performance of the TRWB.

3.3. Design of Adaptive Sliding Mode Controller

Given the fundamental similarity in control principles between lateral and longitudinal translation mechanisms, this study focused specifically on deriving the adaptive sliding mode equation for the lateral translation mechanism. The developed control methodology [31,32] maintained theoretical consistency with the longitudinal translation mechanism's control framework. The sliding surface equation for the lateral positioning mechanism's adaptive sliding mode controller is designed as follows:

$$s_1 = \dot{e} + \rho e = x_2 - \dot{x}_d + \rho e \tag{10}$$

where x_d is the position command issued by the lateral translation motor, $e = x_1 - x_d$ is the lateral position tracking error, and $\rho > 0$. Then:

$$\begin{aligned} \dot{s}_1 &= \dot{x}_2 - \ddot{x}_d + \rho\dot{e} \\ \theta_1\dot{s}_1 &= \theta_1(\dot{x}_2 - \ddot{x}_d + \rho\dot{e}) \end{aligned} \tag{11}$$

Taking $\hat{\theta}_1$ as the estimated value of θ_1, the Lyapunov function is defined as:

$$V_1 = \frac{1}{2}\theta_1 s_1^2 + \frac{1}{2\zeta}\widetilde{\theta}_1 \tag{12}$$

where $\widetilde{\theta}_1 = \hat{\theta}_1 - \theta_1$, and $\zeta > 0$. Then:

$$\begin{aligned} \dot{V}_1 &= \theta_1 s_1 \dot{s}_1 + \tfrac{1}{\zeta}\widetilde{\theta}_1\dot{\widetilde{\theta}}_1 = s_1(\theta_1\dot{x}_2 - \theta_1\ddot{x}_d + \theta_1\rho\dot{e}) + \tfrac{1}{\zeta}\widetilde{\theta}_1\dot{\widetilde{\theta}}_1 \\ &= s_1(u_1 + \Delta_1 - \theta_1(\ddot{x}_d + \rho\dot{e})) + \tfrac{1}{\zeta}\widetilde{\theta}_1\dot{\widetilde{\theta}}_1 \end{aligned} \tag{13}$$

The control law u_1 is designed as:

$$u_1 = \overset{\frown}{\theta}_1(\ddot{x}_d - \rho\dot{e}) - k_s s - \eta\,\text{sign}(s_1) \tag{14}$$

where $k_s > 0$ and $\eta > D$. Combining Equations (13) and (14) gives:

$$\begin{aligned} \dot{V}_1 &= s_1(\overset{\frown}{\theta}_1(\ddot{x}_d - \rho\dot{e}) - k_s s_1 - \eta\,\text{sign}(s_1) + \Delta_1 - \theta_1(\ddot{x}_d - \rho\dot{e})) + \tfrac{1}{\zeta}\widetilde{\theta}_1\dot{\hat{\theta}}_1 \\ &= s_1((\overset{\frown}{\theta}_1 - \theta_1)(\ddot{x}_d - \rho\dot{e}) - k_s s_1 - \eta\,\text{sign}(s_1) + \Delta_1) + \tfrac{1}{\zeta}\widetilde{\theta}_1\dot{\hat{\theta}}_1 \\ &= s_1((\widetilde{\theta}_1(\ddot{x}_d - \rho\dot{e}) - k_s s_1 - \eta\,\text{sign}(s_1) + \Delta_1) + \tfrac{1}{\zeta}\widetilde{\theta}_1\dot{\hat{\theta}}_1 \\ &= -k_s s_1^2 - \eta|s_1| + \Delta_1 \cdot s_1 + \widetilde{\theta}_1(s_1(\ddot{x}_d - \rho\dot{e}) + \tfrac{1}{\zeta}\dot{\hat{\theta}}_1) \end{aligned}$$

The adaptive law is taken as:

$$\dot{\hat{\theta}}_1 = -\zeta s_1(\ddot{x}_d - \rho\dot{e}) \tag{15}$$

Then:

$$\dot{V}_1 = -k_s s_1^2 - \eta|s_1| + \Delta_1 \cdot s_1 \leq -k_s s_1^2 \leq 0$$

Since $\dot{V}_1 = 0$ if and only if $s_1 = 0$, that was, when $\dot{V}_1 \equiv 0, s_1 \equiv 0$. According to LaSalle's invariance principle [33], the closed-loop system was asymptotically stable, meaning that $t_f \to 0, s_1 \to 0$, proving the stability of the adaptive sliding mode controller.

3.4. Crop Detection Test

3.4.1. Maize Seedling Image Segmentation

The host computer performs image segmentation on all maize seedlings detected within the field of view of the seedling positioning sensor, as shown in Figure 13. It prioritized the maize seedling closest to the SEMSW for weeding operations. The host computer was equipped with an NVIDIA GeForce RTX 4090D GPU, and each image segmentation process takes approximately 1 s.

(**a**) (**b**)

Figure 13. Maize image segmentation. (**a**) Original image; (**b**) segmentation effect.

3.4.2. Environment of Weed Detection

The detection model was trained using the dataset described in Section 2.3. The training was conducted on a system with an NVIDIA GeForce RTX 4060 GPU, an Intel Core i5-12600KF CPU (3.70 GHz), 32 GB of RAM, and the Linux Ubuntu 20.04 operating system; the implementation was carried out in Python (v. 3.10) with the PyTorch (v. 2.4.1) library. The dataset was partitioned into training/validation/testing sets at a 0.7:0.2:0.1 ratio. The model underwent 50 epochs of training with a batch size of 16 and an input image size of 640×640.

3.4.3. Result of Weed Detection

As shown in Figure 14, the model's inference accuracy improved consistently with increasing training epochs, indicating effective learning convergence. The detection results are summarized in Table 1, including:

Table 1. Weed detection model metric.

Metrics	Values
Precision	0.951
Recall	0.844
mAP50	0.950
mAP50-95	0.819

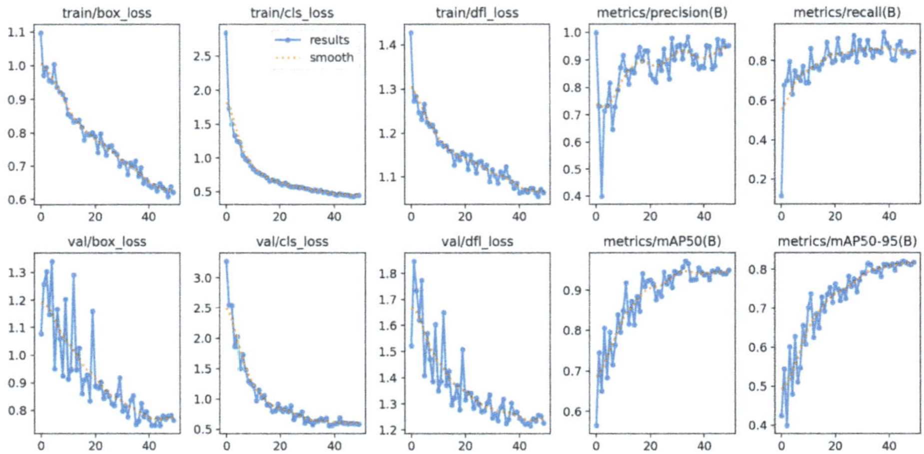

Figure 14. Training results of weed detection model. Note: Precision measures the model's ability to minimize false positives, a critical metric for avoiding crop damage during weeding; Recall reflects the completeness of weed detection, essential for preventing missed weeds; mAP50 (mean average precision at 50% IoU) demonstrates the general robustness of detection; and mAP50-95, a comprehensive evaluation across IoU thresholds ranging from 50% to 95%, indicates the precision of object localization.

The model demonstrates exceptional detection performance across various weed species in maize fields, as shown in Figure 15. Experimental results show the model achieved a precision of 0.951, including mAP50 of 0.95 and mAP50-95 of 0.819. These results demonstrated the model's balance between detection accuracy and precise localization capability, critical for robotic weeding applications. This enabled the SEMSW to accurately identify different weed species, prevent crop misclassification, and minimize crop damage. Consistent performance across weeds with distinct morphological characteristics validated the model's generalization capability under real-field conditions [34].

Figure 15. The results of weed detection.

3.5. *Field Validation Test*

3.5.1. Test Conditions

The experiment was conducted on 24 April 2025 in the experimental field of the Chinese Academy of Agricultural Sciences (39°57′23″ N, 116°19′25″ E). The field, located at 39 m altitude with an atmospheric pressure of 1024.6 hPa, had a silt loam soil type. The maize variety employed was Baitiannuo 708, and the experimental subjects were 18-day-old maize seedlings (post-sowing) and the accompanying weed communities. During the experiment, the daily average temperature ranged from 7 °C to 23 °C, and there was no precipitation.

The experimental field was left fallow for one year, and artificial tillage and land preparation operations were carried out prior to maize sowing. To ensure the survival rate of maize, 4–5 maize seeds were sown per hole. Since all weeds were removed before sowing, there were relatively few weeds on the day of the experiment. As shown in Figure 16a, the weed density was 20–50 plants/m^2, and the main weed species included *Setaria viridis*, mint (*Mentha* sp.), and *Calystegia hederacea*, etc. The actual row-spacing L_s in the experimental field was 60 cm, and the plant-spacing P_s was 27 cm. The environmental parameters of the experimental field are presented in detail in Table 2.

(a) (b)

Figure 16. Field experimental site. (**a**) Field test environment; (**b**) SEMSW's bench platform in the field.

Table 2. Main parameters of field tests.

Items	Parameters	Values
Maize seedlings	Average plant height/mm	203
	Average stem diameter/mm	9.8
Weeds	Height range/mm	2~5
	Stalk diameter/mm	0.1~0.3
	Moisture content/%	66.17
0~100 mm soil layer	Penetration resistance/kPa	326
	Moisture content/%	19.83
	Bulk density/(g·cm^{-3})	1.28
	Temperature/°C	18.3

3.5.2. Test Method and Results

As shown in Figure 16b, the bench platform equipped with the SEMSW was used to conduct weeding performance tests in the experimental field. Since the bench platform lacked autonomous mobility, technicians manually lifted and positioned it above maize seedlings to align with the target plants. The WRR and SDR were adopted as evaluation indices.

Specifically, WRR was defined as the ratio of the number of effectively removed weeds after the bench platform's operation to the total number of weeds before operation. Using a random sampling method, five maize seedling areas were randomly selected in the field for weeding tests. The effective weed counts within the end-effector's coverage area were manually recorded before and after each test. For SDR, it was calculated as the percentage of bruised maize seedlings in the total number of seedlings. Given that visual assessment of seedling damage caused by weeding blades was sometimes challenging, this experiment defined seedling injury [18] as any contact between the TRWB and maize seedlings. After testing the weeding effect on all maize seedlings, the number of injured seedlings was counted to determine SDR. The detailed calculation methods for WRR and SDR are as follows:

$$\begin{cases} \lambda_1 = \frac{F_z - F_s}{F_z} \times 100\% \\ \lambda_2 = \frac{E_s}{E_z} \times 100\% \end{cases} \quad (16)$$

where λ_1 is the WRR, %; λ_2 is the SDR, %; F_z is the number of weeds before operation; F_s is the number of weeds remaining after operation; E_z is the total number of maize seedlings before operation; and E_s is the number of damaged maize seedlings after operation.

Experimental results, as shown in Table 3, demonstrate that the SEMSW achieves simultaneous inter-row and intra-plant weeding, attaining a WRR of 88.3%. The weeding effect is illustrated in Figure 17. The SAM effectively reduced mechanical damage, with two collisions occurring among 91 maize seedlings, yielding an SDR of 2.2%.

Table 3. Result of the WRR tests.

Metrics	Values/%	Range/%	Coefficient of Variation/%	Standard Deviation
WRR	88.3	78.6~94.1	6.2	5.44

(a) (b)

Figure 17. Weeding effect of SEMSW. (**a**) Before weeding operation; (**b**) after weeding operation.

Compared with the 3ZS-2 inter-row cultivation weeder developed by Liu et al. [9] (with an 85% WRR), the SEMSW increased the WRR by 3.9%. In comparison to Chen et al.'s [11] Intra-row Brush Weeding Manipulator, which achieved an 83.4% WRR and 2.5% SDR, the SEMSW improved the WRR by 5.9% while reducing the SDR by 12%. Relative to Jia et al.'s [25] seedling-avoiding weeding control device (5.9% SDR), the SEMSW lowered the SDR by 62.7%. Additionally, compared with the strictest industry benchmark for SDR (3%), the SEMSW reduced the rate by 23.3%, confirming the SAM's superior performance in preventing seedling damage.

These performance metrics confirm that the developed weeding device significantly enhances crop protection while ensuring effective weed control, thus fulfilling the agronomic requirements for maize weeding in northern China.

4. Conclusions

This study proposed a novel shield-cutting end-effector for maize surrounding weeding (SEMSW) that integrated seedling positioning, robotic arm control, and precision weeding functions. The system first localized maize seedlings and weeds via a seedling positioning sensor, then drove the weeding robotic arm to perform XYZ translational motions for precise alignment with the seedlings. The seedling-shielding anti-cutting mechanism (SAM) enclosed crop stems for protection, while the contour-adaptive weeding mechanism (CWM) activated rotational motions of the two-stage retractable weeding blade (TRWB) to remove inter-row and intra-row weeds. Mathematical models were established to determine critical parameters, including the following: the inner diameter D of the seedling-shielding disc was 150 mm; the minimum inscribed-circle diameter d of the RCUs was 30 mm; the ground clearance h of the SAM was 40 mm; the seedling-shielding height L was 170 mm; and the minimum inscribed-circle diameter d_1 of the TRWB was 100 mm. Additionally, the structure and parameters of the contour curve of the SWC and the dimensional chain of the TRWB were designed. Kinematic/dynamic models of the robotic arm were developed, alongside an adaptive sliding mode controller that ensured lateral translation error convergence. A YOLOv8-based weed detection model achieved 0.951 precision, 0.95 mAP50, and 0.819 mAP50-95, striking a balance between detection accuracy and localization precision for robotic weeding. Field tests of the prototype showed an 88.3% weed removal rate (WRR) and a 2.2% seedling damage rate (SDR), meeting the agronomic requirements for maize weeding in northern China.

Future work will optimize the mechanical structure and parameters of SEMSW using EDEM and ANSYS, validate the convergence/stability/robustness of the adaptive sliding mode controller via simulation, and deploy the YOLO algorithm on the weeding robot prototype for field trials to enhance practical applicability.

Author Contributions: Conceptualization, Y.L. (Yunxiang Li) and J.Y.; methodology, Y.L. (Yunxiang Li) and Y.Q.; software, Y.L. (Yunxiang Li) and Y.F.; validation, Y.L. (Yunxiang Li), Y.Q. and J.Y.; writing—original draft preparation, Y.L. (Yunxiang Li); writing—review and editing, Y.L. (Yunxiang Li), Y.Q., J.Y. and Y.L. (Yanfeng Lu); funding acquisition, Y.L. (Yanfeng Lu). All authors have read and agreed to the published version of the manuscript.

Funding: This study is supported by the Strategic Priority Research Program of the Chinese Academy of Sciences under (Grants XDA0450200, XDA0450202), and the Open Projects Program of State Key Laboratory of Multimodal Artificial Intelligence Systems.

Data Availability Statement: The dataset presented in this study are openly available in the weed datasets of Zhangchuanyin (Figure 4), available online at https://github.com/zhangchuanyin/weed-datasets (accessed on 1 May 2025), reference number [28]. The overview of the YOLOv8 structure (Figure 5) is available online: https://github.com/open-mmlab/mmyolo/blob/dev/configs/yolov8/README.md (accessed on 1 May 2025).

Acknowledgments: The authors sincerely thank Wenxia Wu for her valuable assistance with the prototype fabrication and field experiment implementation.

Conflicts of Interest: The authors declare no conflicts of interest.

Nomenclature

WRR/λ_1	Weed removal rate (%)	D	Inner diameter of the seedling-shielding disc (mm)
SDR/λ_2	Seedling damage rate (%)	d	Minimum inscribed-circle diameter of the RCUs (mm)
SEMSW	Shield-cutting end-effector for maize surrounding weeding	d_1	Minimum inscribed-circle diameter of the TRWB (mm)
SAM	Seedling-shielding anti-cutting mechanism	x_1, y_1, z_1	Coordinates of the weeding robotic arm in the spatial coordinate system
CWM	Contour-adaptive weeding mechanism	i	Synchronous transmission ratio between the linear actuator motor and the screw
TRWB	Two-stage retractable weeding blade	L_k	Height of the anti-cutting rotating shield (mm)
RCU	Retractable clamping unit	l_1	Length of the fixed blade body (mm)
SWC	Shape-following weeding cam	l_2	Length of the telescopic blade body (mm)
YOLOv8	You only look once, version 8	W_1	Minimum contraction distance of the TRWB (mm)
CSP	Cross-stage partial	W_2	Maximum stretching distance of the TRWB (mm)
PANet	Path aggregation network	F_z	Number of weeds before operation
L	Seedling-shielding height (mm)	F_s	Number of weeds remaining after operation
r_m	Radius of the maize stalk (cm)	E_z	Total number of maize seedlings before operation
d_m	Diameter of majority maize stalks (mm)	E_s	Number of damaged maize seedlings after operation
h	Ground clearance of the SAM (mm)	ω_x	Angular velocities of the lateral translation motor (rad/s)
h_c	Gap between RCUs and disc (mm)	ω_y	Angular velocities of longitudinal translation motor (rad/s)
h_s	Height of the RCU (mm)	ω_z	Angular velocities of the linear actuator motor (rad/s)
P_s	Plant-spacing of maize sowing (cm)	x_d	Position command issued by lateral translation motor
L_s	Row-spacing of maize sowing (cm)	M_{xa}	Driving torque provided by lateral translation motor (N·m)
T_{za}	Torque of the linear actuator (N·m)	M_{ya}	Driving torque by longitudinal translation motor (N·m)
l_s	Lead of the screw (m)	f_{xa}	Sliding friction resistance of lateral translation (N)
e	Lateral position tracking error	f_{ya}	Sliding friction resistance of longitudinal translation (N)
u_1	Control law equation	r_{xa}	Force arm length of the sliding friction force f_{xa} (m)
s_1	Sliding surface equation for the lateral positioning mechanism	r_{ya}	Force arm length of the sliding friction force f_{ya} (m)
$\hat{\theta}_1$	Estimated value of θ_1	t_f	Time required for the weeding robotic arm to move and align with the target point of the maize seedling (s)
Δ_1	Aggregate uncertainty encompassing external disturbances and unmodeled dynamics	b_g	Thickness of the seedling-shielding fixed disc (mm)
r_w	Radius of the synchronous pulley (m)	$J_1/\theta_1, J_2, J_3$	Moments of inertia of the lateral translation mechanism, longitudinal translation mechanism, and vertical telescopic mechanism during motion, respectively (kg·m^2)

References

1. Wang, A.; Zhang, W.; Wei, X. A review on weed detection using ground-based machine vision and image processing techniques. *Comput. Electron. Agric.* **2019**, *158*, 226–240. [CrossRef]
2. Hamuda, E.; Glavin, M.; Jones, E. A survey of image processing techniques for plant extraction and segmentation in the field. *Comput. Electron. Agric.* **2016**, *125*, 184–199. [CrossRef]
3. Zhang, X.; Wang, Q.; Wang, X.; Li, H.; He, J.; Lu, C.; Yang, Y.; Jiang, S. Automated detection of Crop-Row lines and measurement of maize width for boom spraying. *Comput. Electron. Agric.* **2023**, *215*, 108406. [CrossRef]
4. Wang, X.; Zeng, S.; Luo, J.; Wang, S.; Luo, H. Species and Infestation Level of Weeds in Maize Fields of Bijie Mountain Locality. *Weed Sci.* **2024**, *42*, 17–30. [CrossRef]
5. Oerke, E.C. Crop losses to pests. *J. Agric. Sci.* **2005**, *144*, 31–43. [CrossRef]
6. Zhao, C. Prospect of Agricultural Robot. *China Rural. Sci. Technol.* **2019**, 20–21.
7. Zhao, C. Current situations and prospects of smart agriculture. *J. South China Agric. University* **2021**, *42*, 1–7. [CrossRef]
8. EC-Weeder Intelligent Weeding Machine. Available online: https://www.lemken.cn/Proinfo.aspx?id=154 (accessed on 1 January 2025).
9. Liu, G.; Sun, S.; Cheng, X.; Xu, J. Research and Design of 3ZS-2 Type Plowing Weeding Machine. *J. Agric. Mech. Res.* **1999**, *2*, 49–50 and 119.

10. Li, N.; Chen, Z.; Zhu, C.; Zhang, C.; Sun, Z.; Li, W. System Design and Experiment of Electric Driven Weeding Robot. *Trans. Chin. Soc. Agric. Mach.* **2016**, *47*, 15–20+69. [CrossRef]
11. Chen, Z.; Li, N.; Sun, Z.; Li, T.; Zhang, C.; Li, W. Optimization and experiment of intra-row brush weeding manipulator based on planetary gear train. *Trans. Chin. Soc. Agric. Mach.* **2015**, *46*, 94–99. [CrossRef]
12. Gobor, Z.; Lammers, P.; Martinov, M. Development of a mechatronic intra-row weeding system with rotational hoeing tools: Theoretical approach and simulation. *Comput. Electron. Agric.* **2013**, *98*, 166–174. [CrossRef]
13. Nørremark, M.; Griepentrog, H.W.; Nielsen, J.; Søgaard, H.T. Evaluation of an autonomous GPS-based system for intra-row weed control by assessing the tilled area. *Precis. Agric.* **2011**, *13*, 149–162. [CrossRef]
14. Nørremark, M.; Griepentrog, H.W.; Nielsen, J.; Søgaard, H.T. The development and assessment of the accuracy of an autonomous GPS-based system for intra-row mechanical weed control in row crops. *Biosyst. Eng.* **2008**, *101*, 396–410. [CrossRef]
15. Pérez-Ruíz, M.; Slaughter, D.; Fathallah, F.; Gliever, C.; Miller, B. Co-robotic intra-row weed control system. *Biosyst. Eng.* **2014**, *126*, 45–55. [CrossRef]
16. Hou, X.; Chen, Y.; Guo, W. Machine vision-based navigation for a weeding robot. *Trans. Chin. Soc. Agric. Mach.* **2008**, *39*, 106–108.
17. Guo, W.; Chen, Y. Fuzzy control based autonomous navigation for a weeding robot. *Robot* **2010**, *32*, 204–209. [CrossRef]
18. Jia, H.; Li, S.; Wang, G.; Liu, H. Design and experiment of seedling avoidable weeding control device for intertillage maize (*Zea Mays* L.). *Trans. Chin. Soc. Agric. Eng.* **2018**, *34*, 15–22. [CrossRef]
19. Slaughter, D.; Chen, P.; Curley, R. Vision guided precision cultivation. *Precis. Agric.* **1999**, *1*, 199–216. [CrossRef]
20. Slaughter, D.; Giles, D.; Downey, D. Autonomous robotic weed control systems: A review. *Comput. Electron. Agric.* **2008**, *61*, 63–78. [CrossRef]
21. Pantazi, X.; Moshou, D.; Bravo, C. Active learning system for weed species recognition based on hyperspectral sensing. *Biosyst. Eng.* **2016**, *146*, 193–202. [CrossRef]
22. Sa, I.; Chen, Z.; Popović, M.; Khanna, R.; Liebisch, F.; Nieto, J.; Siegwart, R. weedNet: Dense Semantic Weed Classification Using Multispectral Images and MAV for Smart Farming. *IEEE Robot. Autom. Lett.* **2017**, *3*, 588–595. [CrossRef]
23. Dyrmann, M.; Jørgensen, R.; Midtiby, H. RoboWeedSupport-Detection of weed locations in leaf occluded cereal crops using a fully convolutional neural network. *Adv. Anim. Biosci.* **2017**, *8*, 842–847. [CrossRef]
24. Weed-Datasets of Zhangchuanyin. Available online: https://github.com/zhangchuanyin/weed-datasets (accessed on 1 May 2025).
25. Jia, Z.; Zhang, M.; Yuan, C.; Liu, Q.; Liu, H.; Qiu, X.; Zhao, W.; Shi, J. ADL-YOLOv8: A Field Crop Weed Detection Model Based on Improved YOLOv8. *Agronomy* **2024**, *14*, 2355. [CrossRef]
26. Liao, Y.; Li, C.; Liao, Q.; Wang, L. Research progress of seed guiding technology and device of planter. *Trans. Chin. Soc. Agric. Mach.* **2020**, *51*, 1–14. [CrossRef]
27. Wang, Q.; Cao, X.; Wang, C.; Li, H.; He, J.; Lu, C. Research Progress of No/Minimum Tillage Corn Seeding Technology and Machine in Northeast Black Land of China. *Trans. Chin. Soc. Agric. Mach.* **2021**, *52*, 1–15. [CrossRef]
28. Cao, X.; Wang, Q.; Li, H.; He, J.; Lu, C. Design and experiment of active rotating collective straw-cleaner. *Trans. Chin. Soc. Agric. Eng.* **2021**, *37*, 26–34. [CrossRef]
29. Cao, X.; Wang, Q.; Li, H.; He, J.; Lu, C. Combined row cleaners research with side cutter and stubble clean disk of corn no-till seeder. *Trans. Chin. Soc. Agric. Mach.* **2021**, *52*, 36–44. [CrossRef]
30. Transportation Machinery Design and Selection Manual Committee. *Transport Machinery Design and Selection Manual, the Next Volume*; Chemical Industry Press: Shanghai, China, 1999.
31. Marcos-Andrade, D.; Beltran-Carbajal, F.; Rivas-Cambero, I.; Yañez-Badillo, H.; Favela-Contreras, A.; Rosas-Caro, J.C. Sliding Mode Speed Control in Synchronous Motors for Agriculture Machinery: A Chattering Suppression Approach. *Agriculture* **2024**, *14*, 737. [CrossRef]
32. Peña-Fernández, A.; Colón-Reynoso, M.A.; Mazuela, P. Geometric Analysis of Greenhouse Roofs for Energy Efficiency Optimization and Condensation Drip Reduction. *Agriculture* **2024**, *14*, 216. [CrossRef]
33. Lasalle, J.; Lefschetz, S. *Stability by Lyapunov's Direct Method*; Academic Press: New York, NY, USA, 1961.
34. Chu, B.; Shao, R.; Fang, Y.; Lu, Y. Weed Detection Method Based on Improved YOLOv8 with Neck-Slim. In Proceedings of the 2023 China Automation Congress (CAC), Chongqing, China, 17–19 November 2023; pp. 9378–9382. [CrossRef]

agriculture

MDPI

Article

Evaluation of a Pre-Cut Sugarcane Planter for Seeding Performance

Zhikang Peng [1,2], Fengying Xu [1,2], Pan Xie [1], Jinpeng Chen [1,2], Tao Wu [1,2,*] and Zhen Chen [1,2]

1 College of Engineering, South China Agricultural University, Guangzhou 510642, China; zhikangpeng8@gmail.com (Z.P.); xu_fy@scau.edu.cn (F.X.); 1361334714@stu.scau.edu.cn (P.X.); 19129231971@163.com (J.C.); zhenchenid@scau.edu.cn (Z.C.)
2 Key Laboratory of Key Technology on Agricultural Machine and Equipment, Guangzhou 510642, China
* Correspondence: wt55pub@scau.edu.cn

Abstract: To investigate the relationship between the seeding performance of a novel pre-cut sugarcane planter designed by South China Agricultural University and operational settings, field seeding tests was conducted with the following protocol: First, the John Deere M1654 tractor's forward velocity was calibrated, and the planter's safe loading capacity was determined. Subsequently, eight experimental treatments (A–H) were designed to quantify the relationships between the three performance indicators: seeding density N, the seeding efficiency E and seeding uniformity (coefficient of variation, CV), and three key operational parameters: forward speed of planter v, the discharging sprocket rotational speed n, and the hopper outlet size w. Mathematical models (R20.979) between three key operational parameters with two performance indicators (N, E) was developed through analysis of variance (ANOVA) and regression analysis. The seeding rate per meter was confirmed to follow a Poisson distribution based on Kolmogorov–Smirnov (K–S) tests. When the CV was below 40%, the mean relative error remained within 3%. These findings provide a theoretical foundation for seeding performance prediction under field conditions.

Keywords: sugarcane planter; field test; seeding uniformity; probability distribution; regression analysis

1. Introduction

Sugarcane is a tropical and subtropical crop that serves as a raw material for sugar production. It can also be converted into ethanol as an energy substitute. Global sugarcane production reached 1869 $\times$ 10^6 tons in 2020. Brazil, India and China stand out as the three leading countries in terms of sugarcane production, collectively contributing to 40%, 20%, and 10%, respectively [1]. With the rise in labor costs, there has been a gradual shift from traditional planting to mechanized planting. Sugarcane planters are mainly divided into whole-rode type and billets type.

Pre-cut sugarcane planters currently represent the main development direction. The seeding meter device is the core part of the planter. There are different types of seeding meter devices, such as chain conveyors [2,3], ribbon conveyors [4], fluted-wheels [5,6], etc. To achieve large-scale planting and good seeding uniformity, planters usually use multi-class seed metering device, which are typically a combination of various mechanical structures, such as chain conveyors, ribbon conveyors, fluted wheels, etc. These planters are characterized by their bulky, heavy, and complex structure. These planters are usually pulled directly by a tractor with a ground wheel for support rather than being attached with a three-point linkage, which results in a large turning radius [7–9]. Southern China's sugarcane growing regions are scattered, rough, and small in area, a large turning radius will lead to wasted planting area [10]. In recent decades, as modern agriculture has advanced toward refined planting operations, many researchers have started to focus on precision seeders [11]. Su W, et al. [12] and Ma L, et al. [13] designed double-bud sugarcane seed metering devices. However, their seeding efficiency is low and their seed capacity is small. Currently, many field tests of seed metering devices use the evaluation indexes of multiple -seeds and missing seeds to assess performance. However, the assessment of seed metering device performance lacks comprehensiveness and depth.

To address the problem described above, a new sugarcane planter was designed by South China Agricultural University [14]. In this paper, this planter and its seed metering device structure and feature are first briefly described. Then, the relationships among the gear position, engine speed, and the change in the vehicle speed of the John Deere M1645 tractor, which suspends the planter, and the change in the centroid of the planter with cane seed quantity are studied. Eight field tests were carried out on the planter, and ANOVA and linear regression analysis were performed to examine the factors of the planter's forward speed, the discharging sprocket rotational speed, and the outlet size of the hopper on the indicators of the average seeding rate per meter and the seeding efficiency of the planter. Finally, the distribution of cane seeds per meter was analyzed using the Kolmogorov–Smirnov test.

2. Materials and Methods

2.1. Machine Description and Seeding Operation

This new sugarcane planter designed by South China Agricultural University (SCAU, Guangzhou, Guangdong, China) mainly includes fertilizers, a furrow opener, a ground-wheel control system, a seed metering device, and a soil cover device as shown in Figure 1. Its technical index parameters are shown in Table 1.

Figure 1. Overall structure of sugarcane planter: (1) fertilizers; (2) furrow opener; (3) ground-wheel control system; (4) soil cover device; (5) seed metering device.

Table 1. Specifications of the sugarcane planter [15].

General	Parameter
Number of rows	Two
Type of seeding	Pre-cut seed with two buds
Seeding space (m)	1–1.4
Depth of furrow opener (mm)	250–350
Thickness of covering soil (mm)	80–150
Capacity of fertilizer (L)	341
Matched power (HP)	≥180
Overall weight (t)	About 1.5
Volume of hopper (m^3)	3.246
Size of planter (mm)	3800 × 1900 × 2900 (Length × Width × Height)

The fertilizer discharge device is a rotary scraper fertilizer, featuring a non-clogging structure that ensures optimal performance when applying various fertilizers such as organic, inorganic, and compound fertilizers with excellent adaptability. The furrow opener is the shoe type, the angle between the shoe surface and the main beam is 120°, and the buried angle is 30°, which is acute. This design ensures that the whole ditching part can be sunk into finished cane land to disturb a large amount of soil, and open a wide ditch suitable for planting. Ground-wheel control system is made up of a ground wheel and a speed-dependent seeding control system. The internal gear sensor of the ground wheel detects the planter forward speed and transmits the speed signal to the speed-dependent seeding control system. After the control system processes the calculations, transmits a signal to adjust the discharging sprocket's rotational speed. The seed metering device is made up of seeding cleats, middle cleats, a W-channel, a hopper, a pressure-limiting plate, and an amount-limiting device, among other components. Sugarcane seeds fall into the W-channel from the outlet of hopper and are carried by the seeding cleats out of the seed metering device. The amount-limiting device is positioned at the exit of the tunnel. The film- covering device consists primarily of a frame, an M-shaped soil- covering plate, a film mounting base, a soil- raising disc, and a suppression wheel. The operational process is as follows: once the sugarcane seed falls into the planting ditch, the M-shaped soil-covering plate gathers the soil from both sides of the ditch toward its center. This forms a turtleback shape for the soil layer above the cane seed. This shape offers certain advantages for enhancing the seeds' drought resistance. The thickness of this turtleback-shaped soil layer depends on the height of the M-shaped soil cover plate. The planter can be coupled to the tractor via a hydraulic hitch system, which significantly reduces the turning radius during planting operations compared to conventional direct-towing methods.

The seed metering device is mainly composed of a hopper, a pressure-limiting plate, a W-channel, middle cleats, seeding cleats, and an amount-limiting device. The hopper is used to hold cane seeds, and the pressure-limiting plate at the bottom of the hopper controls the outlet size of the hopper (w) as shown in Figure 2a,b. The size of the hopper outlet (w) determines the rate of cane seed outflow and the jamming of cane seeds.

The cane seeds free fall from the hopper and flow through the W-channel, and then they are transported by the seeding cleats inside the channel. The cane seeds accumulating in the W-channel will move along the channel approaching to the direction of velocity due to the spatial constrain of the channel itself and the law of elongated cylindrical particles tends to be in the direction of motion [16]. The middle cleats can increase the disturbance of the cane seeds in the channel and prevent the cane seeds from avoiding interaction with the seeding cleats. The cane seeds are driven by the seeding cleats, which are transported from the bottom of the channel to the channel outlet. The seeding cleats and the middle cleats are simultaneously driven by the discharging sprocket. The amount-limiting plate is installed at the channel outlet, which can prevent too many cane seeds from being discharged at one time, increasing the seeding uniformity of the planter, as shown in Figure 3.

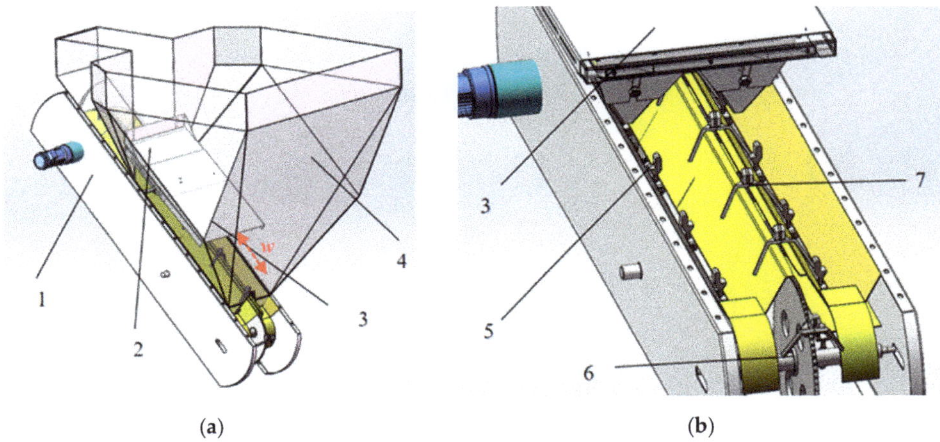

Figure 2. The structure of the seed metering device. (**a**) The overall structure of the seed metering device. (**b**) The interior of the W-channel: (1) W-channel; (2) amount-limiting device; (3) pressure-limiting plate; (4) hopper; (5) seeding cleats; (6) discharging sprocket; (7) middle cleats.

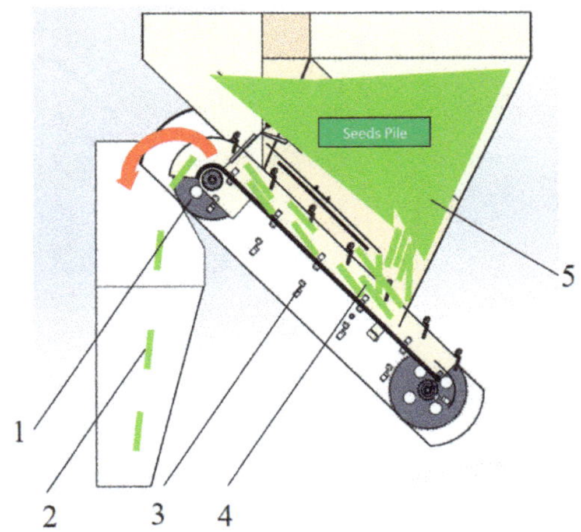

Figure 3. The working principle of the seed-metering device: (1) discharging sprocket; (2) cane seed; (3) seed cleat; (4) W-channel; (5) hopper.

2.2. Evaluation Index

Seeding uniformity and density is key factor in crop growth [17]. Different regions have different requirements for seeding planters. By studying the seeding performance of sugarcane planters, planters can achieve different planting goals. The indices of seeding performance include the seeding density, seeding uniformity, and seeding efficiency.

The planting density of sugarcane planters is defined as the average seeding rate per meter N (seed/m). According to the machine structure of the planter itself, the theoretical value is calculated as follows:

$$N = 0.067\frac{qn}{v} \tag{1}$$

where n is the discharging sprocket rotational speed, r/min; q is the average value of discharging per cleat, seed/cleat; v is the forward speed of the planter, km/h.

Seeding efficiency affects planting efficiency. Without considering the time required for filling the hopper and turning, in this study, the expression the seeding efficiency E (seed/s) is as follows:

$$E = 1.11Nv \quad (2)$$

where N is the theoretical seeding density, seed; v is the forward speed of the planter, km/h.

The coefficient of variation (CV) describes the dispersion of a variable. Therefore, the coefficient of variation of seeding reflects the seeding uniformity. The practical coefficient of variation (CV_p) for the planter is calculated as follows:

$$CV_p = \frac{s_n}{\overline{N}_n} \times 100\% \quad (3)$$

where s_n is the standard deviation of the sample data, and $\overline{N}_n$ is the average value of the sample data, seed/m.

2.3. Probability Distribution and Kolmogorov–Smirnov Test

If the sample distribution of seeds per meter conforms to a theoretical probability distribution where the mean and standard deviation exhibit a specific functional relationship, the seeding uniformity of the seed metering device can be estimated based on the seeding density. To this end, this study performs a goodness-of-fit test between the data distribution of seeds per meter and a specified theoretical distribution. The Kolmogorov–Smirnov (K–S) test, a non-parametric method [18] that does not require prior assumptions about the data distribution [19], is employed.

The K–S test proceeds as follows [20]:

1. Put forward the Hypothesis $H_{0\text{-}1}$:

Hypothesis $H_{0\text{-}1}$. *The cumulative distribution function of the sample data is equal to the cumulative distribution function of the theoretical distribution.*

2. Calculate the absolute differences between the cumulative distribution function of the sample data and that of the theoretical distribution. Take the maximum difference as the test statistic D_s, as shown in Equation (4).

$$D_s = \max_x |F_s(x) - F(x)| \quad (4)$$

where D_s is the test statistic of the K–S test, $F_s(x)$ is the cumulative distribution function of sample data, $F(x)$ is the cumulative distribution function of the theoretical distribution, and s is the sample size.

3. Determine the critical value $D_{s,\alpha}$ based on the sample size s and the significance level α.
4. If $D_s < D_{s,\alpha}$, accept the null Hypothesis $H_{0\text{-}1}$, indicating that the sample data conform to the theoretical distribution; otherwise, reject the null Hypothesis $H_{0\text{-}1}$.

After processing and analyzing the pre-experiment data, this study used the Poisson distribution to fit the distribution of the sample data of the seeding rate, and then used the K–S test method to examine the goodness of fit between the distribution of the sample data and the Poisson distribution. The Poisson distribution is suitable for describing the number of random events occurring within a unit of time (or space). Its cumulative distribution function is expressed as follows:

$$F(x;\lambda) = P(X \le x) = \sum_{i=0}^{x} \frac{\lambda^i e^{-\lambda}}{i!} \quad (5)$$

where x = 0, 1, 2. . ., denotes the number of event occurrences, $\lambda > 0$ represents the average number of events in the Poisson distribution and *e* is the base of the natural logarithm approximately equal to 2.718 [21].

To examine whether the sample data of the seeds per meter conformed to the Poisson distribution, the following null Hypothesis $H_{0\text{-}2}$ was proposed:

Hypothesis $H_{0\text{-}2}$. *The sample data of seeds per meter from different experimental groups, respectively, follow Poisson distributions with different parameters* λ*. The significance level was set at* $\alpha = 0.05$.

In the K–S test, the *p*-value is defined as the probability of observing the current D-value under the condition that the null Hypothesis $H_{0\text{-}2}$ holds. When the H value is 1, it indicates that the corresponding $p \leq \alpha$, and the null Hypothesis $H_{0\text{-}2}$ is rejected. When the H value is 0, it means that the corresponding $p \geq \alpha$, and the null Hypothesis $H_{0\text{-}2}$ is accepted [22]. That is to say, the hypothesis that the sample data of seeds per meter from different experimental groups, respectively, follow Poisson distributions with different parameters λ is valid.

2.4. Experimental Establishment

2.4.1. Experiment Design

In this study, three indexes were used for evaluating the performance of seeding: seeding density, seeding uniformity, and seeding efficiency. To verify the effects of the forward speed, discharging sprocket rotational speed, and the outlet size of the hopper on the index of seeding performance, a field test was conducted as shown in Figure 4.

Figure 4. Field tests were conducted with a sugarcane planter attached to a John Deere M1654 tractor.

Three different forward speeds *v* were used in this experiment: 4, 3, and 2 km/h. Two different hopper outlet sizes (*w*), 400 and 520 mm, were used to ensure that the hopper did not jam. Three different discharging sprocket rotational speeds *n* were used in this experiment: 120, 80, and 40 r/min.

The experiment followed a completely randomized design (CRD) with a fractional factorial design (FFD) approach. Only 8 key treatments were performed. This reduced the number of tests while still capturing the main effects and potential interactions, saving time and sources. The selected treatments are listed in Table 2.

This experiment was conducted at the testing base of Guangken Agricultural Machinery Service Co., Ltd., Zhanjiang City, Guangdong Province, China, on 11 March 2023. To reduce experimental error, the operating region was divided into the start region, test region, and termination region, as shown in Figure 5. The distances of the start, test, and termination regions were 15 m, 50 m, and 15 m, respectively. The number of seeds per meter was recorded in the test region. The planter was mounted on John Deere M1654 tractor and operated during field work. The type of sugarcane used was Guitang 91, grown in Zhanjiang city, Guangdong province. The average length and diameter of the seeds were 250 mm and 30 mm, respectively. In the test region, the seeding rate per meter was manually recorded for a two-row planter, generating 100 data points for treatments A to H. With 3 experimental replications, the total dataset included 300 data points.

Table 2. Completely randomized experiment.

Treatment	Forward Speed v (km/h)	Discharging Sprocket Rotational Speed (r/min)	Hopper Outlet Size (mm)
A	4	80	400
B	3	80	400
C	2	80	400
D	3	120	520
E	2	80	520
F	4	120	520
G	3	80	520
H	2	40	520

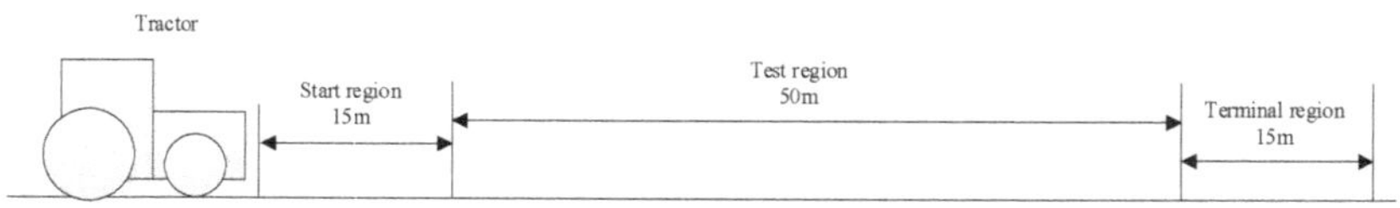

Figure 5. Arrangement of experimental region.

The average seeding rate per meter was calculated from each treatment's data. Using Equations (1) and (2), the average discharge amount per seeding cleat q and seeding efficiency E for each group were then determined. One-way analysis of variance (ANOVA) [23] was performed on q and E to investigate the influence of vehicle speed v, discharging sprocket rotational speed n, and hopper outlet size w on these parameters. Linear regression analysis was employed to establish mathematical relationships among the given operating parameters, the planter's inherent structure, the average seeding rate per meter, and the seeding efficiency. When the distribution of the seeding rate sample data follows a Poisson distribution, the seeding uniformity can be reflected by the average seeding rate per meter. Therefore, the seeding performance of the planter can be directly predicted using its operating parameters and structural characteristics.

2.4.2. Speed Test of John Deere M1654 Tractor

The tractor M1654 used in the experiment lacked a real-time speed display and was equipped only with an engine speed gauge, which presents challenges when conducting speed-related tests. Therefore, this section analyzes the relationships among the vehicle speed, gear position, and engine speed of the John Deere M1654 tractor.

Within the tractor's powertrain, speed transmission begins with the transfer of power from the engine to the main clutch, proceeds through the gearbox, reaches the main reducer, and is finally delivered to the drive wheels via the transfer case [24].

In the transmission system of the M1654 tractor, gear A1 and gear B4 adopt a geometric transmission design. According to reference [25], under the conditions of a constant engine speed, by measuring the vehicle speeds corresponding to any two gears (ignoring actual slippage factors), the common ratio of the transmission system can be calculated.

When operating in the B3 gear at an engine speed of 1500 RPM, the travel time of the John Deere M1654 tractor over a 50-m distance was measured manually. Under these conditions, the tractor's speed was measured to be 5.1 km/h. Similarly, when the John Deere M1654 tractor was operated at 1500 RPM in the A2 gear, it traveled at a speed of 2.0 km/h. So, the drive train ratio for the M1654 tractor was calculated as 1.207. Based on this common ratio, the theoretical vehicle speed of any gear at a specified engine speed can be calculated using Equation (6):

$$v_z = 8.93 \times 10^{-4} \times 1.207^z n_z \quad (6)$$

where: z is the number of gears; n_z is the engine speed, r/min; v_z is the corresponding tractor speed at engine speed n_z for gear z, km/h.

Therefore, when the vehicle speed is set at 3.0 km/h, the corresponding gear and speed would be A3 and 900, respectively; similarly, when the vehicle speed is set at 4.0 km/h, the gear and speed would be B3 and 1200. The study also compiled a selection of tractor models whose first 8 gears within satisfy the requirement for equal ratios. Please refer to Table 3.

Table 3. Tractor models in which the first eight gear ratios within 200 HP are in an equal series.

Brand	Tractor	Gear
John Deere (John Deere, Moline, IL, USA)	940 and 1140	1 to 8
	6M series 1654, 6J series 1854	A1 to B4
CASE (CASE, Racine, WI, USA)	Farmall series 65A, 75A, 85A, 95A, 105A and 115A	2 to 8
	MXM series 120, 130, 140, and 155	A1 to A6
Landini Atlantis (Landini, Fabbrico, Italy)	70, 75, 80, 85 and 95	2 to 8
	105, 115, 130, 145 and 165	3 to 8
McCOR-MICK (McCormick, Basildon, UK)	XTX series 185	1–1 to 1–8

2.4.3. Centroid Analysis

The centroid of the planter, filled with different cane seed amounts, was measured to assess how many seeds the M1654 tractor mount can support. During operation, the total weight of the planter consists primarily of two components: the weight of the unloaded planter and the mass of the loaded seeds.

Firstly, the weight and centroid position of the unloaded planter were measured using the weighbridge, and then the centroid position of the planter with different cane seeds weights was measured using computer technology. When the planter is stationary on the ground, it is supported by the front furrow opener (P1) and two rear legs (P2 and P3). In this study, a weighbridge (range of 0–3 tons, accuracy to 0.5 kg) was used to measure the force on the three points of the unloaded planter, as shown in Figure 6. The weights of the three points P1, P2, and P3 were 756 kg, 414 kg, and 535 kg, respectively. Therefore, the mass M of the unloaded planter was calculated to be 1705 kg, and the horizontal distance X_0 from the center of planter C_P to the suspension point A_t was 1010 mm.

Then, using the software Solidworks (v2016), the material density of the cane seed pile model was set to 415 kg/m^3. By modeling the cane seed pile with different cane seed quantities, the curve showing the change in the centroid position of the cane seed pile with mass was obtained, as shown in Figure 7a,b. The horizontal distance between the center of the mass and the suspension point was 350 mm, and the distance X_h between the front end of the bucket and the suspension point was equal to 1566.49 mm. Using the centroid formula Equation (7) of the assembly, the change curve of the position of the planter's centroid under different cane seed quantities was calculated, as shown in Figure 7b. When the planter is loaded with different masses of seeds, the formula for calculating its centroid position is given by Equation (7):

$$X_1 = \frac{(x_1 + X_h)m_1 + X_0 m_0}{m_1 + m_0} \quad (7)$$

where X_1 is the distance from the centroid to the suspension point after the planter is loaded with seeds, m; X_0 is the distance from the centroid to the suspension point when the planter is not loaded, m; x_1 is the distance from the centroid to the front end, m; X_h is the distance between the front end of the hopper and the suspension point, m; m_1 is the mass of the loaded seeds, kg; m_0 is the self-weight of the unloaded planter, kg.

Figure 6. Weighbridge weighing point P2 of the non-load planter.

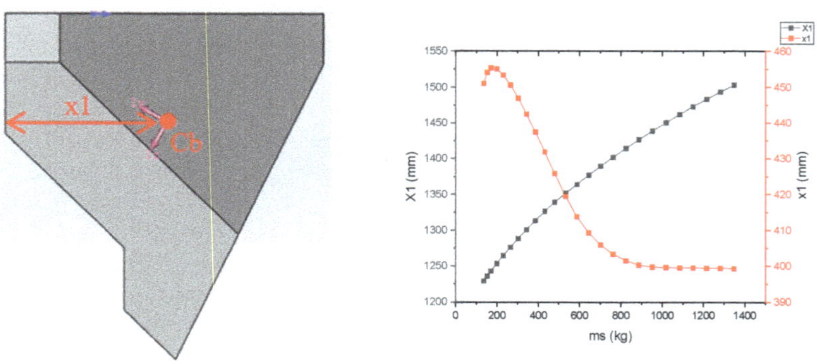

(**a**) (**b**)

Figure 7. Centroid analysis. (**a**) The 1400 kg cane seed pile model in the hopper, where Cb is the centroid of the cane seed pile, and x_1 is the distance from the centroid to the front end. (**b**) X_1 is the curve of the centroid of the planter changing with the cane seed quality, and x_1 is the curve of the centroid of the cane seed pile changing with the cane seed quality.

As the mass of seed pile increased from 200 kg to 800 kg, the distance of the centroid of the seed pile on the x–axis x_1 decreased rapidly, while X_1 gradually increased. When the mass of seed pile was more than 800 kg, x_1 remained almost constant, while X_1 continued to increase. The variation range of $x1$ was within 60 mm, and the variation range of X_1 was within 270 mm. The M1654 tractor has a hydraulic lifting capacity of up to 4000 kg at the 610 mm suspension point [26]. Therefore, based on the data in Figure 7b, the safe loading capacity of the planter was calculated as 360 kg. Under the assumptions of ignoring ground wheel support and uniform mass distribution of seeds within the hopper, the John Deere M1604 tractor can safely suspend the planter for field operations when the loading mass of seeds is less than 360 kg.

2.5. Statistical Analysis

All treatment datasets were imported into the statistical analysis software SPSS (v2017). Within SPSS, a series of analytical operations were carried out, including generating the seeding frequency table, performing the K–S test, conducting one-way analysis of variance, and executing regression analysis.

3. Results

3.1. Analysis of Experimental Data Results

As shown in Figure 8, the field test results were analyzed by counting the number of seeds per meter. After processing the experimental data, the mean and standard deviation were presented in Table 4. Using Equation (1), the average discharge per seed metering cleat was calculated and is also listed in Table 4.

Figure 8. Field test results of sugarcane planting operation.

Table 4. The experimental values of the treatments.

Treatment	Mean (Seeds)	Std. Deviation	Discharging Rate per Cleat q (Seeds)
A	3.25	1.538	2.46
B	3.95	1.751	2.24
C	7.11	2.823	2.69
D	9.32	3.362	3.53
E	7.86	3.294	3.5
F	7.11	2.651	3.5
G	6.03	2.498	3.43
H	5.79	2.754	4.08

3.2. Evaluation of Seeding Density

According to Equation (1), under the given n and v, the average seeding rate per meter N is determined by q. To explore the factors that affect q, one-way ANOVA was carried out on v, n, and w, respectively. The significance of each factor was measured through the F and t-tests, as listed in Table 5.

Table 5. The analysis of variance for factors.

Source of Variance	F-Value	p-Value
Speed	0.291	0.759
Rotation	10.825	0.087
Hopper outlet size	38.272	0.001 *

Note: * Significant impact within a 99% confidence interval.

As seen in Table 5, the outlet size of the hopper had a very significant effect on the discharging rate per cleat. Speed and rotation had an insignificant impact on the discharging rate per cleat. When the outlet sizes of the hopper were 400 mm and 520 mm, the average values of q were 2.55 and 3.61. Assuming that the relationship between q and w is linear, Equation (1) can be written as Equation (8), where k is the proportionality coefficient.

$$N = k\frac{wn}{v} \quad (8)$$

Denoting X as $w \times n/v$, a scatter plot was generated, and linear regression analysis was conducted, excluding a constant between X and the mean N, as shown in Figure 9. The slope of the curve is 4.46×10^{-4} and R-squared is equal to 0.9959.

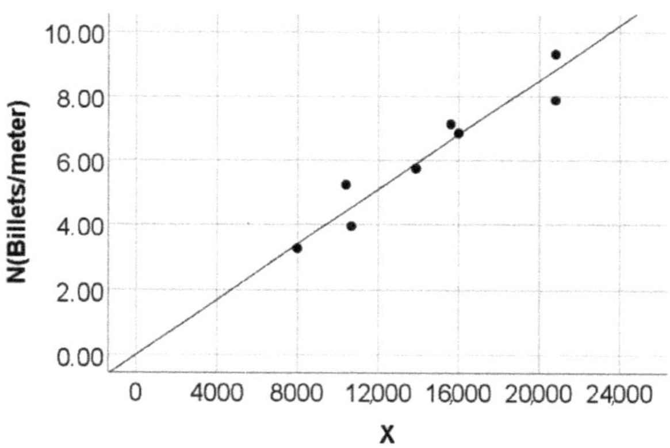

Figure 9. Scatter plot and linear regression curve between X and mean N.

Therefore, the mathematical model of seeding is written as Equation (9).

$$N = 4.46 \times 10^{-4} \cdot \frac{wn}{v} \quad (9)$$

Therefore, the seeding density of the sugarcane planter can be directly calculated using the rotational speed of the discharging sprocket n, the forward speed of the planter v, and the outlet size of the hopper w.

3.3. Evaluation of Effect on Seeding Efficiency

ANOVA was used to test the significant difference in seeding efficiency at different v, w, and n values, as shown in Table 6. The results show that the discharging sprocket rotational speed n and the hopper outlet size w had a significant effect on the seeding efficiency.

Table 6. ANOVA was used to test significant differences in seeding efficiency at different v, w, and n values.

Variance Source	F-Value	p-Value
Speed v	0.106	0.9
Rotation n	5.116	0.019 *
Hopper outlet size w	9.538	0.007 *

Note: * Significant impact within 95% confidence interval.

Linear regression analysis was conducted, taking q and n as independent variables and E as a dependent variable, as shown in Table 6. Therefore, the mathematical linear regression model of seeding efficiency in relation to the outlet size of the hopper and the discharging sprocket rotational speed is established as Equation (10). The model has an R-squared value of 0.979, so the fit is favorable.

$$E = 0.241n + 0.056w - 27.17 \quad (10)$$

Therefore, the seeding efficiency of the sugarcane planter can be directly estimated using the given rotational speed of the seeding sprocket wheel n and the outlet size of the hopper w.

3.4. Evaluation of Seeding Uniformity

To reduce the influence of random errors and improve the reliability of the experiment, the data from three replicate trials in each group were aggregated and processed. This was carried out to analyze the degree of fit between the distribution of seeding quantity per meter and the Poisson distribution. The frequency distribution of seeding of treatment A–H are shown in Figure 10. The frequency distributions of seeding show higher frequencies in the middle and lower frequencies on the sides, forming an approximately bell-shape curve. Skewness and Kurtosis are shown in Table 7. The skewness is all greater than 0, which indicates that the seeding distribution is not symmetric about the mean and majority of the data distribution will be on the left side of the mean. The skewness values were all greater than 0 except for treatment H, which indicates that this seeding distribution was lighter-tailed with a steeper top curve [27].

Table 7. The p-values of the K–S tests and parameter λ of the treatments.

Treatment	Skewness	Kurtosis	p-Value	λ
A	0.151	0.052	0.093	3.25
B	0.623	0.673	0.241	3.95
C	0.579	0.416	0.958	7.11
D	0.478	0.399	0.950	9.32
E	0.738	0.514	0.435	7.86
F	0.476	0.567	0.999	7.11
G	0.452	0.267	1.000	6.03
H	0.331	−0.047	0.535	5.79

The datasets of all treatments were fitted by Poisson distributions, with λ equal to N, using maximum likelihood estimation. The values of λ and the results of the K–S test (under confidence degree $\alpha = 0.05$) are shown in Table 7. All the p-values of K–S tests were larger than $\alpha = 0.05$, so the Poisson distributions fit the datasets of all treatments well [28].

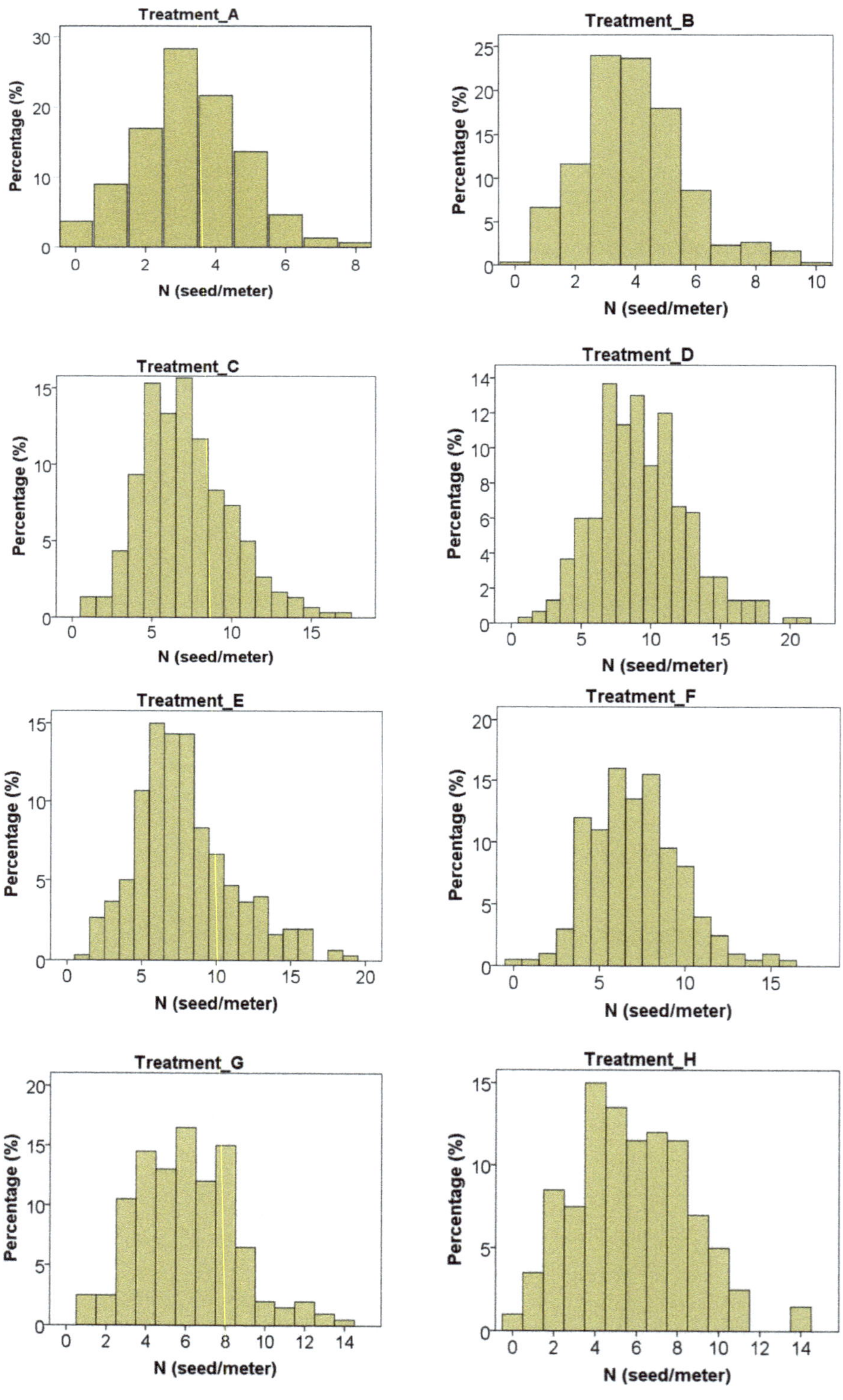

Figure 10. The frequency distribution of seeding for treatments A–H.

As shown in Table 7, the mean and standard deviation of a Poisson distribution are both equal to the parameter λ, which is equal to the mean N. Therefore, the theoretical variable coefficient CV_t can be written as Equation (11) based on Equation (9).

$$CV_t = 47.35\sqrt{\frac{v}{nw}} \tag{11}$$

The practical variable coefficient CV_p can be calculated as the mean divided by the std. deviation of sample data. The ε_a between CV_p and CV_t was 0.94–6.18%. When CV_t was below 40%, the ε_a was less than 3%, as shown in Table 8.

Table 8. Theoretical variable coefficient CV_t, the practical variable coefficient CV_p and the absolute error ε_a (%) between CV_t and CV_p.

Treatment	CV_t (%)	CV_p (%)	Absolute Error ε_a (%)
A	52.94	46.76	6.18
B	45.85	42.72	3.13
C	37.43	35.28	2.15
D	32.83	35.44	2.61
E	35.10	36.37	1.27
F	37.91	37.53	0.38
G	40.21	37.61	2.60
H	46.43	47.37	0.94

Therefore, when the rotational speed of the seeding sprocket wheel n, the forward speed of the planter v, and the outlet size of hopper w of the planter are known, if the calculated theoretical coefficient of variation CV_t is less than 40%, it is approximately equal to the actual coefficient of variation CV_p, with an error of less than 3%. This can relatively accurately reflect the seeding uniformity of the planter.

4. Discussion

This study conducted a comprehensive evaluation of a novel pre-cut sugarcane planter designed by South China Agricultural University. This sugarcane planter represents a breakthrough improvement over current domestic designs [23,29], which are constrained by low efficiency and a limited capacity. Compared to international counterparts (e.g., Gessner's single-row planter [30] or Ury's Manufacturing triple-row planter [31]), this tractor-mounted system achieved a 40% reduction in the turning radius and a superior hopper capacity, making it ideally suited for small-plot cultivation in southern China's hilly regions—a practical solution to China's challenges of high production costs and low mechanization rates in sugarcane farming.

Currently, many scholars analyze the probability of multiple seeding (exceeding a specific seed count) or missed seeding (falling below a specific count) within a unit distance under different working conditions through experiments. However, such an approach is limited because it only provides insights into seeding performance within specific numerical ranges. By conducting large-scale seeding experiments to analyze the probability distribution of seeds per meter, we can comprehensively understand the occurrence probability of each seeding quantity within a unit of distance. Coupled with the regression relationship between operational parameters of planter and seeding density, these findings allow for direct prediction of the occurrence probability for any seeding quantity range under arbitrary operational parameters. Furthermore, these findings can also be used to assess the planter's performance for different regional planting requirements by adjusting the operational parameters of the planter.

The following limitations in this study require further investigation: (1) potential unexamined parameter interactions due to the completely randomized partial factorial design; (2) centroid shift phenomena during uneven seed accumulation and seeding uniformity under varying load conditions; (3) comprehensive economic evaluation (e.g., per-mu operational costs) through large-scale field trials.

5. Conclusions

A new type of pre-cut sugarcane planter designed by South China Agricultural University was used in field seeding experiments. Before the experiment, the vehicle speed of the John Deere M1654 tractor and the safe loading capacity of the planter were analyzed. Then, the distribution of the seeding rate per meter and the seeding performance of the planter at different forward speeds, discharging sprocket rotational speeds, and hopper outlet sizes were investigated through field experiments. The obtained conclusions were as follows:

1. The outlet size of the hopper has a very significant effect on the discharging rate per cleat. When the outlet sizes of the hopper are 400 mm and 520 mm, the average values of q are 2.55 and 3.61. The mathematical model of the average seeding rate per meter of the planter in relation to the speed of the vehicle, the discharging sprocket rotational speed, and the outlet size of hopper is $N = 4.46 \times 10^{-4} \cdot \frac{wn}{v}$ (R^2 = 0.9959).
2. Seeding efficiency is significantly affected by the discharging sprocket rotational speed and the outlet size of the hopper. The mathematical regression model is $E = 0.241n + 0.056w - 27.17$ (R^2 = 0.979).
3. The seeding rate per meter of the planter follows the Poisson distribution, and the Poisson distribution parameter λ is equal to the average seeding rate per meter. Its theoretical coefficient of variation is $CV_t = 47.35\sqrt{\frac{v}{nw}}$. When the theoretical coefficient of variation is less than 40%, the absolute error between the actual and theoretical coefficient of variation is less than 3%.

Author Contributions: Conceptualization, Z.P.; methodology, Z.P.; software, Z.P., J.C. and Z.C.; validation, Z.P. and F.X.; formal analysis, Z.P. and P.X.; investigation, Z.P., F.X., J.C. and Z.C.; resources, Z.P. and J.C.; data curation, Z.P. and P.X.; writing—original draft, Z.P. and P.X.; writing—review and editing, Z.P. and P.X.; visualization, Z.P. and P.X.; supervision, F.X., T.W. and Z.C.; project administration, T.W.; funding acquisition, T.W. All authors have read and agreed to the published version of the manuscript.

Funding: This work was supported by the earmarked fund for CARS-17.

Institutional Review Board Statement: Not applicable.

Data Availability Statement: The data provided in this study are available upon request from the corresponding author.

Conflicts of Interest: The authors declare no conflicts of interest.

References

1. Ungureanu, N.; Vlăduț, V.; Biriș, S.-Ș. Sustainable Valorization of Waste and By-Products from Sugarcane Processing. *Sustainability* **2022**, *14*, 11089. [CrossRef]
2. Liu, F.; Liu, W.; Wen, Z.; Zhang, L.; Wang, X. Design and Development of Chain Plate Precision Seed-Metering Device for Sugarcane. *J. Northeast Agric. Sci.* **2020**, *45*, 103–106+138. [CrossRef]
3. Razavi, J.; Namjoo, M. Determination of forward speed effect on planting uniformity in a sugarcane billet planter. In Proceedings of the XVII World Congress of the International Commission of Agricultural and Biosystems Engineering (CIGR), Quebec City, QC, Canada, 13–17 June 2010. Paper No. CSBE001142.
4. Saengprachatanarug, K.; Chaloemthoi, C.; Kamwilaisak, K.; Kasemsiri, P.; Chaun-Udom, S.; Taira, E. Effect of metering device arrangement to discharge consistency of sugarcane billet planter. *Eng. Agric. Environ. Food* **2018**, *11*, 139–144. [CrossRef]

5. Wu, J.; Guo, X.; Zhang, S. Structure Design of Intelligent Sugarcane Seed Arrangement Device. *Equip. Manuf. Technol.* **2021**, *6*, 64–67.
6. Huang, Y.; Yin, K.; Huang, M.; Wang, X. Single-Bud Cane Seed Discharging Device. Chinese Patent CN104798517A, 24 April 2015.
7. Wen, L.; Li, J. Mechanization of sugarcane field abroad. *Sugarcane Canesugar* **1977**, *Z1*, 85–92.
8. Liu, Q.; Mo, J.; Li, Y.; Ou, Y. Current Situation of Sugarcane Planter and its Key Technical issues in China. *Sugarcane Canesugar* **2011**, *5*, 52–58.
9. Han, J.; Wen, S.; Liu, Q.; Wu, J.; Xu, J.; Wu, J.; Ou, C.; Yan, W. Design and test of pre-cutting type sugarcane planter. *J. South China Agric. Univ.* **2019**, *40*, 109–118.
10. Li, J.; Liu, F.; Liu, W.; Zhang, Y.; Hu, H. Study on the Development of Mechanization of Sugarcane Planting in China: General Situation, Restrictive Factors and countermeasures. *Agric. Mach. Using Maint.* **2018**, *12*, 1–4. [CrossRef]
11. Zhang, H.; Pan, F.; Ji, C.; Chen, J. Research Progress on Automation Key Technology of Precision Seeding in China. *Mod. Agric. Equip.* **2022**, *43*, 2–9+37.
12. Su, W.; Hong, F.; Lai, Q.; Jia, G.; Chen, Z. Design and Test of Pre-cutting and Spoon Seed-metering Device for Sugarcane. *Trans. Chin. Soc. Agric. Mach.* **2020**, *51*, 87–97.
13. Ma, F.; Liang, Y.; Li, S.; Luo, X.; Deng, Z. Improved Design and Experimental Research on Seed Supply Device of Pre-cut Sugarcane Transverse Planter. *J. Agric. Mech. Res.* **2022**, *44*, 176–182.
14. Chen, J. The Invention Relates to a Discharge and Methodology for an Automated Sequential Directional Seed Arrangement Apparatus Used in the Planting of Double Bud Sugarcane. Chinese Patent CN104798517A, 2 April 2015.
15. Chen, J. Design and Experiment of Sugarcane Billet Double Bud Planter with Uniform Quantities. Master's Thesis, South China Agricultural University, Guangzhou, China, 2023.
16. Hao, J. Mechanism Analysis of Polydispersed Shear Flows of Cylindrical Particles Based on DEM. Master's Thesis, Beijing Forestry University, Beijing, China, 2020.
17. Wang, Z.; Dai, S.; Cai, Y.; Mo, X.; Sun, X. Effects of Different Density and Pruning Methods on Seed Production of Seedless Watermelon. *Seed* **2019**, *38*, 138–140. [CrossRef]
18. Pei, J.; Wang, S.; Ren, M.; Peng, J. Research on Reliability of Air Cooler Fan. *Mach. Des. Manuf.* **2020**, *8*, 215–219. [CrossRef]
19. Charles, E. *An Introduction to Reliability and Maintainability Engineering*; Tsinghua University Press: Beijing, China, 2003.
20. Cheng, Y.; Zeng, C.; Miao, H.; Yang, X. Beta Mixture Model Combined with K-S Test for System Harmonic Impedance Estimation. *Electr. Power Syst. Its Autom.* **2025**, *1*, 10. [CrossRef]
21. Zou, D.; Zou, C.; Song, X.; Tian, H.; Zhang, Y.; Zhang, K. Characteristics of the Number of Times and Probability of Rainstorm and Torrential Rain in Zunyi based on Poisson Distribution. *J. Chengdu Univ. Inf. Technol.* **2022**, *37*, 716–720. [CrossRef]
22. Li, C.; Han, Z.; Song, K.; He, L.; Tao, C. Amplitude Extraction Method for Optoelectronic Pulses Based on Synthesis Shaping. *Semicond. Optoelectron.* **2025**, *46*, 149–156. [CrossRef]
23. Zhong, J.; Tao, L.; Li, S.; Che, R.; Li, K.; He, Y. Design and experiments of a seed clearing, reseeding and elevating type seed metering device for sugarcane. *Trans. Chin. Soc. Agric. Eng.* **2023**, *39*, 43–54.
24. Liu, Y. Design and Research of Hilly Mountain Tractor Chassis. Master's Thesis, Shanxi Agricultural University, Shanxi, China, 2021. [CrossRef]
25. Ju, C.; Wang, Y.; Dong, W.; Yang, Y.; Zhai, Y. Research on Adjacent Transmission Ratio Distribution of Tractor Range Transmission. *Tract. Farm Transp.* **2016**, *43*, 28–33.
26. 6M-1654 Tractor | John Deere Official Website. [EB/OL]. 28 June 2022. Available online: https://www.deere.com.cn/zh/%E6%8B%96%E6%8B%89%E6%9C%BA/6m%E7%B3%BB%E5%88%97/1654%E6%8B%96%E6%8B%89%E6%9C%BA/ (accessed on 10 March 2023).
27. Wen, F.; Yang, X. Skewness of Return Distribution and Coefficient of Risk Premium. *J. Syst. Sci. Complex.* **2009**, *22*, 360–371. [CrossRef]
28. Dong, S.; Jiao, C.; Tao, S. Joint return probability analysis of wind speed and rainfall intensity in typhoon-affected sea area. *Nat. Hazards* **2017**, *86*, 1193–1205. [CrossRef]
29. Zhang, B.; Wang, J.; Yang, X.; Chen, B.A. DEM-MBD based method for regulating transfer flux in the supply and discharge of cane seed particles. *Comput. Electron. Agric.* **2024**, *218*, 108732. [CrossRef]
30. Gessner's Single-Row Planter. [EB/OL]. 28 June 2022. Available online: https://gessner-industries.tradeearthmovers.com.au/detail/gessner-single-row-billet-planter-782584 (accessed on 10 March 2025).
31. Ury's Manufacturing Triple-Row Planter. [EB/OL]. 16 August 2024. Available online: https://www.facebook.com/100057479799811/videos/3-row-billet-planter/1490306191857569/ (accessed on 10 March 2025).

agriculture

MDPI

Article

Design and Experimentation of a Height-Adjustable Management Platform for Pineapple Fields

Sili Zhou [1,2,†], Fengguang He [1,2,3,†], Ganran Deng [1,2,*], Ye Dai [1,2], Xilin Wang [1,2], Bin Yan [1,2], Pinlan Chen [1,2], Zehua Liu [1,2], Bin Li [4] and Dexuan Pan [3]

1 Chinese Academy of Tropical Agricultural Sciences, Agricultural Machinery Research Institute, Zhanjiang 524013, China; 19854815520@163.com (S.Z.); hefg1106@163.com (F.H.); 15069116036@163.com (Y.D.); 15756791741@163.com (X.W.); 15069127151@163.com (B.Y.); pl1600875479@gmail.com (P.C.); linzeng14789@outlook.com (Z.L.)

2 Key Laboratory of Tropical Agricultural Machinery, Ministry of Agriculture and Rural Affairs, Zhanjiang 524091, China

3 School of Mechanical and Automotive Engineering, South China University of Technology, Guangzhou 510640, China; 16689740775@163.com

4 Institute of Facility Agriculture, Guangdong Academy of Agricultural Sciences, Guangzhou 510640, China; libin@gdaas.cn

* Correspondence: dengganran@163.com

† These authors contributed equally to this work.

Abstract: To address the challenges of inadequate adaptability, insufficient power, high ground clearance, and limited functionality in existing pineapple field machinery, this study proposes a height-adjustable pineapple field management platform based on previously established cultivation patterns and agronomic requirements. The structural configuration and operational principles of the platform's power chassis are elucidated, with specific emphasis on the development of the traction power system and modular operational systems. Theoretical and experimental analyses of steering parameters, stability, and field performance were conducted. Finite element simulation analysis of the frame revealed that under full-load conditions, the equivalent elastic strains during descent and ascent phases were 0.000317 and 0.00125, respectively. Maximum equivalent stresses (48.27 MPa and 231.6 MPa for descent and ascent, respectively) were localized at the beam–plate junctions, while peak deformations of 1.14 mm (descent) and 4.31 mm (ascent) occurred at mid-frame and posterior–mid regions, respectively. Field validation demonstrated operational velocities of 0.16–1.77 m/s (forward) and 0.11–0.28 m/s (reverse), with a maximum gradability of 20°. The platform exhibited multifunctional capabilities including weeding, spraying, fertilization, flower induction, harvesting, and transportation, demonstrating its potential to fulfill the operational requirements for pineapple field management. Simultaneously, the overall work efficiency is increased by more than 50%, compared to manual labor.

Keywords: pineapple; field management; multi module integration method; height adjustable workstation; design method and process

1. Introduction

As one of the global top ten pineapple producers, China boasts over 400 years of cultivation history, currently accounting for 7% of worldwide pineapple cultivation area and annual production [1]. The pineapple industry plays a pivotal role in tropical agricultural economies, with distinct regional characteristics and substantial market value. However, conventional production models face critical challenges including labor intensiveness (particularly in field management operations), escalating production costs, and limited mechanization adoption. Manual operations still dominate critical processes such as weeding, fertilization, flower induction spraying, and material transportation, constituting over 50% of total production costs [2], with labor expenses showing a persistent upward trend. These traditional practices increasingly fail to meet the demands of large-scale standardized cultivation, necessitating mechanization breakthroughs to overcome industrial constraints.

Current pineapple field management primarily relies on manual or semi-mechanized operations. Research concerning mechanized equipment assisting human labor focuses predominantly on single-function machinery for specific tasks, such as weeders [3,4], fertilizer applicators [5,6], floral induction spraying machines [7,8], and material transport systems [9–11]. While these enhance efficiency for individual operations, they commonly suffer from insufficient adaptability. On one hand, significant variations in planting density and ridge–furrow parameters exist among different pineapple varieties. Furthermore, canopy height can reach 120 cm [12] at maturity, forming densely closed canopy zones that prevent the operation of general-purpose agricultural machinery. On the other hand, the topography of China's primary pineapple-growing regions is complex, with a high proportion of hills and slopes. Existing equipment has limited capability to adjust for ridge spacing and height, hindering stable operation across varied terrain. Additionally, managing multiple operational stages necessitates frequent equipment changes, leading to resource wastage and increased operational costs.

Certainly, some researchers have directly investigated multifunctional pineapple field management machinery. For instance, Wang Binbin [13] designed a compact, adjustable-wheelbase, high-clearance, narrow-tracked crawler chassis. This platform can be equipped with three operational modules: a boom spray component, a solid fertilizer applicator, and an extended platform-transport component. However, it cannot perform floral induction spraying tasks and has a limited payload capacity of only 600 kg. More researchers have explored the application of high-clearance chassis in other contexts. Wan Xingyu [14] developed a high-clearance self-propelled chassis for rapeseed harvesting. Results indicated that the windrower exhibited average straight-line deviations of 0.73% and 1.28% on hard surfaces and soft soil, respectively. Under single-side braking conditions, the average turning radii were 1.91 m and 2.03 m on these respective surfaces, demonstrating stability during ascending/descending slopes, traversing field ridges, and crossing furrows. Du Xiaoqiang [15] developed a tracked high-clearance chassis for vibrating harvesting of camellia fruit. Finite element modal analysis of the striking device frame and harvester frame using ANSYS software (version: 2021R2) determined the first six natural frequencies, confirming resonance avoidance. The camellia fruit harvest rate was 87.56%, with a blossom drop rate of 25.86%, meeting harvesting requirements. Hu Kai [16] and Li Wei [17] each researched high-clearance chassis for field spraying, both employing Proportional–Integral–Derivative (PID) control methods for four-wheel steering, validating the responsiveness, accuracy, and stability of their designed multi-wheel steering systems. However, the high-clearance chassis mentioned in these studies were developed for specific scenarios and lack height adjustability, rendering them unsuitable for pineapple field management operations. Given the demanding operating conditions of agricultural implements in hot, humid, muddy, and bumpy environments, which require characteristics such as high torque at low speeds, stable operation, and low maintenance costs, hydraulic drive systems exhibit significant advantages over electric drives. These advantages include superior torque output characteristics, higher power density, enhanced adaptability to harsh environments, greater layout flexibility, seamless compatibility with diesel engines, and inherent overload protection capabilities, along with proven technical maturity and effective cost control within the agricultural sector.

To address these challenges, our research team proposes an innovative hydraulic-driven modular pineapple field management platform with adjustable ground clearance (60–120 cm), based on our previously established machine-adapted full-process mechanization framework for pineapple production [18]. This integrated system features five key functional modules: (1) a combined weeding-spraying unit; (2) a precision fertilization system; (3) targeted flower induction apparatus; (4) adaptive harvesting mechanism; and (5) a multi-purpose transport platform. Through coordinated module operation and dynamic height adjustment capabilities, the platform enhances operational stability on complex terrain while enabling functional scalability. Field trials demonstrate 40% reduction in labor input and 30% operational cost savings compared to conventional methods, providing a comprehensive mechanization solution for modern pineapple cultivation systems.

2. Methodology

2.1. Agronomic Mode of Pineapple Planting Suitable for Mechanization

The mechanized pineapple production system employs an innovative ridged cultivation pattern characterized by a triangular staggered arrangement of wide–narrow double rows (Figure 1). This optimized configuration features alternating inter-row spacings of 80–90 cm (wide rows) and 50 cm (narrow rows), with intra-row plant spacing maintained at 25 cm. The ridging parameters include a ridge height of 20–25 cm and furrow-to-furrow distance of 130–140 cm, which ensures compatibility with the wheelbases of mainstream tractor brands.

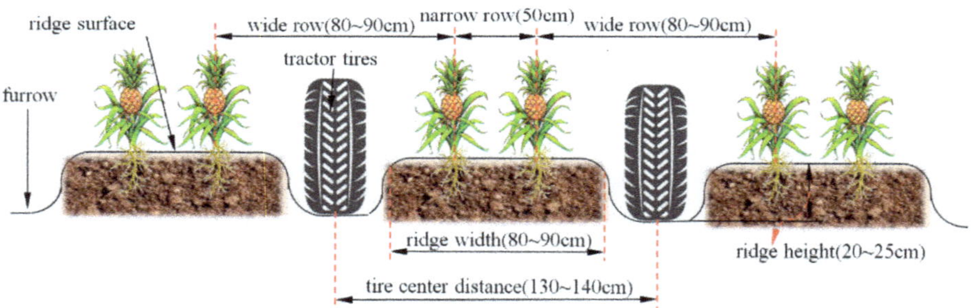

Figure 1. "Pin"-shaped ridging type wide–narrow double-row pineapple planting pattern.

2.2. Complete Machine Structure

The modular height-adjustable pineapple field management platform (Chinese Academy of Tropical Agricultural Sciences, Agricultural Machinery Research Institute, Zhanjiang, China) is primarily composed of an operator cab, power system, locomotion system, hydraulic systems, chassis frame, and control cabinet. As illustrated in Figure 2, the locomotion system integrates four hydraulic motor-driven ground wheels, with synchronized flow control for front and rear wheels on each side through flow-synchronizing valves. The hydraulic systems for both locomotion and operational functions are strategically installed beneath and within the cab compartment.

The chassis configuration employs a two-tier structural design: the upper section houses the operator cab and cargo compartment, while the lower section contains the locomotion mechanism and load-bearing framework. The load-bearing structure adopts a gantry-style design connected via sectionalized support legs, incorporating four hydraulic telescopic columns for vertical adjustment. The elevation of the management platform is controlled by operating the lift control lever, which enables synchronous extension or retraction of the hydraulic telescopic cylinders to achieve platform ascent or descent. The system automatically halts upon reaching its extreme positions, as designated in Figure 2b, providing a height adjustment range of 50 cm to 120 cm. Power generation for both hydraulic circuits is achieved through an integrated diesel-powered system comprising a radiator-equipped diesel engine, hydraulic oil reservoir with dedicated cooling unit, and hydraulic pumps. The operational hydraulic circuit provides motive power for various agricultural implements including weed/pesticide sprayers, flowering induction sprayers, and foliar fertilizer applicators. The locomotion hydraulic system features a four-wheel drive configuration with an integrated safety mechanism, incorporating a self-locking device to prevent unintended vehicle movement on inclines. The system demonstrates exceptional terrain adaptability with a maximum grade ability of 20°, ensuring reliable operation on sloped plantation terrain; the relevant parameters are shown in Table 1.

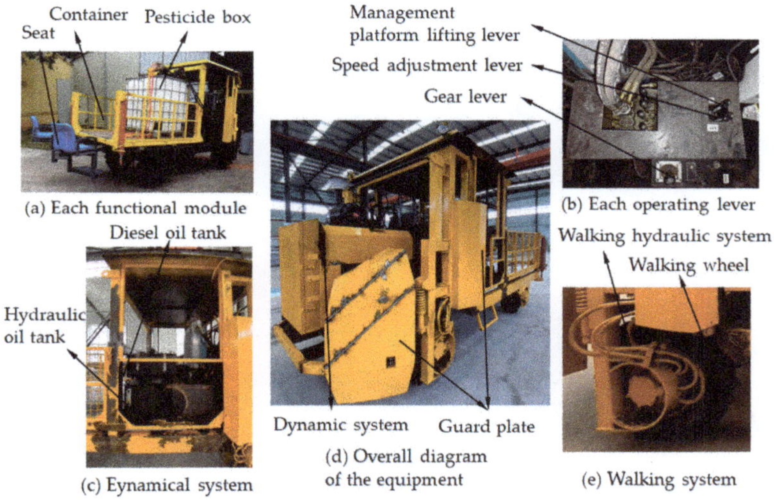

Figure 2. Structure diagram of multi module height-adjustable pineapple field management platform.

Table 1. Main performance parameters of pineapple field management platform.

Parameter	Numerical Value
Power/kW	63
Driving form	Four-wheel hydraulic drive
Overall dimensions (length × width × height) (mm × mm × mm)	4350 × 2100 × 2350
Chassis ground clearance/mm	500~1200
Forward speed/($m \cdot s^{-1}$)	0.16~1.77
Backward speed/($m \cdot s^{-1}$)	0.11~0.28
Rotary speed/($r \cdot min^{-1}$)	2200
Maximum bearing weight/t	2
Turning radius/m	2.9
Maximum torque/(Nm)	≥308
Left- and right-track widths/(mm)	1400
Tire width/(mm)	200
Gear	4 in and 1 out

2.3. Operational Principle

The hydraulic power system is primarily driven by a diesel engine coupled with a hydraulic pump, functionally divided into two distinct circuits: Circuit I for operational hydraulics and Circuit II for travel dynamics. During normal operation, hydraulic fluid is drawn from the reservoir through the pump and subsequently directed to multiple proportional multi-way directional control valves. Among these, one circuit governs four travel motors, enabling precise speed modulation through a high–low speed control lever with analog signal integration, while directional control is achieved via solenoid-operated directional valves.

A separate circuit manages the extension/retraction of hydraulic cylinders for platform ground-clearance adjustment. An additional circuit interfaces with auxiliary hydraulic motors to power multi-functional spray heads capable of performing weed control, flower induction spraying, and foliar fertilizer application.

The system incorporates ergonomic detachable seats at the rear platform, facilitating seated pineapple harvesting operations with direct in situ fruit deposition. Notably, while the hydraulic oil cooler operates via electrical power, all other system components are hydraulically actuated, ensuring energy efficiency and robust performance in agricultural field conditions. This integrated hydraulic architecture (as illustrated in Figure 3) demonstrates optimized power distribution and operational flexibility for specialized horticultural machinery.

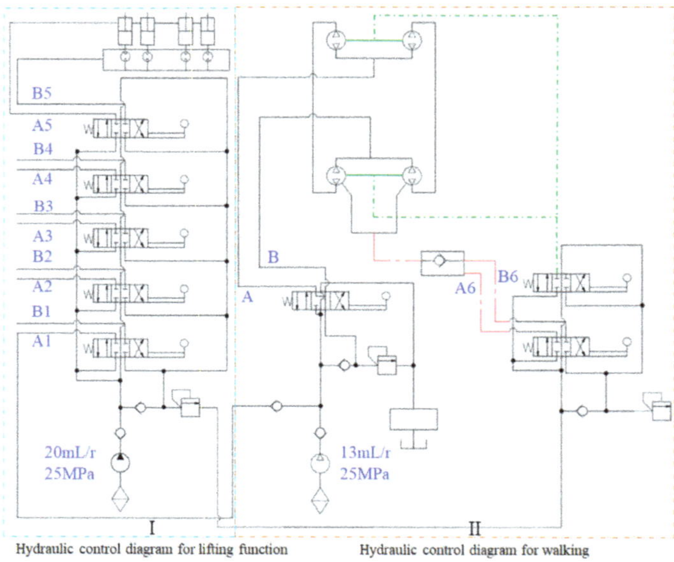

Figure 3. Schematic diagram of hydraulic system.

2.4. *Key Parameter*

Steering-Positioning Parameter Design

Steering-positioning parameters (α) play a critical role in ensuring handling stability, extending tire service life, and guaranteeing operational safety of vehicles [19,20]. This study focuses on the design of these parameters for the high ground-clearance adjustable pineapple field management platform, considering its unique steering system architecture and operational requirements in agricultural applications.

(a) Steering-Axis Inclination

The kingpin inclination angle, defined as the angular deviation between the wheel's central plane and the vehicle's vertical plane (Figure 4), requires careful optimization. An appropriate inclination angle minimizes tire abrasion, reduces hub bearing loads, and enhances directional stability.

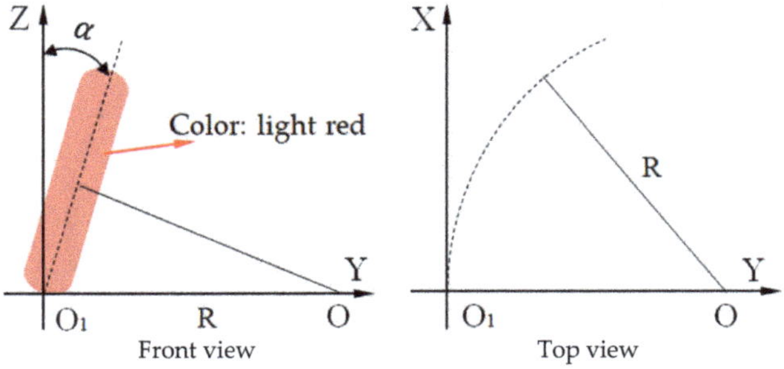

Figure 4. Schematic diagram of steering-axis inclination.

Given the platform's operational characteristics during weed control, flowering induction spraying, and foliar fertilization—where progressive reduction of chemical loads induces significant variations in sprung mass—a minimal variation range of kingpin inclination angle under dynamic loading conditions is essential. Through comprehensive analysis of international counterparts and fundamental steering system design specifications [21,22], the initial kingpin inclination angle was determined as 1.5°.

(b) Front Toe Angle

Toe-in configuration, quantified as the dimensional difference between front- and-rear wheel center planes (Figure 5), serves to eliminate lateral slippage during vehicle motion and mitigate wear. The mathematical expression for toe-in calculation L_s is presented in Equation (1):

$$L_S = \frac{2dSl}{r\,(4L+l)}\sin\beta \tag{1}$$

where d = rim diameter at the toe in, mm;

S = front and rear wheelbase, mm;

l = length of the tire ground imprint, mm;

r = tire rolling radius, mm;

α = front wheel inclination, (°).

The length of tire grounding mark is shown in Formula (2):

$$l = 1.7\sqrt{(D-\Delta x)\Delta x} \tag{2}$$

where D = nominal outer diameter of the front wheel tire, mm:

Δx = radial shape variable of the wheel under vertical load.

The calculation method of Δx is shown in Formula (3) [17]:

$$\Delta x = \frac{CK\,(0.5\,G_1)^{0.85}}{w^{0.7}D^{0.45}p^{0.6}} \tag{3}$$

where C = tire coefficient;

K = coefficient;

w = tire section width, mm;

G_1 = vertical load on the front axle of the platform, N;

P = tire pressure, MPa.

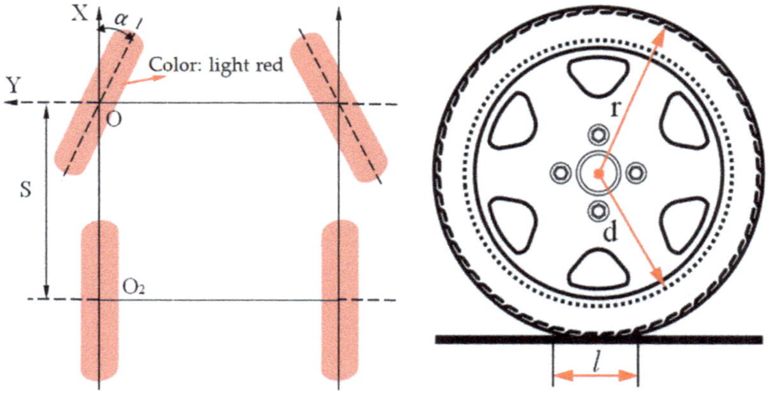

(**a**) Front toe angle in diagram. (**b**) Schematic diagram of single tire analysis.

Figure 5. Schematic diagram of front wheel toe analysis.

According to the actual measured data and Formulas (1)–(3), the toe in angle of the front wheel is about 0.247°.

2.5. Caster Angle

In conventional vehicles, the kingpin is transversely arranged, whereas in high-clearance management platforms, this configuration is replaced by a steering vertical spindle. Consequently, the parameters conventionally referred to as kingpin caster angle and kingpin inclination angle are herein defined as vertical spindle caster angle (β) and vertical spindle inclination angle, respectively. As illustrated in Figure 6, the vertical spindle caster angle α represents the angular deviation between the spindle's central axis and the vertical ground normal. This critical design parameter facilitates automatic wheel-return torque generation, ensuring operational stability during vehicle motion. While increased α values enhance safety through amplified tire deflection counterforce, excessive angles may lead to increased steering resistance, due to elevated reactive forces opposing wheel rotation.

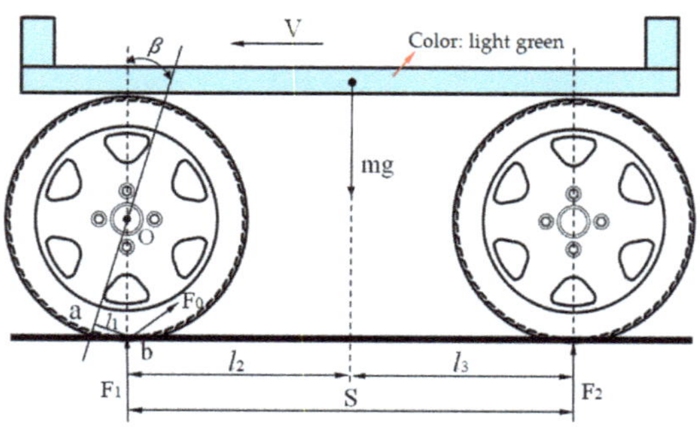

Figure 6. Schematic diagram of caster angle.

Therefore, optimal caster angles are typically constrained within 3° to balance steering efficiency and stability requirements. The geometric configuration creates a mechanical advantage where the intersection point (a) between the spindle axis and ground plane precedes the tire–ground contact point (b). During steering maneuvers, a centrifugal force (F_0) normal to the wheel plane acts at point b, generating a restoring moment (M_β) about the spindle axis. The moment arm (l_1) for this torque corresponds to the perpendicular distance between contact point b and the spindle axis o-a, calculated through Equation (4).

$$l_1 = r \sin\beta \cos\sigma \tag{4}$$

During steering operations where the applied torque remains below the critical restoring moment threshold (M_β), a torque allocation mechanism governs directional control dynamics. In such scenarios, a portion of the restoring moment is dissipated through frictional resistance inherent to the steering system, while the residual torque manifests as an effective restoring-moment component ($M_{\beta 1}$). This mechanical energy distribution establishes functional relationships between centrifugal ground force (F_0), steering input force ($F_{\beta 1}$), and resultant restoring moment ($M_{\beta 1}$), as formulated in Equations (5)–(7) [23].

$$F_0 = \frac{F_1 V^2}{gR} \tag{5}$$

$$F_{\beta 1} = \frac{2F_1 V^2}{\eta_1 D_S i_\omega g R} r \sin\beta \cos\sigma \tag{6}$$

$$M_{\beta 1} = \eta_2 \frac{F_1 V^2}{gR} r \sin\beta \cos\sigma \tag{7}$$

where

g = gravitational acceleration, 9.8 m/s^2;
S = track width, m;
R = turning radius, m;
F_1 = front axle load, N;
V = vehicle velocity, m/s;
D_s = steering wheel diameter, m;
i_ω = angular transmission ratio of steering gear (assigned 10);
η_1 = forward efficiency of steering system (assigned 0.75);
η_2 = reverse efficiency of steering system (assigned 0.60).

2.6. Vertical Spindle Camber Angle

The vertical spindle camber angle (δ) is defined as the angular deviation between the steering spindle axis and the vertical ground normal (Figure 7). This geometric parameter serves a dual function: generating active self-aligning torque to stabilize steering dynamics, and optimizing load distribution across wheel bearings, to mitigate component wear.

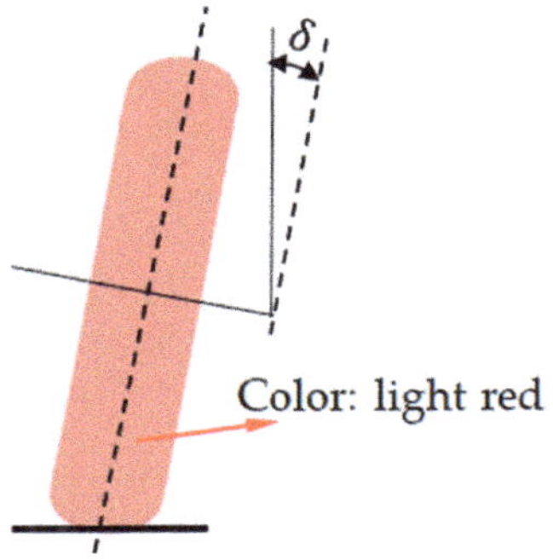

Figure 7. Schematic diagram of vertical spindle camber angle.

During tire deflection under steering forces, the preconfigured vertical spindle caster angle induces vertical displacement (h) of the wheel–body assembly. Upon force removal, ground–tire interaction converts frictional energy into restoring moment (M_β), enabling automatic wheel re-centering. The resultant lift height (h) is quantified by Equation (8).

$$h = l_4 \sin 2\delta \tag{8}$$

where l_4 = steering arm length, m.

When applied steering torque falls below the critical restoring moment threshold ($M_{\delta 1}$), torque allocation follows:

a. Frictional dissipation: Counteracts steering system resistance.
b. Effective restoring component: Generates net self-aligning moment ($M_{\delta 1}$); the governing relationships for ($M_{\delta 1}$) and corresponding steering force ($F_{\delta 1}$) are derived in Equations (9) and (10).

$$M_{\delta 1} = \eta_2 F_1 l_3 \sin 2\beta \frac{\sigma}{180} \tag{9}$$

$$F_{\delta 1} = \frac{2}{\eta_1 D_S i_\omega} F_1 l_3 \sin 2\beta \frac{\sigma}{180} \tag{10}$$

The synergistic effects of vertical spindle caster and camber angles produced are shown as Equations (11) and (12).

$$M = M_{\beta 1} + M_{\delta 1} \tag{11}$$

$$F = F_{\beta 1} + F_{\delta 1} \tag{12}$$

Parametric analysis of Equations (11) and (12) yields optimal spindle angles, balancing steering effort and stability. Computational results recommend initial camber angle $\delta = 5.3^\circ$, subject to iterative refinement through hardware-in-the-loop testing.

2.7. Performance Analysis

The chassis of the pineapple field management platform was designed as a straddle-type configuration, drawing upon existing structural frameworks of high-clearance self-propelled agricultural machinery. The operator's station is strategically positioned at the front of the chassis to ensure optimal visibility, with an ergonomic control panel installed on the left side of the cockpit to enhance operational efficiency and driving safety. A modular pesticide-tank system was implemented, featuring detachable components that permit flexible installation of tanks with varying capacities, according to operational requirements.

To optimize structural integrity while minimizing overall weight, the main chassis beam utilizes 65Mn steel rectangular tubing (180 mm × 65 mm × 8 mm cross-section). The support framework for the operator cabin and lifting platform employs Q355 steel rectangular members (90 mm × 55 mm × 5 mm cross-section), achieving an optimal balance between strength-to-weight ratio and durability. This material selection and structural configuration ensure sufficient load-bearing capacity for field operations, while maintaining maneuverability in pineapple plantation environments. The overall chassis configuration is illustrated in Figure 8.

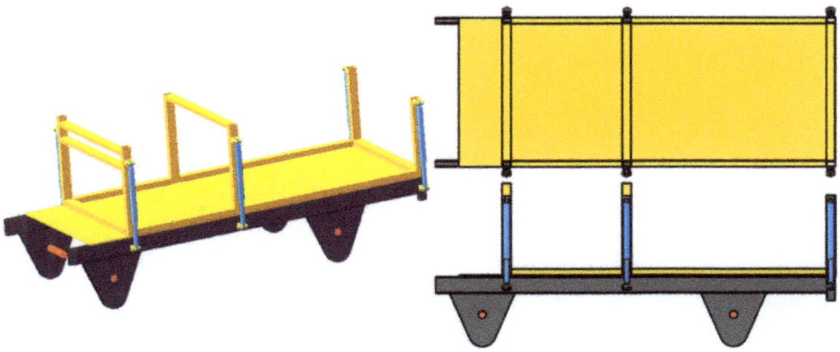

Figure 8. Schematic diagram of chassis frame structure.

Finite Element Simulation Analysis of Chassis Frame

To validate the structural integrity of the designed chassis under anticipated loading conditions, comprehensive strength and stiffness assessments were conducted through finite element analysis (FEA). A refined chassis model was developed using ANSYS Mechanical APDL with t-mm-s-MPa unit consistency, incorporating the material properties of Q355 structural steel: elastic modulus 206 GPa, yield strength 355 MPa, Poisson's ratio 0.30, and mass density 7.85 g/cm^3.

The meshing strategy employed an adaptive meshing technique with system-optimized element sizing, achieving convergence balance between computational accuracy and efficiency. The resulting mesh configuration contained 2,464,340 nodes and 1,129,996 hexahedral elements, with local refinement applied to stress-concentration regions. Boundary conditions simulated realistic field operations through the following:

(1) Full constraint of wheel–ground interaction points.
(2) Distributed static load equivalent to 2-ton payload capacity.
(3) Dynamic load factor of 10 g acceleration for worst-case scenario analysis.

Two critical operational states were investigated:

a. Maximum compressive loading during implement descent.
b. Tensile-dominated loading during platform elevation.

The nonlinear static analysis accounted for both material plasticity and large deformation effects. Stress distribution patterns and deformation characteristics, as illustrated in Figure 9, demonstrated maximum von Mises stresses of 327.4 MPa in the main beam connections during descent phase, corresponding to a safety factor of 1.09, relative to the material yield strength. The maximum displacement of 5.8 mm observed at the cantilevered section remained within permissible deflection limits for agricultural machinery. These results confirm that the chassis design meets structural requirements, while maintaining 22.7% mass reduction compared to conventional configurations, through optimized material distribution.

As illustrated in Figure 9, under full-load conditions, the equivalent elastic-strain values for the chassis frame in descending and ascending states (Figure 9a,d) measure 0.000317 and 0.00125, respectively. The maximum equivalent stress (Figure 9b,e), reaching 48.27 MPa and 231.6 MPa respectively, predominantly occurs at the interface region connecting the crossbeam and plate components. Deformation analysis reveals distinct spatial patterns: the maximum displacement in descending state (1.14 mm, Figure 9c) manifests at the central section, while the ascending state's peak deformation (4.31 mm, Figure 9f) shifts to the middle-rear section. These computational results demonstrate that the chassis configuration satisfactorily meets operational requirements for pineapple field management applications. Notably, the identified stress concentration zones and deformation characteristics provide critical insights for structural reinforcement of vulnerable areas, design optimization strategies, and subsequent lightweight modifications, as referenced in prior research [24].

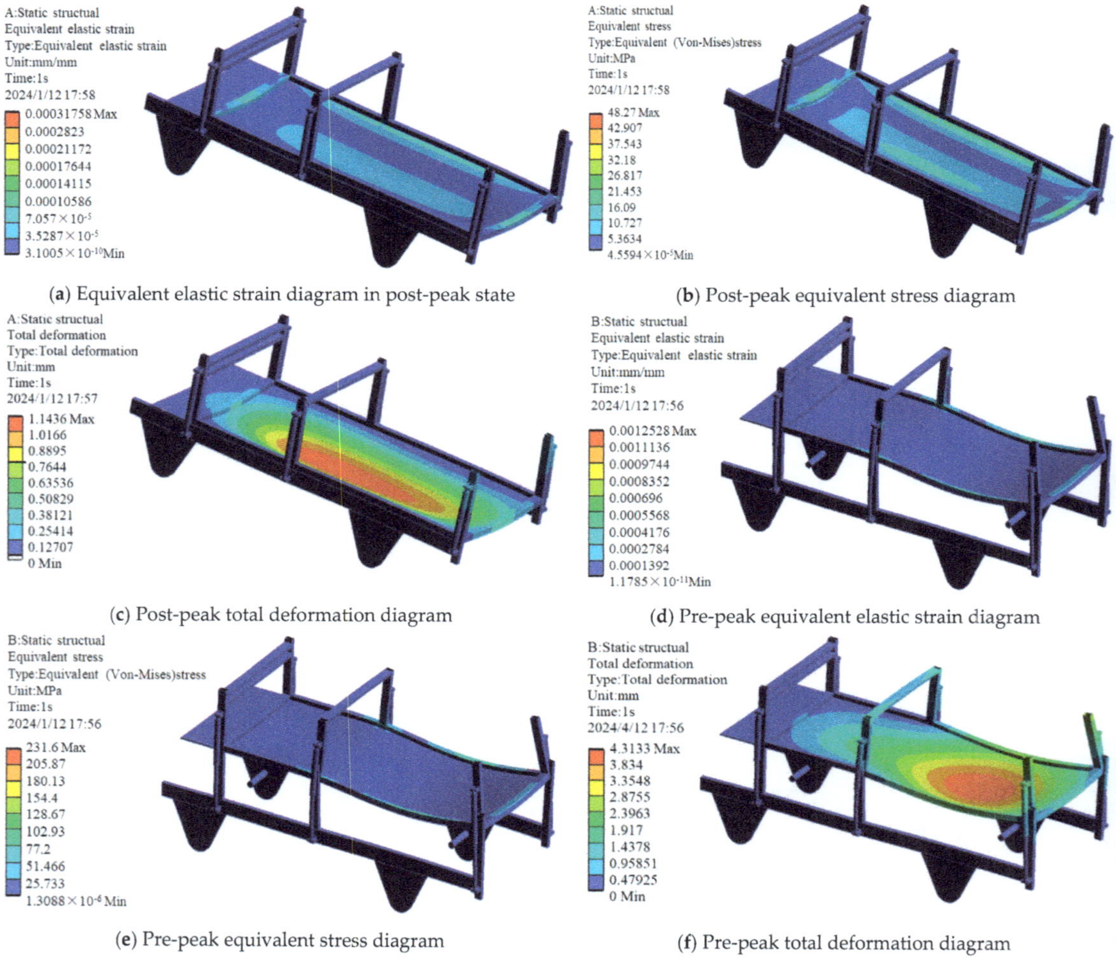

(**a**) Equivalent elastic strain diagram in post-peak state (**b**) Post-peak equivalent stress diagram (**c**) Post-peak total deformation diagram (**d**) Pre-peak equivalent elastic strain diagram (**e**) Pre-peak equivalent stress diagram (**f**) Pre-peak total deformation diagram

Figure 9. FEA-based simulation and analysis diagram of chassis frame structure.

2.8. Stability Performance

During field operations in pineapple plantations, the management platform must maintain sufficient driving stability to accommodate complex terrain conditions. The longitudinal stability of the system (full-load status) is evaluated separately in descending and ascending operational states. Platform stability is quantified by determining the maximum slope angle at which the chassis can remain stationary under braking conditions without experiencing rollover or slippage [25].

For uphill motion analysis under constant velocity (Figure 10a), we establish the mechanical equilibrium equations formulated in Equation (13), incorporating the following assumptions: negligible air resistance, low-speed operation, and rigid-tire modeling.

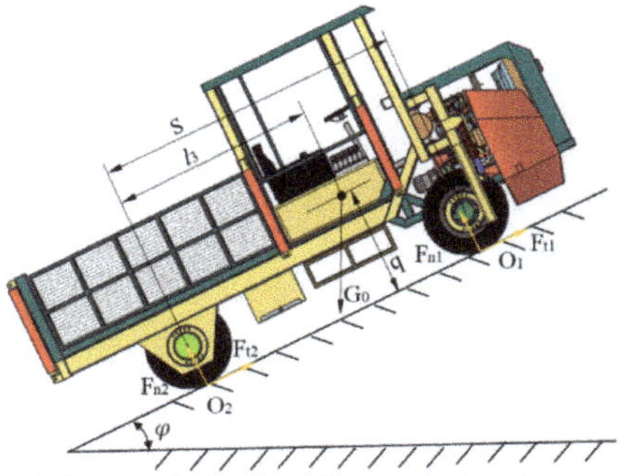

(**a**) Force analysis of uphill state during descent.

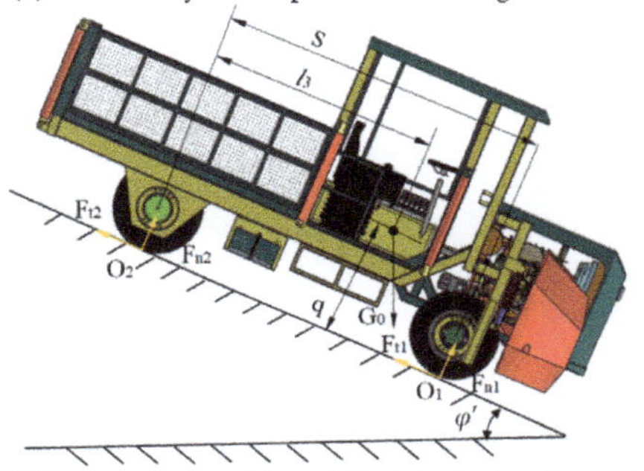

(**b**) Force analysis of downhill state during descent.

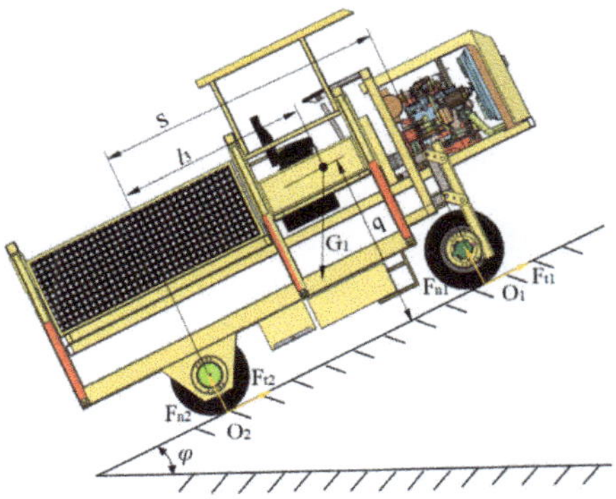

(**c**) Force analysis of uphill state when ascending.

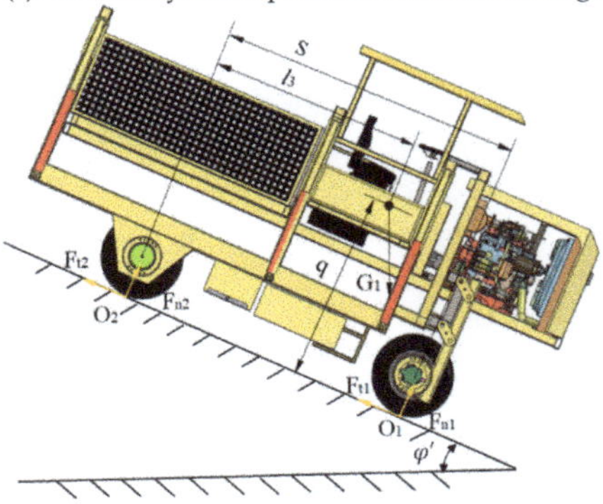

(**d**) Force analysis of downhill state when ascending.

Figure 10. Analysis of longitudinal limit overturning force of pineapple field management-platform chassis.

$$
\begin{aligned}
F_{t1} + F_{t2} &= G_{0,1}\sin\varphi \\
F_{n1} + F_{n2} &= G_{0,1}\sin\varphi \\
Gq\sin\varphi + F_{n1}S &= G_{0,1}l_3\cos\varphi
\end{aligned} \tag{13}
$$

where F_{t1}, F_{t2} = Soil tangential forces on front/rear wheels;

G = Total gravitational force of the platform;

φ = Critical slope inclination angle;

F_{n1}, F_{n2} = Soil normal forces on front/rear wheels;

q = Vertical height from center of gravity to ground;

l_3 = Horizontal distance from center of gravity to rear axle.

The critical rollover threshold occurs when front wheels lose contact with the soil surface (F_{t1} = 0), yielding the maximum sustainable slope angle, as derived in Equation (14).

$$
\varphi = \arctan\frac{l_3}{q} \tag{14}
$$

When analyzing downhill motion of the pineapple management platform chassis under constant velocity conditions (Figure 10b), we similarly establish the mechanical equilibrium equations formulated in Equation (15), maintaining the following validated assumptions:

(1) Aerodynamic resistance exclusion.
(2) Low-speed descent regime.
(3) Non-deformable tire characteristics.

$$
\begin{aligned}
F_{t1} + F_{t2} &= G_{0,1}\sin\varphi' \\
F_{n1} + F_{n2} &= G_{0,1}\sin\varphi' \\
Gq\sin\varphi' + F_{n1}S &= G_{0,1}(S - l_3)\cos\varphi'
\end{aligned} \tag{15}
$$

The critical downhill rollover threshold occurs when the rear wheels lose ground contact, corresponding to the vanishing of rear-wheel tangential force (F_{t2} = 0). Through equilibrium resolution of the governing mechanical equations, we derive the maximum sustainable downhill inclination angle, as formulated in Equation (16).

$$
\varphi' = \arctan\frac{S - l_3}{q} \tag{16}
$$

Analogously, the mechanical equilibrium framework was extended to analyze ascending-state stability, through complementary free-body diagrams (Figure 10c,d). The elevated COG (center of gravity) position during platform elevation introduces distinct parametric variations in rollover thresholds.

From Equations (14) and (16), the critical longitudinal rollover angles demonstrate explicit dependence on COG spatial parameters. Platform stability on slopes is maintained when the gravitational vector remains within the polygon defined by wheel–ground contact points, preventing moment-arm development about the critical axis. Parametric analysis was conducted through CAD/CAE integration using CAXA 3D modeling software, extracting precise mass distribution characteristics:

a. Vertical COG height: q = 728 mm
b. Horizontal COG-rear axle distance: l_3 = 1670 mm

   Substitution into the derived equations yields the following:

c. Descending State

   Uphill limit: 57.85°

   Downhill limit: 68.43°

d. Ascending State
Uphill limit: 46.29°
Downhill limit: 53.67°

In addition, pineapple may also be planted along the slope to reduce rainwater scouring. Therefore, the calculation method of the roll angle of the pineapple field management platform is as follows:

$$G_{0,1}cos\xi \cdot \frac{S}{2} = G_{0,1}sin\xi \cdot \mathrm{q} \tag{17}$$

$$\xi \leq \arctan\left(\frac{S}{2q}\right) \tag{18}$$

where ξ = roll angle.

It is calculated that the maximum inclination when moving along the slope is 43.78°.

3. Results and Discussion

Field experiments were conducted at the Institute of Agricultural Machinery, Chinese Academy of Tropical Agricultural Sciences (21.159° N, 110.269° E) during 2024, to evaluate agricultural operations in pineapple cultivation. The comprehensive trials comprised three key components: (1) flower-promoting spray application, (2) fruit harvesting operations, and (3) postharvest transportation protocols. Figure 11 illustrates the experimental setup and field implementation of these mechanized agricultural processes. All trials were performed under standardized tropical conditions characteristic of the Leizhou Peninsula region, with temperature and humidity parameters recorded throughout the experimental periods to ensure environmental consistency. The average temperature is 17 ° C in February and 30 ° C in July.

(**a**) Flower-inducing and spraying operation.

(**b**) Pineapple harvesting operation.

Figure 11. Field operation scenario.

Field performance assessments were conducted using standardized measurement tools, including a tape measure (SJG-100, Zhengzhou, Henan, China, resolution: 1 mm), platform scale (YihengTCS-03, Hengyang, Hunan, China, capacity: 500 kg), digital stopwatch M.Air-J893, Zhejiang, China, and hydraulic pressure gauge (BZ-GM523, Foshan, Guangdong, China, range: 0–10 MPa). Particle-free water under ambient conditions (25 ± 2 °C) served as the test medium for hydraulic system evaluations. The pineapple field management platform prototype underwent systematic testing across three critical operational parameters [26,27]:

(1) Travel velocity: triplicate trials were performed on dry, level paved surfaces (concrete/asphalt) under maximum throttle settings. Velocity (v, m/s) was calculated via Equation (19).

$$\overline{V} = \frac{s}{t} \tag{19}$$

where s = the walking distance; t = the walking time.

(2) Turning radius: steering performance was quantified through three full 360° rotations at minimal forward speed (<0.2 m/s) with steering wheels locked at extreme left/right positions. The inner trajectory radius of front tires was measured post-rotation using laser rangefinder-assisted ground marking.

(3) Grade ability: slope negotiation capability was evaluated on engineered inclines (5–25° gradient) [28] using onboard inertial measurement units to record actual tilt angles during ascent at creep speed (0.1 m/s).

The test was repeated three times for each index and the average value taken. The test results are shown in Table 2.

Table 2. Field test results.

Parameter	Detection Result	Technical Requirement
Forward speed/($m{\cdot}s^{-1}$)	0.16~1.77	0.16~1.77
Reverse speed/($m{\cdot}s^{-1}$)	0.11~0.28	0.11~0.28
Minimum turning radius/m	2.9	≤3
Maximum climbing angle/(°)	20	≥15
Height adjustment range/mm	52–1194	50–1200
Spray efficiency/(hm^2/h)	0.23	≥0.20
Fruit picking efficiency/(hm^2/h)	0.11	≥0.12

(4) Operational efficiency: flower-inducing efficiency was determined through timed trials using 35–40 mL spray volumes per plant. Mean operation duration per plant (0.7 ± 0.05 s) yielded a theoretical field capacity of 0.23 hm^2/h at standard planting density (45,000 plants/hm^2). Fruit detachment-to-transport cycle time (2.1 ± 0.3 s per fruit) was measured via synchronized video analysis, translating to a harvesting efficiency of 0.11 hm^2 /h under equivalent density conditions. Compared with manual operations, the overall operational efficiency has been significantly improved, while substantially reducing workers' labor intensity.

Field test results demonstrate that the pineapple field management platform achieves a forward travel speed ranging from 0.16 to 1.77 m/s and a reverse travel speed ranging from 0.11 to 0.28 m/s. Key design parameters, such as the wheel camber angle, kingpin inclination, and kingpin caster angle, were appropriately selected, resulting in a minimum turning radius of 2.9 m. The structural strength validation results for the cargo box aligned with simulation data; when fully loaded with pesticide, the platform attained a maximum climbing angle of 20°. Overall operational efficiency was comparable to that of skilled workers, meeting the anticipated design requirements. Moreover, the platform significantly reduces operator physical exertion and lowers operational costs. The field test results largely met the design specifications, confirming the platform's capability to fulfill diverse operational demands in pineapple field management.

4. Conclusions

Based on the wide–narrow double-row pineapple planting pattern, a multi-module height-adjustable field management platform was designed, incorporating its working principle, key steering parameters (minimum turning radius: 2.9 m), and powertrain system. Theoretical analysis of the chassis indicated that the ultimate overturning angles were 57.85° uphill and 46.29° downhill during descent, and 68.43° uphill and 53.67° downhill during ascent. Three-dimensional modeling and ANSYS finite element analysis of the frame revealed that under full load, the equivalent elastic strains for descent and ascent states were 0.000317 and 0.00125, respectively. The maximum equivalent stresses (located at beam–plate joints) were 48.27 MPa and 231.6 MPa for descent and ascent, respectively. Maximum deformation occurred centrally during descent (1.14 mm total) and mid-rear during ascent (4.31 mm total). Field trials demonstrated forward and reverse speeds ranging from 0.16–1.77 m/s to 0.11–0.28 m/s, respectively, with a maximum gradability of 20°. The platform effectively performs weeding, spraying, fertilization, flower induction, harvesting, and transportation, meeting operational requirements for pineapple field management.

Author Contributions: S.Z. and F.H.: formal analysis, writing—original draft, funding acquisition. G.D. and Y.D.: writing—review and editing. X.W.: conceptualization, methodology. B.Y.: investigation, visualization. P.C., Z.L., B.L. and D.P.: data curation. All authors have read and agreed to the published version of the manuscript.

Funding: This work was supported by the Foundation for Innovative Research Groups of the National Natural Science Foundation of China (Grant NO. 52405300); the Hainan Provincial Natural Science Foundation Youth Fund (Grant NO. 323QN284).

Institutional Review Board Statement: Not applicable.

Data Availability Statement: The data used to support the findings of this study are available from the corresponding author upon request.

Acknowledgments: The authors would like to thank their college and laboratory, as well as the reviewers who provided helpful suggestions for this manuscript.

Conflicts of Interest: The authors declare that they have no known competing financial interests or personal relationships that could have appeared to influence the work reported in this paper.

References

1. Liu, T.; Mai, B.; Zhang, J.; Liu, S.; Chen, J.; Sun, W. Design and experiment of chain feeding combined with roller and flexible rod breaking pineapple picking mechanism. *Trans. Chin. Soc. Agric. Mach.* **2024**, *55*, 116–125.
2. Ni, Y.; Dai, Y.; Deng, G.; Qin, J.; Li, L.; Zheng, S. Overview of the development of mechanized technology for pineapple production in China. *Mod. Agric. Equip.* **2024**, *45*, 17–23.
3. Zheng, H.; Li, Z.; Shan, Y.; Zhu, H.; Xiao, M. Current status and development trends of agricultural machinery weeding technologies and equipment. *J. Nanjing Agric. Univ.* **2025**, *48*, 285–295.
4. Fu, L.; Li, S.; Zhang, L.; Meng, Y. Research progress on field weeding robots: A review. *Robot* **2021**, *43*, 751–768.
5. Zhang, M.; Zhang, H.; Dong, X.; Bai, H.; Yao, J.; Lan, H. Research progress of fertilizing machinery based on bibliometrics. *J. Chin. Agric. Mech.* **2024**, *45*, 91–100.
6. Liu, S.; Xu, C.; Zhang, H.; Jiang, H.; Quan, Z.; Wang, J. Research status and development analysis of base—Frtilizer application equipment of orchard. *Trans. Chin. Soc. Agric. Mach.* **2020**, *51*, 99–108.
7. Zhang, Z.; Huang, C.; Lan, H. Flower core recognition and location of pineapple flower inducing robot based on RGBD. *Control Theory Appl.* **2025**, *4*, 1–9.

8. Zhang, R.; Zhang, Q.; Ou, J.; He, K.; Wang, Z.; Lu, D. Design and control of self-propelled pineapple flower induction machine with high Gap. *Mech. Electr. Eng. Technol.* **2022**, *51*, 18–21.
9. Bui, L.C.Q.; Hoang, S.A.; Nguyen, P.T.A.; Tran, C.C. Dynamics and motion control of a pineapple harvesting robotic system. In Proceedings of the 2022 6th International Conference on Robotics and Automation Sciences (ICRAS), Wuhan, China, 9–11 June 2022; pp. 132–137.
10. Hoang, S.A.; Van Tai, D.; Le Cuong Quoc, B. Developing robotic system for harvesting pineapples. In Proceedings of the 2020 International Conference on Advanced Mechatronic Systems (ICAMechS), Hanoi, Vietnam, 10–13 December 2020; pp. 39–44.
11. Zheng, S.; Deng, G.; Cui, Z.; Li, L.; Zhou, W.; Lv, Y. Design of self-propelled pineapple harvest truck with high gap crawler. *Mod. Agric. Equip.* **2020**, *41*, 9–32.
12. Sugita, N.H.; Herrera, M.D.G.; Marano, R.P.; Luaces, P.A. Evaluation of irrigation frequency on pineapple plants grown in greenhouse conditions. *Sci. Hortic.* **2022**, *302*, 111186. [CrossRef]
13. Wang, B.; Liu, W.; Deng, W.; Chen, Z. A multifunctional field management machine based on pineapple mechanized field management technology design and testing. *China South. Agric. Mach.* **2024**, *55*, 1-5+25.
14. Wan, X.; Shu, C.; Liao, Q.; Fan, W.; Zhou, Q.; Liao, Y. Design and experiment of self- propelled middle- placement rape windrower with high ground elearance crawler. *Trans. Chin. Soc. Agric. Mach.* **2022**, *53*, 109–121.
15. Du, X.; Ning, C.; He, L.; Qian, Y.; Zhang, G.; Yao, X. Design and test of crawler-type high clearance camellia oleifera fruit vibratory harvester. *Trans. Chin. Soc. Agric. Mach.* **2022**, *53*, 113–121.
16. Hu, K.; Zhang, W. Design and experiment of multi-mode hydraulic steering system of high clearance self-propelled sprayer. *Trans. Chin. Soc. Agric. Mach.* **2020**, *51*, 366–373.
17. Li, W.; Xue, T.; Mao, E.; Shao, M.; Li, Z.; Du, Y. Design and experiment of multifunctional steering system for high clearance self-propelled sprayer. *Trans. Chin. Soc. Agric. Mach.* **2019**, *50*, 141–151.
18. Cui, Z.; Chen, M.; Xie, J.; Deng, G.; Zhou, W.; Zheng, S. Preliminary study on the cultivation mode of mechanized pineapple and its mechanization in Zhanjiang. *Mod. Agric. Equip.* **2022**, *43*, 2–9.
19. Wang, B.; Peng, W.; Liang, Y.; Cheng, J.; Hu, H.; Xu, D. Characteristics analysis and optimization design of suspension mechanism of all-terrain mobile robot. *J. Mech. Eng.* **2022**, *58*, 71–86.
20. Yang, Y.; Xie, M. An analytical study of alignment parameters and ground intersection offset of steering kingpin. *Automot. Engineering.* **2021**, *43*, 692–696.
21. Shi, B.; Ye, X.; Lv, H.; Lai, F. Research on the mechanism of differential steering based on tire cornering. *China Mech. Eng.* **2025**, *4*, 1–10.
22. Liu, H.; Long, Y.; He, S.; Cui, Y.; Shen, Y. Design and experiment of the auxiliary steering system for a four-wheel independent electrically driven high clearance sprayer. *Trans. Chin. Soc. Agric. Eng.* **2021**, *37*, 30–37.
23. Zhou, Q.; Wen, H.; Li, Z.; Zhang, W.; Yi, Z.; Wang, G. Design and test of fuzzy anti-skid control of four-wheel drive hydraulically driven sprayer. *Trans. Chin. Soc. Agric. Mach.* **2020**, *51*, 283–288.
24. Hao, Z.; Yang, X.; Liu, L.; Yin, S.; Liu, S.; Zhao, J. Design and experiment of multifunctionaldynamic chassis for orchard. *Trans. Chin. Soc. Agric. Mach.* **2018**, *49*, 66–73+92.
25. Bhat, K.; Chayalakshmi, C.L. Microcontroller-based semiautomated pineapple harvesting system. In *International Conference on Mobile Computing and Sustainable Informatics*; Springer: Berlin/Heidelberg, Germany, 2020; pp. 383–392.
26. Sun, J.; Meng, X.; Zeng, L.; Zheng, H.; Ying, J.; Zhang, H.; Xu, G. Design and performance test of experimental platform for omnidirectional control of agricultural chassis center of gravity in hilly and mountainous areas. *Trans. Chin. Soc. Agric. Mach.* **2025**, *56*, 511–522.
27. Zhu, Y.; Sang, Y.; Hu, M.; Tan, W.; Shen, K. Design and tability analysis of sprayer with adjustable height gap. *J. Agric. Mech. Res.* **2020**, *42*, 68–73.
28. Deng, Z.; Cao, X.; Hu, G.; Xie, G.; Wang, P. Study on the climbing characteristics of tracked all-terrain vehicle based on DEM-MBD coupling method. *J. Chin. Agric. Mech.* **2024**, *45*, 144–151.

agriculture

MDPI

Article

Intelligent Path Tracking for Single-Track Agricultural Machinery Based on Variable Universe Fuzzy Control and PSO-SVR Steering Compensation

Huanyu Liu [1], Zhihang Han [1], Junwei Bao [2], Jiahao Luo [1], Hao Yu [1], Shuang Wang [1] and Xiangnan Liu [1,*]

[1] Institute of Modern Agricultural Equipment, Xihua University, Chengdu 610039, China; liuhy0528@163.com (H.L.); 17262126203@163.com (Z.H.); 212023085500057@stu.xhu.edu.cn (J.L.); yh-email@stu.xhu.edu.cn (H.Y.)

[2] Inner Mongolia Academy of Agricultural and Animal Husbandry Sciences, Hohhot 010000, China; 2@imaaahs.ac.cn

* Correspondence: 0120240191@xhu.edu.cn; Tel.: +86-028-87723323

Abstract: Single-track electric agricultural chassis plays a vital role in autonomous navigation and driving operations in hilly and mountainous regions, where its path tracking performance directly affects the operational accuracy and stability. However, in complex farmland environments, traditional methods often suffer from frequent turning and large tracking errors due to variable path curvature, uneven terrain, and track slippage. To address these issues, this paper proposes a path tracking algorithm combining a segmented preview model with variable universe fuzzy control, enabling dynamic adjustment of the preview distance for better curvature adaptation. Additionally, a heading deviation prediction model based on Support Vector Regression (SVR) optimized by Particle Swarm Optimization (PSO) is introduced, and a steering angle compensation controller is designed to improve the turning accuracy. Simulation and field experiments show that, compared with fixed preview distance and fixed-universe fuzzy control methods, the proposed algorithm reduces the average number of turns per control cycle by 30.19% and 18.23% and decreases the average lateral error by 34.29% and 46.96%, respectively. These results confirm that the proposed method significantly enhances path tracking stability and accuracy in complex terrains, providing an effective solution for autonomous navigation of agricultural machinery.

Keywords: electric tracked agricultural chassis; single-track braking steering; path tracking; variable universe fuzzy control; support vector regression; particle swarm optimization

1. Introduction

Single-track electric agricultural chassis holds great promise for crop production in hilly and mountainous regions due to its compact structure, large ground contact area, low soil pressure, small turning radius, and strong off-road capability. Implementing path tracking control for such agricultural machinery is essential for improving the operational efficiency and quality in these terrains while also reducing the labor intensity and operational costs [1–3].

At present, the Pure Pursuit algorithm is widely used in agricultural path tracking due to its simple structure and high computational efficiency [4,5]. Some studies have introduced the bicycle model into the kinematic modeling of tracked vehicles and realized steering control via PWM signals [6]. He et al. [7] developed a tracked peanut harvester equipped with a dual-PID preview controller, achieving an average deviation of 2.2 cm during straight-line motion at 2 km/h under sandy soil conditions. Qin et al. [8] proposed a state feedback-based differential steering control system, which demonstrated better performance than conventional single-side brake steering in various environments, improving both the turning accuracy and stability. Although such methods perform well in simulations, they still suffer from large tracking errors and reduced performance in real-world scenarios with high-curvature turning points, such as orchards. He et al. [9] proposed a "preview–tangent" local path tracking method based on a two-wheel differential model, which maintained tracking errors within 0.05 m in field experiments. However, traditional path tracking algorithms generally assume that the vehicle can execute the required steering angle accurately while neglecting slippage, ground friction, and other factors on unstructured terrains. These unmodeled disturbances may lead to steering execution errors, thereby causing system instability and cumulative tracking deviations [10]. Moreover, although robust control approaches, such as sliding mode variable structure control, offer certain adaptability, they require high model accuracy and careful parameter tuning [11]. Wang et al. [12] constructed a motion model for a tracked platform using an improved A-Star algorithm combined with Support Vector Regression (SVR) to predict the optimal theoretical speeds of the left and right tracks. The results demonstrated its feasibility for autonomous navigation in orchard environments.

As a key parameter in the Pure Pursuit algorithm, the preview distance has a significant impact on the path tracking accuracy [13]. Hu et al. [6] optimized the preview distance under different path curvature segments using a genetic algorithm, thereby improving tracking precision. Zhang et al. [14] employed a Particle Swarm Optimization (PSO) algorithm to dynamically adjust the preview distance, using lateral and heading errors as the fitness function, and improve the convergence speed by refining the inertia weight. Netto et al. [15] further suggested that the preview distance should vary with vehicle velocity and proposed a quadratic fitting model for adjustment. To address the issue of reduced tracking accuracy caused by varying curvature in autonomous navigation, Zhou et al. [16] proposed a fuzzy control-based path tracking method, which effectively met the operational demands of agricultural machinery under curved paths. Song et al. [17] adopted the Sparrow Search Algorithm (SSA) to construct a fitness function based on predicted deviation and searched for the optimal goal point; field experiments showed high adaptability to time-varying path curvature. Guan et al. [18] sought the optimal preview point by considering both the longitudinal look-ahead distance and lateral position. They used the Deep Deterministic Policy Gradient (DDPG) to train the optimal longitudinal preview distance, constructed a lateral error search region, and applied an evaluation function to determine the best prediction point. The results indicated a significant improvement in tracking accuracy under marginal curvature conditions. Zhou et al. [16] designed a fuzzy controller with three inputs and one output using lateral deviation, heading deviation, and the curvature of the preview path as the input variables. A control function was then constructed by combining pose deviation and preview path curvature to realize variable speed control. The results showed that the maximum lateral error was reduced by more than 50%, significantly improving the tracking accuracy under varying curvatures. Wang et al. [19] proposed a preview-based tracking algorithm and applied an improved sparrow search algorithm to adaptively solve for the preview distance. The results demonstrated that the number of turns per control cycle was reduced by up to 13.59%, and the error area was decreased by as much as 54.59%. Although the aforementioned methods have achieved certain effectiveness in specific scenarios, most of them fail to fully account for the coupling relationship between path curvature and the vehicle's dynamic response. This limitation is particularly evident in hilly regions with frequent turning and significant terrain undulations, where control accuracy and system stability still require further improvement.

In summary, research on path tracking control for single-track electric agricultural chassis remains limited. The existing methods often exhibit shortcomings in adapting to path curvature, ensuring steering accuracy, and minimizing the number of turns. This study focuses on single-track electric agricultural chassis and proposes a path tracking algorithm that integrates a segmented preview model with variable universe fuzzy control. Additionally, a steering angle prediction compensator is developed based on Support Vector Regression (SVR), optimized by Particle Swarm Optimization (PSO), with the goal of reducing the turning frequency and enhancing the vehicle stability and tracking accuracy. The main contributions of this work are as follows: (1) the construction of a path tracking experimental platform and the development of an autonomous navigation control system; (2) the design of a segmented preview algorithm based on variable universe fuzzy control and a steering angle compensation controller using the SVR-PSO framework; and (3) the validation of the proposed algorithm's effectiveness and robustness through both simulation and field experiments.

2. Experimental Equipment and Materials

2.1. Experimental Vehicle

The experimental platform employed in this study is a single-motor-driven tracked agricultural vehicle designed for operation in hilly terrains and small-field environments, as shown in Figure 1. This vehicle features a compact structure and is equipped with a tracked propulsion system that provides excellent terrain adaptability and obstacle-crossing capability, enabling stable performance under complex road conditions. The vehicle platform supports a modular assembly, allowing for flexible mounting of attachments such as spraying systems and weeding devices to perform a variety of agricultural tasks. It is also suitable for applications such as autonomous inspection in unmanned farms. The main technical specifications of the vehicle are listed in Table 1.

Figure 1. Single-track agricultural vehicle.

Table 1. Technical specifications of the vehicle.

Parameter	Specification
Power system	Brushless DC motor
Rated motor power (kW)	24
Maximum motor shaft speed (RPM)	4000
Dimensions (mm)	700 (L) × 470 (H) × 360 (W)
Track width (mm)	100
Ground clearance (mm)	170
Turning radius (mm)	250
Maximum travel speed (km/h)	5

2.2. Steering System and Operating Principle

The vehicle's steering and transmission system is primarily composed of a relay unit, electromagnetic clutches, a motor controller, and a brushless DC motor. As illustrated in Figure 2, the industrial PC controls the state of the electromagnetic clutches by operating the relay unit, thereby enabling the vehicle to move forward, stop, or steer. The motor driver regulates the vehicle's speed and direction of motion, while the electromagnetic clutches control the turning and braking actions.

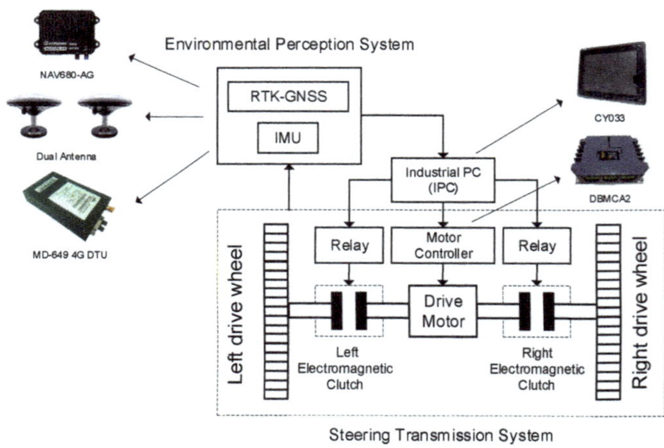

Figure 2. Design of the autonomous navigation system.

When the left clutch is disengaged and the right clutch is engaged, the left drive wheel is braked while the right drive wheel rotates, causing the vehicle to turn left. Conversely, when the left clutch is engaged and the right clutch is disengaged, the vehicle turns right. When both clutches are engaged, the vehicle moves forward; when both are disengaged, the vehicle stops.

In addition, the vehicle supports reverse motion. By controlling the motor driver to reverse the direction of rotation of the brushless DC motor, both tracks can be driven in the opposite direction simultaneously, thereby enabling backward movement. The engagement logic of the clutches remains the same as in forward driving, ensuring consistent control behavior.

This drive system features a simple structure and reliable control, offering advantages such as a small turning radius and suitability for confined operational areas, like small plots. It is also easy to implement in engineering applications [20]. However, since the steering control relies on the engagement and disengagement of clutches, the vehicle can only adjust its heading with a fixed turning radius. This necessitates frequent switching between left and right steering states during path tracking, resulting in a degree of motion discreteness. Such characteristics can affect the smoothness and continuity of path tracking, thereby placing higher demands on the control algorithm.

2.3. Navigation System

To enable autonomous path tracking control, the experimental platform was retrofitted with an automatic navigation system. This navigation system consists of three primary modules: external environment perception, control decision-making, and steering control.

The external environment perception module includes an RTK-GNSS and IMU integrated navigation unit capable of high-precision positioning. Differential signals are received via a DTU connected to the base station, allowing for accurate and continuous positioning even in lightly obstructed environments. The control decision-making module is based on an industrial PC, which receives and parses sensor data, executes the path tracking algorithm, and sends control commands to the steering and drive systems. The drive system comprises a relay unit, electromagnetic clutches, and a brushless DC motor. The hardware components involved in the navigation system are illustrated in Figure 2.

3. Path Tracking Algorithm Design

3.1. Kinematic Model of the Tracked Vehicle

By analyzing the locomotion mechanism of a single-track vehicle, it is observed that during steering, the track on the braked side remains stationary, while the track on the unbraked side drives the vehicle to rotate. The turning center is located at the contact point of the braked track. This motion behavior can be simplified as a classical differential-drive system.

To facilitate kinematic modeling, the following assumptions are made in this study:

1. The vehicle moves on a planar (two-dimensional) surface, with the vertical displacement neglected.
2. The contact between the tracks and the ground is uniform, and no slippage occurs.
3. The center of mass coincides with the geometric center of the vehicle.

Based on these assumptions, a kinematic model of the tracked vehicle is established in the navigation coordinate frame *XOY*, as illustrated in Figure 3.

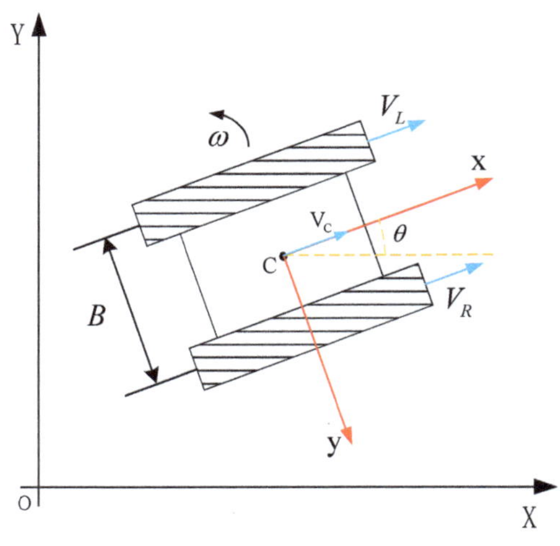

Figure 3. Kinematic model of the tracked vehicle.

To enhance clarity, the definitions of the key symbols used in this model are summarized in Table 2 below.

Table 2. Description of the key symbols in the kinematic model.

Symbol	Description
(X, O, Y)	Global coordinate system fixed to the ground, used as the reference frame.
(x, C, y)	Vehicle coordinate system with origin at the center of mass.
C	Position of the vehicle's center of mass.
B	Distance between the centers of the two tracks (in m).
V_R	Velocity of the right track (in m·s^{-1}).
V_L	Velocity of the left track (in m·s^{-1}).
V_C	Velocity of the vehicle's center of mass (in m·s^{-1}).
θ	Angle between the vehicle's x-axis and the global X-axis (rad), positive in the counterclockwise direction.
ω	Angular velocity of the vehicle around its center (in rad·s^{-1}).

According to the kinematic equations of the tracked vehicle, the following can be derived:

$$\begin{cases} \dot{x} = V_C \cdot \cos\theta \\ \dot{y} = V_C \cdot \sin\theta \\ \dot{\theta} = \omega \end{cases} \tag{1}$$

The forward velocity V_C of the vehicle, defined as the linear velocity of the vehicle's center of mass in the forward direction, is given as the average velocity of the two tracks:

$$V_C = \frac{V_R + V_L}{2} \tag{2}$$

When the vehicle turns by braking one track, the angular velocity ω and turning radius R are determined by the track width B and the velocity difference between the two tracks. The turning center is located at the midpoint of the braked track:

$$R = \frac{V_C}{\omega} = \frac{(V_L + V_R)B}{2(V_L - V_R)} = \frac{B}{2} \tag{3}$$

$$\omega = \frac{V_R - V_L}{B} \tag{4}$$

When the vehicle turns left, the left track is braked, i.e., $V_L = 0$, then:

$$\begin{cases} V_C = \frac{V_R + 0}{2} = \frac{V_R}{2} \\ \omega = \frac{V_R - 0}{B} = \frac{V_R}{B} \end{cases} \tag{5}$$

When the vehicle turns right, the right track is braked, i.e., $V_R = 0$, then:

$$\begin{cases} V_C = \frac{V_L + 0}{2} = \frac{V_L}{2} \\ \omega = \frac{0 - V_L}{B} = -\frac{V_L}{B} \end{cases} \tag{6}$$

3.2. *Segmented Preview Tracking Model*

This study adopts a segmented preview tracking model, which divides the path tracking process into two stages: "steering adjustment" and "straight-line motion" (as shown in

Figure 4), in order to accommodate the structural characteristic of fixed turning radius in single-track vehicles. The vehicle first utilizes its single-side braking steering mechanism to adjust its heading by calculating the heading deviation and determining the required turning angle, thereby aligning the vehicle toward the preview point. Subsequently, the vehicle moves along a straight path toward the preview point. Upon reaching it, the next preview point is computed and selected, and the process is repeated until the entire path is successfully tracked.

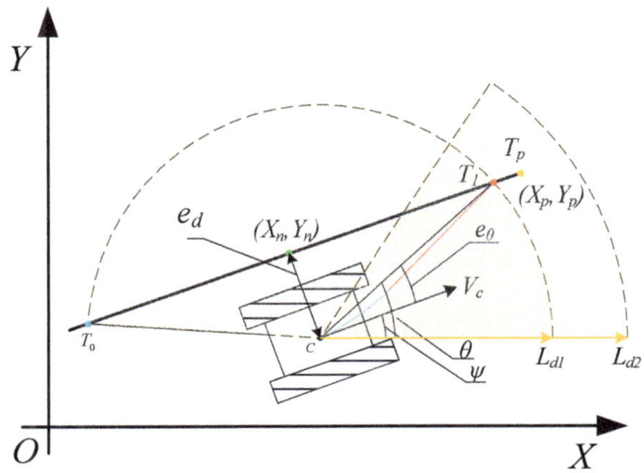

Figure 4. Segmented preview tracking model.

To realize path tracking control, it is necessary to perform real-time computation and updating of key parameters throughout the tracking process. The lateral deviation e_d represents the perpendicular distance between the vehicle's current position and the target path, indicating the extent of lateral offset. To ensure computational accuracy, the lateral deviation is calculated based on the local direction of the path. By identifying the nearest index point (X_n, Y_n), the corresponding path segment (e.g., from (X_{n-1}, Y_{n-1}) to (X_{n+1}, Y_{n+1})) is obtained, and the tangent direction of the path is calculated accordingly:

$$\theta_p = arctan2(Y_{n+1} - Y_{n-1}, X_{n+1} - X_{n-1}) \tag{7}$$

where θ_p denotes the local heading angle of the path.

Subsequently, the vehicle position vector $\vec{P} = (X - X_n,\ Y - Y_n)$ is projected onto the direction perpendicular to the path, and the lateral error of the vehicle is calculated as follows:

$$e_d = (X - X_n)sin\theta_p - (Y - Y_n)cos\theta_p \tag{8}$$

Note: the sign of ed indicates the vehicle's relative position with respect to the path; $e_d > 0$ means that the vehicle is on the left side of the path, while $e_d < 0$ means that it is on the right side.

The current preview point of the vehicle is determined based on the preview distance by calculating the Euclidean distance between the vehicle's current position and the path points and searching for the point whose distance equals or is closest to the preview distance. However, the vehicle may sometimes select a point located behind its current position as the preview point (e.g., selecting point T_0 in Figure 4), which can lead to path tracking failure. Therefore, a forward-looking constraint must be introduced to ensure that the vehicle progresses along the forward direction of the path. Assuming the path consists of discrete points $\{(X_i, Y_i)\}$ for $i = 0, 1, 2, \cdots, N$, a forward constraint is first imposed as follows:

1. Find the nearest index point (X_n, Y_n) relative to the vehicle. Compute the Euclidean distance d_i between the vehicle's current position (X, Y) and each path point.

$$d_{\mathrm{i}} = \sqrt{(\mathrm{X_i} - \mathrm{X})^2 + (\mathrm{Y_i} - \mathrm{Y})^2} \quad (9)$$

2. Identify the point corresponding to the minimum d_i as the nearest index point (X_n, Y_n), with the index value n (i.e., $d_n = min(d_i)$).
3. Select the path point that satisfies the preview distance constraint while ensuring that its index satisfies $I > n$, thereby guaranteeing that the preview point is located ahead of the vehicle. The resulting point is defined as the current preview point.

Meanwhile, to ensure complete path tracking and avoid failures in the final segment caused by the inability to locate a preview point (i.e., $(d_i < L_{d2})$), it is necessary to assign the endpoint of the planned path as the preview point when it falls within the current preview distance. In such cases, the endpoint T_p is taken as the vehicle's preview tracking point, thereby ensuring complete path coverage.

After determining the preview point, the current heading deviation e_θ of the vehicle is calculated to guide the steering adjustment. The target heading angle θ is computed based on the vehicle's current position (X, Y) and the coordinates of the preview point (X_T, Y_T) as follows:

$$\theta = arctan2(\mathrm{Y}_T - Y, \mathrm{X}_T - X) \quad (10)$$

The heading deviation e_θ is defined as the difference between the target heading angle θ and the current heading angle ψ:

$$e_\theta = \theta - \psi \quad (11)$$

To avoid angular discontinuity, e_θ should be normalized to the interval $[-\pi, \pi]$.

$$e_\theta = atan2(sin(\theta - \psi), cos(\theta - \psi)) \quad (12)$$

Note: $e_\theta > 0$ indicates that the vehicle needs to turn left, while $e_\theta < 0$ indicates a right turn.

To provide a quantitative measure of the dynamic characteristics of the path, the average curvature of the forward path (within the preview range) is calculated. The path curvature k represents the degree of bending of the path and is mathematically defined as the rate of change of the tangent angle with respect to arc length at a given point on the curve. For circular arcs, where the curvature is constant, the curvature k is given by the following:

$$k = \frac{\delta}{s} \quad (13)$$

Here, δ represents the central angle corresponding to the arc, and s is the arc length of the arc.

From the nearest point (X_n, Y_n) of the vehicle, a set of points within the preview distance L_d is selected along the direction of the path, denoted as $(X_{n+1}, Y_{n+1}), \ldots, (X_{n+m}, Y_{n+m})$, where m is determined by the preview distance L_d. The curvature at each point on the path is estimated using the three-point method. As shown in Figure 5, three consecutive points (X_{i-1}, Y_{i-1}), (X_i, Y_i), and (X_{i+1}, Y_{i+1}) are selected. The central angle $\Delta\delta_i$ is approximated by the included angle between the two line segments:

$$\Delta\delta_i = arccos\left(\frac{\vec{v_1} \cdot \vec{v_2}}{\left|\vec{v_1}\right|\left|\vec{v_2}\right|}\right) \quad (14)$$

where $\vec{v_1} = (X_i - X_{i-1}, Y_i - Y_{i-1})$, $\vec{v_2} = (X_{i+1} - X_i, Y_{i+1} - Y_i)$

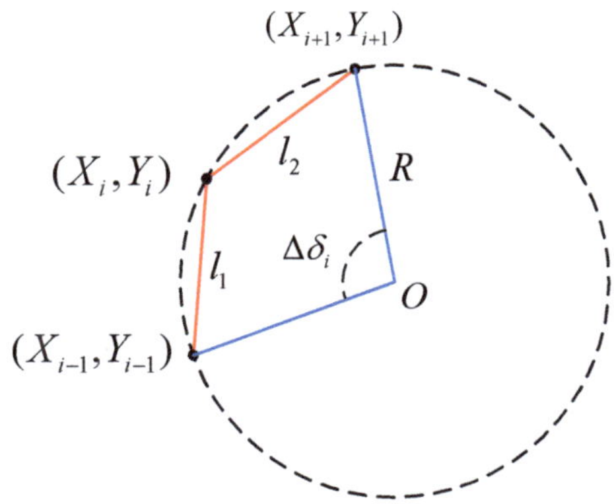

Figure 5. Curvature estimation using the three-point method.

The arc length Δs_i is approximated based on the three-point arc assumption by taking the average of the lengths of the two segments:

$$\Delta s_i = \frac{\sqrt{(X_i - X_{i-1})^2 + (Y_i - Y_{i-1})^2} + \sqrt{(X_{i+1} - X_i)^2 + (Y_{i+1} - Y)^2}}{2} \tag{15}$$

This approximation originates from a geometric simplification in curvature calculations. For a circular arc formed by three points, the true arc length is $\Delta s_i = R \cdot |\Delta\delta_i|$, where R is the arc radius. However, calculating R is computationally complex and difficult to obtain in real time. Therefore, the average of the two segment lengths, l_1 and l_2, is used to approximate the arc length. The resulting error is negligible for small angles, ensuring a balance between real-time performance and accuracy.

The curvature at a single point is given by the following:

$$k_i = \frac{|\Delta\delta_i|}{\Delta s_i} \tag{16}$$

The average curvature is calculated over the m points ahead as follows:

$$k = \frac{1}{m} \sum_{i=n}^{n+m-1} k_i \tag{17}$$

To facilitate the interpretation of Figure 4, a summary of the associated variables and their definitions is provided in Table 3.

Table 3. Symbol definitions related to the segmented preview tracking model in Figure 4.

Symbol	Description
(X, O, Y)	Global coordinate system fixed to the ground.
L_{d1}, L_{d2}	Preview distances (in m).
C	Center of mass of the vehicle.
V_c	Velocity direction of the center of mass (in m·s^{-1}).
T_1	Preview point of the tracked vehicle.

Table 3. *Cont.*

Symbol	Description
T_p	Endpoint of the path.
T_0	Point located behind the vehicle.
e_d	Lateral deviation of the vehicle (in m).
θ	Target heading angle (in rad).
ψ	Current heading angle of the vehicle (in rad).
e_θ	Heading deviation (in rad).
blue line	Tracked vehicle's turning trajectory.
red line	Straight-line tracking trajectory.
(X_n, Y_n)	Point on the path closest to the vehicle.

3.3. Dynamic Adjustment Strategy for Preview Distance

In the preview tracking algorithm for tracked vehicles, the preview distance L_d is a key parameter that determines path tracking performance and turning behavior. However, a fixed preview distance often results in frequent turning or excessive tracking deviation under specific conditions, thereby affecting the smoothness of navigation and operational efficiency.

As shown in Figure 6, using a large preview distance may cause the vehicle to skip certain path segments, leading to path tracking failure. Conversely, a small preview distance can cause the vehicle to oscillate back and forth near the reference path. Therefore, to enhance the effectiveness of the preview tracking algorithm, it is essential to dynamically select an appropriate preview distance based on road information and vehicle state.

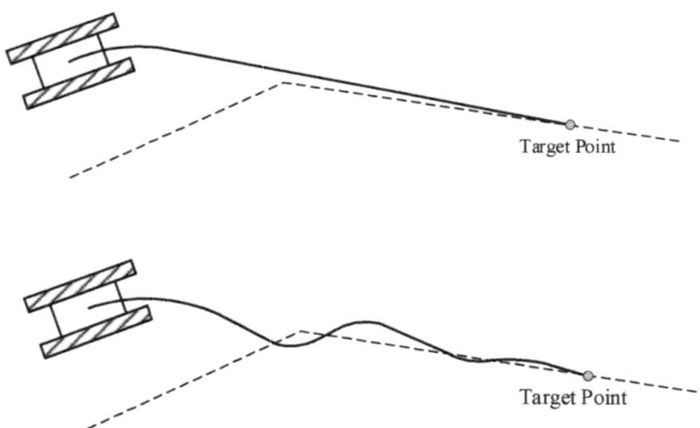

Figure 6. Tracking behavior under inappropriate preview distances.

To ensure stable path tracking, the selection of the preview distance must also guarantee the stability of the tracking system [21]. When the reference path is fully followed, the system must remain in a stable state. The local stability of the nonlinear system under equilibrium conditions can be analyzed by linearizing the system around the equilibrium point.

Based on the geometric preview control model for tracked vehicles, the evolution of lateral error and heading error over time can be described by the following kinematic relationships:

$$\begin{cases} \dot{e}_d = -v \cdot e_\theta \\ \dot{e}_\theta = v \cdot \gamma \end{cases} \tag{18}$$

where e_d is the lateral deviation, e_θ is the heading deviation, v is the vehicle speed, T is the time constant, and γ is the vehicle rotation angle.

To ensure closed-loop stability, the controller applies a dynamic regulation rate to adjust γ based on the observed tracking errors. The adopted regulation law is given as follows:

$$\dot{\gamma} = \frac{1}{T}(\phi_{e_\theta} \cdot e_d + \phi_{e_d} \cdot e_\theta + \phi_\gamma \cdot \gamma - \gamma) \tag{19}$$

where $\Phi_{e\theta}$, Φ_{ed}, and Φ_γ are the linearization coefficients of the path tracking system.

Combining the above expressions, the closed-loop state-space system can be written as follows:

$$\begin{bmatrix} \dot{e}_d \\ \dot{e}_\theta \\ \dot{\gamma} \end{bmatrix} = \begin{bmatrix} 0 & -v & 0 \\ 0 & 0 & v \\ \frac{1}{T}\phi_{e_\theta} & \frac{1}{T}\phi_{e_d} & \frac{1}{T}(\phi_\gamma - 1) \end{bmatrix} \begin{bmatrix} e_d \\ e_\theta \\ \gamma \end{bmatrix} \tag{20}$$

The characteristic polynomial of the system is calculated as follows:

$$f(\lambda) = \lambda^3 - \frac{1}{T}(\phi_\gamma - 1)\lambda^2 - \frac{v}{T}\phi_{e_\theta}\lambda + \frac{v^2}{T}\phi_{e_d} \tag{21}$$

Based on the Routh–Hurwitz stability criterion, the stability conditions of the system can be obtained as follows:

$$\begin{cases} \phi_\gamma - 1 \leq 0 \\ \phi_{e_\theta} \leq 0 \\ -(\phi_\gamma - 1)\phi_{e_\theta} + Tv\phi_{e_d} \leq 0 \end{cases} \tag{22}$$

If the above conditions are satisfied, the system is locally stable at the equilibrium point. Based on the stability conditions, the following expression is derived:

$$L_d \geq Tv \tag{23}$$

where L_d is the preview distance, T is the system response time, and v is the vehicle speed.

Through experimental analysis, the system response time of the vehicle was determined to be 0.6 s, and the vehicle speed was set to 2 m/s. As a result, the minimum preview distance was calculated to be 1.2 m, indicating that the system can remain stable when the preview distance satisfies $L_d > 1.2$ m.

Secondly, this study designs a preview distance adjustment controller based on variable universe fuzzy control. The controller takes the average curvature k of the forward path and the vehicle speed v at a given moment as inputs to an adaptive scaling factor module, which calculates and outputs a universe scaling factor α. Subsequently, the variable universe fuzzy control module adjusts the input universe based on α and performs inference using the current lateral deviation and heading deviation to obtain the optimal preview distance at that moment. Finally, the calculated preview distance is passed into the segmented preview tracking module, which, together with the vehicle's steering angle prediction compensator, achieves path tracking control.

The algorithm implementation flowchart is shown in Figure 7. The functions and roles of the key modules involved in the control algorithm, as shown in Figure 7, are summarized in Table 4.

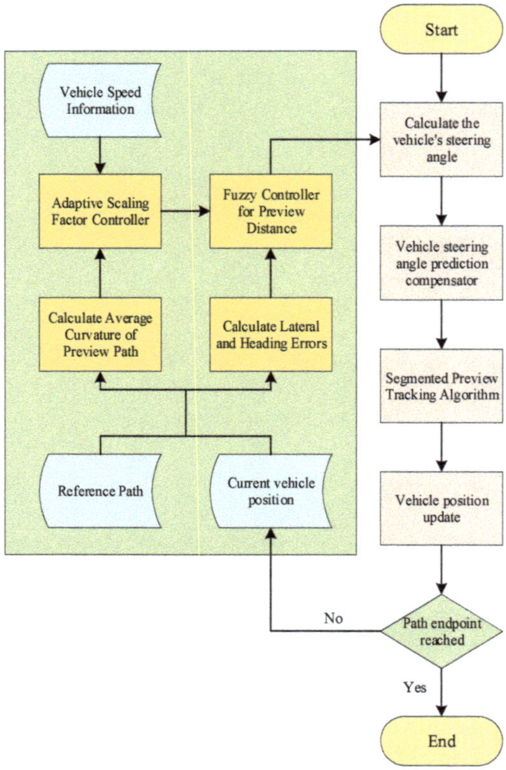

Figure 7. Algorithm flowchart.

Table 4. Functional descriptions of modules used in the control algorithm (Figure 7).

Module/Block	Description
Vehicle Speed Information	Provides the vehicle's current speed for use in adaptive scaling.
Adaptive Scaling Factor Controller	Computes the scaling factor based on average path curvature and speed.
Fuzzy Controller for Preview Distance	Infers optimal preview distance using lateral and heading errors.
PSO-SVR Steering Angle Compensator	Predicts and corrects heading deviation to improve steering accuracy.
Segmented Preview Tracking Algorithm	Computes the target heading and guides vehicle movement.
Vehicle Position Update	Updates vehicle state based on GNSS/IMU feedback and trajectory execution.

3.3.1. Design of the Variable Universe Fuzzy Controller

The essence of the variable universe fuzzy control lies in a dynamically pointwise-converging interpolation method. While keeping the original fuzzy rules of the controller unchanged, it adjusts the scaling factors online, thereby enabling the dynamic modification of the fuzzy universe [21].

For a dual-input single-output fuzzy preview tracking system, let the fuzzy universe of the input variables be defined as follows:

$$\begin{cases} X_1 = [-E_1, E_1] \\ X_2 = [-E_2, E_2] \end{cases} \quad (24)$$

and the fuzzy universe of the output variable as follows:

$$Y = [-U, U] \quad (25)$$

The adjusted fuzzy universes after applying the scaling factors are given by the following:

$$\begin{cases} X_1 = [-\alpha E_1, \alpha E_1] \\ X_2 = [-\alpha E_2, \alpha E_2] \\ Y = [-\alpha U, \alpha U] \end{cases} \quad (26)$$

where α is the universe scaling factor.

The ranges of the input and output fuzzy universes dynamically contract or expand according to the system's real-time state, such as the average curvature k of the forward path and the vehicle speed v. Although the total number of fuzzy rules remains unchanged, when the universe contracts, the fuzzy partitions become finer, thereby increasing the number of effective rules and improving the control accuracy. Conversely, when the universe expands, the fuzzy partitions become coarser, reducing the number of effective rules while accommodating other control requirements. This mechanism enables the controller to apply more appropriate fuzzy rules during different control phases, thereby enhancing the system's adaptability [22,23].

In this study, a dual-input single-output (DISO) fuzzy controller is designed with the objective of enabling the vehicle to follow a predefined trajectory. The initial parameters of the fuzzy controller, including the membership function ranges and rule base, were set based on control experience and standard fuzzy control principles. These parameters were subsequently fine-tuned through offline simulation experiments to enhance the tracking responsiveness and stability of the algorithm under various deviation conditions.

The input variables are the current lateral deviation e_d and the heading deviation e_θ, while the output variable is the preview distance L_d. Since the segmented preview tracking model divides the vehicle tracking into turning and execution phases, and the turning direction can be determined by the sign of the heading deviation, the design of the preview distance focuses more on the magnitude rather than the direction of the input errors. Therefore, the absolute values of the input parameters are used as inputs to the controller.

The basic universe of discourse for the input and output variables is defined as follows: $e_d \in [0, 50]$ cm, $e_\theta \in [0, 90]^\circ$, and $L_d \in [1, 5]$ m. The initial value of the scaling factor is set to 1, with its specific value to be determined through system tuning during operation. The input variables are described using three levels of fuzzy linguistic terms: small (*S*), medium (*M*), and large (*L*); the output variable is defined using four levels: very small (*VS*), small (*S*), medium (*M*), and large (*L*). To ensure smoothness in the algorithm's output, all membership functions are modeled using Gaussian membership functions. The membership functions of each variable are shown in Figure 8.

The mathematical expression is given by the following:

$$\mu(x) = e^{-(x-c)^2/2\sigma^2} \quad (27)$$

where μ_i is the membership degree, and α_i is the center value of the fuzzy set.

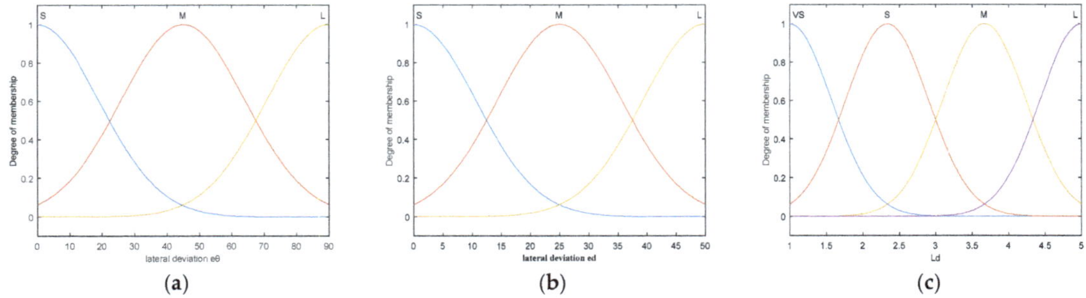

Figure 8. (**a**) Heading deviation (e_θ) membership function. (**b**) Lateral deviation membership function (e_d). (**c**) Preview distance membership function (L_d).

Reasonable fuzzy rules are established as follows: when both e_d and e_θ are large, the preview distance L_d should be reduced to quickly correct the error and improve the tracking accuracy. When e_d and e_θ are small, L_d should be increased to extend the straight-line distance and reduce unnecessary turning. The fuzzy control rules are listed in Table 5.

Table 5. Variable universe fuzzy control rules.

e_d	e_θ		
	S	***M***	***L***
S	*L*	*M*	*VS*
M	*M*	*S*	*VS*
L	*VS*	*VS*	*VS*

3.3.2. Design of the Adaptive Scaling Factor

Based on the concept of a variable universe, it is necessary to design a reasonable and effective scaling factor. Existing studies commonly adopt mathematical functions to compute the scaling factor, where its magnitude directly depends on the input variables. These function-based approaches have explicit mathematical models, allowing for fast computation and implementation within control systems. However, such methods often rely on empirically defined parameters [24], which are difficult to tune. Moreover, fixed mathematical formulations cannot fully capture the nonlinear and variable characteristics of complex systems, thereby affecting control accuracy. As a result, they may perform poorly under the diverse operating conditions of agricultural machinery. To enhance the adaptability of the system, this study adopts a fuzzy control approach for scaling factor adjustment. By applying fuzzy inference, the scaling factor is dynamically adjusted in real time based on the average curvature of the path ahead and the vehicle speed. This allows the fuzzy universe to contract more quickly when errors are large—thereby improving the system's responsiveness—and to expand when errors are small, reducing unnecessary turns and improving the system's stability.

The adaptive scaling factor controller is designed as a dual-input single-output fuzzy controller. The input variables are the average curvature k of the forward path and the vehicle speed v, while the output variable is the scaling factor α. The adaptive scaling factor controller consists of the following three main steps:

(1) Fuzzification. The fuzzy universe of input variable k is set to [0, 0.5] m^{-1}, and the fuzzy range of vehicle speed v is defined as [2, 12] km/h. Both input variables adopt three levels of fuzzy linguistic terms: small (S), medium (M), and large (L). The fuzzy universe of the output variable α is set to [0.1, 1.3], with four fuzzy linguistic terms: very small

(VS), small (S), medium (M), and large (L). The membership functions of both input and output variables use triangular functions, which are computationally simple and suitable for real-time control. The general form is as follows: ($\mu(x) = max(0, 1 - |x - c|/b)$), where c is the center, and b is the width. The membership functions are illustrated in Figure 9.

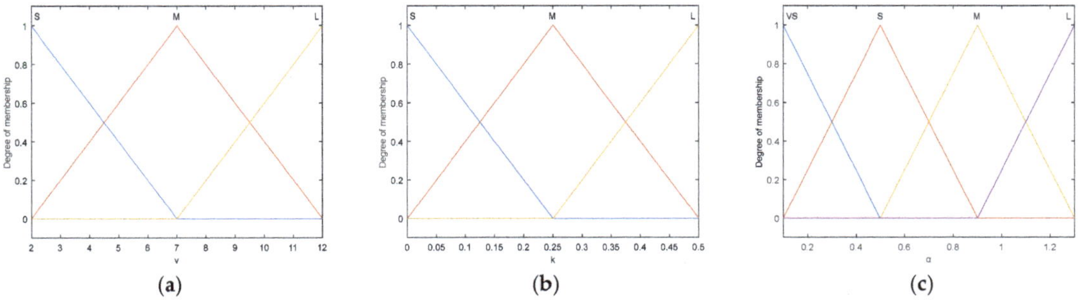

Figure 9. (**a**) Membership function diagram of vehicle speed (v). (**b**) Membership function diagram of average forward path curvature (k). (**c**) Membership function diagram of scaling factor (α).

(2) Establishment of the fuzzy rule base. The design principle of the fuzzy rules for the adaptive scaling factor is as follows: When both k and v are large, a smaller scaling factor is applied to compress the universe, increase sensitivity to errors, and enhance the path tracking capability. When both k and v are small, a larger scaling factor (close to or slightly greater than 1) is used to appropriately expand the universe, reducing excessive adjustments and improving smoothness. When k is large but v is small, a moderately small scaling factor is used to improve the tracking accuracy while avoiding unnecessary sharp corrections. When k is small but v is large, a moderately large scaling factor is adopted to enhance the vehicle stability while maintaining the required tracking precision. The designed fuzzy rule base is shown in Table 6.

Table 6. Fuzzy control rule table for the universe scaling factor.

K	*V*		
	S	*M*	*L*
S	*L*	*L*	*M*
M	*S*	*S*	*VS*
L	*VS*	*VS*	*VS*

(3) Defuzzification and fuzzy inference. Fuzzy inference is performed using the Mamdani method. Based on the membership degrees of inputs k and v, the fuzzy rule base is used to compute the fuzzy output of α. The fuzzy output is then defuzzified using the centroid method to obtain a crisp value, calculated as follows: $\alpha = \sum(\mu_i \cdot \alpha_i)/\sum\mu_i$, where μ_i is the membership degree, and α_i is the center value of the fuzzy set. Finally, the computed scaling factor is fed into the variable universe fuzzy controller and applied in the calculation of the vehicle's preview distance.

3.4. Vehicle Steering Angle Prediction and Compensation Based on SVR and PSO

Heading deviation refers to the difference between the current heading of the vehicle and the desired heading, which determines the vehicle's turning direction and turning angle. In practical applications, however, due to factors such as muddy terrain and track slippage, the actual angular velocity during single-track steering tends to be discrete or exhibit nonlinear jumps. As a result, the control process shows discontinuous steering behavior, making it difficult to achieve precise control. This often leads to steady-state errors and an inability to maintain low lateral deviation [25]. Therefore, it is necessary to predict the vehicle's heading deviation and compensate for the steering angle accordingly so that the vehicle can advance or delay steering actions as needed. Given the highly nonlinear disturbances present in complex farmland environments, it is challenging to develop an accurate model to estimate the actual heading deviation error. In this context, machine learning offers a feasible solution.

Support Vector Regression (SVR), as a nonlinear regression model based on the principle of structural risk minimization, demonstrates strong fitting capabilities for small-sample and high-dimensional data [26]. It can analyze and learn from actual observational data to infer and predict the vehicle's state at the next moment. Based on the prediction results, compensation can be applied to certain parameter values to achieve more accurate tracking. Therefore, this study adopts the SVR algorithm to predict the vehicle's heading at the next time step in real time based on the current heading deviation.

During navigation, due to muddy and uneven terrain, the vehicle's navigation data D (including current GPS coordinates, heading angle, and track speed) exhibit continuous fluctuations, containing both sudden changes and noise. To mitigate the impact of noise and abrupt variations, this study employs the exponential moving average (EMA) method:

$$D_i = \begin{cases} 0 & i = 0 \\ \beta D_{i-1} - (1-\beta)\theta_i & i \geq 1 \end{cases} \tag{28}$$

where β is the weighting coefficient, which is set to 0.6 in order to enhance the stability of the algorithm and better handle sudden changes. D_i denotes the exponential moving average of the i-th sample, and θ_i represents the original value of the i-th sample. Each round of exponential smoothing is performed using 300 consecutive sampling points.

Considering that the exponential moving average may misestimate the true value due to insufficient initial data, resulting in a systematic bias, a bias correction formula is introduced for compensation:

$$D_i = \frac{D_i}{1-\beta^{\mathrm{i}}} \tag{29}$$

SVR can be used to construct a nonlinear mapping relationship between input and output variables. Unlike traditional regression methods, SVR maps input samples into a high-dimensional space using a kernel function and determines the regression hyperplane based on the support vectors [27]. Its mathematical expression is given as follows:

$$f(x) = \sum_{i=1}^{m} (\hat{\alpha}_i - \alpha_i) \cdot k(x_i, x_j) + b \tag{30}$$

where $k(x_i, x_j) = \Phi(x_i)^T \Phi(x_i)$ is the kernel function.

During navigation, the 300 most recent navigation data points are processed using exponential moving averaging. The first 70% of the smoothed results are used as the training set, and the remaining 30% as the test set. The input features include vehicle position information, heading deviation, lateral deviation, and vehicle speed. The target output is the predicted heading deviation, which is used to guide steering control. To avoid falling into local optima and to improve the performance of the SVR model, this study introduces the Particle Swarm Optimization (PSO) algorithm to automatically optimize the key SVR parameters, including the penalty coefficient C and the kernel parameter g. In the *PSO* algorithm, each particle represents a candidate parameter pair (C, g), where

$C \in [0.1, 100]$ and $g \in [0.1, 100]$. The mean squared error (MSE) of the SVR is adopted as the fitness function in PSO, and the fitness of each particle is evaluated through cross-validation errors.

PSO dynamically updates each particle's velocity and position by referring to its own historical best position and the current global best position of the entire swarm [28,29]. Through multiple iterations, the algorithm converges to an optimal set of parameters that minimizes the prediction error of the SVR model, thereby improving its prediction accuracy and generalization capability. In this study, the PSO hyperparameters—including population size, inertia weight, and learning factors—are empirically set to achieve a good balance between convergence speed and prediction performance.

The equations for updating the particle velocity and position are as follows:

$$\begin{cases} v_{i,d}(t+1) = \omega v_i + c_1 r_1 (P_{i,d} - x_{i,d}(t)) + c_2 r_2 (P_{g,d} - x_{i,d}(t)) \\ x_{i,d}(t+1) = x_{i,d}(t) + v_{i,d}(t+1) \end{cases} \tag{31}$$

Note: i is the index of the particle; t is the current iteration number; d is the dimension index of the search space (i.e., the d-th parameter); $v_{i,d}(t)$ is the velocity of particle i in dimension d at iteration t; $x_{i,d}(t)$ is the position (i.e., parameter value) of particle i in dimension d at iteration t; $P_{i,d}$ is the historical best position of particle i in dimension d; $\mathrm{P_{g,d}}$ is the global best position found so far in dimension d; ω is the inertia weight, set to 1; c_1 and c_2 are learning factors, set to 1.5 and 1.7, respectively; and r_1 and r_2 are random numbers uniformly distributed in the interval [0, 1].

After training is completed, new navigation data are input into the trained SVR model, which uses the learned weights and bias to compute the predicted heading deviation of the vehicle. For clarity of representation, the heading deviation is denoted as φ.

To improve the prediction accuracy, this study employs the Mean Bias Error (MBE) to correct the prediction results. MBE measures the average deviation between the predicted heading deviation and the actual heading deviation. The calculation formula is as follows:

$$MBE = \frac{1}{n}\sum_{i=1}^{n}(\varphi_{\alpha_i} - \varphi_{p_i}) \tag{32}$$

The predicted heading deviation after MBE correction is denoted as φ'_{p_i}, and its calculation formula is as follows:

$$\varphi'_{p_i} = MBE + \varphi_{p_i} \tag{33}$$

To further compensate for the heading deviation, a deviation compensation coefficient μ_{ci} is defined, and its expression is given as follows:

$$\mu_{ci} = \frac{\varphi_{e_{i-1}} - \varphi_{e_i}}{\varphi'_{p_{i-1}} - \varphi'_{p_i}} = \frac{\Delta\varphi_e}{\Delta\varphi'_p} \tag{34}$$

where φ_e is the expected heading deviation, and φ'_p is the MBE-corrected predicted heading deviation.

The prediction error is defined as follows:

$$\varphi_{error,i} = \varphi_{e_i} - \varphi'_{p_i} \tag{35}$$

The corresponding compensation amount is calculated as follows:

$$\varphi_{c_i} = \mu_{c_i} \cdot \varphi_{error,i} \tag{36}$$

The final actual output heading deviation is as follows:

$$\varphi_i = \varphi_{c_i} + \mu_{c_i} \cdot \varphi_{error,i} \tag{37}$$

By introducing the heading deviation compensation term to adjust the output of the controller, the heading deviation becomes more consistent with the actual steering behavior of the vehicle, thereby improving the path tracking accuracy of autonomous navigation. The flow of this feedback control algorithm is illustrated in Figure 10.

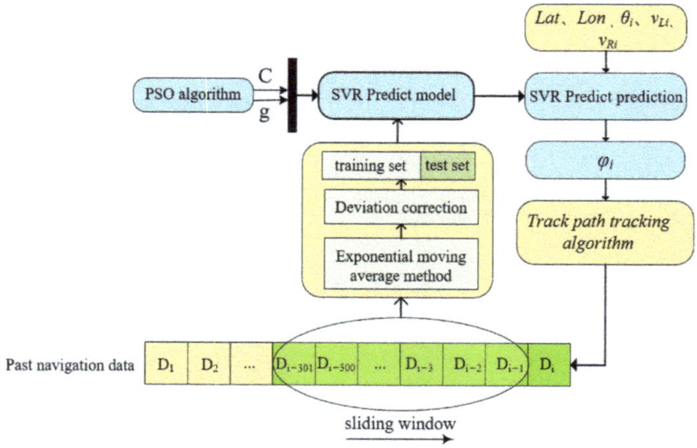

Figure 10. Flowchart of the feedback control algorithm.

4. Experiments and Analysis

4.1. Simulation Experiments

To verify the effectiveness of the single-track autonomous navigation system based on variable universe fuzzy control, a simulation platform for single-side braking tracked vehicles was developed using MATLAB 2022b. This platform is used to conduct comparative evaluations against fixed preview distance and fixed-universe fuzzy control methods. In the simulation, the vehicle moves at a constant speed of 2 km/h, with a track width of B = 1 m. The target path is a 40 m straight line oriented at a 45° angle to the x-axis. A comparison of the simulated trajectories is shown in Figure 11, and the corresponding experimental data are provided in Table 7.

The distance required for the vehicle to align with the reference path decreases as the preview distance decreases. However, this is accompanied by a significant increase in the number of turns within a control cycle. When the preview distance becomes too small, the vehicle may even deviate beyond the reference path. The algorithm proposed in this study appropriately reduces the fuzzy universe when the initial deviation is large, resulting in more responsive control and a shorter alignment distance. As shown in the table, the alignment distance is reduced by 12.99% compared to fixed-universe fuzzy control, and the number of turns per control cycle is reduced by 9.09%. This demonstrates that the proposed algorithm not only achieves rapid path alignment but also effectively reduces the number of turns during control cycles, thereby enhancing vehicle stability.

Although the proposed algorithm achieves a lower turning frequency, its average lateral error is slightly higher. This is attributed to the control strategy prioritizing rapid alignment in the initial stage while accepting slightly larger oscillations during steady-state tracking. Since the preview distance increases adaptively once the vehicle is aligned, minor lateral deviations tend to persist longer, especially in low-curvature segments. However, the resulting lateral deviation remains within an acceptable range and does not compromise the effectiveness or stability of the control algorithm.

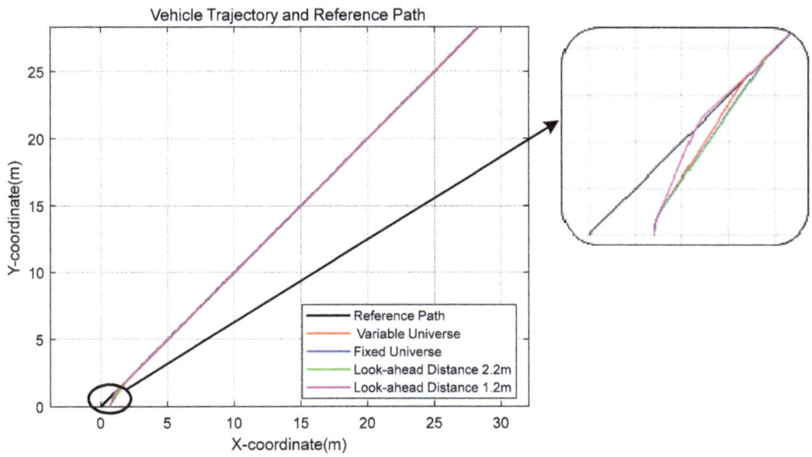

Figure 11. Verification of path alignment distance.

Table 7. Simulation data for path alignment.

Preview Distance (m)/Control Method	Alignment Distance (m)	Number of Turns per Cycle	Average Lateral Error (cm)
2.2	2.0139	18	2.0637
1.2	1.0944	33	1.3160
Fixed-universe fuzzy control algorithm	1.9667	11	2.4727
Variable-universe fuzzy control algorithm (proposed)	1.7111	10	2.7514

To further verify the adaptability of the proposed path tracking control algorithm under different vehicle speeds and forward path curvature conditions, a test path was designed. The path includes three turning segments with curvatures of 0.2, 0.3, and 0.4, respectively, corresponding to semicircular curves, as well as four straight-line segments, each 40 m in length, simulating various operational conditions. The path tracking result is shown in Figure 12. It can be observed that the variable universe fuzzy control algorithm is capable of dynamically adjusting the preview distance when the path curvature changes, enabling effective path tracking. Compared to the fixed-universe fuzzy control method, the total number of turns over the entire path is reduced by 32.94%, and the average lateral deviation is reduced by 47.11%. Compared to the fixed preview distance method, the total number of turns over the full trajectory is reduced by 20.52%, and the average lateral deviation is reduced by 42.56%. Moreover, no distortion occurs in high-curvature path segments, indicating strong adaptability of the proposed method.

It should be noted that although the Variable Universe Fuzzy Control algorithm may exhibit denser steering actions in curved segments—as visually illustrated in Figure 12—this does not necessarily indicate a higher total number of turns. This phenomenon results from the algorithm's adaptive behavior: it increases the turning frequency in high-curvature regions to minimize the lateral deviation while reducing unnecessary corrections in straight segments. Consequently, the overall number of turns over the entire trajectory is lower than that of the fixed-universe fuzzy control, as reflected in Table 8.

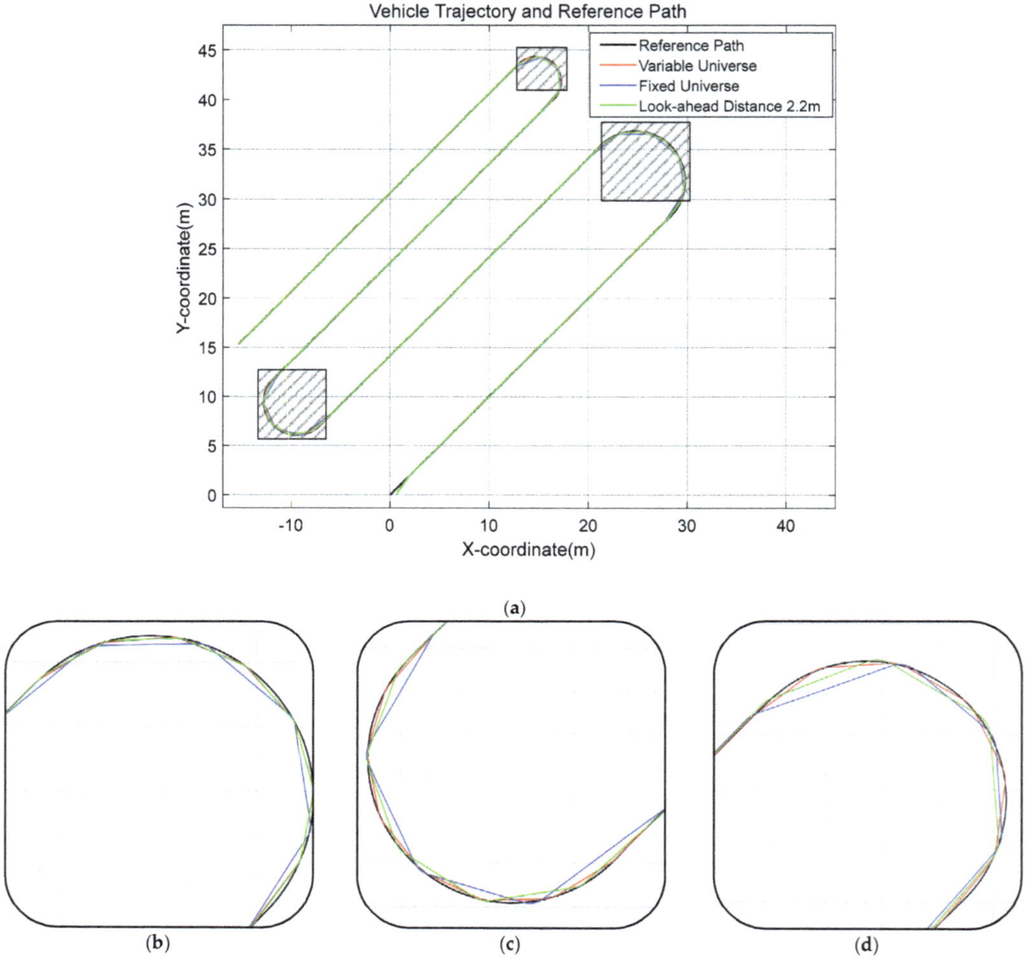

Figure 12. (**a**) Overall path tracking diagram. (**b–d**) are the local enlarged views.

Table 8. Comprehensive path simulation data.

Preview Distance/Control Method	Number of Turns	Average Lateral Error (cm)
2.2	85	4.4425
Fixed-universe fuzzy control algorithm	71	3.5682
Variable-universe fuzzy control algorithm (proposed)	57	2.3498

4.2. Field Experiments

To further validate the proposed path tracking algorithm, a field test was conducted on an open farmland path. The test site was located at the experimental field of Xihua University in Chengdu, China, with coordinates of longitude 103.9571° and latitude 30.7781°.

The operating environment is shown in Figure 13, where uneven terrain and the presence of weeds can be observed. To evaluate the effectiveness of the designed vehicle steering angle prediction compensator, multiple reference points were set along the path. The vehicle was commanded to track these reference points sequentially at a speed of 2 km/h, and a comparative analysis was performed between the cases with and without the prediction compensator enabled. The results showed that the mean absolute heading angle deviation without the compensator was 10.06°, while with the compensator enabled, it was reduced to 3.78°, representing a 32.34% improvement. As illustrated in Figure 14, in the absence of the prediction compensator, the vehicle exhibited significant deviation from the reference points after turning, requiring multiple adjustments to realign with the path. In contrast, when the compensator was activated, the steering control was significantly more accurate, effectively reducing the need for repeated heading adjustments.

In addition, a gradually inward-curving path was designed in the field to verify the adaptability of the proposed path tracking algorithm to curvature variations. The field experiments were conducted to evaluate the overall tracking performance of the proposed algorithm in a continuous and realistic operating environment. To ensure stability and avoid unnecessary disturbances in the experimental data, no artificial lateral offset was introduced at the starting point.

The vehicle was tested three times under each of the three control strategies: variable-universe fuzzy control, fixed-universe fuzzy control, and fixed preview distance. The path tracking result of one representative trial is shown in Figure 15, while the distribution of lateral tracking errors is illustrated in Figure 16. The complete results from the three trials are summarized in Table 9.

To assess whether the differences in path tracking performance among the control methods were statistically significant, a significance analysis was conducted based on the data from three repeated field trials. One-way ANOVA and pairwise independent-sample *t*-tests were applied to key performance indicators [30], including the number of turns per control cycle, average lateral deviation, and its standard deviation. The results indicate that the proposed variable-universe fuzzy control method exhibited statistically significant improvements over the other two methods in multiple metrics ($p < 0.05$), demonstrating its superior tracking performance under complex terrain conditions. Significant differences are marked in Table 10 using superscript symbols.

Figure 13. Field test environment.

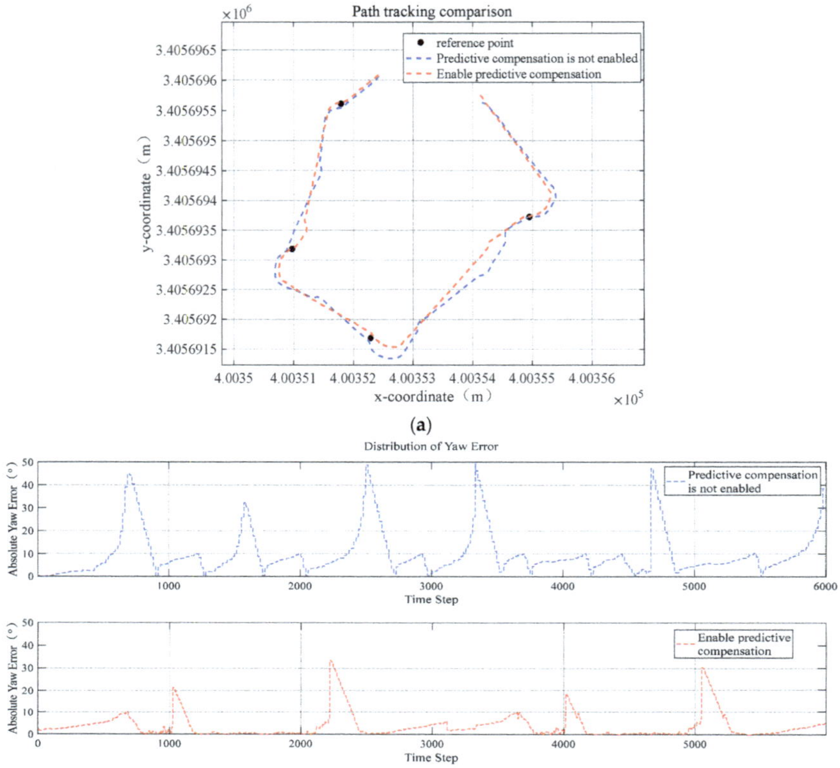

(**a**)

(**b**)

Figure 14. (**a**) Comparison of vehicle trajectories when tracking reference points. (**b**) Comparison of heading deviation during tracking.

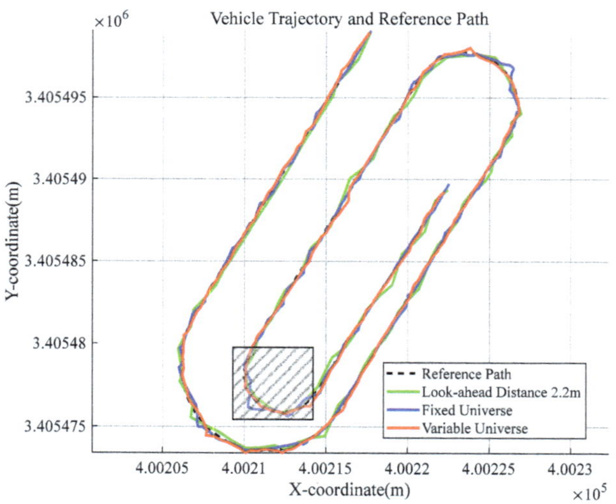

(**a**)

Figure 15. *Cont.*

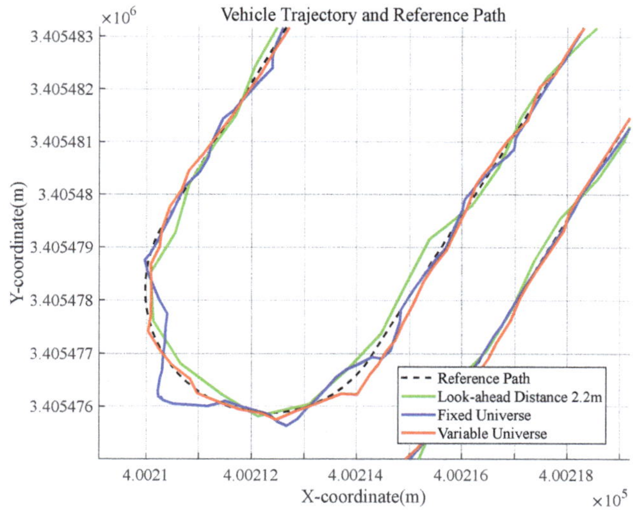

(**b**)

Figure 15. (**a**) Overall vehicle trajectory tracking diagram. (**b**) Local enlarged view.

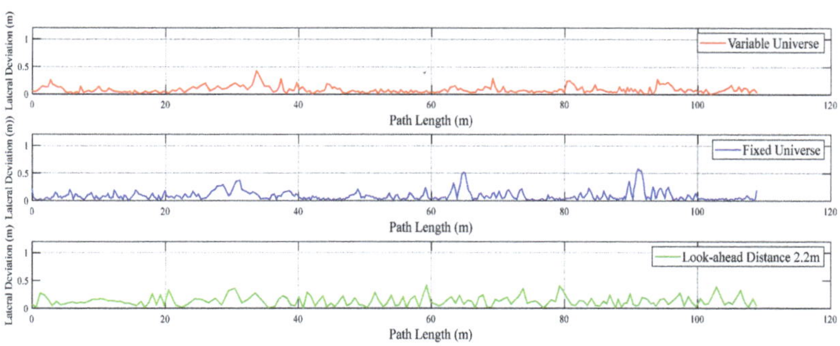

Figure 16. Distribution of lateral tracking errors.

Table 9. Comparison of path tracking performance among different algorithms in three trials.

	Path Tracking Algorithm	Number of Turns per Cycle	Average Lateral Error (m)	Maximum Lateral Error (m)	Standard Deviation of Lateral Error (m)
Trial 1	Variable-universe (proposed)	142	0.0485	0.4266	0.0610
	Fixed-universe	173	0.0840	0.5790	0.0940
	Fixed preview distance	213	0.1070	0.4186	0.0790
Trial 2	Variable-universe (proposed)	153	0.0493	0.5140	0.0592
	Fixed-universe	178	0.0712	0.6584	0.0843
	Fixed preview distance	220	0.0894	0.5136	0.0814
Trial 3	Variable-universe (proposed)	149	0.0516	0.6120	0.0602
	Fixed-universe	191	0.0723	0.5123	0.0927
	Fixed preview distance	205	0.0853	0.4541	0.0876

Table 10. Statistical comparison of overall path tracking performance among different algorithms (mean ± SEM).

Path Tracking Algorithm	Number of Turns per Cycle	Average Lateral Error (m)	Standard Deviation of Lateral Error (m)
Variable-universe (proposed)	148 ± 5.5 (*)	0.0498 ± 0.0016 (*)	0.0601 ± 0.0009 (*)
Fixed-universe	181 ± 9.0 (*)	0.0758 ± 0.0071 (*)	0.0903 ± 0.0042 (*)
Fixed preview distance	212 ± 7.6 (*)	0.0939 ± 0.0115 (*)	0.0827 ± 0.0043

Note: (*) indicates that the result differs significantly from other control methods ($p < 0.05$) based on one-way ANOVA and independent-sample t-tests conducted over three repeated field trials.

Due to the presence of uneven terrain, lateral deviations exhibited some oscillations. However, the proposed algorithm still maintained good stability. As shown in the data in Table 10, the number of turns per control cycle was reduced by 30.19% and 18.23% compared to the fixed preview distance and fixed-universe fuzzy control methods, respectively. The average lateral deviation was reduced by 34.29% and 46.96%, respectively.

5. Discussion

To better clarify the technical innovation and engineering feasibility of this study, a comparative analysis is conducted against representative existing methods, and the applicability of the proposed approach in complex environments is further explored.

Jaramillo-Martínez et al. [31] proposed a reinforcement learning-based path planning method that combines Deep Q-Learning with a Lyapunov-based path tracking controller for autonomous mobile robots (AMRs). This method improves the trajectory smoothness and reduces the number of turns through a carefully designed reward function. Although it performs well in minimizing path length and turning frequency, its heavy reliance on discrete grid modeling and large-scale training iterations limits its adaptability to real-time variations and nonlinear disturbances commonly found in unstructured agricultural environments.

Compared with reinforcement learning-based methods, the approach proposed in this study offers advantages such as interpretable control behavior, high computational efficiency, and deployment-friendliness on embedded platforms. Specifically, the variable universe fuzzy controller adopts Gaussian membership functions and a three-level linguistic classification, maintaining a compact rule base (typically 3×3 or 3×4). The inference and defuzzification process can be implemented using lookup tables and multiply–accumulate operations, making it well-suited for deployment on low- to mid-end microcontrollers (MCUs) using fixed-point or integer operations. It can operate stably under sampling rates of 100–200 Hz without noticeable delay.

Meanwhile, the SVR-based compensation module adopts a sliding window update strategy with one-step-ahead prediction and offline PSO optimization of parameters, which avoids the need for online kernel learning or complex regression matrix solutions. Once deployed, the model only requires evaluating a small number of kernel functions (e.g., RBF) and computing their linear combination. In our experiments, the input dimension was around 4–5, with fewer than 50 support vectors, and the inference time per prediction was less than 2 ms—demonstrating stable real-time performance on embedded Linux platforms, such as ARM A53. These features make the method particularly suitable for resource-constrained agricultural platforms [32].

Pérez-Juárez et al. [33] developed a fuzzy PID-based trajectory tracking controller for wheeled robots operating in virtual environments. Their fuzzy system symmetrically adjusts PID parameters to achieve good error convergence and control robustness. While this method demonstrates certain adaptive capabilities, it is tailored for continuous control in holonomic systems and is not suitable for the nonholonomic tracked vehicles considered in this study, which use unilateral braking and exhibit a fixed turning radius. To bridge this gap, our work introduces a preview point segmentation model based on tracked vehicle dynamics, along with a PSO-optimized SVR-based yaw compensation strategy that significantly enhances turning accuracy and path tracking continuity.

In addition, considering that agricultural machinery often operates under challenging conditions—such as high travel speeds, steep slopes, loose gravel, and irregular terrain—existing control methods require further improvements in environmental adaptability. At higher speeds, vehicle response delays and increased inertia demand more responsive preview distance estimation and real-time error prediction. Therefore, future work may incorporate speed-segmented control strategies or dynamically weighted fuzzy compensation mechanisms to enhance responsiveness and accuracy under high-speed scenarios.

Furthermore, in environments with severe terrain disturbances—such as loose soil, gravel roads, or inclined fields—tracked vehicles may experience slipping, sinking, or fluctuating traction, which can exacerbate the deviation between the model and actual behavior. To improve the system's robustness under such conditions, future work can consider integrating multi-source sensor information (e.g., IMU, terrain adhesion estimation), incorporating terrain factors into the fuzzy inference system, or introducing disturbance observers and hybrid neuro-fuzzy networks for real-time modeling and compensation of nonlinear interferences.

Overall, the proposed method has demonstrated strong adaptability and control performance under current speeds and typical hilly terrains. Future extensions toward higher speeds, stronger disturbances, and more diverse field conditions may focus on adaptive fuzzy rule expansion, terrain-aware disturbance modeling, and integration with predictive control to further improve the system's intelligent decision-making and control capabilities in complex agricultural environments.

6. Conclusions

This study focuses on the path tracking problem of single-track electric agricultural chassis in farmland environments. Based on preview tracking, a variable-universe fuzzy control method was employed to dynamically adjust the preview distance. This approach effectively reduces the number of turns during tracking and enhances the system's stability. In addition, a vehicle steering angle prediction compensator was developed using a combination of Support Vector Regression (SVR) and Particle Swarm Optimization (PSO), enabling precise control of turning maneuvers for single-track agricultural vehicles and reducing steady-state errors. Comprehensive path tracking experiments were conducted in field conditions. The results showed that, compared with fixed preview distance and fixed-universe fuzzy control methods, the proposed algorithm reduced the average number of turns by 30.19% and 18.23% and decreased the average lateral error by 34.29% and 46.96%, respectively, thus verifying the effectiveness of the proposed approach. The following conclusions are drawn:

(1) A universe scaling factor fuzzy controller and a preview distance fuzzy controller were designed. The average path curvature k and vehicle speed v were used as inputs to compute the universe scaling factor α. The current lateral deviation e_d and heading deviation e_θ were then input into the controller, with the fuzzy universe adjusted dynamically based on α to determine the preview distance L_d. This approach enables the preview distance to be adaptively adjusted according to environmental information, thereby reducing the number of turns per control cycle.

(2) The PSO-SVR method was employed to predict the vehicle's heading deviation and to compensate for the steering angle, which effectively improved the vehicle's turning

control capability under external environmental disturbances, thereby reducing the path tracking error of agricultural equipment. Additionally, this method enables real-time adjustment based on navigation information, allowing the system to rapidly adapt to different operating environments.

In future work, the proposed algorithm will be further enhanced to improve its adaptability to complex and slippery farmland environments. Additionally, specific issues related to single-track turning—such as severe soil disruption and soil accumulation caused by excessive turning of one track—will be considered in the path tracking algorithm design to improve the environmental adaptability of agricultural equipment.

Author Contributions: Conceptualization, H.L. and X.L.; methodology, H.L.; software, H.L.; validation, Z.H., J.L. and S.W.; formal analysis, H.L.; investigation, Z.H. and J.L.; resources, H.Y.; data curation, Z.H.; writing—original draft preparation, H.L.; writing—review and editing, X.L. and J.B.; visualization, H.L.; supervision, X.L.; project administration, J.B.; funding acquisition, J.B. All authors have read and agreed to the published version of the manuscript.

Funding: This research was funded by the Natural Science Foundation of Sichuan Province, China, under project number 2025ZNSFSC1004. The project title is "Research on Dynamic Scheduling Optimization Method for Multiple Agricultural Machines Oriented to Cooperative Tasks". The APC was funded by the Natural Science Foundation of Sichuan Province, China.

Institutional Review Board Statement: Not applicable.

Data Availability Statement: Data is contained within the article.

Conflicts of Interest: The authors declare no conflict of interest.

References

1. Sun, J.B.; Ren, G.Y.; Zhang, Y.; Liu, Y.H. Research status and prospects of key technologies for agricultural crawler chassis. *Trans. Chin. Soc. Agric. Mach.* **2024**, *53*, 202–220. (In Chinese)
2. Yao, C.; Yuan, J.; Pan, J. Design and simulation analysis of the terrain adaptive chassis's walking system of rubber track combine harvester. *J. Chin. Agric. Mech.* **2020**, *41*, 107–113. (In Chinese)
3. Ma, Z.; Yin, C.; Du, X.; Zhao, L.; Lin, L.; Zhang, G.; Wu, C. Rice row tracking control of crawler tractor based on the satellite and visual integrated navigation. *Comput. Electron. Agric.* **2022**, *197*, 106935. [CrossRef]
4. Yang, Y.; Li, Y.; Wen, X.; Zhang, G.; Ma, Q.; Cheng, S.; Qi, J.; Xu, L.; Chen, L. An Optimal Goal Point Determination Algorithm for Automatic Navigation of Agricultural Machinery: Improving the Tracking Accuracy of the Pure Pursuit Algorithm. *Comput. Electron. Agric.* **2022**, *194*, 106760. [CrossRef]
5. Wang, Q.; He, J.; Lu, C.; Wang, C.; Lin, H.; Yang, H.; Li, H.; Wu, Z. Modelling and Control Methods in Path Tracking Control for Autonomous Agricultural Vehicles: A Review of State of the Art and Challenges. *Appl. Sci.* **2023**, *13*, 7155. [CrossRef]
6. Hu, C.; Ru, Y.; Li, X.; Fang, S.; Zhou, H.; Yan, X.; Liu, M.; Xie, R. Path tracking control for brake-steering tracked vehicles based on an improved pure pursuit algorithm. *Biosyst. Eng.* **2024**, *242*, 1–15. [CrossRef]
7. He, J.; Man, Z.; Hu, L.; Luo, X.; Wang, P.; Li, M.; Li, W. Path tracking control method and experiments for the crawler-mounted peanut combine harvester. *Trans. Chin. Soc. Agric. Eng.* **2023**, *39*, 9–17. (In Chinese with English Abstract) [CrossRef]
8. Qin, W.; Zhang, G.; Hu, S.; Zhou, Y.; Wen, C.; Fu, W.; Meng, Z. Differential steering control system for single HST tracked tractors. *Trans. Chin. Soc. Agric. Mach.* **2024**, *55* (Suppl. S1), 405–411. (In Chinese with English Abstract)
9. He, Y.; Zhou, J.; Yuan, L.; Zheng, P.; Liang, Z. Local tracking path planning based on steering characteristics of crawler-type combine harvester. *Trans. Chin. Soc. Agric. Mach.* **2022**, *53*, 13–21. (In Chinese with English Abstract)
10. Li, Y.; Zhao, Z.; Huang, P.; Guan, W.; Wu, X. Automatic navigation system of tractor based on DGPS and double closed-loop steering control. *Trans. Chin. Soc. Agric. Mach.* **2017**, *48*, 11–19. (In Chinese with English Abstract) [CrossRef]
11. Sun, J.; Wang, Z.; Ding, S.; Xia, J.; Xing, G. Adaptive disturbance observer-based fixed time nonsingular terminal sliding mode control for path-tracking of unmanned agricultural tractors. *Biosyst. Eng.* **2024**, *246*, 96–109. [CrossRef]
12. Wang, M.; Xu, J.; Zhang, J.; Cui, Y. An autonomous navigation method for orchard rows based on a combination of an improved A-star algorithm and SVR. *Precis. Agric.* **2024**, *25*, 1429–1453. [CrossRef]
13. Zhang, W.; Hu, L.; Wang, H.; Zhang, G.; Luo, X.; Zhang, Z. Improved Pure Pursuit Agricultural Machinery Navigation Curve Path Tracking Method Based on B-spline Optimization. *Trans. Chin. Soc. Agric. Mach.* **2024**, *55*, 42–51. (In Chinese with English Abstract) [CrossRef]

14. Zhang, H.; Wang, G.; Lü, Y.; Qin, C.; Liu, L.; Gong, J. Research on Path Tracking Algorithm for Agricultural Machinery Based on Improved Pure Pursuit Model. *Trans. Chin. Soc. Agric. Mach.* **2020**, *51*, 18–25. (In Chinese with English Abstract) [CrossRef]
15. Netto, M.; Blosseville, J.-M.; Lusetti, B.; Mammar, S. A new robust control system with optimized use of the lane detection data for vehicle full lateral control under strong curvatures. In Proceedings of the 2006 IEEE Intelligent Transportation Systems Conference (ITSC), Toronto, ON, Canada, 17–20 September 2006; pp. 1382–1387.
16. Zhou, J.; Wen, J.; Yao, L.; Yang, Z.; Xu, L.; Yao, L. Agricultural Machinery Path Tracking with Varying Curvatures Based on an Improved Pure-Pursuit Method. *Agriculture* **2025**, *15*, 266. [CrossRef]
17. Song, Y.; Xue, J.; Zhang, T.; Sun, X.; Sun, H.; Gao, W.; Chen, Q. Path tracking control of crawler tractor based on adaptive adjustment of lookahead distance using sparrow search algorithm. *Comput. Electron. Agric.* **2025**, *234*, 110219. [CrossRef]
18. Guan, Y.; Li, N.; Chen, P.; Zhang, Y. Research on Path Tracking Control Based on Optimal Look-Ahead Points. *Int. J. Automot. Technol.* **2024**, *25*, 1355–1374. [CrossRef]
19. Wang, F.; Wang, B.; Zhang, Z.; Xie, K.; Li, A.; Ni, C.; Fang, H. Adaptive preview tracking fuzzy control algorithm for tracked vehicles. *Trans. Chin. Soc. Agric. Eng.* **2024**, *40*, 32–43. (In Chinese with English Abstract) [CrossRef]
20. Wang, W.; Luo, C.; Zhang, G.; Fu, J.; Dong, Z.; Ji, C.; Zhao, Z. Design and experiment of automatic navigation system for rice harvester with dual-motor crawler chassis. *J. Huazhong Agric. Univ.* **2022**, *41*, 199–207. (In Chinese with English Abstract) [CrossRef]
21. Chai, S.; Yao, L.; Xu, L.; Chen, Q.; Xu, T.; Yang, Y. Research on greenhouse agricultural machinery path tracking based on dynamic look-ahead distance pure pursuit model. *J. Chin. Agric. Mech.* **2021**, *42*, 58–64. (In Chinese with English Abstract) [CrossRef]
22. Zhu, D.; Shi, M.; Wang, Y.; Xue, K.; Liao, J.; Xiong, W.; Kuang, F.; Zhang, S. Path tracking control method for automatic navigation rice transplanters based on VUFC and improved BAS algorithm. *Robotica* **2023**, *41*, 3116–3136. [CrossRef]
23. Ji, G.; Li, S.; Feng, G.; Wang, H. Enhanced Variable Universe Fuzzy Control of Vehicle Active Suspension Based on Adaptive Contracting–Expanding Factors. *Int. J. Fuzzy Syst.* **2023**, *25*, 2986–3000. [CrossRef]
24. Zhang, J.; Lin, H.; Gao, J.; Zhou, X.; Zhou, J. Intelligent vehicle path tracking control considering variable universe fuzzy steering angle compensation and network time delay. *J. Chongqing Jiaotong Univ. Nat. Sci. Ed.* **2024**, *43*, 103–113. (In Chinese with English Abstract) [CrossRef]
25. Li, S.; Zhang, M.; Cao, R.; Ji, Y.; Zhang, Z.; Li, H.; Yin, Y. Development of the automatic navigation system for combine harvester based on GNSS. *Int. J. Agric. Biol. Eng.* **2021**, *14*, 163–171. [CrossRef]
26. Easterly, D.R.; Adamchuk, V.I.; Kocher, M.F.; Hoy, R.M. Using a vision sensor system for performance testing of satellite-based tractor auto-guidance. *Comput. Electron. Agric.* **2010**, *72*, 107–118. [CrossRef]
27. Ahanda, J.J.-B.M.; Mbede, J.B.; Melingui, A.; Essimbi, B. Robust adaptive control for robot manipulators: Support vector regression-based command filtered adaptive backstepping approach. *Robotica* **2018**, *36*, 516–534. [CrossRef]
28. Tian, Q.; Wang, T.; Song, Y.; Wang, Y.; Liu, B. Autonomous Underwater Vehicle Path Tracking Based on the Optimal Fuzzy Controller with Multiple Performance Indexes. *J. Mar. Sci. Eng.* **2023**, *11*, 463. [CrossRef]
29. Tolossa, T.D.; Gunasekaran, M.; Halder, K.; Verma, H.K.; Parswal, S.S.; Jorwal, N.; Joseph, F.O.M.; Hote, Y.V. Trajectory tracking control of a mobile robot using fuzzy logic controller with optimal parameters. *Robotica* **2024**, *42*, 2801–2824. [CrossRef]
30. Liu, H.; Luo, J.; Zhang, L.; Yu, H.; Liu, X.; Wang, S. Research on Traversal Path Planning and Collaborative Scheduling for Corn Harvesting and Transportation in Hilly Areas Based on Dijkstra's Algorithm and Improved Harris Hawk Optimization. *Agriculture* **2025**, *15*, 233. [CrossRef]
31. Jaramillo-Martínez, R.; Chavero-Navarrete, E.; Ibarra-Pérez, T. Reinforcement-Learning-Based Path Planning: A Reward Function Strategy. *Appl. Sci.* **2024**, *14*, 7654. [CrossRef]
32. Zhao, B.; Zheng, D.; Yang, C.; Wang, S.; Mansurova, M.; Jomartova, S.; Kunicina, N.; Zabasta, A.; Beliaev, V.; Caiko, J.; et al. Design and Optimization of an Internet of Things-Based Cloud Platform for Autonomous Agricultural Machinery Using Narrowband Internet of Things and 5G Dual-Channel Communication. *Electronics* **2025**, *14*, 1672. [CrossRef]
33. Pérez-Juárez, J.G.; García-Martínez, J.R.; Medina Santiago, A.; Cruz-Miguel, E.E.; Olmedo-García, L.F.; Barra-Vázquez, O.A.; Rojas-Hernández, M.A. Kinematic Fuzzy Logic-Based Controller for Trajectory Tracking of Wheeled Mobile Robots in Virtual Environments. *Symmetry* **2025**, *17*, 301. [CrossRef]

agriculture

MDPI

Article

Configurational Comparison of a Binary Logic Transmission Unit Applicable to Agricultural Tractor Hydro-Mechanical Continuously Variable Transmissions and Its Wet Clutch Optimization Design Based on an Improved General Regression Neural Network

Wenjie Li, Zhun Cheng * and Mengchen Yang

College of Automobile and Traffic Engineering, Nanjing Forestry University, Nanjing 210037, China; wy13813938850@163.com (W.L.); a_ymcc@163.com (M.Y.)

* Correspondence: cz38@njfu.edu.cn

Abstract: Binary logic transmission (BLT), a stepped transmission system, has been utilized in military vehicles and heavy-duty commercial vehicles due to its high transmission efficiency, strong load-bearing capacity, and compact structure. Its adaptability to agricultural tractor operations is notable. This study modularizes BLT into a binary logic transmission unit (BLT-U) for application in agricultural tractor Hydro-Mechanical Continuously Variable Transmission (HMCVT), optimizing its wet clutch to enhance HMCVT shifting performance. This provides a basis for BLT-U's application in other transmission systems and subsequent optimization. A wet clutch test bench was employed to validate the modeling approach. The optimal BLT-U configuration was selected using both light/heavy load conditions and subjective–objective evaluation criteria. The WOA improved the spread value in the GRNN algorithm, establishing a GRNN to predict the optimal range for wet clutch design values in BLT-U; the model validation showed an average correlation coefficient of 0.92 for speed curves and an average relative error of 5.58% for dynamic loads. Under light-load conditions, the optimal configuration improved average and maximum scores by 13.38% and 11.53%, respectively, while under heavy-load conditions, the corresponding improvements were 9.38% and 5.86%. Under light-load conditions, the optimized GRNN reduced total relative error by 39.6%, while under heavy-load conditions, it achieved a 61% reduction. This study confirms the rationality of the modeling method, identifies Configuration 1 as optimal, and determines the optimal range for clutch design values under light-load and heavy-load conditions, respectively.

Keywords: agricultural tractor; HMCVT; binary logic transmission unit; improved neural network algorithm; wet clutch optimization design

1. Introduction

Tractors, as common agricultural vehicles, can undertake various tasks by mounting different agricultural equipment, such as rotary tillage, plowing, ditching, sowing, harvesting, and transportation. These agricultural operations impose specific requirements on the power and transmission systems of tractors. For instance, low speeds are needed for tillage, medium speeds for sowing and harvesting, and higher speeds for transportation. Therefore, the power sources and transmission systems of agricultural tractors are critical components for achieving their functions. Currently, single-power-source agricultural tractors on the market commonly adopt transmission schemes such as PST (Power Shift Transmission), HST (Hydrostatic Transmission), and HMCVT. For example, John Deere's 8R series tractors use a 16–48 stage PST, while LCAG's 4H series employs HMCVT. The described transmission system exhibits distinct technical trade-offs: Hydrostatic Transmission (HST) suffers from suboptimal drivetrain efficiency inherent to hydraulic power transfer, while Power Shift Transmission (PST), as a discrete-ratio gearbox, fundamentally lacks continuous speed regulation capability. Although the HMCVT achieves superior efficiency and true stepless operation, its structural complexity imposes design constraints—unlike the modular gear-stage configurability of PST systems, the HMCVT's stage division options remain comparatively limited due to mechanical integration requirements. On the other hand, multi-power-source agricultural tractors often utilize hybrid power schemes. For instance, Dou Haishi's team at Beijing Institute of Technology [1] proposed a coupled-split dynamic configuration with an engine and dual motors as power sources to reduce the impact of load shocks on tractor power. By optimizing the torque distribution strategy, the engine torque fluctuation range was minimized. Zhang Kai's team at Nanjing Agricultural University [2] proposed an output-split electric continuously variable transmission (OS-ECVT) configuration based on planetary gears. By utilizing different power combinations for different operating conditions, they achieved power matching and improved economic efficiency. For the newly proposed hybrid powertrain solution, the coexistence of multiple power sources makes the coordination between them during power output the current research focus. Due to the complexity of this coordination, commercially available hybrid products remain rare. Furthermore, since planetary gear mechanisms typically serve as the power-coupling devices in hybrid systems, research on BLT-U can also advance hybrid powertrain development. From this perspective, the design and optimization of agricultural tractor transmission systems hold significant research importance.

HMCVT, as one of the more advanced transmission systems for agricultural tractors currently on the market, offers advantages such as strong load capacity, a wide speed regulation range, high transmission efficiency, continuous variable speed control, uninterrupted power during stage changes, and smooth power delivery. There has been some research progress on the configuration design and parameter optimization of HMCVT. For example, Macor Alarico [3] highlighted the need and necessity for optimizing the design of HMCVT. Sunghyun Ahn [4] proposed a comprehensive control algorithm for HMCVT systems, with a particular focus on transmission efficiency. This approach identified the optimal operating line for the system, improved overall efficiency, and developed a simulation model based on AME Sim-Simulink, ultimately achieving a 7.5% improvement in fuel economy. Zhun Cheng [5] introduced a geometric design method based on I-GA. The HMCVT design scheme proposed using this method effectively matches the required speed of tractors. Current research on HMCVT design emphasizes transmission economic performance, though the efficiency improvement remains limited due to inherent transmission characteristics. The study of stage change performance not only affects economic metrics but also significantly characterizes dynamic performance and ride comfort.

Binary logic transmission (BLT) is commonly used in military vehicles and heavy-duty commercial vehicles. This transmission system features high transmission efficiency, high power density, strong load-bearing capacity, and a compact structure, making it highly adaptable to the working conditions of agricultural tractors. However, due to its stepped characteristics, it has not yet been applied in the agricultural tractor sector. BLT was first detailed by Kerr [6], who explained its fundamental principles and explored its application value in pure electric vehicle transmission systems. Subsequently, the Ker-Train Research Institute was established, collaborating with the Toronto Transit Commission and the U.S.

Army's Tank-Automotive Research, Development, and Engineering Center (TARDEC) to provide BLT systems for buses [7] and military vehicles [8,9]. Additionally, the institute, along with Professors Ferguson and Morrison, conducted in-depth analyses of the efficiency of BLT systems based on experimental data, investigating the causes of shock during dynamic shifting and the impact of hydraulic system time delays on shift shock [10,11]. Apart from the Ker-Train Research Institute, Spain's SAPA Placencia company is another key research institution in this field. Since 2002, it has been investing in the research and development of BLT-related technologies and has successfully commercialized mature products [12]. Wang Yanzhong's team conducted exploratory research on BLT technology, proposing a "multi-state modular" concept that goes beyond "binary logic" [13]. The research on the BLT-U has been exclusively developed based on Kerr's innovative design incorporating double intermeshing planetary gear sets [14] and concentric V-groove clutches [15]. While the double intermeshing planetary gear sets demonstrate superior efficiency characteristics, they inherently suffer from eccentric rotation dynamics and potential lock-up risks [16]. The corresponding V-groove clutch design presents significant structural complexity, exhibiting fundamental architectural differences from conventional wet clutch systems. These specialized planetary gear mechanisms and clutch configurations impose stringent manufacturing requirements and substantial production costs for equipment suppliers, consequently limiting their practical adoption in other transmission system applications. Chang Yile's research [17] demonstrated that a binary logic transmission device composed of a single planetary gear mechanism can also function as a two-stage automatic transmission.

The core components for stage shifting in the aforementioned HMCVT and BLT systems are clutch devices. Therefore, the performance of wet clutches has a significant impact on the transmission system and even overall vehicle performance. Current research on wet clutches primarily focuses on thermal characteristics and control measurement studies. For example, Shin Seungin [18] proposed a thermal dynamics-based friction control modeling method for wet clutches with paper-based friction linings. This study thoroughly explored the maximum model uncertainty in sliding control clutch friction. The proposed model integrates the Coulomb friction coefficient and includes variables such as clamping force, friction speed, and friction surface temperature, with model validation conducted. Strobl Patrick [19] introduced a comprehensive measurement method for the friction behavior of wet clutches, which has broad applicability and provides in-depth insights into friction characteristics within the sliding speed range during brake shifting operations, including forced sliding operations up to static torque transfer. Existing research on wet clutches has predominantly focused on their intrinsic performance characteristics. However, as wet clutches serve as critical components for stage shifting and gear changing in many transmission systems, comprehensive investigation into their impact on overall system performance is equally essential.

Currently, research on HMCVT design often involves complex configurations. Expanding the transmission ratio range typically requires significant changes to the structure, and there is limited research on how configuration changes affect the system performance or on optimizing the wet clutch, a core component for stage shifting, based on the system performance. For binary logic transmission systems, the main challenges in experimental research and application lie in the high production and processing requirements for double intermeshing planetary gear sets and concentric V-groove clutches, as well as the associated high production costs.

To address the aforementioned issues, broaden the application scenarios of binary logic transmission systems, improve the performance of HMCVT, and optimize the parameters of wet clutches, this paper proposes a BLT-U applicable to a wide range of systems. This unit is applied to the HMCVT system to enhance the performance of the transmission system. Furthermore, based on the system's performance, the WOA-GRNN optimization algorithm is utilized to optimize the wet clutch within the BLT-U. In addition, this paper utilizes Simulation X to build a simulation model and combines it with a wet clutch test bench to validate the actual oil pressure in the simulation model. The research presented in this paper provides valuable references for the applications of BLT-Us and the optimization of wet clutches within them.

2. Materials and Methods

The specific research process of this paper mainly includes three parts: modeling method validation, BLT-U configuration comparison, and wet clutch optimization design. The detailed workflow is shown in the Figure 1. The simulation models in this study were established using Simulation X software (Version 3.8), while data analysis and algorithm optimization were performed using MATLAB software (R2019b).

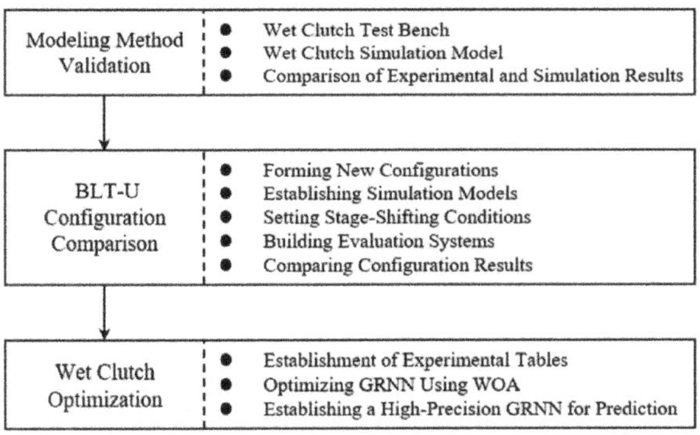

Figure 1. Workflow of this research.

2.1. Modeling Method Validation

2.1.1. Wet Clutch Test Bench

The primary working principle of a wet clutch involves using the oil pressure in the hydraulic cylinder to compress the friction plates and steel plates within the wet clutch, eliminating the gap and causing them to engage. This process gradually transitions from sliding friction to static friction, thereby transmitting torque, speed, and power from one end of the wet clutch to the other. Therefore, the oil pressure curve of the wet clutch has a significant impact on its performance. Due to the numerous factors influencing the oil pressure curve, there are still some discrepancies in the descriptions of wet clutch oil pressure curve variations in the research of different scholars [20,21]. To avoid the influence of the oil pressure curve generated by the simulation program on the experimental results, this paper directly adopts the measured oil pressure curve of the wet clutch as the loading oil pressure for the simulation model.

To validate the feasibility of the modeling method and oil pressure loading approach adopted in this paper, the wet clutch engagement test bench shown in the Figure 2 was used:

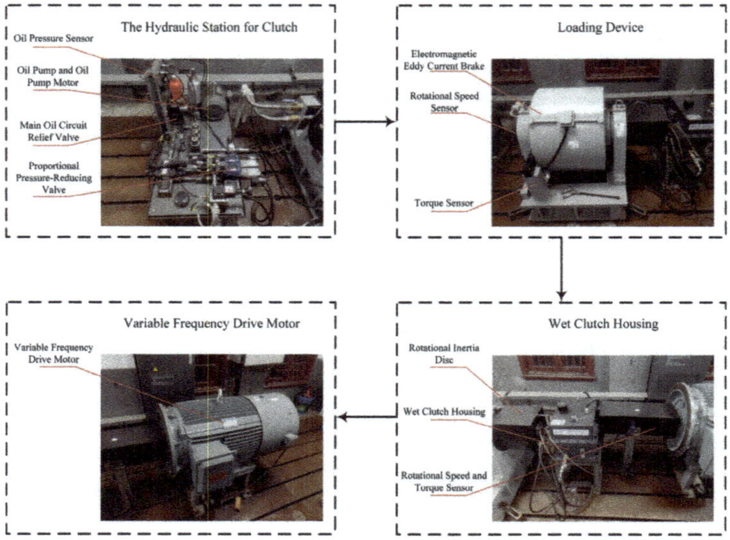

Figure 2. Wet clutch engagement test bench.

The variable frequency drive motor used in this study was the YXVF315L2-4, which was selected from Delixi Electric Co., Ltd. and procured from Hangzhou, China, and the selected electromagnetic eddy current brake was the CW2000B, which was selected from Jiangsu Lanmec Technology Co., Ltd. and procured from Nantong, China. This test bench is equipped with two wet clutches, but only one clutch was utilized for the engagement tests in this paper.

For the aforementioned test bench, the following two sets of test parameters were selected to align with the low-speed, high-load operating conditions of agricultural tractors, as shown in Table 1:

Table 1. Test parameters for the wet clutch engagement test bench.

Test Group	Input Speed (rpm)	Load Torque (N·m)	Oil Pressure (bar)	Flow Rate (r)
1	800	800	25	4
2	500	800	30	4

Note: the flow rate adjustment on the test bench in this paper is controlled by a knob, where "r" represents the number of rotations.

2.1.2. Wet Clutch Simulation Model

The simulation model established based on the working principles of the aforementioned test bench is shown in the Figure 3:

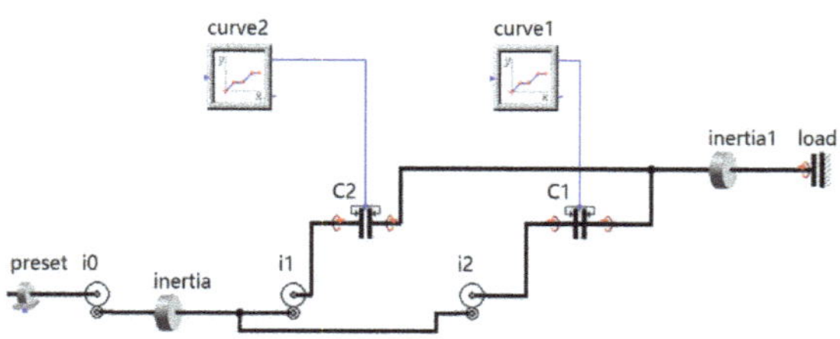

Figure 3. Simulation model of the wet clutch engagement test bench.

2.2. BLT-U Configuration Comparison

2.2.1. Basic Parameters of the Tractor and HMCVT Used in This Study

The transmission system proposed in this study is designed for tractors with power sources of 90 horsepower (hp) and above. The basic parameters of the agricultural tractor adopted in this paper are shown in the Table 2 [22]:

Table 2. Basic parameters of the agricultural tractor.

Parameters		Value
Vehicle dimension (mm)	Length	4600
	Width	1985
	Height	2764
Vehicle cross-sectional area (m^2)		6.24
Coefficient of air resistance		0.6
Complete vehicle kerb mass (kg)		3640
Tire specification	Driving wheel	13.6~24
Engine speed (rpm)		800~2200
Final driver ratio		24
Tire moment of inertia ($kg{\cdot}m^2$)		4
Differential shaft moment of inertia ($kg{\cdot}m^2$)	Length	0.01

The original HMCVT configuration adopted in this paper is shown in the Figure 4.

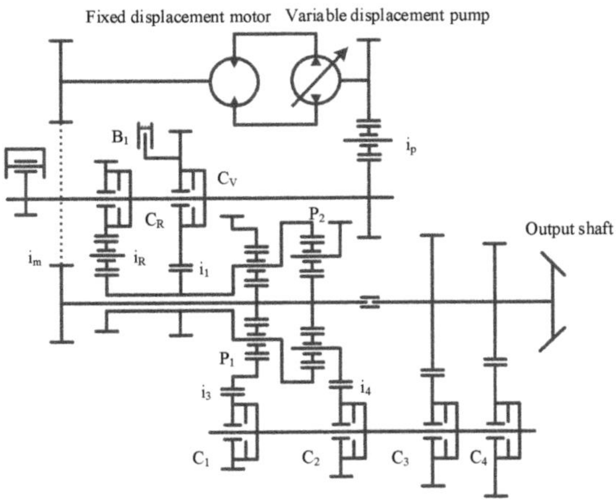

Figure 4. Five-stage HMCVT adopted in this paper.

The HMCVT adopted in this paper consists of five forward stages: one pure hydraulic working stage, H0, and four hydraulic–mechanical power flow coupling working stages, H1 to H4. In the H0 stage, brake B1 locks the planet carrier of planetary gear P1. The engine's output is transmitted through the hydraulic transmission part and then input to the sun gear of planetary gear P1. The power is then output through the gear ring of planetary gear P1 and a pair of gear sets. In the H1 to H4 stages, the wet clutch CV engages, allowing the engine's power to be transmitted through a gear set to the planet carrier of P1 and the ring gear of P2. Another part of the engine's power is input to the variable displacement pump and then transmitted through a fixed displacement motor to the sun gears of P1 and P2. The power is then output through the engagement and disengagement of wet clutches C1 or C2 and C3 or C4. Additionally, this HMCVT has five reverse stages with the same working principle as the forward stages; CV disengages and CR engages. The engagement status of clutches and brakes during operation is shown in Table 3:

Table 3. Engagement status of clutches and brakes.

Stage	Brakes and Wet Clutches					
	C_V	B_1	C_1	C_2	C_3	C_4
H_0	×	•	•	×	•	×
HM_1	•	×	×	•	•	×
HM_2	•	×	•	×	•	×
HM_3	•	×	×	•	×	•
HM_4	•	×	•	×	×	•

Note: • represents that the wet clutch (or brake) is working, × represents that the wet clutch (or brake) is out of working.

The transmission ratio variation characteristics of the HMCVT adopted in this paper are shown in the Figure 5.

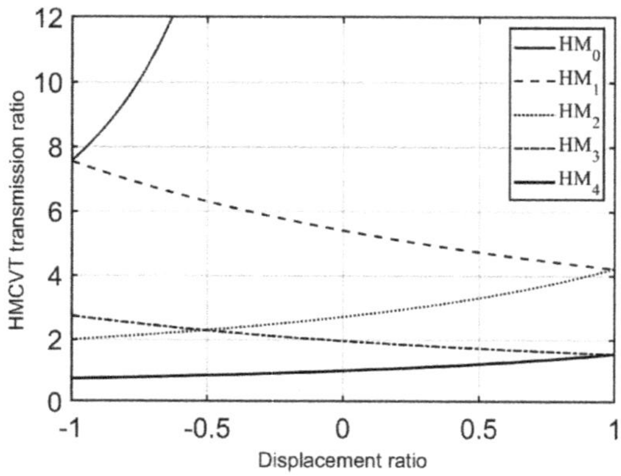

Figure 5. Transmission ratio variation characteristics of HMCVT.

This study proposes replacing clutches C_3 and C_4 in Figure 4 with a BLT-U, followed by stage-shifting simulation tests and shift quality evaluation. Therefore, the selected stage-shifting conditions should align with the actuation requirements of the BLT-U's brake-switching clutches. From Table 3, the applicable stage transition is HM_2 to HM_3, where from Figure 5 the displacement ratio during shifting is -0.511, and the transmission ratio change is 2.29. Combining these with the parameters in Table 2, the vehicle speed range during shifting should be 3.57–9.8 km/h. Subsequent operating conditions must be selected within this range.

2.2.2. Forming BLT-U Configurations

Currently, binary logic transmission devices often use double intermeshing planetary gear as the planetary gear set. Double intermeshing planetary gears differ from traditional planetary gear structures in that the planetary gears are eccentrically arranged. When designing the planet carrier, certain space must be reserved for adding or removing weight to achieve dynamic balance for the planet carrier and planetary gears as a whole. This increases the difficulty and cost of manufacturing. Although the dual internal meshing planetary gear set has advantages such as high efficiency and compact size, its structure results in significant meshing forces at the engagement points between the planetary gears and the central gears. This leads to efficiency reduction and potential self-locking issues under high transmission ratios [16].

This paper selects the NGW planetary gear mechanism as the planetary gear set, where the ring gear and planetary gears are internally meshed, and the planetary gears and sun gear are externally meshed. This mechanism transmits different gear ratios when different components are locked. The specific details are shown in the Table 4:

Table 4. Transmission ratios of NGW under different locking conditions.

Scheme	Input Part	Output Part	Locked Part	Transmission Ratio
1	Sun Gear	Planet Carrier	Ring Gear	$1+k$
2	Sun Gear	Ring Gear	Planet Carrier	$-k$
3	Ring Gear	Sun Gear	Planet Carrier	$-1/k$
4	Ring Gear	Planet Carrier	Sun Gear	$(1+k)/k$
5	Planet Carrier	Sun Gear	Ring Gear	$1/(1+k)$
6	Planet Carrier	Ring Gear	Sun Gear	$k/(1+k)$
7	Any two components engaged			1
8	Neither locked nor engaged			Free

Note: k is the characteristic parameter of the planetary gear mechanism, defined as the ratio of the number of ring gear teeth to the number of sun gear teeth.

Considering the intervals between transmission ratios and the requirements for working conditions, this paper selected the sun gear as the input, the planet carrier as the output, and the locked ring gear as State 1. Engaging any two components was defined as State 2, resulting in the three configurations shown in the Figure 6:

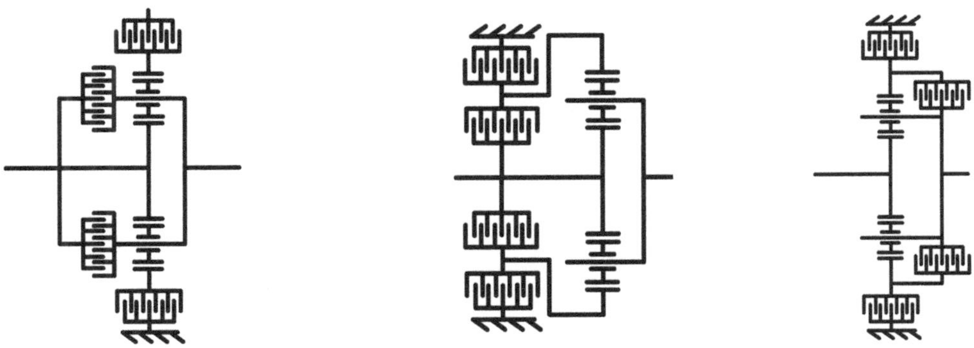

(a) (b) (c)

Figure 6. Configuration diagram of the BLT-U. (**a**) Configuration 1; (**b**) Configuration 2; (**c**) Configuration 3.

2.2.3. Establishment of the Simulation Model

Based on the working principles of the HMCVT described above, the three configurations were integrated with the HMCVT without affecting its stage-shifting functionality, resulting in new configuration schemes. The new HMCVT schemes formed after replacement are shown in the Figure 7:

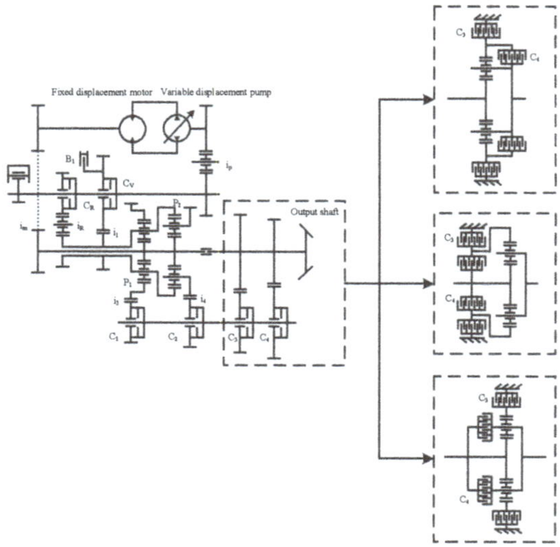

Figure 7. New HMCVT schemes integrated with BLT-U configurations.

Based on the above schemes, Simulation X was used to build simulation models for the original scheme and the three new configuration schemes. The pump–motor system in the simulation model is shown in the Figure 8:

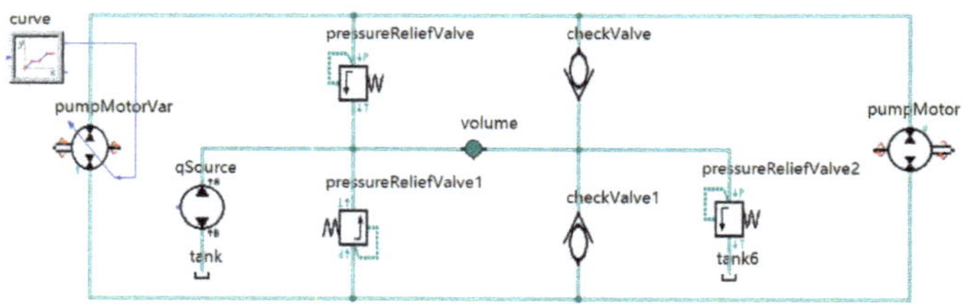

Figure 8. Pump–motor system in the simulation model.

The oil pressure simulation system for the constant-pressure clutch and pressure-release clutch under simulation conditions in the simulation model is shown in the Figure 9:

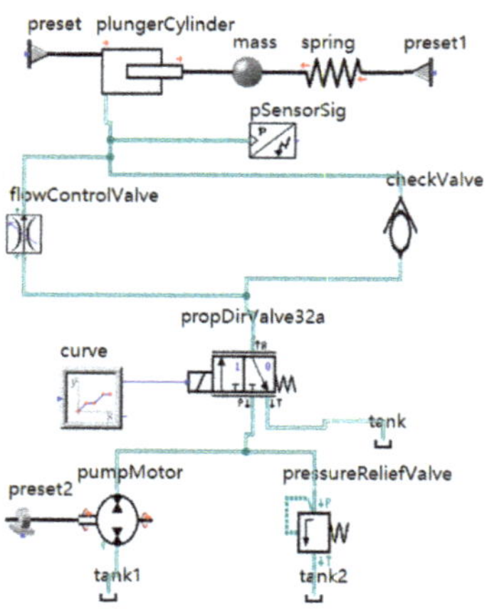

Figure 9. Oil pressure simulation system in the simulation model.

The mechanical transmission part of the simulation model is shown in the Figure 10:

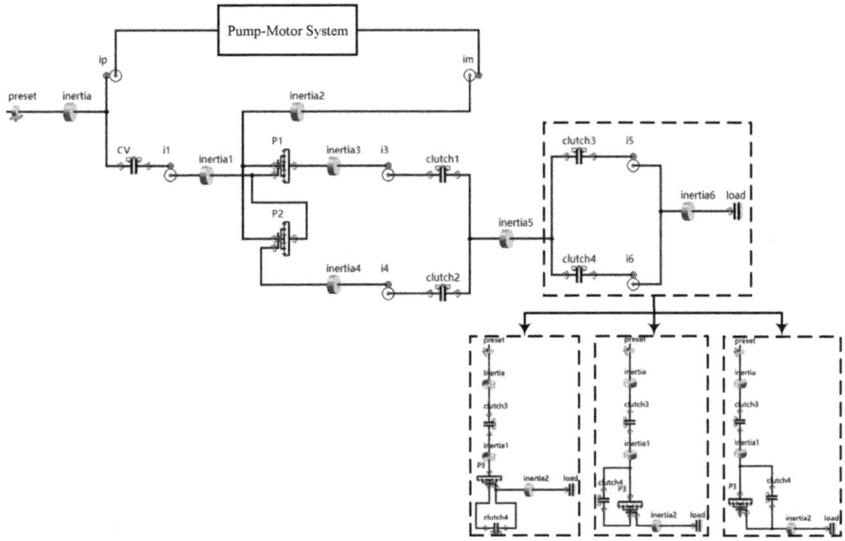

Figure 10. Mechanical transmission part of the simulation model.

To compare the performance with the original HMCVT configuration, it is necessary to switch between C3 and C4. Therefore, the stage-shifting condition selected in this paper was shifted from the HM2 stage to the HM3 stage. The simulation was set to perform the stage shift at the 2 s mark, during which the wet clutches C2 and C4 need to engage. Since the other wet clutches were either constantly engaged or disengaged, and their impact on the overall system performance during stage shifting was minimal, the simulated oil pressure curve was used for engagement.

2.2.4. Selection of Influencing Factors and Establishment of the Evaluation System

Agricultural tractors are required to withstand heavy loads and frequently shift stages to meet power demands during operation. During stage shifting, issues such as jerking, impact, and power interruption leading to stalling often occur. Therefore, research on the shifting quality of tractors primarily focuses on these two aspects.

The stage shifting of HMCVT is achieved through the engagement and disengagement of wet clutches. Shifting quality refers to performing smooth and rapid stage shifts while ensuring vehicle driving power. During this process, the oil pressure chamber of the engaging clutch gradually fills with oil pressure, pushing the piston to perform axial movement. The piston gradually compresses the steel plates and friction plates connected to the driving end and driven end, respectively. As the gap between the steel plates and friction plates gradually decreases until they are fully in contact and rotate at the same speed, the driving end and driven end transmit torque T through the static friction of the steel plates and friction plates. The calculation formula is:

$$T = \mu_1 Q R_e Z K \quad (1)$$

where μ_1 is the friction coefficient of the friction plates, Q is the total clamping force provided by the oil pressure, R_e is the equivalent friction radius of the friction plates, Z is the number of friction pairs, and K is the compression force loss coefficient.

To ensure that the clutch can reliably transmit torque, the outer diameter D of the friction plates must satisfy the following formula:

$$D = \sqrt[3]{\frac{16\alpha T_{e\,max}}{\pi\mu ZKq(1-c^2)(1+c)}} \tag{2}$$

where α is the reserve coefficient, c is the ratio of the inner to outer radius of the clutch, and q is the clamping oil pressure of the clutch.

Additionally, the outer diameter of the clutch must satisfy the following constraints:

$$D \le \frac{v_1 \times 60 \times 10^3}{\pi \times n_{emax}} \tag{3}$$

where v_1 represents the maximum permissible circumferential speed, and n_{emax} denotes the maximum rotational speed of the friction plates.

Based on the above formula and adjustments in the simulation software, this paper selected D, c, Z, and q as the influencing factors.

The aforementioned influencing factors have different impacts on the performance of wet clutches. An excessively large outer diameter of the friction plates can cause significant centrifugal hydraulic pressure, leading to slower disengagement, while an excessively small outer diameter may result in insufficient friction to meet load requirements. Based on Equations (2) and (3) above, this study calculated the minimum non-slipping outer diameter for the original configuration under 20 bar oil pressure. Incorporating the clutch outer diameter constraints (Equation (4)) and the referenced literature, the permissible range for D was 280 mm to 330 mm [23]. The ratio of the inner to outer radius of the friction plates is an important parameter in friction plate design, affecting the service life of the friction plates and the overall performance of the clutch. A smaller c value results in a larger gap between the inner and outer radii. With a constant outer diameter, a smaller inner diameter is unfavorable for the overall design of the clutch and accelerates the wear and damage of the friction plates. If the inner diameter remains constant and c is too small, the equivalent friction radius of the friction plates decreases, reducing the torque transmission capacity. Conversely, if c is too large, the effective pressure area of the friction plates decreases, resulting in higher pressure that may exceed the allowable limit. The reasonable ratio of the inner to outer radius of the clutch friction plates is 0.4 to 0.8 [24,25]. Under the condition of meeting torque transmission requirements, the number of friction plates and steel plates should be minimized. A smaller number of plates results in a lighter clutch, higher transmission efficiency, faster cooling, and less wear. Additionally, fewer friction plates facilitate easier disengagement of the clutch. The reasonable number of clutch friction pairs Z is 5 to 14. The value of q should not be too large, as an excessively large q will cause the friction plates to generate heat that cannot be dissipated in time, leading to rapid temperature rise and damage to the friction plates. If q is too small, the clutch size will need to increase to meet operational requirements. The reasonable range for clutch clamping oil pressure q is 20 to 50 bar [20,21].

The HMCVT studied in this paper can effectively address the power interruption issue caused by stage shifting during operation. Therefore, the primary optimization goals are to resolve jerking and impact during stage shifting. The following evaluation metrics were selected: speed drop, dynamic load, sliding time, and maximum sliding power of the shifting clutches (C2, C4).

The speed drop refers to the fluctuation range of the transmission output shaft speed after the stage-shifting command is issued. The calculation formula is as follows:

$$\beta_1 = |\omega_0 - \omega_{min}| \tag{4}$$

where ω_0 represents the transmission output shaft speed under stable conditions, and ω_{min} represents the minimum transmission output shaft speed during the stage-shifting process.

The dynamic load refers to the impact generated during the stage-shifting process when the clutch transitions from sliding friction to static friction and achieves speed synchronization in a short time. The calculation formula is shown in Equation (12).

The sliding time is the duration from when the stage-shifting command is issued until the transmission output shaft speed drops and recovers to 95% of its stable speed value. The maximum sliding power is the maximum instantaneous sliding power of the clutch from the start to the end of the stage-shifting process.

Among the above evaluation metrics, the speed drop and sliding time reflect the jerking and its duration during stage shifting, while the dynamic load reflects the impact generated during stage shifting. Minimizing the sliding time and maximum sliding power helps avoid thermal failure caused by excessive friction in the clutch.

To obtain a comprehensive optimization result, this paper used the Analytic Hierarchy Process (AHP) and variance analysis to establish a comprehensive evaluation index. For the two distinct load conditions, domain experts provided differentiated AHP judgment matrices based on simulation data analysis and empirical knowledge: under light-load operations, shift quality was prioritized, whereas under heavy-load conditions, system reliability—particularly clutch engagement performance—dominated the criteria. The judgment matrix provided by experts in the AHP is shown in the Table 5:

Table 5. AHP judgment matrix under light-load conditions.

	Metric 1	Metric 2	Metric 3	Metric 4
Metric 1	1	2	1	4
Metric 2	1/2	1	1	2
Metric 3	1	1	1	3
Metric 4	1/4	1/2	1/3	1

Note: Metric 1 represents speed drop, Metric 2 represents sliding time, Metric 3 represents dynamic load, and Metric 4 represents maximum sliding power. The consistency ratio (CR) of the above judgment matrix is 0.0172, which is acceptable. The AHP weights obtained are [0.3774, 0.2254, 0.2968, 0.1005].

The consistency ratio (CR) of the Table 6 was 0.0172, which is acceptable. The AHP weights obtained were [0.1484, 0.1945, 0.2312, 0.4258]. Variance analysis involved normalizing all simulated values of Metric 1, Metric 2, Metric 3, and Metric 4 based on their respective maximum and minimum values, followed by single-factor variance analysis. The weights were then assigned based on the variance analysis results.

Table 6. AHP judgment matrix for heavy-load conditions.

	Metric 1	Metric 2	Metric 3	Metric 4
Metric 1	1	1	1/2	1/3
Metric 2	1	1	1	1/2
Metric 3	2	1	1	1/2
Metric 4	3	2	2	1

2.2.5. Establishment of Simulation Conditions and Test Tables

Based on the above, the stage-shifting transmission ratio from HM_2 to HM_3 was calculated as 2.29. At this ratio, the vehicle speed ranged from 3.57 km/h to 9.8 km/h, which aligns with the following:

(1) Light-load operations: rotary tillage (selected: 5 km/h).

(2) Heavy-load operations: plowing (selected: 8 km/h).

These conditions were adopted as the working scenarios for HMCVT stage-shifting simulation tests [26].

The driving resistance of the agricultural tractor in this paper is simplified as the sum of rolling resistance and tillage resistance. The calculation formula is as follows:

$$F_t = F_f + F_T \tag{5}$$

$$F_f = \mu_2 \times mg \tag{6}$$

where F_t is the driving resistance, F_f is the rolling resistance, F_T is the tillage resistance, which includes rotary tillage resistance F_{Ro} and plowing resistance F_{Pl}, μ_2 is the rolling resistance coefficient of the farmland, with a value of 0.1 [27], and m is the total weight of the tillage implement weight and the agricultural tractor.

For light-load conditions, the rotary tillage resistance is calculated as follows [28]:

$$F_{Ro} = 10k_\lambda \cdot k_j \cdot B_{Ro} \cdot h_{Ro} \tag{7}$$

where k_λ is the soil-specific correction factor for rotary tillage resistance, k_j is the constant term of rotary tillage resistance, B_{Ro} is the working width of the rotary tiller, and h_{Ro} is the rotary tilling depth.

Based on the given formula, the required driving force of the vehicle was calculated to be 12,886.52 N at a speed of 5 km/h. The corresponding engine power demand was 23.52 hp.

The input speed of the HMCVT and the required output torque were ultimately calculated using the following equations:

$$n_e = \frac{u_a \times i_g \times i_0}{0.377 r_d} \tag{8}$$

$$T_{tq} = \frac{F_t}{i_0} \tag{9}$$

where n_e represents the input speed of the HMCVT, u_a represents the vehicle speed under tillage conditions, i_g represents the stage-shifting transmission ratio of the HMCVT, and T_{tq} represents the required output torque of the HMCVT.

According to Table 2 and Equations (2) and (3), the driving wheel radius was 0.65 m. The following parameters were derived: transmission input speed was 1099.5 rpm, and the transmission output torque was 349.01 N·m.

For the measurement of plowing resistance under heavy-load conditions in tillage operations, this paper employed the test bench for data acquisition shown in the Figure 11:

Figure 11. Plowing resistance test bench.

In the above test bench, a speed of 8 km/h, a plowing depth of 50 mm, and a one-furrow plow were selected as the plowing conditions (due to the limitations of the test bench and field conditions). The plowing resistance obtained using this plowing resistance test bench is shown in the Figure 12 (the test bench experimental data were collected in July 2023):

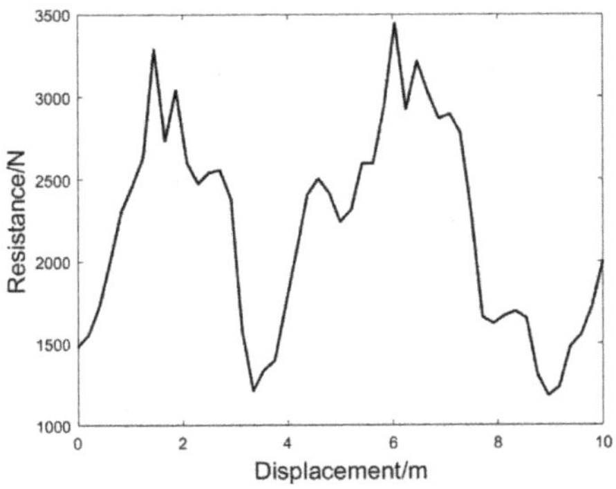

Figure 12. Plowing resistance.

Under these conditions, the calculated average plowing resistance is 2242.19 N. The empirical formula for calculating plowing resistance is as follows [2]:

$$F_{Pl} = Z_{Pl} \cdot B_{Pl} \cdot h_{Pl} \cdot k_0 \tag{10}$$

where k_0 is the specific soil resistance, Z_{Pl} is the number of plowshares, B_{Pl} is the working width of plowing, and h_{Pl} is the plowing depth.

According to the above equation, the plowing resistance is positively correlated with both working depth and the number of plowshares. This study selected a heavy-duty plowing condition with a 20 cm working depth, three-bottom plow, and a wheel speed of 8 km/h. Based on Equations (5)–(7), the required tractive force was calculated as 30,826.35 N at 8 km/h, requiring an engine power output of 91.86 HP. Field research on plowing resistance indicates that under a 20 cm working depth, the resistance for 3–5 bottom plows typically ranges between 16,000 N and 30,000 N [2,29,30]. The calculated plowing resistance in this study (26,906.35 N) falls within this expected range.

According to Table 2, Equations (2) and (3), the driving wheel radius is 0.65 m. The following parameters were derived: the transmission input speed was 1794.2 rpm, and the transmission output torque was 834.88 N·m.

This paper uses the Latin Hypercube Sampling (LHS) method for experiments. Latin Hypercube Sampling is an improved multidimensional stratified Monte Carlo sampling method. Compared to traditional Monte Carlo random sampling, LHS provides more uniform and extensive coverage of the sampling space with fewer samples, effectively utilizing sampled values to describe the distribution function of random variables.

The LHS method for sampling random variables involves dividing the range of the variable into x non-overlapping intervals. Then, for each of these intervals, a random value is sampled within the interval. Finally, the sampled values are shuffled to form the test groups. In this paper, x was selected as 10. Since q represents the actual oil pressure curve and cannot obtain so many groups within the value range, under light-load conditions, hydraulic pressure adjustment exhibited minimal impact on clutch failure. This allowed for the investigation of shift performance across varying pressures. Therefore, this study selected 20 bar, 25 bar, and 30 bar as the test pressures for light-load operation. For heavy-load conditions, pressure variation may induce clutch slippage, potentially leading to transmission system failure. To prevent this, a higher pressure of 50 bar was adopted for heavy-load testing. The experimental table obtained by applying the above sampling method to the remaining parameters is shown in the Table 7:

Table 7. Latin Hypercube experimental table.

Test Group	D (mm)	c	Z
1	307.3	0.511	14
2	328.5	0.786	13
3	296.3	0.407	10
4	304.1	0.639	12
5	291.6	0.718	11
6	284.1	0.753	5
7	323.2	0.454	8
8	316.8	0.571	6
9	311.1	0.534	7
10	289.6	0.662	9

Note: The Z value in the table represents the number of friction pairs, which must be an integer; it is obtained by rounding down the LHS sampled values.

2.3. *Wet Clutch Optimization*

2.3.1. Establishment of Experimental Tables

To determine the optimal design values for the clutch under this HMCVT system, the oil pressure and configuration were fixed at 25 bar and Configuration 1, respectively. Since the value of Z was fixed during the configuration comparison, under the fixed oil pressure and configuration, LHS with $x = 10$ was performed again for $Z = 5, 7, 9, 11, 13$, and D and c were resampled five times, resulting in five experimental groups. Simulation tests were conducted for these five groups, and the results were incorporated into the comprehensive evaluation system to obtain the final comprehensive evaluation scores.

After obtaining the comprehensive evaluation scores for the above 60 groups (including the experimental groups in Table 6), the 60 experimental datasets (the experimental data were obtained under light-load conditions) were randomly shuffled, with the first 48 sets allocated to the training set and the remaining 12 to the test set. Three neural network models were each trained 10 times under the following configurations: GRNN (Generalized Regression Neural Network): spread = 1; RBF (Radial Basis Function): spread = 1, 5 neurons, Gaussian basis function; BP (Backpropagation): 10 hidden-layer neurons, 1000 epochs, learning rate = 0.01. The total relative error (i.e., the mean relative error between predicted values and the training dataset) was calculated for each iteration, with results presented in the Table 8.

Table 8. The total relative error of the model.

	1	2	3	4	5	6	7	8	9	10
GRNN	0.916	1.209	0.918	1.121	1.030	1.579	0.859	1.120	0.990	0.806
RBF	2.542	1.510	0.568	2.229	2.180	1.106	1.292	0.605	0.758	1.753
BP	1.896	0.066	1.993	1.592	0.904	2.623	1.436	2.313	1.632	0.777

Note: The same serial number does not indicate identical experimental groups, as all 30 result sets were randomly generated.

Based on the experimental results presented in the table above, the average total relative errors for GRNN, RBF, and BP networks were calculated as 1.055%, 1.454%, and 1.523%, respectively. Given that GRNN requires fewer adjustable parameters and demonstrates faster computational performance, this study selected the GRNN for subsequent optimization and modeling.

2.3.2. Optimizing GRNN Using WOA

GRNN is a variant of the Radial Basis Function (RBF) neural network. GRNN is based on non-parametric regression, using sample data for testing and performing Parzen non-parametric estimation. It calculates the output according to the principle of maximum probability. This algorithm has excellent nonlinear approximation capabilities and, compared to RBF, is more convenient for training, making it particularly suitable for solving curve-fitting problems [31].

The network structure of GRNN is similar to that of RBF, consisting of four layers: the input layer, pattern layer, summation layer, and output layer. The corresponding network input is $X = [x_1, x_2, \cdots x_n]^T$, and the output is $Y = [y_1, y_2, \cdots y_k]^T$. The specific structure is shown in the Figure 13:

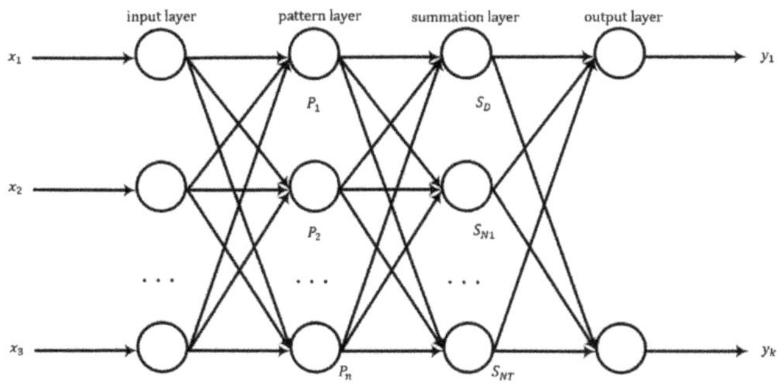

Figure 13. Network structure of GRNN.

In the process of building the GRNN, the only parameter that affects the accuracy of the model is the spread value of the Radial Basis Function. If this value is too large, the approximation process will be smoother, but the error will also increase accordingly. Conversely, if the value is too small, the function approximation will be more accurate, but the process will be less smooth, and it may easily lead to overfitting issues.

This paper used the Whale Optimization Algorithm (WOA) to optimize the spread value, thereby improving the accuracy of the established GRNN and ensuring the reliability of the prediction results.

The WOA is a heuristic optimization algorithm that simulates the hunting behavior of humpback whales. Its main features include using random or optimal search agents to simulate hunting behavior and mimicking the bubble-net feeding mechanism of humpback whales. This algorithm has the advantages of simple mechanisms, few parameters, and strong optimization capabilities. Its basic principle originates from the unique hunting method of humpback whales, known as bubble-net feeding, which mainly includes three stages: encircling prey, bubble-net attacking, and searching for prey [32].

Combining the above algorithm principles, the flowchart for optimizing the GRNN's spread value using the WOA is shown in the Figure 14. (During the supplementary heavy-duty testing, 2 out of the 50 additional test groups were discarded due to clutch slippage-induced data loss, resulting in 58 valid test groups under high-load conditions. The dataset was partitioned into a 46-group training set and a 12-group validation set for subsequent analysis.)

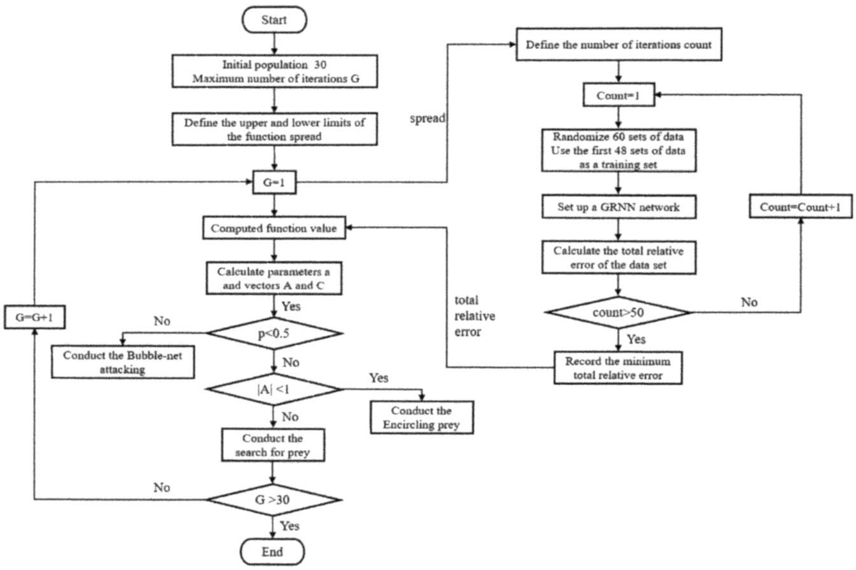

Figure 14. Flowchart of WOA optimizing GRNN spread value.

2.3.3. Establishing a High-Precision GRNN for Prediction

After obtaining the optimal spread value, the optimized GRNN was established using the program shown in the Figure 15, and the NET of the GRNN at the minimum total relative error was recorded (for the heavy-load conditions, a total of 58 experimental datasets were obtained, with 46 datasets allocated to the test set and 12 datasets reserved for validation purposes).

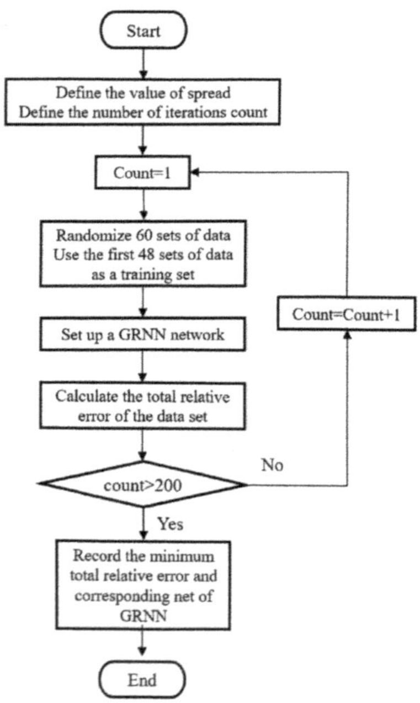

Figure 15. Flowchart for establishing GRNN.

3. Results

3.1. Modeling Method Validation

3.1.1. The Results of Wet Clutch Test Bench

Based on the test parameters in Table 1, the test bench was adjusted, and the maximum output torques obtained were 1480 N·m and 1570.5 N·m, respectively. The corresponding oil pressure curves and speed curves are shown in the Figure 16 (the test bench experimental data were collected in December 2024):

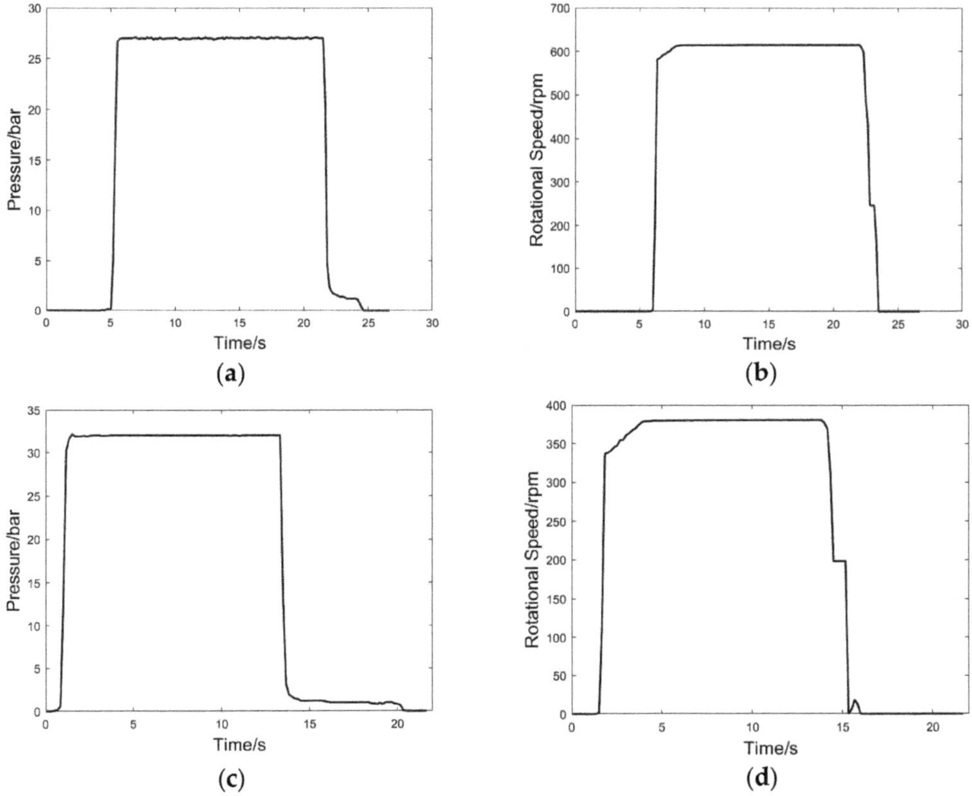

Figure 16. Oil pressure curves and output speed curves obtained from the test bench. (**a**) Oil pressure curve for test group 1; (**b**) output speed curve for test group 1; (**c**) oil pressure curve for test group 2; (**d**) output speed curve for test group 2.

3.1.2. Comparison of Experimental and Simulation Results

By inputting the test parameters from Table 1 into the simulation model and loading the experimentally obtained oil pressure into Curve 1 of the model, this paper only utilized Clutch 1 for the engagement tests. The maximum torque values obtained from the simulation results are 1335.56 N·m and 1548.479 N·m, respectively. The simulated speed curves were compared with the experimentally obtained speed curves, and the results are shown in the Figure 17:

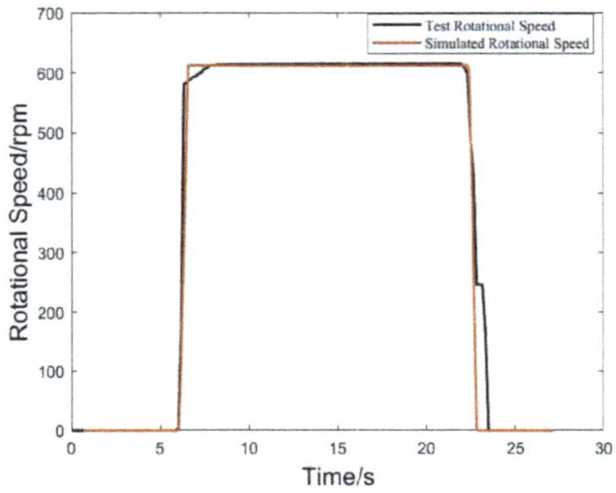

(a)

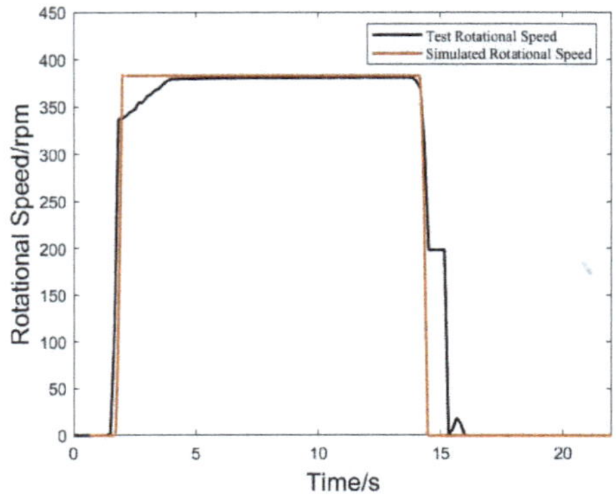

(b)

Figure 17. Comparison of experimental and simulated output speed curves. (**a**) Comparison of output speed curves for test group 1; (**b**) comparison of output speed curves for test group 2.

A correlation analysis was conducted between the experimental and simulation results for the two test groups. To avoid affecting the analysis results, the parts of the curves where the speed was zero were removed. This paper uses the Pearson correlation coefficient to characterize the correlation between the two curves. For the torque analysis, the maximum torque was converted into dynamic load, which is one of the evaluation indicators mentioned below, for relative error analysis. The formulas for the Pearson correlation coefficient and dynamic load are as follows:

$$r(X,Y) = \frac{Cov(X,Y)}{\sqrt{Var[X]Var[Y]}} \quad (11)$$

$$\beta_2 = \frac{T_{max}}{T_{sto}} \quad (12)$$

where T_{max} represents the maximum output torque, and T_{sto} represents the stable output torque.

The Pearson correlation coefficients for the speed curves and the relative errors of the dynamic loads for the two test groups are shown in the Table 9:

Table 9. Comparison of results.

Test Group	Correlation Coefficient	Dynamic Load		Relative Error
		Experimental	Simulated	
1	0.9519	1.85	1.67	9.76
2	0.8860	1.96	1.94	1.4

The average output speed correlation coefficient is 0.92, and the average relative error of the dynamic load is 5.58%, indicating that the modeling method and oil pressure loading approach have a small error range and meet the requirements.

3.2. BLT-U Configuration Comparison

3.2.1. Analysis of Raw Simulation Experiment Data

For light-load operations, according to the above experimental table, simulation tests were conducted for the original configuration, Configuration 1, Configuration 2, and Configuration 3 at clutch clamping oil pressures of 20 bar, 25 bar, and 30 bar (i.e., 12 test groups). Each group was tested according to the LHS experimental table. The results of these tests, totaling 120 sets, were converted into corresponding evaluation metrics. It was found that the average fluctuation range (i.e., the fluctuation range/average value) of the maximum slip power of Clutch 2 across the 12 test groups under various oil pressures and configurations was 1.22%. Therefore, this evaluation metric was excluded. The experimental results of 10 sets for four configurations under oil pressures of 20 bar, 25 bar, and 30 bar are shown in Figures 18, 19, and 20, respectively. In the figure, all four evaluation metrics indicate that lower values correspond to better performance. Values exceeding the upper limit in the figure indicate that C_4 slipped during the simulation experiment under the corresponding oil pressure and experimental group.

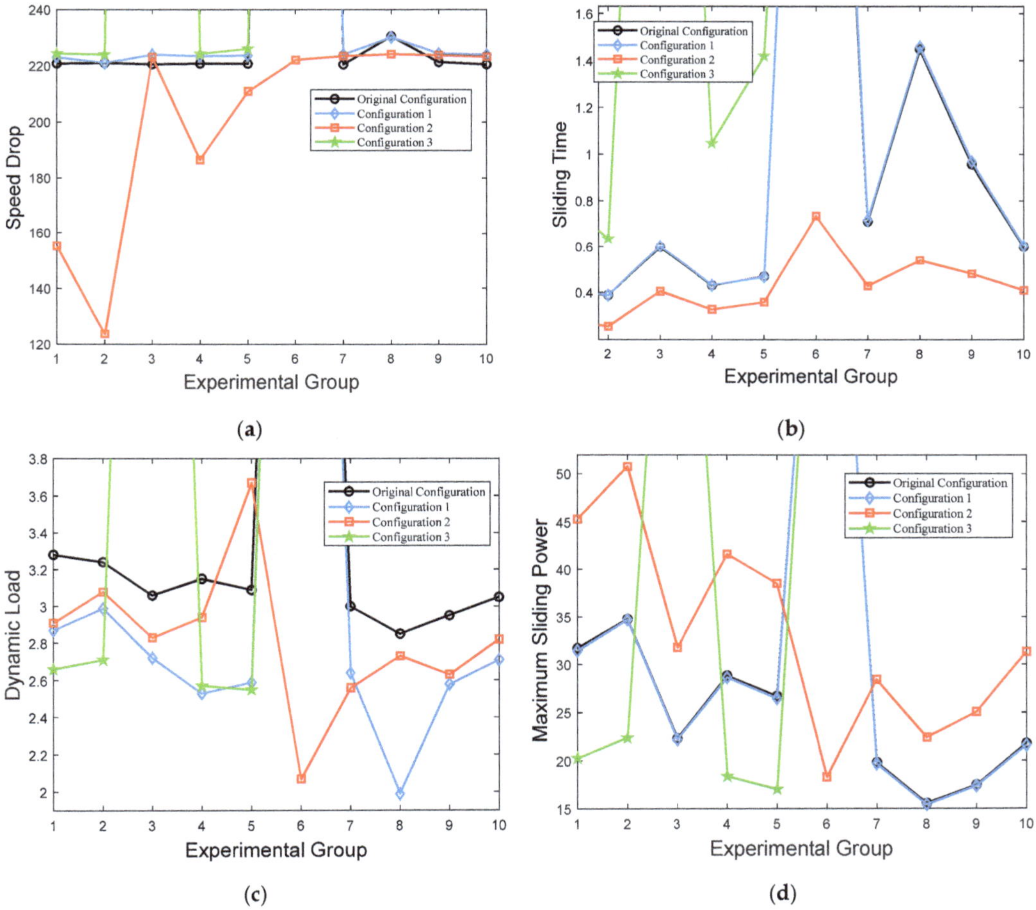

Figure 18. Simulation experimental data table for 20 bar. (**a**) Speed drop; (**b**) sliding time; (**c**) dynamic load; (**d**) maximum sliding power of Clutch 4.

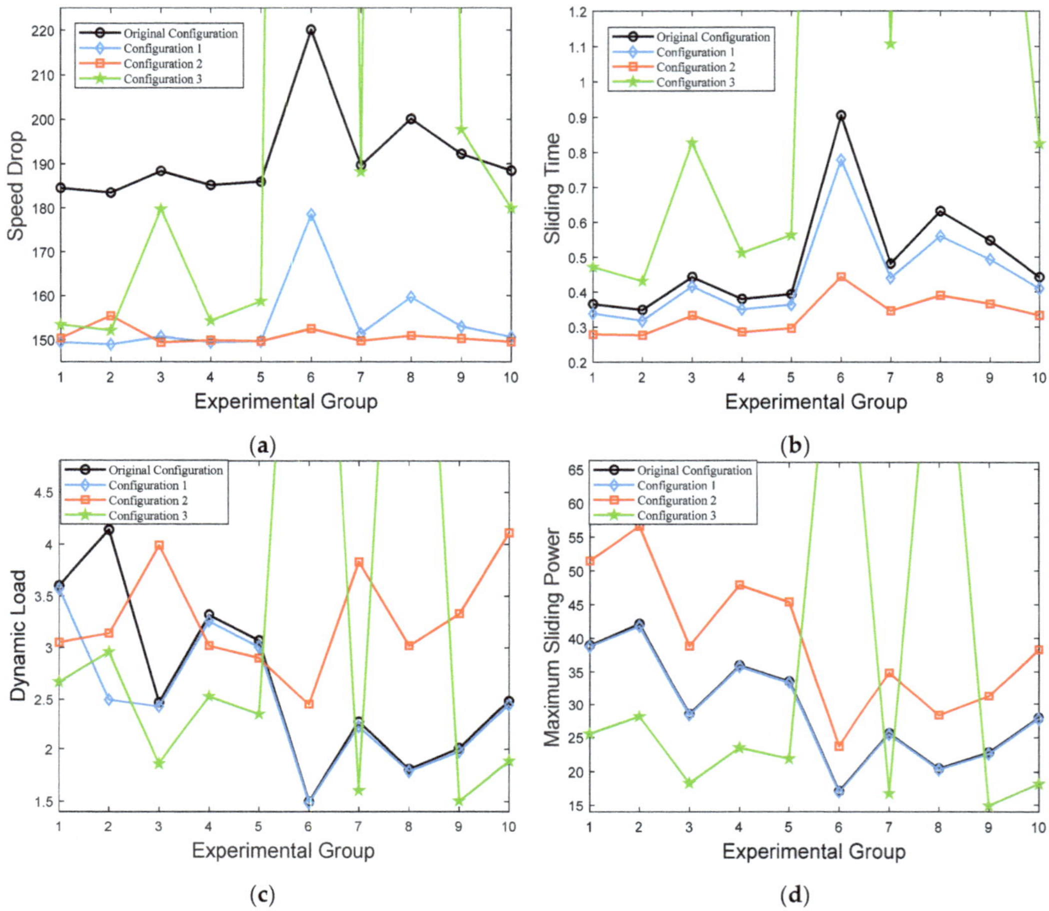

Figure 19. Simulation experimental data table for 25 bar. (**a**) Speed drop; (**b**) sliding time; (**c**) dynamic load; (**d**) maximum sliding power of Clutch 4.

As shown in Figures 18–20, at 20 bar, except for Configuration 2, the other three configurations showed experimental groups with slippage. Configuration 2 is the least prone to slippage, with no slippage observed under all three oil pressures, while Configuration 3 is the most prone to slippage, with slippage still occurring at 30 bar. In terms of evaluation metrics, Configuration 2 has smaller speed drops and slip times but larger dynamic loads and maximum slip power. The trends of the four evaluation metrics for the original configuration and Configuration 1 are very similar under the three oil pressures, but Configuration 1 outperforms the original configuration in some aspects. Configuration 3 performs poorly in speed drop and slip time but performs well in dynamic load and maximum slip power. From the perspective of oil pressure, the speed drop and slip time of all four configurations are smaller at 25 bar. The overall dynamic load is better at 20 bar, and the maximum slip power is better at 30 bar.

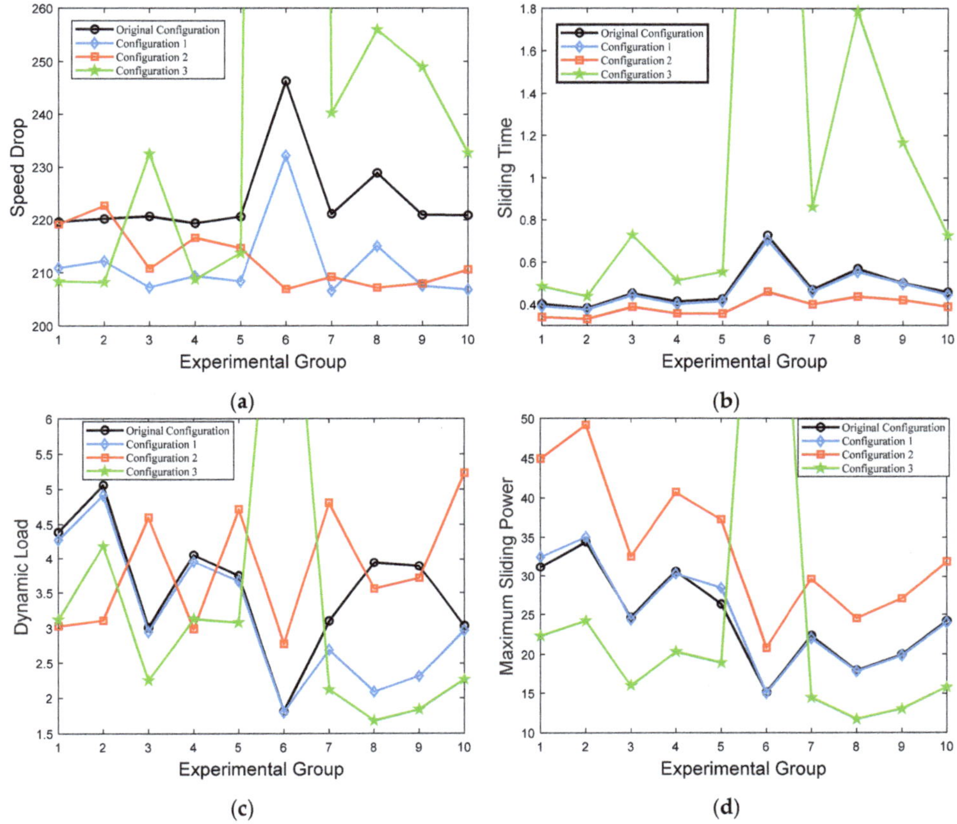

Figure 20. Simulation experimental data table for 25 bar. (**a**) Speed drop; (**b**) sliding time; (**c**) dynamic load; (**d**) maximum sliding power of Clutch 4.

Under heavy-load conditions, the analysis revealed that among the four configurations, clutch 2 exhibited an average fluctuation range (i.e., fluctuation range/mean value) of 1.52% for maximum slippage power. Consequently, the maximum slippage power of clutch 2 may also be excluded from the evaluation metrics under heavy-load operation. The corresponding test results at 50 bar are presented below.

As shown in Figure 21, under 50 bar pressure, all configurations except Configuration 3 exhibited no slippage incidents during testing, demonstrating that Configuration 3 remains the most susceptible to slippage under both light-load and heavy-load operating conditions. The evaluation metrics indicate that Configuration 2 continues to show optimal performance characteristics, including minimal speed reduction, shortest slippage duration, and maximum slippage power output, while also displaying improved dynamic load performance. Comparative analysis of the three light-load pressure conditions reveals significant improvements in three of the four key evaluation metrics (speed reduction, slippage duration, and maximum slippage power), though dynamic load performance showed only marginal improvement with some instances of slight degradation.

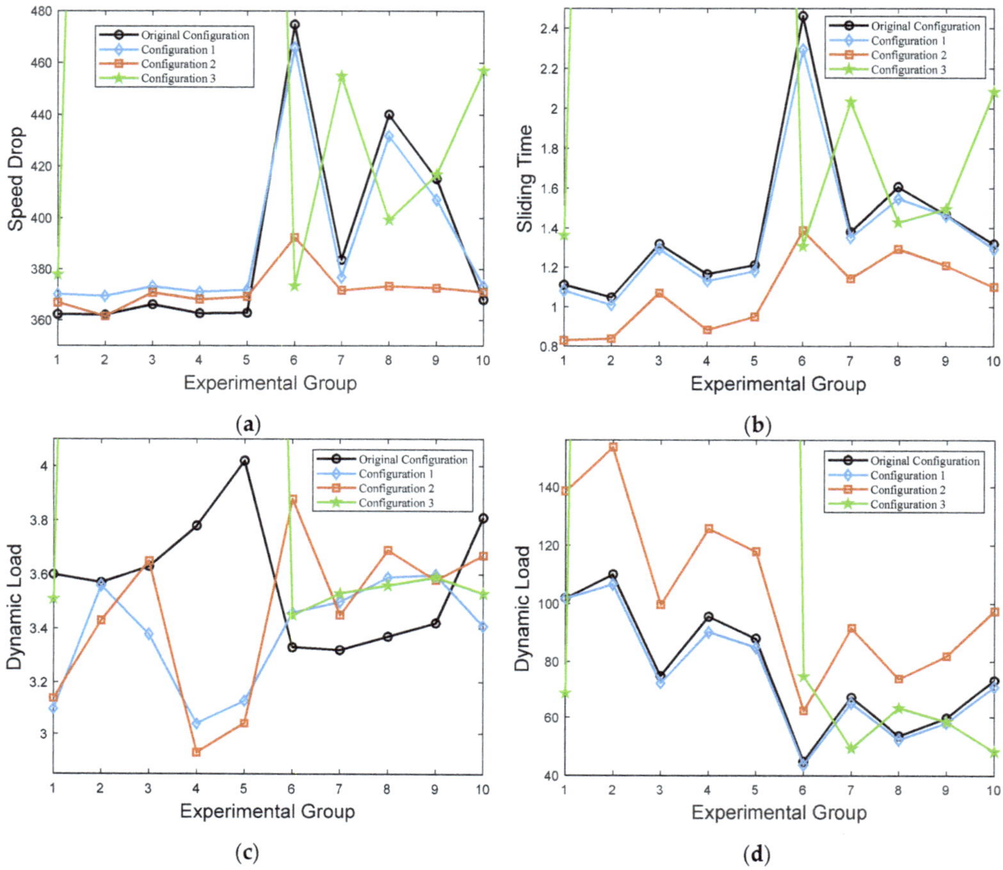

Figure 21. Simulation experimental data table for 50 bar. (**a**) Speed drop; (**b**) sliding time; (**c**) dynamic load; (**d**) maximum sliding power of Clutch 4.

3.2.2. Analysis of Evaluation System Results

For the light-load operations, the ANOVA weights obtained from the 120 sets of experiments are [0.3912, 0.1057, 0.2655, 0.2376]. By assigning a weight coefficient of 1/2 to both the AHP weights and the ANOVA weights, the final weights become [0.3843, 0.1655, 0.2812, 0.1690]. Based on these weights, the highest scores and average scores for each experimental group are shown in the Figure 22.

As shown in Figure 22, the average and maximum scores of Configuration 1, Configuration 2, and Configuration 3 have improved compared to the original configuration. Under the three oil pressures, the average scores of Configuration 1, Configuration 2, and Configuration 3 increased by 13.38%, 5.94%, and 10.54%, respectively, compared to the original configuration. Under the three oil pressures, the average maximum scores of Configuration 1, Configuration 2, and Configuration 3 increased by 11.53%, 19.17%, and 4.64%, respectively, compared to the original configuration.

For the heavy-load operations, the ANOVA weights obtained from the 40 sets of experiments are [0.1736, 0.2225, 0.3603, 0.2435]. By assigning a weight coefficient of 1/2 to both the AHP weights and the ANOVA weights, the final weights become [0.1610, 0.2085, 0.2958, 0.3347]. Based on these weights, the highest scores and average scores for each experimental group are shown in the Figure 23.

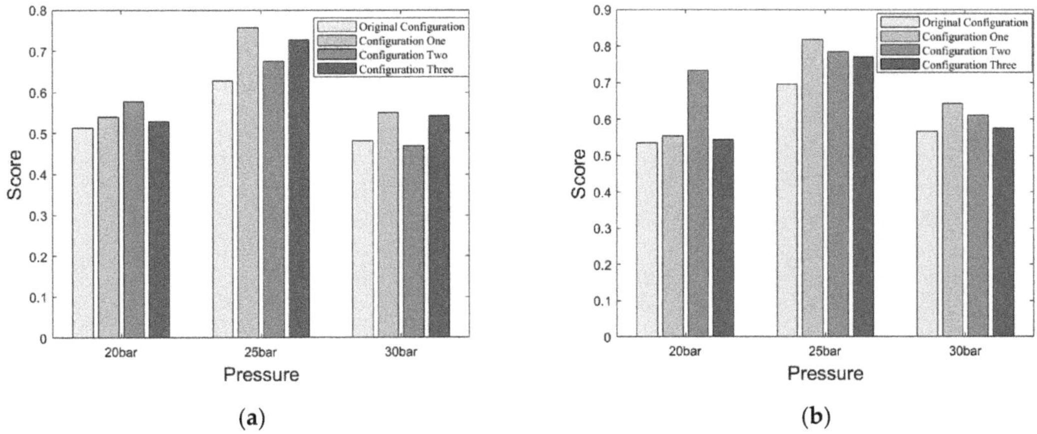

Figure 22. Scores of each experimental group. (**a**) Average scores of each experimental group; (**b**) maximum scores of each experimental group.

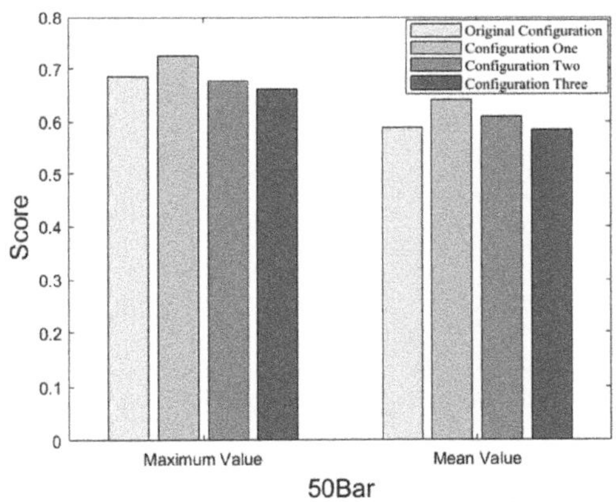

Figure 23. Scores of 50 bar experimental group.

Figure 23 demonstrates that at 50 bar operating pressure, Configuration 1 achieved comprehensive performance enhancements over the original configuration, showing improvements in both average and maximum evaluation scores. In contrast, Configuration 2 exhibited mixed results—while registering gains in average performance metrics, it failed to surpass the original configuration's peak score capability. Configuration 3 displayed a similar but more pronounced performance degradation pattern compared to Configuration 2. Under 50 bar oil pressures, Configuration 1 delivered a 5.86% maximum score improvement and 9.38% average score increase, whereas Configurations 2 and 3 showed respective maximum score reductions of 1.50% and 3.87%, despite managing modest average score improvements of 2.63% and 0.27% over the original configuration.

Comprehensive testing results demonstrate that Configuration 1 delivered optimal performance across both operational regimes, achieving superior average and peak evaluation scores under light-load conditions at 25 bar pressure while maintaining this performance advantage in heavy-load scenarios. This empirical validation confirms Configuration 1 as the preferred design solution for both load conditions. The implementation of an optimized clutch mechanism within the BLT-U significantly enhances the Hydro-Mechanical Continuously Variable Transmission system's overall efficiency, establishing the technical feasibility of modular BLT-U architectures for diverse transmission system applications. These findings substantiate the engineering value of standardized, unitized design approaches incorporating dual-mode logic control principles.

3.3. *Wet Clutch Optimization*

3.3.1. Analysis of Raw Simulation Experiment Data

For light-load conditions, under Configuration 1 and 25 bar, the five experimental groups with $Z = 5, 7, 9, 11, 13$ combined with the experimental group in Table 6 resulted in a total of 60 experimental results. These 60 results are represented in Figures 24, 25, and 26 based on D, c, and Z, respectively.

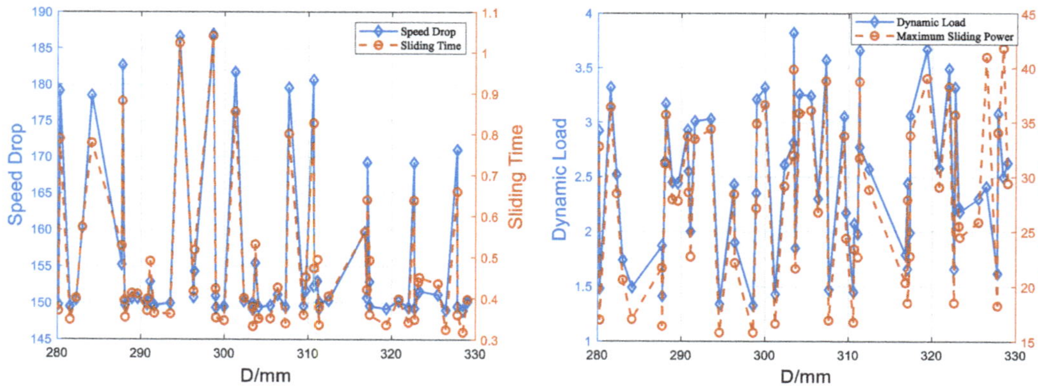

(a) (b)

Figure 24. Simulation experimental data based on D. (**a**) Speed drop and sliding time; (**b**) dynamic load and maximum sliding power of Clutch 4.

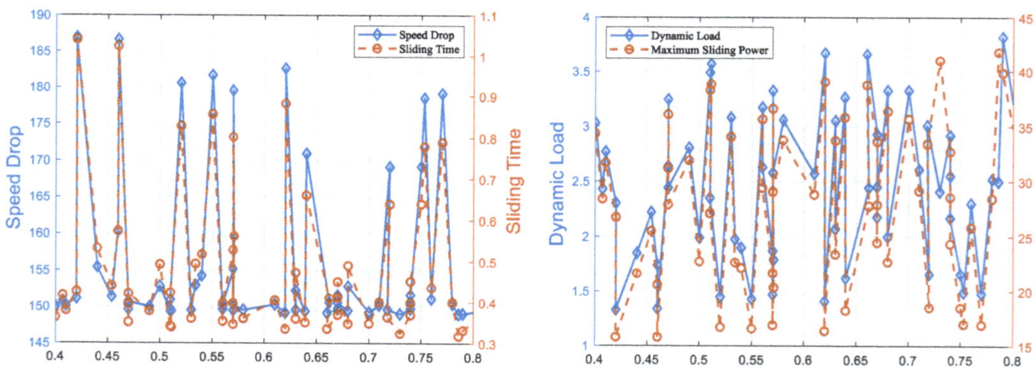

(a) (b)

Figure 25. Simulation experimental data based on c. (**a**) Speed drop and sliding time; (**b**) dynamic load and maximum sliding power of Clutch 4.

As shown in Figures 24–26, the trends of speed drop and slip time are similar, and the trends of dynamic load and maximum slip power are also similar. Speed drop and slip time tend to decrease as D, c, and Z increase, while dynamic load and slip time tend to increase as Z increases. However, the above trends are weakly correlated with changes in D and c but strongly correlated with changes in Z.

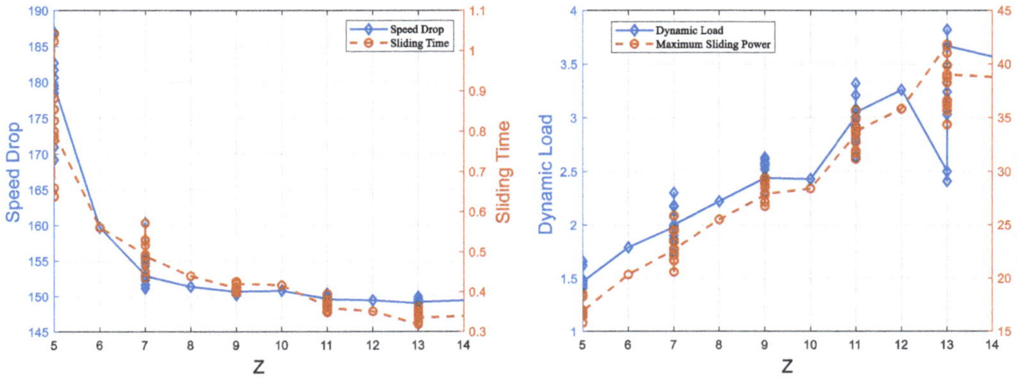

(a) (b)

Figure 26. Simulation experimental data based on Z. (**a**) Speed drop and sliding time; (**b**) dynamic load and maximum sliding power of Clutch 4.

For heavy-load conditions, under Configuration 1 and 25 bar, the five experimental groups with $Z = 5, 7, 9, 11, 13$ combined with the experimental group in Table 6 resulted in a total of 58 experimental results (2 out of the 50 additional test groups were discarded due to clutch slippage-induced data loss). These 58 results are represented in Figures 22, 23, and 24 based on D, c, and Z, respectively.

As shown in Figures 27–29, under heavy-load conditions, the trends of speed reduction and slippage duration with parameter changes remain relatively consistent, while the trends of dynamic load and maximum slippage power differ significantly, even showing opposite patterns in certain parameter ranges. Under heavy load, the trends of the indicators with parameters D and c are not obvious, with reductions only occurring in some local parameter regions. For the parameter Z, speed reduction and slippage duration still gradually decrease as Z increases, dynamic load decreases somewhat with increasing Z, and maximum slippage power continues to increase with increasing Z.

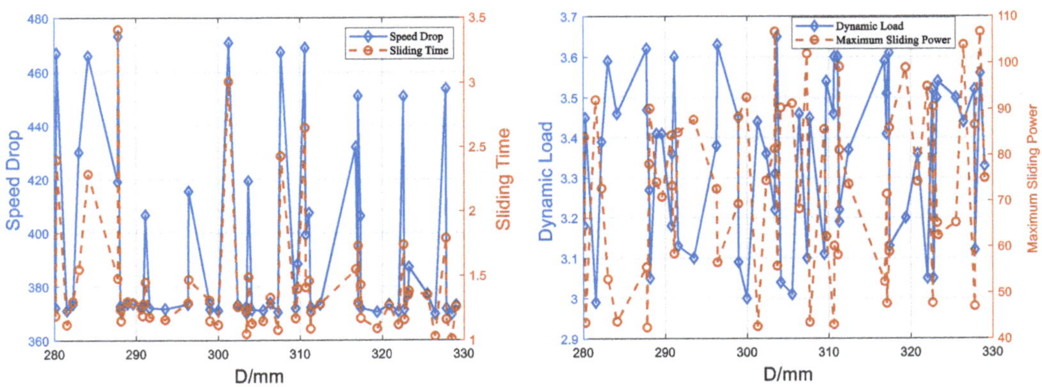

(a) (b)

Figure 27. Simulation experimental data based on D. (**a**) Speed drop and sliding time; (**b**) dynamic load and maximum sliding power of Clutch 4.

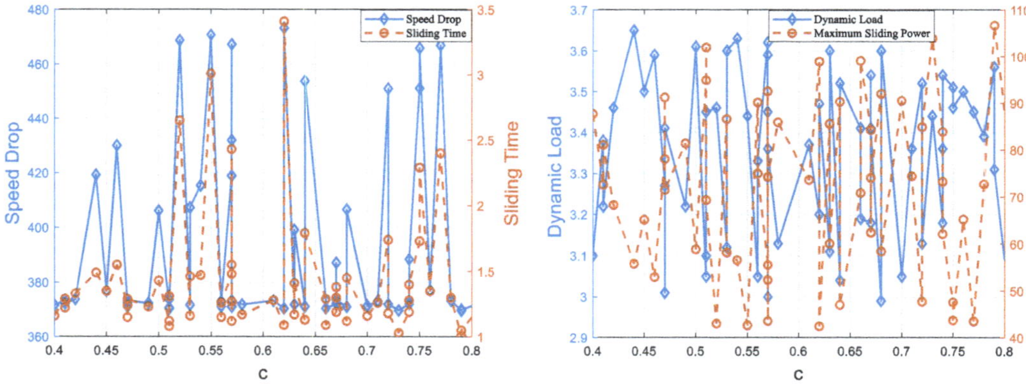

(**a**) (**b**)

Figure 28. Simulation experimental data based on *c*. (**a**) Speed drop and sliding time; (**b**) dynamic load and maximum sliding power of Clutch 4.

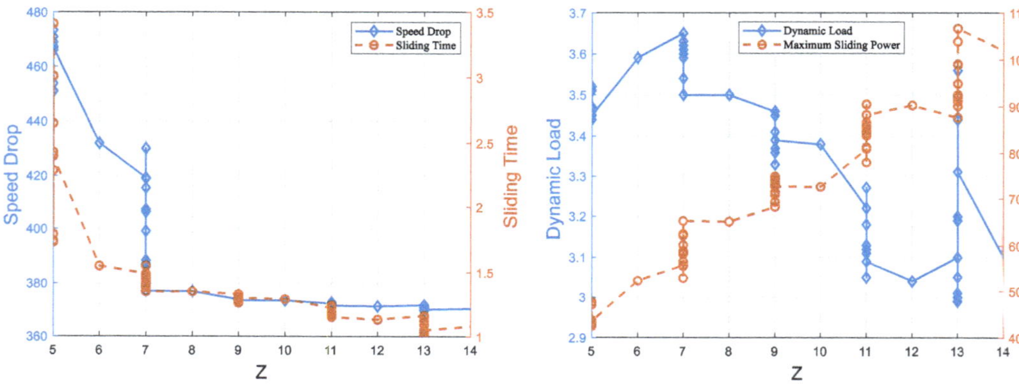

(**a**) (**b**)

Figure 29. Simulation experimental data based on *Z*. (**a**) Speed drop and sliding time; (**b**) dynamic load and maximum sliding power of Clutch 4.

3.3.2. GRNN Improvement Results

The optimization process illustrated in Figure 14 employed the Whale Optimization Algorithm (WOA) with a population size of 30 over 30 iterations. This achieved optimized dispersion values of 0.3416 for light-load conditions and 0.3108 for heavy-load conditions. Under these dispersion values, the minimum total relative error reached 0.7459% (light load) and 1.527% (heavy load). The recorded total relative error is shown in the Figure 30:

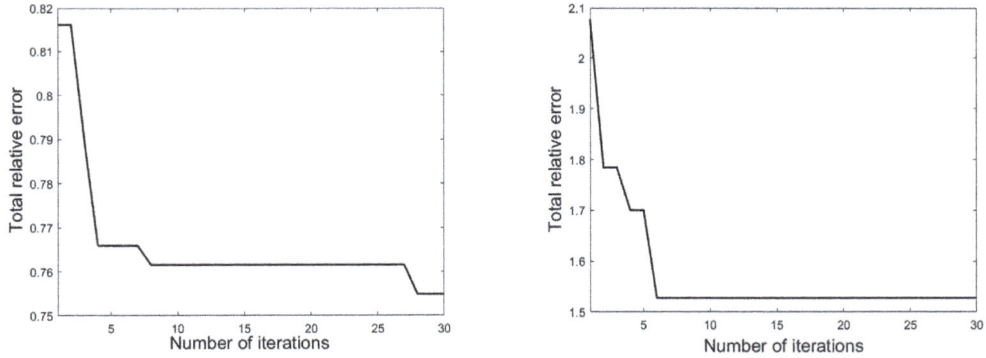

(a) (b)

Figure 30. Curve of total relative error recorded by WOA. (**a**) Light-load conditions; (**b**) heavy-load conditions.

By incorporating the obtained spread values into the GRNN (Generalized Regression Neural Network) and conducting 200 iterations, the recorded minimum total relative errors were reduced to 0.4508% (light-load) and 0.5952% (heavy-load), representing 39.6% and 61% improvements, respectively, compared to the previously trained GRNN model.

3.3.3. Analysis of Results from the Evaluation System and Optimized GRNN

After obtaining the above GRNN's NET, 100 points were linearly taken within the value ranges of *D* and *c*. Since *Z* can only take positive integers, 10 values within its range were still selected. This expanded the variables to 100,000 groups, which were then input into the recorded GRNN's NET to predict the corresponding scores for the two distinct load conditions. The scores were fitted with *D* and *Z* parameters, and the *c* and *Z* parameters using a second-order polynomial, and the resulting fitted surfaces are shown in the Figure 31:

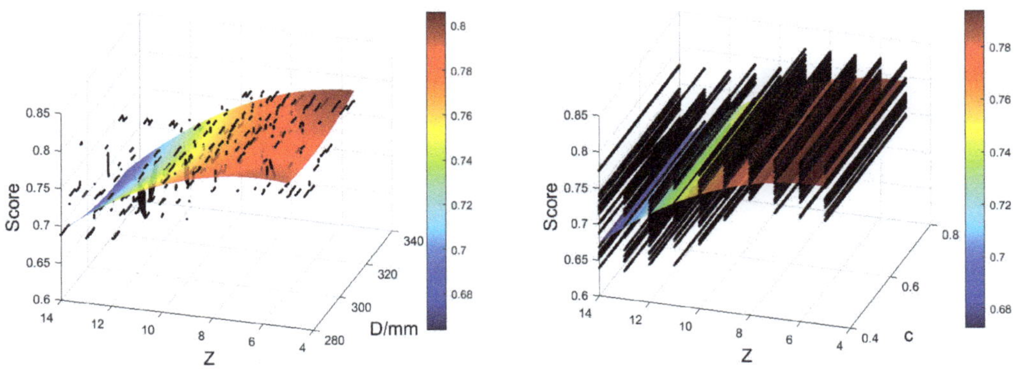

(a) (b)

Figure 31. *Cont.*

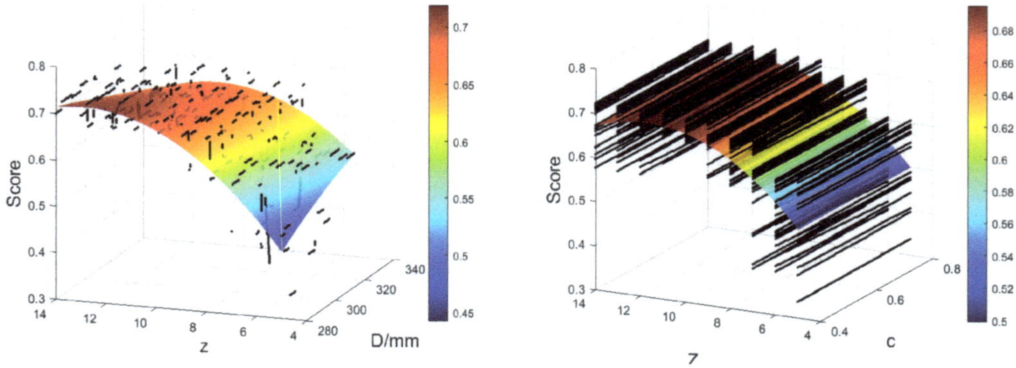

(c) (d)

Figure 31. Fitted surfaces. (**a**) Second-order polynomial fitting of D and Z for the light-load conditions; (**b**) second-order polynomial fitting of c and Z for the light-load conditions; (**c**) second-order polynomial fitting of D and Z for the heavy-load conditions; (**d**) second-order polynomial fitting of c and Z for the heavy -load conditions.

In the fitted surface plot, since there are 1000 scores for the same D and c values and 10,000 scores for the same Z value, it is difficult to observe and obtain the corresponding optimization results. Therefore, the predicted scores were analyzed based on the single factors D, c, and Z. The maximum, minimum, and average scores for each factor's individual values were recorded, and the results for factor D are shown in the Figure 32:

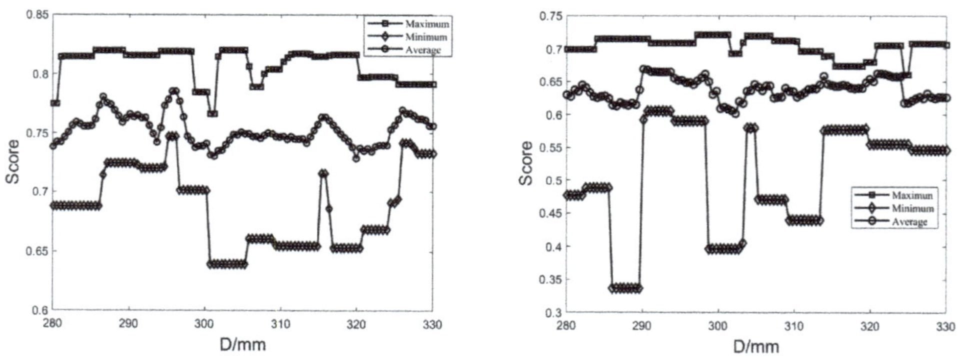

(a) (b)

Figure 32. Analysis results of factor D. (**a**) The light-load conditions; (**b**) the heavy-load conditions.

From Figure 32, under light-load conditions, to achieve the optimal design results, the optimal range for D should be between (302, 305), combined with appropriate c and Z. To ensure a certain level of performance while other parameters vary, the range for D can be set to (287, 296). If a larger radius for D is required by design needs, (315, 316) can be chosen to ensure the system's performance. Under heavy-load conditions, to achieve optimal design results, the parameter D should be set within the range of (297, 301) while determining corresponding values for parameters c and Z. To maintain certain performance levels when other parameters vary, an alternative range of (290, 298) may be adopted. For design requirements involving larger radii of D, the range (320, 324) can be selected to ensure system performance.

The results for factor c are shown in the Figure 33:

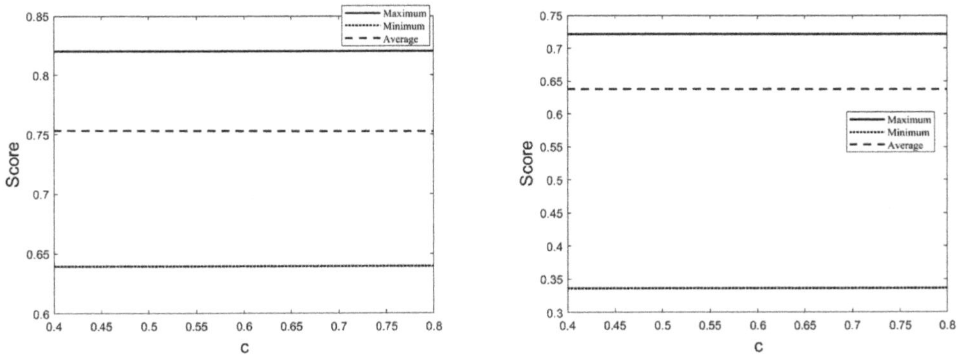

(**a**) (**b**)

Figure 33. Analysis results of factor c. (**a**) The light-load conditions; (**b**) the heavy-load conditions.

Figure 33 demonstrates that under both light-load and heavy-load conditions, the maximum and minimum scores remain consistently stable across different values of the parameter c. The overall trend of the average score is that it decreases as c increases, but the fluctuation range of the average score (i.e., the range of average scores divided by the minimum average score) is 0.1% and 0.8%. Therefore, it can be concluded that to achieve the optimal design results, an appropriate c value should be selected under the determined D and Z values. However, analyzing c alone shows that it has almost no impact on the overall system performance.

The results for factor Z are shown in the Figure 34:

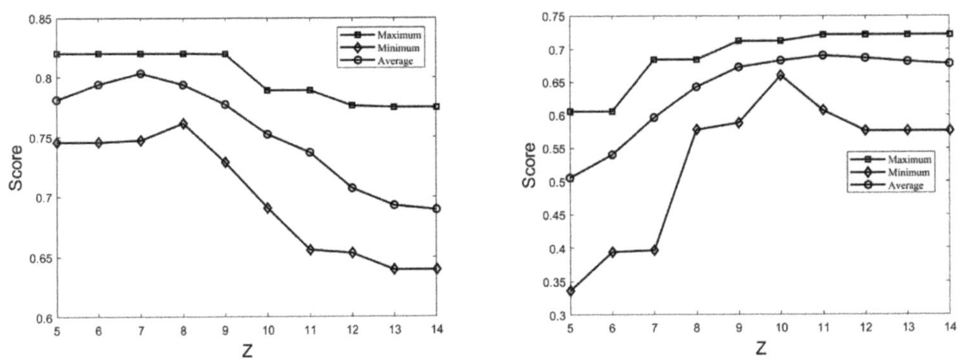

(**a**) (**b**)

Figure 34. Analysis results of factor Z. (**a**) The light-load conditions; (**b**) the heavy-load conditions.

From Figure 34, under light-load conditions, the scores show a clear pattern as Z changes, with scores gradually decreasing as Z increases. To achieve the optimal design results, the range for Z should be (7, 8). To ensure a certain level of performance while other parameters vary, the range for Z can be set to (5, 9). When Z exceeds 9, the score will significantly decline. Under heavy-load conditions, the scoring trend exhibits an inverse relationship compared to light-load operation, demonstrating gradual score improvement with increasing Z values. For optimal design outcomes, the parameter Z should be selected within the (11, 14) range. To maintain baseline performance stability during variations in other parameters, an alternative Z range of (9, 11) proves effective. Notably, when Z values fall below 8, significant score degradation occurs.

4. Discussion

The main problems addressed in this paper are the optimization of stage-shifting performance and structural design for the Hydro-Mechanical Continuously Variable Transmission (HMCVT) of agricultural tractors. Due to the complex operating conditions of agricultural tractors, optimizing stage-shifting performance can enhance their power performance, fuel economy, and comfort, and extend the transmission's service life. For HMCVT, increasing the number of stages can effectively expand its working range and improve power performance and fuel economy. Therefore, introducing a transmission system that improves stage-shifting performance and increases the number of stages is essential. This paper selects the binary logic transmission system as the introduced transmission system, which is a stepped transmission with a compact structure, strong load-bearing capacity, and potential for modular and unitized design. When unitized, it can be connected in series to increase the number of gears and stages.

The first problem to be solved is conducting transmission system research without a complete transmission system test bench. For wet clutches and some transmission systems, building a complete test bench is costly and challenging. A hardware-in-the-loop and digital twin system development approach can address this issue by combining hardware and software, ensuring test accuracy while reducing costs. This method involves using test results from core components of the transmission system as inputs to the overall system simulation model.

The second problem is the unitization, modularization, and performance comparison of the binary logic transmission system. Currently, binary logic transmission systems in use often consist of planetary gear sets with different structural sizes, primarily using double intermeshing planetary gear sets and concentric V-groove clutches. These systems have high production and processing requirements, high costs, and issues with eccentric transmission. Therefore, this paper selects the simpler NGW planetary gear reducer and discusses different locking methods to achieve binary logic functionality. The final choice is to lock the ring gear and connect any two components with clutches, forming transmission ratios of $1 + k$ and 1, which better match the transmission ratio intervals required for different stages of agricultural tractors. This results in three different configurations of the BLT-U. During configuration comparison, four evaluation metrics were selected, but the importance of these metrics and the principles for selecting configurations were unclear. Thus, an evaluation system combining expert-evaluated Analytic Hierarchy Process (AHP) and variance analysis was established, proposing a scoring standard that includes both subjective and data-driven characteristics. Future work can adapt different planetary gear mechanisms and locking/connection methods for various transmission systems. The transmission unit proposed in this paper has potential for application in other agricultural tractor transmission systems.

The third problem is how to improve the stage-shifting performance of the transmission after selecting the appropriate configuration and clutch charging pressure. To address this issue, considering the adjustable simulation parameters in Simulation X software for the BLT-U system—such as the planetary gear mechanism characteristic parameter k, clutch friction plate area, number of clutch friction pairs, and clutch engagement pressure—this paper optimizes stage-shifting performance by improving the design parameters of the engaging clutch. Since the oil pressure and planetary gear mechanism characteristic parameter k were determined earlier, the optimization focused on the engaging clutch's design parameters. Specific metrics for clutch performance include speed drop and dynamic load to quantify jerking and impact, as well as slip time and maximum slip power to quantify clutch engagement performance. These metrics characterize the transmission's stage-shifting performance. Future work can propose new evaluation metrics and analyze their relationships and differences.

5. Conclusions

This paper proposes an improved Hydro-Mechanical Continuously Variable Transmission (HMCVT) for agricultural tractors using a BLT-U. By integrating stage-shifting performance and clutch engagement performance, a corresponding evaluation system is established. A Whale Optimization Algorithm (WOA)-improved Generalized Regression Neural Network (GRNN) is employed for modeling and training, and a method for optimizing the wet clutch in the BLT-U is introduced.

The accuracy of the modeling method was validated using a clutch test bench. The average correlation coefficient between the simulated and experimental speed curves was 0.92, and the average relative error of the dynamic load was 5.58%. Comparative analysis revealed the clutch's simulation data achieved peak rotational speed faster than bench test results, exhibiting a steeper acceleration curve, whereas bench tests showed gradual slope variations in the speed ascent profile. This finding suggests implementing controlled time delays for the simulated speed ramp-up phase to better approximate physical test conditions.

This paper proposes an improved HMCVT using a BLT-U. By comparing the new configurations with the original configuration, the performance of different configurations under the same oil pressure and the same configuration under different oil pressures is analyzed. The optimal configuration is determined using the evaluation system. Under two load conditions, the average and maximum scores of the selected optimal configuration increased by 13.38%, 11.53% and 9.38%,5.86%, respectively, compared to the original configuration. This demonstrates that a well-designed BLT-U can enhance the overall performance of the HMCVT system.

The analysis reveals that when any two components of the planetary gear mechanism engage at a 1:1 transmission ratio, different engagement combinations significantly impact overall system performance. In Configuration 1, the clutch essentially directly connects the input and output shafts, resulting in performance trends similar to the original configuration while substantially improving both speed drop during gear shifts and dynamic load capacity. This demonstrates that replacing conventional clutches with such planetary gear arrangements can offer distinct performance advantages for certain systems. Configuration 2 connects the ring gear and sun gear via clutch, requiring lower clamping force for equivalent torque transmission compared to the original configuration, albeit with slightly compromised shift quality. This design proves particularly advantageous for systems with strict clutch size limitations. Configuration 3 links the ring gear and planetary carrier, necessitating greater clamping force to ensure torque transfer while providing smoother engagement characteristics with reduced dynamic loads and slippage power—making it suitable for shock-sensitive applications. Notably, across four pressure levels in both light and heavy-load conditions, all four configurations exhibit fundamentally similar trends in evaluation metrics, suggesting Configuration 1 likely maintains its performance advantages at other pressure settings as well.

This study demonstrates that optimizing the GRNN model using the WOA significantly improves prediction accuracy, confirming the dual value of optimization algorithms in both model parameter tuning and post-model optimization. Experimental data reveal that under identical oil pressures and configurations, speed drop and slippage duration exhibit highly consistent variation trends across light-/heavy-load conditions, allowing these metrics to serve as interchangeable indicators. In contrast, peak slippage power and dynamic load display load-dependent discrepancies—particularly for parameter Z, which exerts completely opposite effects on dynamic load under different operating conditions due to strong load-case dependencies. Performance metrics show inherent trade-offs: improvements in speed drop/slippage duration often coincide with deterioration in dynamic load/peak slippage power, resulting in opposing Z-score evaluations across conditions. Optimization must balance these conflicting metrics. Parameter D shows cross-condition optimization potential with overlapping optimal range (301, 302), while parameter c, despite weaker influence, requires proper matching to avoid performance degradation. The determined optimal design space is as follows: D in the range (301, 302), and Z in the range (8, 10) with corresponding c-values—ensuring optimization feasibility while preventing significant system performance deterioration.

Author Contributions: Methodology, Z.C., W.L. and M.Y.; software, W.L.; validation, W.L. and Z.C.; investigation, W.L.; resources, Z.C.; writing—original draft preparation, W.L.; writing—review and editing, W.L. and M.Y.; supervision, Z.C.; and project administration, Z.C. All authors have read and agreed to the published version of the manuscript.

Funding: This research was funded by the China Postdoctoral Science Foundation (grant number: 2023T160780), the Postgraduate Research & Practice Innovation Program of Jiangsu Province (grant number: KYCX24_1348), and the National Natural Science Foundation of China (grant number: 52105063).

Institutional Review Board Statement: Not applicable.

Data Availability Statement: The data presented in this study are available on request from the corresponding author or first author.

Acknowledgments: We sincerely acknowledge the financial support from the aforementioned funding agencies. Our sincere thanks also go to Zhixiong Lu and his research team at Nanjing Agricultural University for their valuable assistance in the plow tillage resistance tests and wet clutch experiments. Additionally, we also thank the anonymous reviewers for providing critical comments and suggestions that improved the manuscript.

Conflicts of Interest: The authors declare no conflicts of interest.

References

1. Dou, H.S.; Wei, H.Q.; Ai, Q.; Zhang, Y. Control Strategy for Rotary Tillage Condition of Hybrid Electric Tractor with Coupled-split Dynamic Configuration. *Trans. Chin. Soc. Agric. Mach.* **2024**, *55*, 393–400+414.
2. Zhang, K.; Wang, L.; Deng, X.T.; Lu, Z. Research on energy-saving control strategy of OS-ECVT configuration of hybrid tractor based on improved ECMS. *J. Nanjing Agric. Univ.* **2025**, 1–14.
3. Macor, A.; Rossetti, A. Optimization of hydro-mechanical power split transmissions. *Mech. Mach. Theory* **2011**, *46*, 1901–1919. [CrossRef]
4. Ahn, S.; Choi, J.; Kim, S.; Lee, J.; Choi, C.; Kim, H. Development of an integrated engine-hydro-mechanical transmission control algorithm for a tractor. *Adv. Mech. Eng.* **2015**, *7*, 1687814015593870. [CrossRef]
5. Cheng, Z.; Chen, Y.; Li, W.; Zhou, P.; Liu, J.; Li, L.; Chang, W.; Qian, Y. Optimization Design Based on I-GA and Simulation Test Verification of 5-Stage Hydraulic Mechanical Continuously Variable Transmission Used for Tractor. *Agriculture* **2022**, *12*, 807. [CrossRef]
6. Kerr, J.H.; Ferguson, R.J. The binary logic incremental transmission. *Mech. Mach. Theory* **1983**, *18*, 357–361. [CrossRef]
7. Kerr, J.H. Development of a 64-speed binary logic transmission for transit buses. *Transp. Can. Publ.* **1994**, 1–18.
8. Marquardt, B.; Brown, M. Development of a Super Compact, High Efficiency, 32-Speed Transmission for Tracked Vehicles. In Proceedings of the 2016 Ground Vehicle Systems Engineering and Technology Symposium (NDIA), Michigan, MI, USA, 2–4 August 2016; pp. 1–6.
9. Brown, M.; Marquardt, B. Ker-Train High Efficiency Drive-by-Wire Transmission Systems. In Proceedings of the 2017 Ground Vehicle Systems Engineering and Technology Symposium (NDIA), Michigan, MI, USA, 8–10 August 2017; pp. 1–7.
10. Ferguson, R.J.; Morrison, J.R.G. The efficiency of the 64-ratio Ker-Train transmission. *SAE Trans.* **1986**, *95*, 61–68.

11. Morrison, J.R.G.; Ferguson, R.J. Control of shift disturbances in the 64-ratio Ker-Train transmission. *SAE Trans.* **1986**, *95*, 69–77.
12. Defense World. BAE Systems Selects SG 850C Automatic Transmission for Next Gen Bradley Fighting Vehicles. Available online: https://defensemirror.com/news/17293/BAE_Systems_Selects_SG_850C_Automatic_Transmission_For_Next_Gen_Bradley_Fighting_Vehicles (accessed on 25 September 2021).
13. Wang, Y.Z.; Wang, D.P.; Li, Y.; Ren, S.Y.; Dou, D.L.; He, X.J. Simulation and test on shift process of multi-state modular automatic transmission. *J. Harbin Inst. Technol.* **2018**, *50*, 103–110.
14. Kerr, J.H. Coplanar Reverted Gear Train Loop. US6669594, 30 December 2003.
15. Kerr, J.H. Concentric V-Groove Coupling. US0053525, 26 February 2015.
16. Wang, Y.Z.; Wang, P.K.; Li, Y.; Wang, D.P.; Jia, S. Design and Experimental Study of Double Intermeshing Planetary Gear Transmission. *J. Mech. Transm.* **2017**, *41*, 146–151.
17. Chang, Y.L.; Yang, K.; Wang, J.; Chen, Y.; Wang, Y. Matching and Structure Design Verification for Planetary Gear Two Speed Transmission. *J. Mech. Transm.* **2021**, *45*, 106–113.
18. Shin, S.; Choi, S.B. Control-oriented modeling of wet clutch friction considering thermal dynamics. *Mechatronics* **2024**, *99*, 103146. [CrossRef]
19. Strobl, P.; Meingassner, G.J.; Pflaum, H.; Voelkel, K.; Schneider, T.; Stahl, K. Holistic Measurement of the Friction Behavior of Wet Clutches. *Lubricants* **2024**, *12*, 235. [CrossRef]
20. Yang, Z.H.; Du, Y.H.; Tan, Y.Y.; Li, Q.H.; Chen, X.H. Analysis of the Influence of Continuous Variable Characteristic Parameters on the Quality of HMCVT Transition. *J. Mech. Transm.* **2025**, 1–9.
21. Wu, Z.G.; Wang, D.Q.; Zhang, Y.A.; Du, Y.F.; Gao, L.Y.; Guo, X.B. Modeling and Verification of Wet Clutch Hydraulic System for High-Power Tractor. *Tract. Farm Transp.* **2024**, *51*, 32–38.
22. Ji, D.X.; Chen, X.; Yan, S.M. Calculation Method for Equivalent Moment of Inertia of Different Gear Trains in Gearbox. *Automob. Applier Technol.* **2024**, *51*, 32–38.
23. Tang, P.X.; Liu, Y.F.; Wang, S.H.; Xu, X.Y. Optimal design of clutch in automatic transmission. *J. Beijing Univ. Aeronaut. Astronaut.* **2010**, *36*, 291–294. [CrossRef]
24. Gong, R.; Shen, Z.Z.; Zhang, Y.D.; Zhan, C. Calculation method and experimental verification of centrifugal pressure in flow field of wet clutch cylinder. *Agric. Equip. Veh. Eng.* **2023**, *61*, 56–61.
25. Jiang, K.; Li, H.; Chen, S.; Li, S.L.; Ren, R.Z. Multi-objective Optimization Design of Wet Friction Clutches. *J. Mech. Transm.* **2023**, *61*, 56–61.
26. Cheng, S.M.; Zhong, Y.; Wang, Y. Working condition matching and gear shift optimization of intelligent agricultural machinery power system. *J. Fujian Univ. Technol.* **2023**, *21*, 55–62.
27. Zhao, J.H.; Zhao, T.L.; Xu, L.Y.; Li, Y.Y.; Zhang, J.Y. General Model and Optimization Experiment on Traction Performance Prediction Based on APSO Algorithm. *Trans. Chin. Soc. Agric. Mach.* **2025**, 1–11.
28. Cheng, S.M. Research on AMT Shifting Strategy of Intelligent tractor Based on Working Condition Recognition. Master's Thesis, Fujian University of Technology, Fujian, China, 2023.
29. Wang, Q. Research on Plowing Drag-adaptive Operation Control Method of Electric Tractor. Master's Thesis, Northwest A&F University, Shaanxi, China, 2023.
30. Cheng, B.Y. Design and experimental study of tillage resistance acquisition system. Master's Thesis, South China Agricultural University, Guangdong, China, 2019.
31. Specht, D.F. A general regression neural network. *IEEE Trans. Neural Netw.* **1991**, *2*, 568–576. [CrossRef]
32. Mirjalili, S.; Lewis, A. The Whale Optimization Algorithm. *Adv. Eng. Softw.* **2016**, *95*, 51–67. [CrossRef]

agriculture

MDPI

Article

Design of Dynamic Deep Sowing System for Peanut Planter with Double-Loop Feedback Fuzzy PID Control

Moxian Li [1,*], Xueliang Chang [2,*], Yaqing Gu [1], Ping Wang [1] and Shuqi Shang [2]

[1] College of Electrical Engineering, Shandong Huayu University of Technology, Dezhou 253034, China; guyaqing6666@163.com (Y.G.); wp@huayu.edu.cn (P.W.)

[2] College of Electrical and Mechanical Engineering, Qingdao Agricultural University, Qingdao 266109, China; sqshang@qau.edu.cn

* Correspondence: 17863998618@163.com (M.L.); cxl@stu.qau.edu.cn (X.C.); Tel.: +86-17863998618 (M.L.)

Abstract: To enhance peanut sowing depth consistency, an active depth adjustment planter was designed. This study employs inclination and pressure sensors for ridge surface detection, coupled with a hydraulic cylinder and profiling mechanism to dynamically adjust furrow depth according to ground variations. A mathematical model integrating detection, adjustment, and execution processes was established. The control system adopts an improved DLF-Fuzzy PID (double-loop feedback fuzzy PID) control strategy, with co-simulation in MATLAB/AMESIM for performance comparison. The results demonstrate the improved algorithm's superiority in sowing depth accuracy. Field experiments evaluated three operational parameters (vehicle speed, pressure, and sowing depth) with the qualification rate as the metric. At 50 mm sowing depth and 3 km/h speed, the system achieved a 94.6% dynamic qualification rate and 2.38% maximum depth variation coefficient. Compared with existing methods, this approach enhances sowing depth control effectiveness by 6.05% and reduces variation by 2.85%.

Keywords: peanut seeder; control system; seeding depth; improved double-loop feedback fuzzy PID

1. Introduction

Peanuts, known for their high nutritional and economic value, are a crucial oil crop in China [1,2]. With the current saturation in cultivation area, the focus has shifted to enhancing yield within the existing acreage. The uniformity of sowing depth significantly impacts both the yield and quality of peanuts [3]. Consistent sowing depth alleviates seedling emergence difficulties, boosts germination rates, and increases yield, thereby serving as a vital strategy for economic growth [4–6]. However, at present, there are fewer studies on peanut sowing depth control systems in China, and most peanut planters do not have a sowing depth adjustment function. The development of a peanut sowing depth control system is of necessity and practical significance.

Currently, two main types of control systems on peanut sowing depth control systems are existing in China. One is the mechanical sowing depth adjustment device developed by Chang Xueliang et al. The consistency of sowing depth is achieved by pushing the soil at the top of the ridge surface flat [7]. Geng Duanyang [8] developed a ridge profiling sowing depth control device, which adopts a similar mechanical structure. The other is the sowing depth control system based on fuzzy control developed by Wang Linsheng et al. This system detects the height of the ridge surface through sensors and further regulates the sowing depth using an electric actuator [9]. These two approaches improved the consistency of sowing depth of peanut but still need further improvement compared to other major crops. Research by domestic and international scholars highlights two primary phases for ensuring sowing depth consistency: measurement and execution. Depending on whether or not the measuring device is in contact with the ridge surface, the measurement method can be classified as contact or non-contact measurement. Non-contact methods, as utilized by Søren Kirkegaard Nielsen [10–12] and Han Bao [13], involve technologies like ultrasonic or laser sensors for ground height measurement. For instance, Sajad Kiani [14] adjusted sowing depth based on trench depth measurements using ultrasonic sensors. Contact methods, adopted by Jensen Lynndavid [15], Weatherly E.T. [16], Huang Dongyan [17], and Fu Weiqiang [18], typically use angle, displacement, or pressure sensors. Li Yuhuan [19], for example, calculated ground height based on pressure sensor readings from compaction wheels. The execution phase involves researchers like Fu Weiqiang [18] and Xie Yangmin [20–22] employing electro-hydraulic or electric systems to control microcontrollers or PLCs (Programmable Logic Controller). Zhao Jinhui [23] utilized displacement sensors to detect movement and adjusted the sowing depth via a hydraulic system, achieving a stability coefficient over 90%. Control strategies, such as PID algorithms and fuzzy PID controls, were applied by Wen Liping [24], Xia Junfang [25], and Qi Wenchao [26] for depth regulation. Wang Can [27] integrated a multi-channel neural network intelligent algorithm with PLC to adjust sowing depth in garlic planters. From the above description, it can be seen that compared with the existing peanut sowing depth control system, the control system for other crops is more deeply researched and has higher control accuracy. However, they are not directly usable for peanut planting due to the difference in sowing depth and plant spacing. Therefore, it is necessary to develop a sowing depth control system tailored to the characteristics of peanuts.

Based on the ridging characteristics of peanut cultivation, this study proposes an active sowing depth control system integrating tilt/pressure sensor detection, electro-hydraulic actuators, and a feedforward-improved double-loop feedback fuzzy PID algorithm [28]. An integrated mathematical model encompassing detection, adjustment, and execution processes was established to address the insufficient adjustment accuracy of traditional mechanical devices and poor adaptability of existing control strategies. The system features a terrain-adaptive profiling mechanism and implements the double-closed-loop control of pressure and actuator height through the enhanced control strategy, with parameter optimization achieved via MATLAB-R2020b/AMESIM-2020.2 co-simulation. Compared to existing methods, the proposed system demonstrates superior performance under operational conditions of 50 mm sowing depth and 3 km/h speed, achieving a 94.6% dynamic qualification rate (a 6.05% improvement over comparable systems) and a maximum depth variation coefficient of 2.38% (a 2.85% reduction). This study realizes the precise control of peanut sowing depth, offering both theoretical references and technical foundations for improving sowing depth consistency in peanut cultivation.

2. Materials and Methods

2.1. Mechanical Structure and Working Principle

2.1.1. The Mechanical Structure of the Sowing Depth Control System

The sowing depth control device of the peanut planter consists of three main components: the detection unit, the execution unit, and the control unit. The detection unit includes a profiling wheel, a profiling wheel frame, a pressure sensor, and an inclination sensor. The execution unit comprises a parallel four-bar linkage, a hydraulic cylinder, a electro-hydraulic proportional valve, a pressure sensor, and a furrow opener. The control unit is composed of a control box. The overall structure of the sowing depth control device is illustrated in Figure 1.

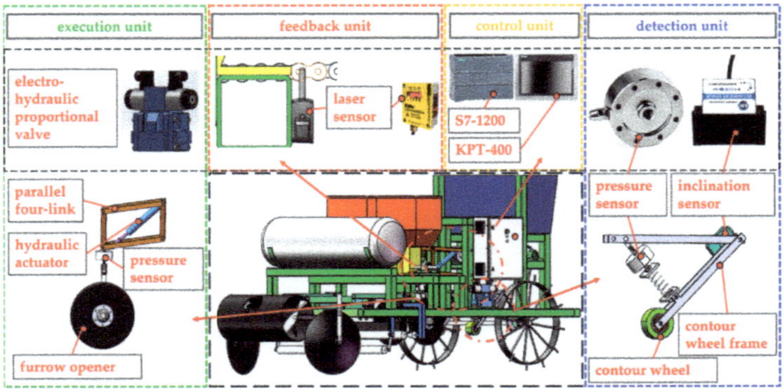

Figure 1. Structural design drawing of detection and control device.

2.1.2. Working Principle of the Sowing Depth Control System

During operation, the system is started and initialized. Control parameters are input via the HMI (Human–Machine Interface). The inclination sensor measures angular changes in the profiling wheel frame. The pressure sensor detects pressure between the profiling wheel and its frame, enabling ridge height detection. Data are sent to the PLC in the control box. The PLC calculates the height difference from the previous unit value. It determines if a step change or feedforward system intervention is needed. The PLC computes and outputs a PWM (Pulse Width Modulation) waveform current value. This controls the electro-hydraulic proportional valve, adjusting the hydraulic cylinder to move the parallel four-bar linkage. Simultaneously, the hydraulic cylinder's pressure sensor measures pressure and sends data to the PLC. The laser sensor also transmits data to the PLC, providing a reference for the next sowing point. The system records all data and calculates furrow opener performance using predefined formulas. The results are displayed in real time at the HMI. The working principle is shown in Figure 2.

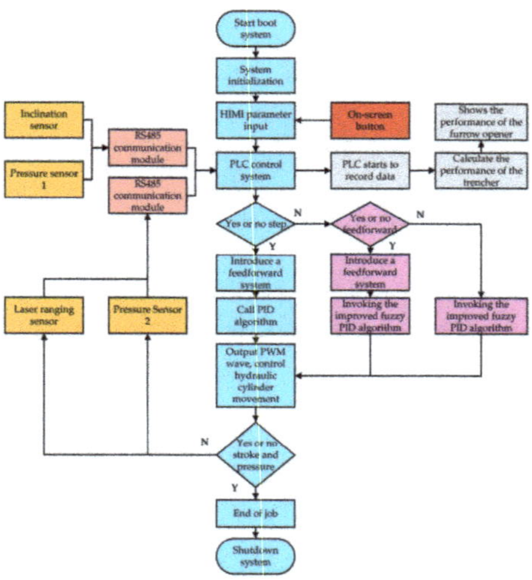

Figure 2. A schematic diagram of the detection and control device.

2.2. Key Structures and Working Mechanism

2.2.1. The Design of the Ridge Surface Profiling Mechanism

A working schematic of the detection device is shown in Figure 3. The profiling wheel contacts the ground. When ground elevation changes, the annular pressure sensor captures pressure values. The inclination sensor detects angular changes in the profiling wheel-connecting frame.

Figure 3. Schematic diagram of theoretical calculation of detection device.

In Figure 3, H_1 is the vertical distance between the transverse bar of the detection device and the center of the profiling wheel; L_1 is the center distance between the center of the profiling wheel and the center of the upper center of the inclination sensor; L_T is the vertical distance between the center of the upper center of the inclination sensor and the spring; and H_y is the vertical distance between the center of the profiling wheel with radius R and the contact surface. C_h is the distance between the contact surface of the profiling wheel and the vertical ridge surface of the center line of the profiling wheel, H_L is the distance between the ridge surface, the angle between the profiling wheel and the transverse bar is α, and the angle between the vertical and vertical directions of the profiling wheel is γ. F_{DS} is the force of the ground on the profiling wheel; G is the gravity of the profiling wheel and the bracket; and F_n is the pressure applied to the detection device at moment n.

The actual measurement of the imitation wheel beam frame is 300 mm from the ground, L_1 is 300 mm, R is 60 mm, and L_T is 200 mm. To detect the height of the monopoly surface H_L, substituting Equations (1) and (2) into Equation (3) is carried out to obtain Equation (4):

$$H_1 = L_1 sin\alpha \tag{1}$$

$$H_y + C_h = \frac{R}{cos\gamma} \tag{2}$$

$$H_L = 300 - (H_1 + H_y + C_h) \tag{3}$$

$$H_L = 300 - L_1 sin\alpha - \frac{R}{cos\gamma} \tag{4}$$

Further, the following was obtained to measure the height of the ridge at that point during the operation of the planter:

$$H_{L_n} = 300 - L_1 sin\alpha_n - \frac{R}{cos\gamma_n} \quad (n = 1, 2, 3 \ldots\ldots) \tag{5}$$

where α is measured and collected by the inclination sensor and γ is calculated by Equation (9); the relationship between α and γ is

$$\Delta H_{\text{Ln}} = H_{\text{Ln}} - H_{\text{Ln}-1} = (L_1 sin\alpha_{n-1} + \frac{R}{cos\gamma_{n-1}}) - (L_1 sin\alpha_n + \frac{R}{cos\gamma_n}) \tag{6}$$

We reduce Equation (6) to Equation (7).

$$\Delta H_{Ln} = L_1(sin\alpha_{n-1} - sin\alpha_n) + \frac{R(cos\gamma_n - cos\gamma_{n-1})}{cos\gamma_{n-1} cos\gamma_n} \quad (n = 1, 2, 3 \ldots\ldots) \tag{7}$$

The value of the initial state is calculated through Equation (8).

$$\begin{cases} \alpha_0 = \arcsin\frac{H_1}{L_1} \\ \gamma_0 = 0 \\ H_{L_1} = 300 - L_1 sin\alpha_0 - \frac{R}{cos\gamma_0} \end{cases} \quad (8)$$

The relationship between pressure and angle in the detection device is discerned via Equations (9) and (10).

$$F_n = G\cos\alpha - \frac{L_1 F_{DS}\sin\gamma}{L_T} = k\Delta x_n = kL_T(tan\alpha_n - tan\alpha_{n-1}) \quad (9)$$

$$\Delta F_n = F_n - F_{n-1} = k(\Delta x_n - \Delta x_{n-1}) = kL_T(tan\alpha_{n-2} - 2tan\alpha_{n-1}) \quad (n = 1,2,3\ldots\ldots) \quad (10)$$

The value of the change in the displacement of the profiling device in the vertical direction is

$$\Delta H_N = \frac{\Delta x_n - \Delta x_{n-1}}{\cos\alpha_n} = \frac{\Delta F_n}{k\cos\alpha_n} \quad (11)$$

In the formula, the following applies:

ΔH_N—The value of the change in the displacement of the profiling device in the vertical direction, m.

The initial values of the variables in Equations (9) and (10) are calculated by Equation (12). F_{DS} (the ground reaction force) is determined through calibration, involving the vertical damping coefficient and motion damping characteristics of the profiling wheel. Based on rheological theory, the soil is modeled as a nonlinear elastic material. Under the assumption that soil deformation dominates during wheel operation (negligible wheel structural deformation), F_{DS} can be expressed as

$$\begin{cases} F_{n0} = k\Delta x_0 = G\cos\alpha_0 \\ F_{DS} = K_1 H + C\dot{H} \end{cases} \quad (12)$$

In the formula, the following applies:

K_1—The resistance coefficient of the ditching machine in the vertical direction, N/m.

C—The damping of the trenching machine's movement in the vertical direction, N/(m/s).

H—The depth of the trenching machine compared to the ridge surface, m.

$\dot{H}$—The first-order time derivative of H, m/s.

Equation (5) is the relationship between the angle measured and collected by the inclination sensor and the height of the ridge surface. Equation (9) is the relationship between the pressure measured and collected by the pressure sensor and the angle measured and collected by the inclination sensor. The definitions of the variables in this chapter are shown in Table 1.

Table 1. Parameter definition table.

Parameters	Connotation	Unit
H_1	the vertical distance between the transverse bar of the detection device and the center of the profiling wheel	m
L_1	the center distance between the center of the profiling wheel and the center of the upper center of the inclination sensor	m
R	the radius of the profiling wheel	m
L_T	the vertical distance between the center of the upper center of the inclination sensor and the spring	m

Table 1. *Cont.*

Parameters	Connotation	Unit
F_n	the pressure applied to the detection device at moment n	N
ΔF_n	the change value of F_n in the adjacent detection point	N
G	the gravity of the profiling wheel and the bracket	N
F_{DS}	the force of the ground on the profiling wheel	N
Δx_0	the initial change value of the spring	m
Δx_n	the change value of the spring at moment n	m
k	the elastic coefficient of the spring	-
H_y	the vertical distance between the center of the profiling wheel with radius R and the contact surface	m
C_h	the distance between the contact surface of the profiling wheel and the vertical ridge surface of the center line of the profiling wheel	m
H_L	the distance between the ridge surface	m
ΔH_{Ln}	the change value of H_L in the adjacent detection point	m
ΔH_N	the value of the change in the displacement of the profiling device in the vertical direction	m
α	the angle between the profiling wheel and the transverse bar	°
γ	the angle between the vertical and vertical directions of the profiling wheel	°
K_1	the resistance coefficient of the ditching machine in the vertical direction	N/m
C	the damping of the trenching machine's movement in the vertical direction	N/(m/s)
H	the depth of the trenching machine compared to the ridge surface	m
$\dot{H}$	The first-order time derivative of H	m/s

2.2.2. Design of Sowing Depth Adjustment Mechanism

Due to variations in peanut rotary tillage equipment or field soil types, the ridge surface may become uneven after ridging, causing ground fluctuations. The furrow opener requires floating adjustment to adapt to these changes. By adjusting the hydraulic cylinder and the position of the parallel four-bar linkage, the system accommodates these variations. The schematic diagram is shown in Figure 4.

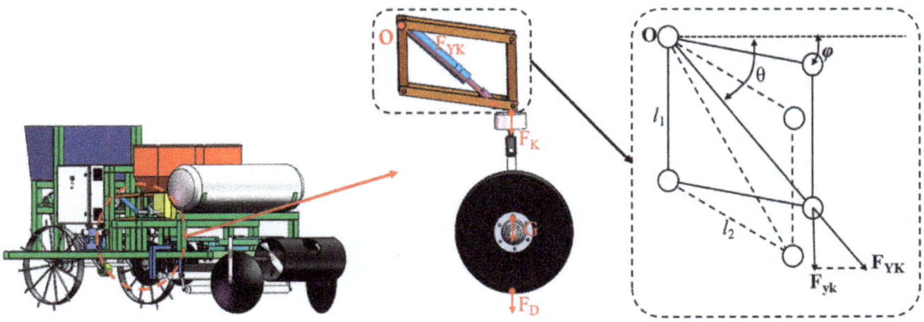

Figure 4. Theoretical calculation schematic diagram of executive regulating device.

In Figure 4, F_{YK} is the downforce of the hydraulic cylinder on the parallel four-bar linkage rod, N; F_{yk} is the hydraulic cylinder in the vertical direction component of N; G is the gravity of the floating part, N; φ is the swinging angle of the four-bar connecting rod, °; and θ is the angle of swing of the hydraulic cylinder, °.

The downward pressure of the hydraulic cylinder on the parallel four-bar linkage is in the vertical direction:

$$\begin{cases} F_{yk} = F_K = F_{YK} sin\theta \\ \tan\theta = \frac{l_1 + l_2 \sin\varphi}{l_2 \cos\varphi} \end{cases} \tag{13}$$

$$F_D = F_{yk} + G \tag{14}$$

In the formula, the following applies:
F_K—The force detected by the sensor.
F_D—The downforce of the execution adjustment device on the ground.
The downward pressure of the execution adjustment device on the sensor is

$$F_{yk} = F_{YK} sin\left(\arctan\frac{l_1 + l_2 \sin\varphi}{l_2 \cos\varphi}\right) \quad (15)$$

Because the execution adjustment device acts on the trenching machine with all the underground pressure, the following applies:

$$F_D = F_{YK} sin\left(\arctan\frac{l_1 + l_2 \sin\varphi}{l_2 \cos\varphi}\right) + G \quad (16)$$

The downforce of the execution adjustment device on the ground at moment n can be shown as

$$F_{Dn} = F_{Kn} + G = F_{YKn} sin\left(\arctan\frac{l_1 + l_2 \sin\varphi_n}{l_2 \cos\varphi_n}\right) + G \quad (17)$$

In the formula, the following applies:
F_{Dn}—The downforce of the execution adjustment device on the ground at moment n.
F_{Kn}—The force detected by the sensor at moment n.
Then, we can obtain the value of change in pressure at two neighboring points as

$$\Delta F_{Dn} = \Delta F_{Kn} = F_{Kn} - F_{Kn-1} = F_{YKn} sin\theta_n - F_{YKn-1} sin\theta_{n-1} \quad (18)$$

By structural design, the initial state of the four-link is horizontal, and the initial value of the above equation can be expressed as

$$\begin{cases} \varphi_0 = 0^\circ \\ \theta_0 = \arctan(l_1/l_2) \\ F_{K0} = G \\ F_{D0} = 0 \end{cases} \quad (19)$$

The value of the change in the displacement of the trenching device in the vertical direction is

$$\Delta H_{Dn} = l_1(\sin\varphi_n - \sin\varphi_{n-1}) \quad (20)$$

In the formula, the following applies:
ΔH_{Dn}—the value of the change in the displacement of the trenching device in the vertical direction.

The reaction force of the ground on the executive adjusting device comes from the deformation resistance and soil deformation friction when the two are in contact; according to the rheology theory, the soil is approximated to be a nonlinear elastic material, assuming that the deformation variable of the executive adjusting device and the ground is mainly the soil in the process of the operation, and then the ground's reaction force on the executive adjusting device is

$$F_D = K_1 H + C\dot{H} \quad (21)$$

F_D is also obtained by calibration, and the parameters required by its calibration formula are shown in Equation (12). Equation (21) shows the relationship between the pressure between the actuating regulator and the ground, which can be measured by a pressure sensor. The above parameters are defined as shown in Table 2.

Table 2. Parameter definition table.

Parameters	Connotation	Unit
F_{YK}	the downforce of the hydraulic cylinder on the parallel four-bar linkage rod	N
F_{yk}	the hydraulic cylinder in the vertical direction component of N	N
G	the gravity of the floating part below the sensor	N
l_1	the length of the vertical connecting rod	m
l_2	the length of the horizontal connecting rod	m
φ	the swinging angle of the four-bar connecting rod	°
θ	the angle of swing of the hydraulic cylinder	°
F_D	the downforce of the execution adjustment device on the ground	N
F_{Dn}	the downforce of the execution adjustment device on the ground at moment n	N
F_K	the force detected by the sensor	N
F_{Kn}	the force detected by the sensor at moment n	N
ΔF_{Dn}	the value of the change in pressure at two neighboring points on the ground at moment n	N
ΔF_{Kn}	the value of the change in pressure at two neighboring points by the sensor at moment n	N
ΔH_{Dn}	the value of the change in the displacement of the trenching device in the vertical direction	m

2.3. Hardware Design of Sowing Depth Control System

During field operation, the peanut planter encounters soil agglomeration due to residual crop debris, resulting in uneven ridge surfaces post ridging. This study employs a contact measurement method to precisely detect ridge surface irregularities.

The sowing depth control system hardware consists of three units: detection, execution, and control. The detection unit includes a profiling wheel, profiling wheel frame, pressure sensor, and inclination sensor to monitor ridge height changes in real time. The execution unit comprises a parallel four-bar linkage, hydraulic cylinder, electro-hydraulic proportional valve, and furrow opener to adjust sowing depth based on control signals. The control unit, centered on a PLC, integrates an HMI interface and laser sensor for data processing, parameter input, and system status display. A power module ensures the stable operation of all components.

The pressure sensor and inclination sensor communicate with the PLC via RS485, transmitting ground height data in real time. The PLC calculates height differences and outputs PWM signals to control the electro-hydraulic proportional valve. This drives the hydraulic cylinder to adjust the parallel four-bar linkage, changing the furrow opener depth. The HMI connects to the PLC via Ethernet for parameter setting and status monitoring. The laser sensor measures ground height and provides the difference between the execution and detection units. The hardware connection diagram is shown in Figure 5.

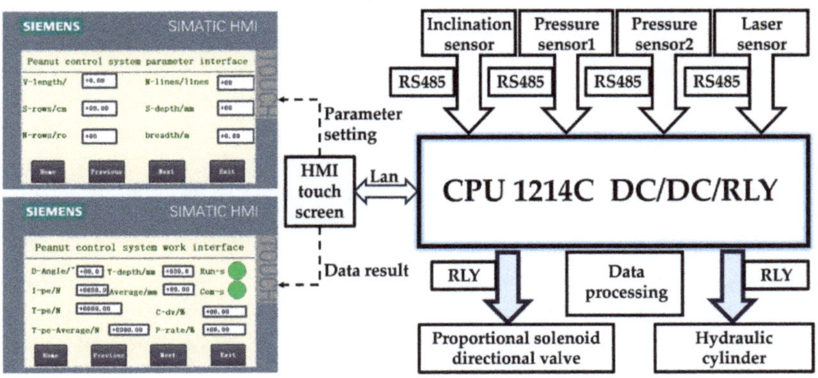

Figure 5. Hardware connection diagram of depth control system.

The detection unit is located at the front of the planter. The pressure sensor and inclination sensor are mounted on the profiling wheel frame, with the profiling wheel at the bottom of the frame. The PLC and HMI are installed in an easily accessible position, and the power module is placed near the PLC. The execution unit is positioned in the middle of the planter. The electro-hydraulic proportional valve, hydraulic cylinder, and parallel four-bar linkage are located at the front of the seed box. The furrow opener is placed at the bottom of the execution unit. The hardware parameters of the control system are listed in Table 3.

Table 3. Hardware parameters of the control system.

Component	Model	Key Parameters	Function
Pressure Sensor	DVLF-102	Range: 0–500 N, Accuracy: 2.0 ± 10%	Measures the pressure of the profiling wheel and the furrow opener
Inclination Sensor	LVT518T	Range: ±90°, Accuracy: ±0.1°	Detects profiling wheel frame angle
Hydraulic Cylinder	ROX28 × 125	Stroke: 60 mm, Force: 1000 N	Drives parallel four-bar linkage
Electro-Hydraulic Proportional Valve	EFBG-03-H	Response Time: <50 ms, Control Accuracy: ±0.1%	Controls hydraulic cylinder movement
PLC	Siemens S7-1214C	Input: 16 Channels, Output: 12 Channels, Communication: RS485	Data processing and control
HMI	Siemens KTP-400	Size: 4 inches, Resolution: 800 × 480	Parameter input and status display
Laser Sensor	BL-400NZ	Measurement Range: 200–260 mm, Accuracy: ±0.8 mm	Measures ground height for feedback

2.4. Control Strategy of Sowing Depth Control System

The software of the peanut planter's sowing depth control system includes data acquisition, processing, control algorithm, and human–machine interaction modules. The data acquisition module reads data from pressure sensors, inclination sensors, and laser sensors via RS485 and analog interfaces. The data processing module filters, calibrates, and calculates height differences. The control algorithm module employs a DLF-Fuzzy PID (double-loop feedback fuzzy PID) strategy to dynamically adjust PID parameters and output PWM signals to control the electro-hydraulic proportional valve. The execution control module drives the hydraulic cylinder to adjust the furrow opener depth. The human–machine interaction module enables parameter setting, status display, and fault alarm through the HMI. All modules work together to ensure precise sowing depth control. The system is programmed using Portal v16 software, with a workflow including initialization, main loop (data acquisition, processing, and control execution), and fault handling (anomaly detection and alarm).

The DLF-Fuzzy PID control strategy uses an outer loop to control height difference and an inner loop to regulate the hydraulic cylinder position. When the height difference from the target depth is significant, a feedforward system is introduced to enhance stability and precision. The response speed of the execution unit directly affects sowing depth consistency during ridge surface fluctuations. A mathematical model is established, with target depth as the factor and execution unit response time and downward pressure as variables, providing theoretical support for the DLF-Fuzzy PID control strategy.

2.4.1. Mathematical Model of DLF-Fuzzy PID Control Strategy

1. Mathematical Model Without Step Changes

The actuator uses a fuzzy controller algorithm to obtain a stable pressure output to ensure a consistent target depth if the change in depth between the target depth and the previous target point does not exceed 10 mm. Through Equations (11) and (20), the following can be obtained.

$$\begin{cases} \Delta H_{Dn} = \Delta H_N \\ kL_T(tan\alpha_{n-2} - 2tan\alpha_{n-1}) = F_{YKn}sin\theta_n - F_{YKn-1}sin\theta_{n-1} \end{cases} \quad (22)$$

2. Mathematical Model with Step Changes

When the depth change between the target depth and the previous target point is more than 10 mm, a feedforward system is introduced to input a reserve amount of pressure, and the PID controller is used to adjust the algorithm in order to obtain a stable pressure output and to ensure that the target depth is consistent. Through Equations (11) and (20), the following can be obtained.

$$\begin{cases} \Delta H_{Dn} = \Delta H_N + 5 \\ kL_T(tan\alpha_{n-2} - 2tan\alpha_{n-1}) = F_{YKn}sin\theta_n - F_{YKn-1}sin\theta_{n-1} + F_b \end{cases} \quad (23)$$

In the formula, the following applies:

F_b—the step pressure input when ΔH_{Dn} is equal to (ΔH_N + 5 mm).

Based on the change in H_L of the monopoly surface, the value of the amount of change between that point and the previous point is calculated, and the condition for whether a step occurs is set to a difference of more than 10 mm between the two points;

When 0 < ΔH_{Ln} < 5 mm, i.e., the variation between two points does not exceed 5 mm, PWMi is adaptively adjusted by fuzzy PID;

When 5 < ΔH_{Ln} < 10 mm, when the change between the two points exceeds 5 mm but does not exceed 10 mm, the feedforward system PWMi is introduced to be adjusted adaptively by fuzzy PID;

When ΔH_{Ln} > 10 mm, the amount of change between the two points exceeds 10 mm, and it undergoes a step change and is introduced into the feedforward system; PWMi is adjusted using a PID controller.

2.4.2. Design of DLF-Fuzzy PID Control Strategy

DLF-Fuzzy PID combines the advantages of fuzzy PID [29,30] control and conventional PID control, ensuring real-time performance and accuracy through double-loop feedback. When a step change in target depth occurs, the system automatically switches to conventional PID control to avoid overshoot caused by fuzzy PID control. During stable target depth conditions, fuzzy PID control is employed to enhance dynamic response and steady-state precision. Additionally, feedforward control is introduced during step changes to further reduce response time and steady-state error. The depth control program diagram is shown in Figure 6.

DLF-Fuzzy PID monitors and adjusts system states in real time through inner and outer loops. The inner loop ensures rapid response, while the outer loop achieves the precise tracking of the target depth. The pressure of the proportional pressure-reducing valve is regulated by outputting PWM signals of varying frequencies, enabling precise pressure control and ensuring sowing depth stability. By intelligently switching control modes and incorporating feedforward control, DLF-Fuzzy PID significantly improves the system's dynamic performance and steady-state accuracy, making it suitable for high-precision sowing operations.

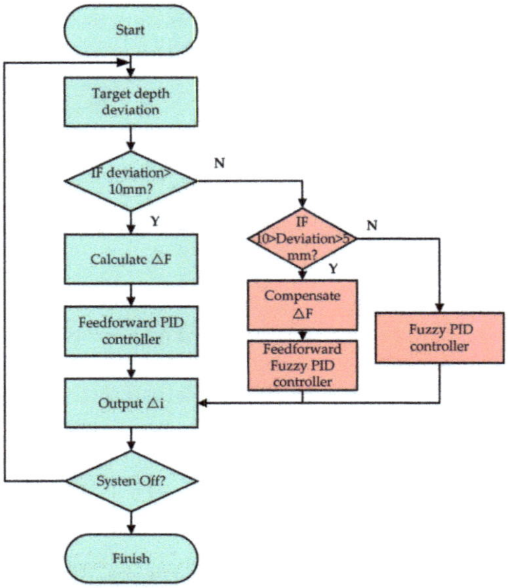

Figure 6. Depth control program diagram.

3. Fuzzy PID algorithm

(1) Design of fuzzy PID control system

The pressure value at the target depth is compared with the feedback value from the pressure sensor, and the error e and the error change rate e_c are used as input quantities. The PID parameters are adjusted through the fuzzy PID controller outputs Kp, Ki, and Kd to correct the PWM generation frequency, and at the same time, the ground undulation perturbation information is transferred into the feedforward controller to be analyzed and processed, and the output PWM control signals are used to control the opening size of the proportional valve and the lifting and lowering of the hydraulic cylinder. The current loop is used as the inner loop, with the pressure loop as the output, and the current information feedback and proportional valve information is used as the input to offset the perturbation and ensure the stability of the dynamic performance of the hydraulic cylinder. As inputs are the difference e between the target pressure of the hydraulic cylinder of the actuator and the pressure feedback and the rate of change in the difference e_c, e and e_c are exact quantities, which are fuzzified to obtain the fuzzy quantities E and Ec. The three parametric quantities are ΔKp, ΔKi, and ΔKd, which represent the correction parameters of the controller. The three correction parameters are automatically optimized and adjusted in real time to achieve the self-tuning of the parameters. The depth of peanut planting in Shandong is generally between 4 and 6 cm, and the height variation range of monopoly height is 12–16 cm from Equation (5), and the pressure variation range is 0–800 N from Equation (10), Equation (22), and Equation (23), and the fuzzy theoretical domains of the input variables e and e_c are $[-6, 6]$ through several tests and parameter adjustments.

For the output variables ΔKp, ΔKi and ΔKd, the fuzzy theoretical domains are [8.5, 9.1], [0.2, 0.8] and [0.09, 0.15]. The fuzzy subsets of fuzzy linguistic variables are defined as *FA*, *FB*, *FC*, *DL*, *ZE*, *ZF*, and *ZG*, meaning Negative Big, Negative Medium, Negative Small,

Zero, Positive Small, Positive Medium, and Positive Big. The parameter settings of the fuzzy PID algorithm are shown in Table 4.

Table 4. Parameter settings of fuzzy PID algorithm.

Input and output variables	e	e_c	ΔKp	ΔKi	ΔKd
Fuzzy language variable	E	E_C	ΔKp	ΔKi	ΔKd
Fuzzy subset			[*FA FB FC DL ZE ZF ZG*]		
Fuzzy domain	[−6, 6]	[−6, 6]	[8.5, 9.1]	[0.2, 0.8]	[0.09, 0.15]

(2) Determination of affiliation function

In this study, the triangular affiliation function is selected for input variables and the Gaussian-type affiliation function is selected for output variables. The fuzzy controller is of the Mamdani type. Its membership function is shown in Figure 7.

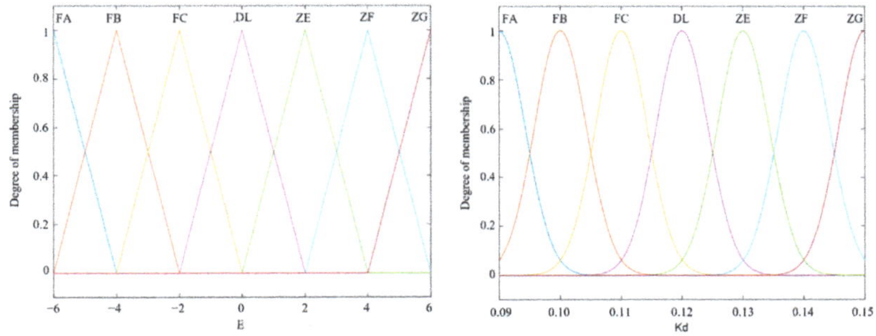

Figure 7. Graph of membership function.

(3) Fuzzy control rule setting

The fuzzy rule-based in-depth control system is a set of control rules accumulated by the practical experience and expert knowledge of engineers and technicians. The fuzzy design rules in this paper are jointly analyzed by programmers and field operators and formulated on the basis of professional technology and operational experience, as shown in Table 5.

Table 5. Control parameter self-tuning rule table.

Kp/Ki/Kd	e_c						
e	***FA***	***FB***	***FC***	***DL***	***ZE***	***ZF***	***ZG***
FA	*ZG/FA/ZE*	*ZG/FA/FC*	*ZF/FB/FA*	*ZF/FB/FA*	*ZE/FC/FA*	*DL/DL/FB*	*DL/DL/ZE*
FB	*ZG/FA/ZE*	*ZG/FA/FC*	*ZF/FB/FA*	*ZE/FC/FB*	*ZE/FC/FB*	*DL/DL/FC*	*FC/DL/DL*
FC	*ZF/FA/DL*	*ZF/FB/FC*	*ZF/FC/FB*	*ZE/FC/FB*	*DL/DL/FC*	*FC/ZE/FC*	*FC/ZE/DL*
DL	*ZF/FB/DL*	*ZF/FB/FC*	*ZE/FC/FC*	*DL/DL/FC*	*FC/ZE/FC*	*FB/ZF/FC*	*FB/ZF/DL*
ZE	*ZE/FB/DL*	*ZE/FC/DL*	*DL/DL/DL*	*FC/ZE/DL*	*FC/ZE/DL*	*FB/ZF/DL*	*FB/ZG/DL*
ZF	*ZE/DL/ZG*	*DL/DL/FC*	*FC/ZE/ZE*	*FB/ZE/ZE*	*FB/ZF/ZE*	*FB/ZG/ZE*	*FA/ZG/ZG*
ZG	*DL/DL/ZG*	*DL/DL/ZF*	*FB/ZE/ZF*	*FB/ZF/ZF*	*FB/ZF/ZE*	*FA/ZG/ZE*	*FA/ZG/ZG*

The parameters (*Kp*, *Ki*, and *Kd*) of the PID controller are adaptively tuned online through fuzzy control rules based on real-time measurements of pressure error (e) and the error rate (e_c). The control strategy operates in three phases: When $|e|$ is large, *Kp* is maximized with minimized *Kd* and zero *Ki* to enhance response speed while preventing derivative-induced overshoot. For intermediate $|e|$ and $|e_c|$ values, all parameters maintain moderate magnitudes. When $|e|$ becomes small, *Kp* and *Ki* are increased to improve steady-state accuracy, with *Kd* appropriately adjusted to suppress oscillations. The adjustment algorithm formulas for *Kp*, *Kd*, and *Ki* are as follows:

$$\begin{cases} Kp = Kp' + \{E, E_C\}Kp = Kp' + \Delta Kp \\ Ki = Ki' + \{E, E_C\}Ki = Ki' + \Delta Ki \\ Kd = Kd' + \{E, E_C\}Kd = Kd' + \Delta Kd \end{cases} \tag{24}$$

In the formula, *Kp*, *Ki*, and *Kd* are PID control parameters, with their initial values $K_p{}'$, $K_i{}'$, and $K_d{}'$ derived from conventional methods. During operation, the system continuously monitors the output, calculates the error e and its rate of change e_c, and fuzzifies them into linguistic variables E and Ec. Based on the fuzzy adjustment matrix, *Kp*, *Ki*, and *Kd* are dynamically adjusted to achieve adaptive fuzzy PID control.

(4) Defuzzification methods

The output of the fuzzy controller in this control system is a fuzzy dataset after inference. To obtain a precise numerical value, defuzzification is required. To ensure real-time performance, this study adopts the centroid method via the offline lookup of the fuzzy controller's control rule table as the output strategy. The centroid of the fuzzy set is selected as the crisp output, defined by the following function:

$$df_g(u) = \frac{\int_a^b u \cdot \mu_c(u)du}{\int_a^b \mu_c(u)du} \tag{25}$$

In the formula, $df_g(u)$ is the clear value and $\mu_c(u)$ is the closeness function of u.

4. PID algorithm with introduction of feedforward term

When the ridge surface detection device detects a step change in target depth, the feedforward term is introduced to reduce system response time. The deviation e between the target pressure and the pressure sensor value, calculated based on depth theory, serves as the input. The relationship between input e and output u is defined as follows:

$$u(t) = \Delta K_p e(t) + \Delta K_i \int_0^t e(t)dt + \Delta K_d \frac{de(t)}{dt} \tag{26}$$

In the formula, *Kp* is the proportional coefficient of the system; *Ki* is the integral coefficient; and *Kd* is the differential coefficient. The PID parameters are adjusted through trial and error. Finally, the proportional coefficient *Kp* is determined to be 1.8, the integral coefficient *Ki* is 1.6, and the differential coefficient *Kd* is 0.1. A flowchart of the DLF-Fuzzy PID control process is shown in Figure 8.

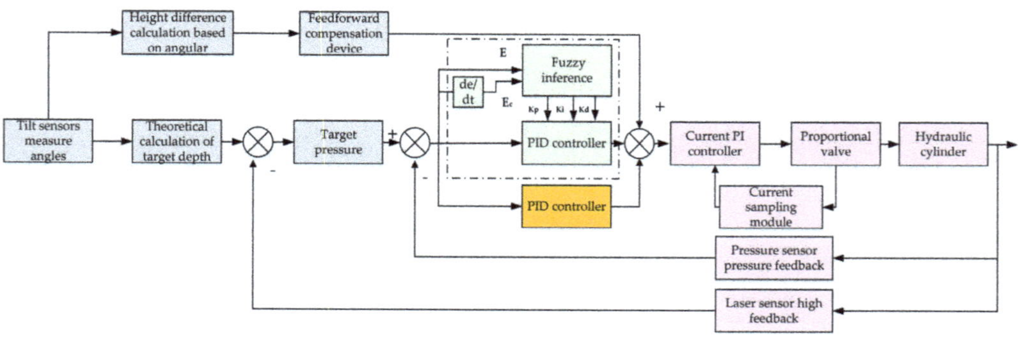

Figure 8. The flowchart of the DLF-Fuzzy PID control process.

3. Results

3.1. Simulation Verification and Analysis Based on MATLAB and AMESIM

To validate the algorithm's advantages, simulation models of double-loop feedback fuzzy PID (DLF-Fuzzy PID), traditional PID, and fuzzy PID control were developed in Matlab-R2020b/Simulink, with a hydraulic simulation model created in Amesim-2020.2 for co-simulation (Figure 9).

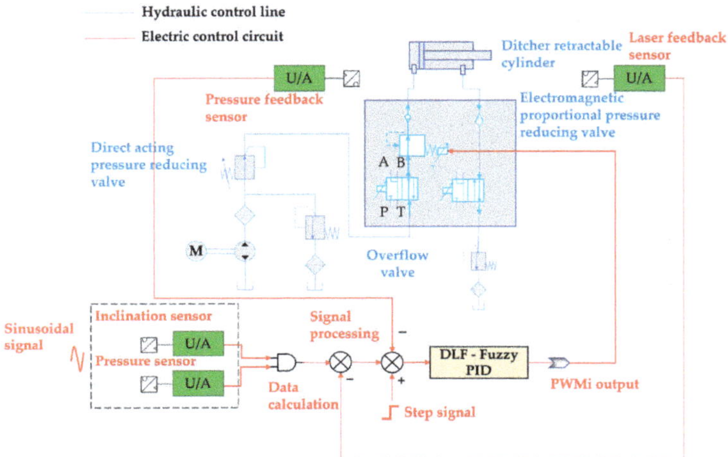

Figure 9. Simulink and Amesim co-simulation.

The simulation used a sinusoidal signal to mimic ground ridges and a step signal to simulate ground disturbances, replicating peanut planter operations on gentle ridge slopes. Key parameters included a control system sampling time of 0.025 s, proportional solenoid valve current range of 100–2600 mA, hydraulic cylinder total length of 300 mm, piston rod stroke of 60 mm, maximum oil supply pressure of 5 Mpa, and hydraulic cylinder inner and outer diameters of 28 mm and 45 mm, respectively. Partial electro-hydraulic system parameters are listed in Table 6.

Table 6. Partial electro-hydraulic system parameter settings.

Parameters	Connotation	Value	Unit
m	Armature mass	0.5	kg
K	Electromagnetic force conversion coefficient	160.5	m/s×A
K_y	Force conversion coefficient	470	$N \times m^{-1}$
E	Armature assembly spring stiffness	1×10^4	$N \times m^{-1}$
R	Internal resistance of proportional valve electromagnet	25	Ω
L	Proportional valve electromagnet inductance	6.6×10^{-2}	H
K_q	Proportional valve flow gain is sufficient	2.4	-
K_{ce}	Total flow pressure coefficient	4.5×10^{-12}	-
M	The load is converted to the equivalent mass on the piston	150	kg
β	Effective bulk modulus of elasticity port	1×10^9	Pa
V_t	Total volume of hydraulic cylinder	6×10^{-4}	m^3
A_p	Equivalent area of load flow	3.75×10^{-3}	m^2

A co-simulation was conducted 8000 times, with Figure 10a,b presenting partial magnified views of input simulation signals from representative cases. Figure 10a illustrates scenarios with significant differentials between two measurement points on the ridge surface, while Figure 10b demonstrates minimal differentials corresponding to smoother ridge profiles. To validate control performance under both flat and undulating ridge conditions, Figure 10c,d comparatively present results from conventional PID, Fuzzy PID, and the proposed double-loop feedback Fuzzy PID control methods. All three control strategies exhibit adaptive parameter adjustment capabilities in response to ridge profile variations.

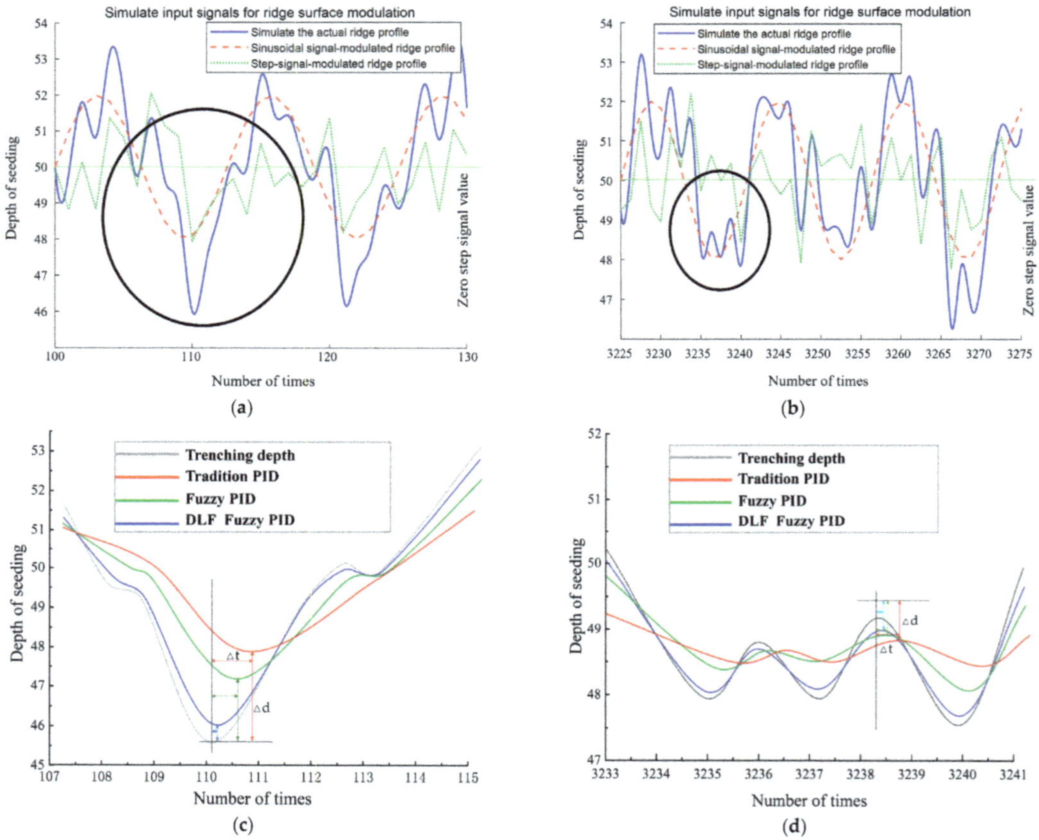

Figure 10. A comparative analysis of simulated signal input plots and control strategy effectiveness in a seedbed depth control system. (**a**,**b**) Zoomed-in views of partial simulated input signals; (**c**,**d**) control strategy performance curves at 50 mm sowing depth.

During sudden ridge deformation, all strategies demonstrate comparable response times with significant parameter adjustments yet exhibit substantial accuracy discrepancies. Under stable ridge conditions, improved tracking accuracy correlates with prolonged response times across the control schemes. The parameter Δt denotes the deviation between the algorithm's response time and the prescribed standard duration, reflecting system responsiveness. Δd represents the discrepancy between actual actuator adjustment depth and theoretical requirements, indicating control accuracy and tracking performance. Under operational parameters of 3 km/h sowing speed and 50 mm target depth, the enhanced DLF Fuzzy PID algorithm demonstrates marked improvements in both response speed and control accuracy compared to conventional PID and standard Fuzzy PID implementations. It enhances dynamic performance and stability, making it more suitable for high-precision sowing operations.

3.2. Field Trials

To verify the performance of the peanut planter with a dynamic sowing depth control system, field tests were conducted using the 2MB-1/2 peanut planter manufactured by Shandong Yuanquan Machinery Co., Ltd. (Yishui County, Linyi City, China). The tests used Luhua peanut seeds and were carried out in May 2023 in Zhujiazhuang Village, northwest of Yishui County, Linyi City, Shandong Province. The test plot measured 100 m in length and 40 m in width, with soil primarily consisting of cinnamon soil (brown earth) characterized by a loam or light loam texture, which had undergone rotary tillage treatment. The field trial is shown in Figure 11.

(**a**) (**b**)

Figure 11. Field trial diagram. (**a**) Field trial; (**b**) measurement data.

In order to make the comparison test more accurate, the traditional sowing control method was adopted on the peanut planter side, and the rest of the sowing devices were simplified to avoid the deviation of results caused by human factors. On the basis of GB-T6973-2005 Test Methods for Single Seeder (Precision Seeder) [31], the qualification rate of sowing depth and coefficient of variation in sowing depth were tested under the influence of different speeds.

3.2.1. Test Methods

The test parameters were set according to Shandong peanut sowing standards, with a typical sowing depth of 40–60 mm and a working speed of 3–5 km/h. Based on the mechanical plant spacing of the peanut planter, three sowing depths (40 mm, 50 mm, and 60 mm) and three speeds (3 km/h, 4 km/h, and 5 km/h) were tested. To eliminate acceleration and deceleration effects, measurements were taken in the middle area, 20 m from each end of the plot. The ground surface height was calibrated, and the soil was scraped to expose peanut seeds. The distance between the seeds and the ground surface was measured. Six groups of measurements were conducted, with 100 seeds per group. The test data are shown in Figure 12.

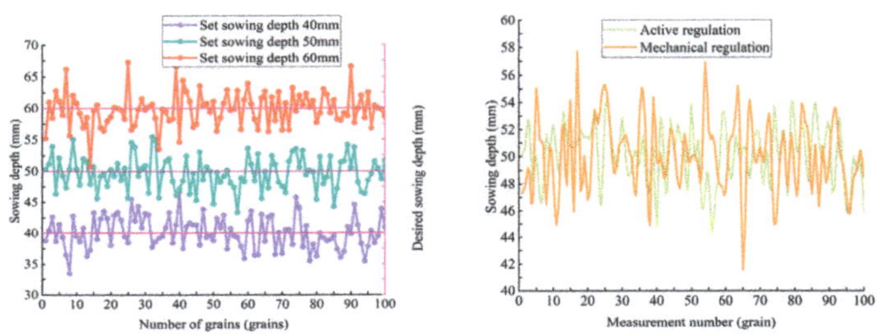

(a) (b)

Figure 12. *Cont.*

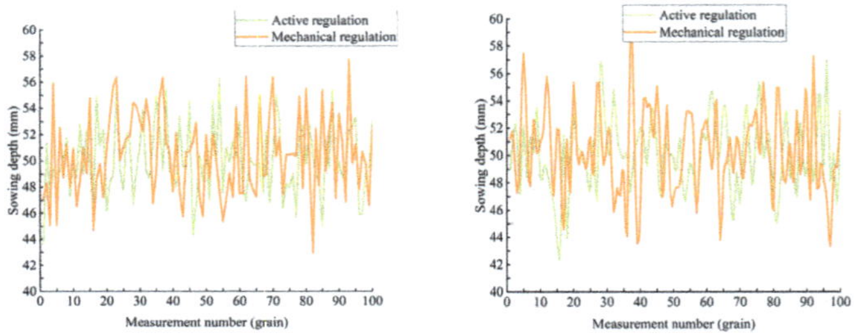

(c) (d)

Figure 12. Experimental data. (**a**) When the speed is 3 km/h, the field data of active adjustment methods with different sowing depths are set. (**b**) The field data of two adjustment methods when the sowing depth is set to 50 mm at a speed of 3 km/h. (**c**) The field data of two adjustment methods when the sowing depth is set to 50 mm at a speed of 4 km/h. (**d**) The field data of two adjustment methods when the sowing depth is set to 50 mm at a speed of 5 km/h.

3.2.2. Calculation of Test Results

The experimental data of the two different methods were analyzed according to the formula of each calculated parameter, and the results are shown in Table 7.

$$\overline{h} = \frac{\Sigma h_i}{N} \tag{27}$$

$$\eta = \frac{n}{N} \times 100\% \tag{28}$$

$$S_h = \sqrt{\frac{\Sigma(h_i - h)^2}{N}} \tag{29}$$

$$V_h = \frac{S_h}{\overline{h}} \times 100\% \tag{30}$$

In the formula, H—the sowing depth pass rate, %; n—the number of seeding depth passes, piece; N—the total number of sowing depth measurement points, piece; h_i—seeding depth measurement, mm; S_h—the standard deviation of sowing depth, mm; V_h—the coefficient of variation of sowing depth, %; and $\overline{h}$—mean value of sowing depth, mm.

Table 7. Comparison of results of each test's data.

Measuring Parameters	Operating Speed/(km/h)			Sowing Depth (mm)			Adjustment Mode
Working speed/km/h	3	4	5	3	3	3	-
Sowing depth/mm	50	50	50	40	50	60	-
Average sowing depth/mm	50.4	47.6	44.04	41.27	50.4	44.32	Dynamic
	46.02	52.9	41.02	46.32	46.02	44.97	Mechanical
Sowing depth pass rate/%	94.6	90.4	86.19	94.03	94.6	92.35	Dynamic
	89.21	83.7	75.78	89.67	89.21	87.52	Mechanical
Standard deviation of sowing depth/mm	2.38	3.74	6.19	2.36	2.38	4.67	Dynamic
	4.53	5.67	7.95	5.02	4.53	5.39	Mechanical
Sowing depth coefficient of variation	3.99	5.76	6.94	5.32	3.99	5.84	Dynamic
	7.94	10.95	14.28	7.06	7.94	12.69	Mechanical

3.2.3. Analysis of Test Results

A comprehensive analysis of the 3 km/h and seeding depth at 50 mm, compared with the setting depth, dynamic depth average difference of 0.4 mm, and the traditional average depth of 3.98 mm, was conducted. The qualified rate of dynamic sowing depth was 94.6%, while that of traditional sowing depth was 89.21%. The standard deviation of dynamic sowing depth was 2.38 mm and that of traditional sowing depth was 4.53 mm. At 4 km/h and 5 km/h, the mean sowing depth, pass rate, standard deviation and coefficient of variation decreased when the sowing depth was 50 mm. At 3 km/h, when the sowing depth is 40 mm and 60 mm, and when the sowing depth is 50 mm, the coefficients reach the best results. When the speed rises from 3 km/h to 5 km/h, the performance of the dynamically adjusted peanut planter is better than that of the traditional peanut planter. The complex shows the better performance of the stable and consistent adjustment of the sowing depth of the hydraulic adjustment mode, achieving the purpose of the design.

4. Conclusions

(1) Compared to traditional peanut planters, this study developed a dynamic sowing depth control system based on an existing planter. It uses hydraulic control to adjust downward pressure and includes a ridge height detection device.
(2) A dynamic control mathematical model was established, incorporating angle–pressure–depth dynamics. A double-loop feedback fuzzy PID (DLF-Fuzzy PID) control strategy was designed. Co-simulation using Matlab-R2020b and Amesim-2020.2, along with a downward pressure control platform, validated the algorithm. It demonstrated significant advantages over traditional methods.
(3) Field tests showed that the dynamic sowing depth adjustment outperformed traditional methods at speeds of 3–5 km/h. The dynamic system exhibited stable and reliable control performance. At 5 km/h, the sowing depth qualification rate exceeded 85%, laying the foundation for precision peanut planting.

Author Contributions: Conceptualization, M.L.; methodology, M.L.; software, M.L.; validation, X.C., Y.G. and P.W.; formal analysis, Y.G.; investigation, M.L.; resources, M.L.; data curation, M.L.; writing—original draft preparation, M.L.; writing—review and editing, M.L.; visualization, P.W.; supervision, S.S.; project administration, M.L.; funding acquisition, M.L. All authors have read and agreed to the published version of the manuscript.

Funding: This research was supported by Dezhou Intelligent Equipment Research and Development Center (PT2025KJT004) and Shandong Province Higher Education Undergraduate Teaching Reform Project' Research and Practice on the Construction of Efficient Classroom of Automation Specialty under the Background of New Engineering'(Z2021212).

Institutional Review Board Statement: Not applicable.

Data Availability Statement: The original contributions presented in the study are included in the article material; further inquiries can be directed to the corresponding authors.

Conflicts of Interest: The authors declare no conflicts of interest.

References

1. Sun, C.R. Analysis of peanut planting technology and measures to improve planting efficiency. *Hebei Agric. Mach.* **2024**, *22*, 94–96. [CrossRef]
2. Zhu, X.Q. Analysis of peanut cultivation techniques and measures to improve planting benefits. *Hebei Agric. Mach.* **2023**, *21*, 136–138. [CrossRef]
3. Li, D.D.; Tang, K.; Zeng, N.B. Effects of imbibition characteristics and sowing depth on interspecific soil temperature and humidity and seedling emergence of peanut seeds with different grain types. *Shandong Agric. Sci.* **2022**, *06*, 27–34. [CrossRef]
4. Zhang, D.X.; Liu, J.; Yang, L. Research on soil moisture measurement sensor in seeding ditch based on VIS-NIR. *Trans. Chin. Soc. Agric. Mach.* **2021**, *52*, 218–226.
5. Gao, Y.Y.; Wang, X.; Yang, S. Design and experiment of Pneumatic Downforce Control System for Seeder. *Trans. Chin. Soc. Agric. Mach.* **2019**, *50*, 19–29+83.
6. Guan, X.K.; Yang, M.D.; Bai, T.T. Suitable deep sowing to improve the seedling emergence rate of summer maize by subsurface drip irrigation and promote seedling growth. *Trans. Chin. Soc. Agric. Eng.* **2016**, *32*, 75–80.
7. Chang, X.L.; Shang, S.Q.; He, X.N. Design and Test of Precision Planter for Peanut Community Breeding. *J. Agric. Mech. Res.* **2025**, *47*, 121–126. [CrossRef]

8. Geng, D.Y.; He, K.; Yin, X. Design and test of corn-peanut intercropping seeding-fertilizing machine. *Trans. Chin. Soc. Agric. Eng.* **2017**, *33*, 34–41. [CrossRef]
9. Wang, L.S.; Ma, Y.; Wang, X.Z. Simulation on Precision Single Sowing Depth Control of Seeder Based on Fuzzy Control. *J. Agric. Mech. Res.* **2016**, *38*, 86–89. [CrossRef]
10. Nielsen, S.K.; Nørremark, M.; Green, O. Sensor and control for consistent seed drill coulter depth. *Comput. Electron. Agric.* **2016**, *127*, 690–698. [CrossRef]
11. Nielsen, S.K.; Munkholm, L.J.; Lamandé, M. Seed drill instrumentation for spatial coulter depth measurements. *Comput. Electron. Agric.* **2017**, *141*, 207–214. [CrossRef]
12. Nielsen, S.K.; Munkholm, L.J.; Lamandé, M. Seed drill depth control system for precision seeding. *Comput. Electron. Agric.* **2018**, *144*, 174–180. [CrossRef]
13. Han, B.; Yang, Y.N.; Wang, H.W. Design and Bench Test of PID automatic Control System for embedded depth of weeding Parts between seedlings. *Trans. Chin. Soc. Agric. Eng.* **2018**, *34*, 68–77.
14. Kiani, S. Automatic on-line depth control for seeding units using a non-contacting ultrasonic sensor. *Int. J. Nat. Eng. Sci.* **2012**, *6*, 39–42.
15. Jesen, L.D.; Nelson, C.; Leclaire, J.P. Depth Control Device for Planting Implement. U.S. Patent No. 6,701,857, 03 September 2004.
16. Weatherly, E.T.; Bowers, C.G. Automatic depth control of a seed planter based on soil drying front sensing. *Trans. ASAE* **1997**, *40*, 295–305. [CrossRef]
17. Huang, D.Y.; Jia, H.L.; Qi, Y. Seed discharge monitoring System of planter based on polyvinylidene fluoride piezoelectric film. *Trans. Chin. Soc. Agric. Eng.* **2013**, *29*, 15–22.
18. Fu, W.Q.; Dong, J.J.; Mei, H.B. Design and experiment of Lower Pressure Control System for Maize seed monomer. *Trans. Chin. Soc. Agric. Mach.* **2018**, *49*, 68–77. [CrossRef]
19. Li, Y.H.; Meng, P.X.; Geng, D.Y. Research on intelligent control system of maize sowing depth. *Trans. Chin. Soc. Agric. Mach.* **2016**, *47*, 62–68+42.
20. Xie, Y.M.; Alleyne, A.; Greer, A. Fundamental limits in combine harvester header height control. *J. Dyn. Syst. Meas. Control-Trans. ASME* **2013**, *135*, 034503. [CrossRef]
21. Xie, Y.M.; Alleyne, A. Two degree of freedom control synthesis with applications to agricultural systems. *J. Dyn. Syst. Meas. Control* **2014**, *136*, 051006. [CrossRef]
22. Xie, Y.M.; Alleyne, A. A robust two degree-of-freedom controller for systems with both model and measurement uncertainty. *Control Eng. Pract.* **2014**, *25*, 55–65. [CrossRef]
23. Zhao, J.H.; Liu, L.J.; Yang, X.J. Design and test of teff planter based on PLC. *Trans. Chin. Soc. Agric. Mach.* **2016**, *47*, 84–89. [CrossRef]
24. Wen, L.P.; Zhang, Y.; Fan, X.F. Precision seeder sowing depth control system based on PLC study. *J. Agric. Mech. Res.* **2014**, *4*, 98–101.
25. Xia, J.F.; Li, D.; Liu, G.Y. Research on electro-hydraulic monitoring System of tractor suspension depth based on Angle detection. *Trans. Chin. Soc. Agric. Mach.* **2021**, *52*, 386–395. [CrossRef]
26. Qi, W.C.; Li, Y.M.; Zhang, J.H. Double closed-loop fuzzy PID Control Method for Body Leveling of Tractors in hilly areas. *Trans. Chin. Soc. Agric. Mach.* **2019**, *50*, 17–23+34. [CrossRef]
27. Wang, C. Research on intelligent control of garlic sowing depth. *Agric. Mech. Res.* **2018**, *40*, 185–188, 193. [CrossRef]
28. Ding, X.; Li, R.; Cheng, Y.; Liu, Q.; Liu, J. Design of and Research into a Multiple-Fuzzy PID Suspension Control System Based on Road Recognition. *Processes* **2021**, *9*, 2190. [CrossRef]
29. Chen, L.Q.; Xie, B.B.; Li, Z.D. Design of precision seeding control system for maize based on Double closed-loop PID fuzzy algorithm. *Trans. Chin. Soc. Agric. Eng.* **2018**, *34*, 33–41.
30. Song, S.R.; Ruan, Y.C.; Hong, T.S. Self-tuning fuzzy PID control of liquid Pressure in orchard pipe spray System. *Trans. Chin. Soc. Agric. Eng.* **2011**, *27*, 157–161.
31. *GB-T6973-2005*; Test Methods for Single Seeder (Precision Seeder). Machinery Industry Publishing House: Beijing, China, 2005.

agriculture

MDPI

Article

Study on Structure Design and Parameter Optimization of Diversion Rifled Feeder Based on CFD-DEM

Wancheng Dong [1], Xiongye Zhang [1], Zhen Jiang [1], Xue Hu [1,2], Yun Ge [1] and Lixin Zhang [1,2,*]

1 College of Mechanical and Electrical Engineering, Shihezi University, Shihezi 832003, China; dongwancheng@shzu.edu.cn (W.D.); 20212309406@stu.shzu.edu.cn (X.Z.); 20222509002@stu.shzu.edu.cn (Z.J.); 201601142@shzu.edu.cn (X.H.); 20222009022@stu.shzu.edu.cn (Y.G.)
2 School of Energy and Materials, Shihezi University, Shihezi 832003, China
* Correspondence: zlx_mac@shzu.edu.cn; Tel.: +86-18-899-596-017

Abstract: The feeder, a critical component of pneumatic conveying and distributing fertilizer machines for cotton, significantly impacts cotton quality and yield. This study addresses the limitations of conventional feeders, namely poor delivery stability and insufficient air–fertilizer mixture uniformity, by introducing a novel feeder design incorporating guide-flow chamber vanes. Based on the structural principles of pneumatic fertilizer distribution and the theory of vortex flow feeding, the new feeder design was developed and analyzed using a CFD-DEM coupled model to simulate its operational process. Response surface methodology was employed to optimize key structural parameters, aiming to enhance both air–fertilizer mixture uniformity and fertilizer delivery stability. The optimized design features two guide vanes, a fertilizer discharge contraction angle of 23.2°, and eight chamber lines. Validation through simulation confirmed the model's reliability, with a discrepancy of less than 5% between simulated and predicted values. Experimental validation on a pneumatic fertilizing test prototype further demonstrated the effectiveness of the optimization, resulting in a feeder stability coefficient of variation of 13.30% and a coefficient of variation in fertilizer distribution across the system's branches of 3.96%, satisfying practical operational requirements. After optimization, the stability and uniformity of the feeder fertilizer are significantly improved. The optimized feeder design effectively improves the stability and air–fertilizer mixing uniformity of cotton pneumatic fertilizing machines, providing valuable theoretical and technical support for their design optimization and performance enhancement.

Keywords: pneumatic conveying and distributing fertilizer machine; feeder; vortex flow feeding; gas–solid coupling; parameter optimization

1. Introduction

The application of pneumatic conveying technology in agriculture, especially in fertilizer distribution systems, has exhibited a notable upward trend in recent years [1–3]. With the advantage that it can pneumatically transport the fertilizer to multiple fertilizing points, the pneumatic seeding and fertilizing compound machine has become an important development direction of fertilizing machinery in the agricultural field at home and abroad [4–6]. At present, pneumatic fertigation technology is mainly used in the fertigation of rice, corn, wheat, and other crops [7,8]. The feeder, as the core component of the pneumatic conveying and distributing fertilizer machine, is responsible for the crucial function of transporting fertilizer from the storage bin to the fertilizer distributor via airflow [9]. Its performance directly affects the precision, efficiency, and uniformity of fertilization. The physical characteristics of seeds or fertilizers vary among cotton and other crops. Consequently, the performance requirements for feeders differ significantly. Therefore, it is crucial to consider the specific crop characteristics and application scenarios comprehensively.

The existing pneumatic fertilizing device adopts airflow to transport the fertilizer particles and distribute them to each branch for fertilizing. When the air–fertilizer mixture is distributed to each branch, the discharge of fertilizer is greatly affected by the structure of the pneumatic conveying system. This can lead to fertilizer crushing and bounce at the outlet, causing significant variations in the fertilizer discharge across branches. However, the fertilizing process of cotton usually adopts large area wide multi-row operation. When the uniformity of fertilizer discharge of pneumatic collective discharge fertilizer is poor, it is more difficult to control the amount of fertilizer discharged by each row, which has a greater impact on the yield and quality of cotton. Therefore, optimizing the structure of the fertilizer feeder in the pneumatic conveying system is of great significance to improve the fertilizer uniformity and stability of each branch of the cotton pneumatic collecting and discharging fertilizer. With the development of discrete element and finite element simulation technologies, the EDEM-Fluent coupled simulation method has been widely applied in the simulation and optimization of particulate material conveying systems [10,11]. It provides a powerful tool for optimizing the structural design optimization of the feeder. Currently, there is relatively little research on the feeders of cotton pneumatic fertilizing devices, especially regarding issues such as air–fertilizer mixing uniformity and delivery stability [12]. Most studies have focused on gas–solid coupling in crop conveying systems, liquid fertilizer mixing, and powder material transport [13–15]. For example, Liao et al. [16] numerically simulated the gas–solid flow of mixed oat and pea seeds in a venturi feeder, analyzing the effects of seed entry angle and airflow velocity on mixing uniformity and determining the optimal structure for seed entry angle and airflow velocity under optimal mixing conditions. Gao et al. [17] used the CFD-DEM simulation method to model the motion of seed particles, discharge uniformity, and airflow in a quantitative feeding system, analyzing the influence of various parameters and identifying the optimal settings through a comparison and analysis of the simulation results. Chandramohan et al. [18] designed the structure of a water–fertilizer venturi tube and performed CFD simulations to determine the optimal parameters, thus improving fertilizer intake under uniform water flow conditions. Li et al. [19] conducted gas–solid coupling simulations of the fertilizer mixing chamber in a pneumatic rice fertilizing machine, optimizing the fertilizer feeding system to improve discharge performance. Guo et al. [20] studied the fluid dynamics and discharge principles inside a fertilizer distributor, determined the basic size parameters and coupling model, and performed CFD-DEM coupling simulations to identify the optimal distributor structure.

In conclusion, relevant scholars at home and abroad used CFD-DEM and other methods to carry out the design and study of a pneumatic collection and discharge delivery system feeder for material grains, oat, corn, rice, and other crops, and optimized their different structural properties. However, few studies have focused on feeders for pneumatic fertilizer collection in cotton. Additionally, research on the cavity flow field, fertilizer particle motion, and feeder structure optimization for cotton pneumatic conveying is scarce.

Therefore, based on the working principle of the pneumatic collector feeder, this study designs a diversion linear feeder to optimize the uniformity of gas–fertilizer mixing and fertilizer stability. The goal is to optimize the feeder design for cotton pneumatic collection, providing a theoretical foundation and technical support for improving cotton fertilization machinery.

2. Materials and Methods

2.1. Feeder Operating Principle

During the operation of the pneumatic fertilizer distribution system, fertilizer particles from the fertilizer tank are discharged through the variable-rate fertilizer applicator to the upper inlet of the feeder. The left-side inlet of the feeder is connected to a fan, and the airflow from the fan mixes with the fertilizer inside the feeder to form the mixture of gas and fertilizer. This mixture of gas and fertilizer are then conveyed through horizontal and vertical delivery pipelines to the fertilizer distributor. The distributor features a centrally symmetric structure, ensuring that the gas–fertilizer mixture is evenly split and directed into the fertilizer application openers at the ends of each branch pipeline. The operating principle is illustrated in Figure 1.

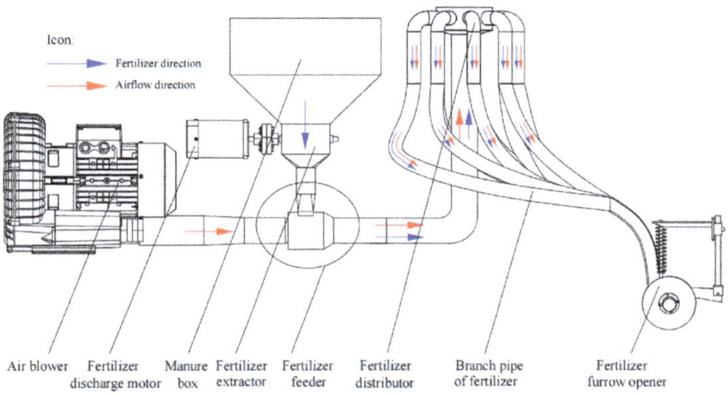

Figure 1. Schematic diagram of the working principle of the pneumatic fertilizer distribution system.

The feeder is primarily composed of an intake section, fertilizer inlet section, gas–fertilizer mixing zone, and output section. Figure 2 illustrated the basic structure and operating principle. The feeder's intake is a narrow nozzle, and when the fan airflow enters the feeder, high-speed airflow is generated at the nozzle. According to Bernoulli's principle, this high-speed airflow creates a pressure drop within the gas–fertilizer mixing chamber of the feeder, causing the fertilizer particles discharged from the variable-rate fertilizer applicator to enter the feeder due to the pressure difference and gravity. The airflow mixes with the fertilizer particles inside the feeder to form a gas–fertilizer mixture. This mixed gas–fertilizer flow is then conveyed through the output section into the pneumatic fertilizer delivery system pipelines.

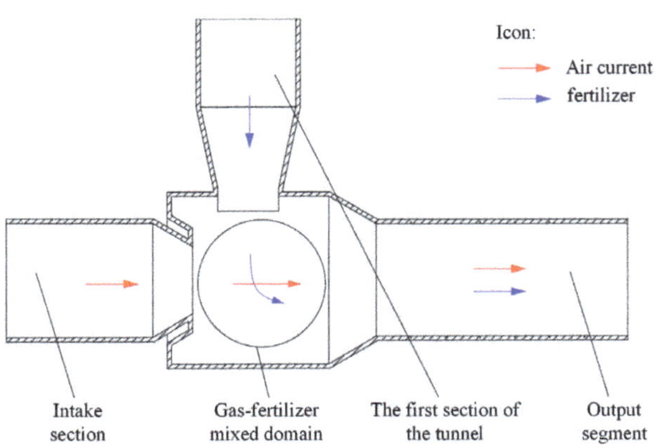

Figure 2. Schematic diagram of the feeder structure principle.

2.2. Feeder Structural

In the pneumatic fertilizer distribution system, the feeder is crucial for the uniform delivery of fertilizer and the operational performance of the fertilizer applicator. In this study, both cyclone conveying technology and steady-flow conveying technology are introduced into the feeder design [21,22]. Specifically, the plan involves applying cyclone conveying technology in the feeder's inlet pipeline, which generates a swirling airflow in the gas–fertilizer mixing chamber of the feeder. This swirling airflow exerts a vortex effect on the fertilizer particles, addressing the issue of fertilizer settling under the influence of gravity when transported through the straight pipelines after leaving the feeder. This approach improves the uniformity of fertilizer delivery, reduces airflow delivery energy consumption, and indirectly enhances the operational performance of the pneumatic fertilizer applicator.

Additionally, steady-flow conveying technology is considered for implementation in the pneumatic fertilizer system. This technology induces a self-rotating motion of the fertilizer particles within the delivery pipeline, ensuring stable movement. This helps resolve issues such as increased fertilizer particle bouncing and damage when the fertilizer flow velocity is too high during delivery from the feeder.

Based on the aforementioned cyclone airflow initiation structure and steady-flow conveying structure, and in accordance with the feeder design principles, a vane-type feeder is designed to create a spiral airflow at the intake guide, as shown in Figure 3a. To address the potential issue of flow turbulence caused by the introduction of the cyclone structure, a steady-flow structure is incorporated into the output section of the feeder. This steady-flow design, based on the principles of steady-flow conveying technology, helps stabilize the fertilizer discharge process, as shown in Figure 3b. Therefore, this paper selects a combination of the vane-type intake airflow initiation structure and the output chamber-line steady-flow structure, resulting in the design of the flow-guide spiral feeder.

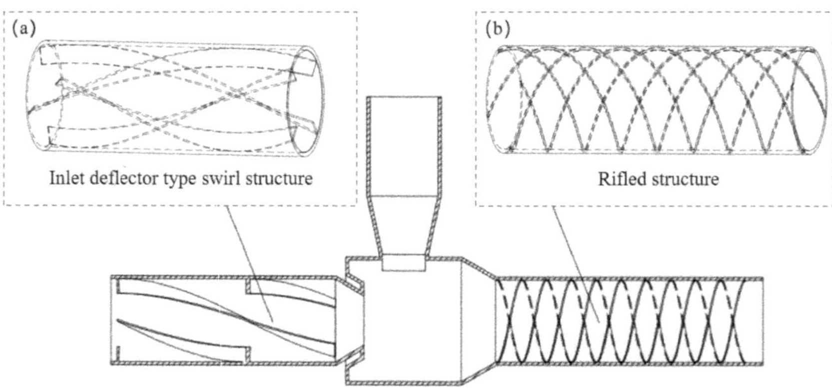

Figure 3. Structure scheme of the diversion rifled feeder. (**a**) Inlet deflector type swirl of feeder structure; (**b**) Rifled structure of feeder.

2.3. Establishment of Feeder Flow–Solid Coupling Simulation Model Based on DEM-CFD

2.3.1. Model Grid Division

Based on the structural design comparison of the feeder, the 3D model of the feeder is created at a 1:1 scale for use in discrete element method (DEM) and computational fluid dynamics (CFD) simulations. In the fluid simulation, the internal space of the feeder is enclosed at both the outlet and inlet end faces of the distributor. The fluid domain model of the feeder is extracted using ANSYS 2020R2 software. Next, we proceed to mesh the fluid domain, referencing the actual diameter size of the fertilizer particles (1.5–3 mm) to ensure the suitability of the mesh division. To verify grid independence, i.e., the extent to which grid size affects simulation results, we conducted appropriate grid independence tests. During the testing, we took into account both the requirements for computational accuracy and the availability of computational resources, ultimately setting the minimum grid size to 3 mm and the maximum to 8 mm. This size range is capable of effectively capturing the flow characteristics within the fluid domain while avoiding excessive computational load due to overly dense grids. The volumetric grid size needs to be about three times the diameter of the fertilizer particles to simulate their movement and distribution within the fluid. Based on this requirement, we set the maximum grid size to 9 mm. Additionally, three layers were added to the boundary layer. The fluid domain model and volume mesh model for the flow-guiding rib-style feeder simulation process are shown in Figure 4a,b.

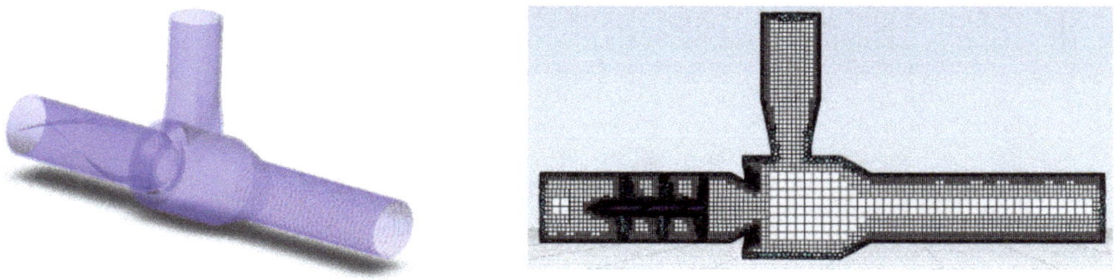

(**a**) (**b**)

Figure 4. Feeder simulation model diagram. (**a**) Fluid domain model diagram; (**b**) body mesh division.

2.3.2. Setting of the Simulation Parameters

(1) Finite Element Simulation Parameters

In the simulation, the EDEM and Fluent software are used for gas–solid coupling simulation of the gas–fertilizer mixing process in the feeder. The airflow in the feeder is treated as incompressible turbulent flow, while the motion of the fertilizer particles is considered as the collective motion of individual particles. There is no adhesive force between the surfaces of the fertilizer particles, and the system follows Newton's second law. For the gas phase in the feeder, an incompressible fluid model using the Eulerian–Lagrangian method is selected, and the standard k-ε model is used for solving. The solid phase uses the Hertz–Mindlin no-slip contact model. The forces acting on the particles from the airflow in the DEM-CFD coupling simulation module are based on the Freestream Equation fluid drag, Saffman lift, and Magnus lift models [23,24]. The time step for EDEM and CFD is set to 5×10^{-6} s and 1×10^{-3} s, respectively. The number of steps in Fluent is set to 5000, with a total simulation time of 5 s. Each time step can be iterated up to 30 times. The boundary conditions for the model are set as follows: inlet velocity at 40 m/s and pressure outlet. The fertilizer inlet flow rate is set to 490 g/s, with the fertilizer generation time set to 3 s and the total simulation duration to 5 s. The gravitational acceleration is set to 9.81 m/s^2, and the fertilizer particle inlet speed is set to 5 m/s.

(2) DEM Simulation Parameters

Additionally, in the EDEM particle flow simulation, the physical property parameters of the fertilizer particles are crucial for conducting discrete element simulations. Existing discrete element parameters for fertilizer particles are used in the simulation. The specific parameters are as follows: urea particles with 46.4% nitrogen content are used as the fertilizer, with a sphericity of 93.6%, and a particle size range of 1.2 mm to 3.0 mm. The particle diameters are randomly distributed [25]. In the simulation model, the fertilizer particles are simplified as spheres with a diameter of 2.10 mm. The feeder material is chosen as PVC, and mesh generation is performed using the Fluent-Meshing module. The specific parameters for the fertilizer particles and the feeder model are summarized in Table 1 [26,27].

Table 1. Parameter of the simulation model of fertilizer particles.

Parameter	Numerical Value
Poisson's ratio of the pneumatic conveying system material	0.32
Elastic modulus of the pneumatic conveying system material (Pa)	4.22×10^9
Density of the pneumatic conveying system material ($kg{\cdot}m^{-3}$)	1418
Poisson's ratio of urea particles	0.3
Elastic modulus of urea particles (Pa)	6.72×10^7
Density of urea particles ($kg{\cdot}m^{-3}$)	1341
Coefficient of restitution for collisions between urea particles and the pneumatic conveying system	0.35
Coefficient of static friction between urea particles and the pneumatic conveying system	0.32
Coefficient of rolling friction between urea particles and the pneumatic conveying system	0.04
Coefficient of restitution for collisions between urea particles	0.26
Coefficient of static friction between urea particles	0.27
Coefficient of rolling friction between urea particles	0.11

To obtain information on the maximum velocity variation and the particle count, as well as to calculate subsequent evaluation metrics, a sampling section A-A is set at the fertilizer discharge outlet at the end of the feeder in the EDEM post-processing module. The center coordinates of the sampling region are (265, 0, 0) mm, and the sampling area is defined as a rectangle, as shown in Figure 5.

Figure 5. Schematic diagram of the feeder simulation outlet sampling section A-A.

2.4. Experimental Evaluation Index

To accurately and realistically evaluate the fertilizer distribution performance of the feeder, the coefficient of variation of stability and the gas–fertilizer mixing uniformity coefficient are introduced for assessment [28,29]. The coefficient of variation of stability is an important indicator of the uniformity of the feeder's material supply. A lower coefficient of variation of stability indicates less fluctuation in the fertilizer flow rate within the pipeline over time, resulting in a more stable output from the feeder. The coefficient of variation of stability is defined as the ratio of the standard deviation of the fertilizer output over a given time period to the average fertilizer output during that period, calculated for the sampling section A-A. The specific calculation method is as follows:

Additionally, in pneumatic fertilization systems, the uniformity of fertilizer particle distribution is typically quantified using a uniformity index. The uniformity index is a quantitative measure used to describe the uniformity of particle distribution, which can be evaluated by observing changes in particle concentration along the pipeline. The value of the uniformity index ranges from 0 to 1, with values closer to 1 indicating more uniform particle distribution. The uniformity index of the feeder's fertilizer output can be calculated using the following formula:

$$Y_1 = 1 - \frac{Z_1}{\overline{y_1}} \tag{1}$$

where Z_1 is the standard deviation of the output pipeline particle concentration in periods 2, 3, 4, 5 of the simulation section A-A, $Z_1 = \sqrt{\frac{\sum_{i=1}^{n}(y_1-\overline{y_1})^2}{n_1-1}}$, the y_i is the fertilizer concentration of each i sampling point, the mean value of each fertilizer concentration in y_2, g/cm^3, and n_1 is the number of samples.

$$y_i = \frac{\rho_a \times (1-\vartheta_i)}{1-\vartheta_i\frac{\rho_c}{\rho_a}} \tag{2}$$

where ρ_a and ρ_c are the density of particles and air, respectively, and ϑ_i is the void ratio of the fertilizer particles at each sampling point. In the simulation test, the real-time void rate value of fertilizer particles can be directly derived in EDEM post-treatment.

Fertilizer particle count in the export area A-A is sampled during the time periods 1.8–2.2 s, 2.8–3.2 s, 3.8–4.2 s, and 4.6–5.0 s. A comparative analysis of the variability in the number of fertilizer particles at the export section is performed across the four time periods. The formula for calculating the coefficient of variation of stability is shown in Equation (1):

$$Y_2 = \frac{Z_2}{\overline{x}} \times 100\% \tag{3}$$

where Z_2 is the standard deviation of sample samples in each period of the simulation section A-A, $Z_2 = \sqrt{\frac{\sum_{i=1}^{n}(x_i-\overline{x})^2}{n_2-1}}$, the x_i is the fertilizer discharge of each sampling point in i, $\overline{x}$ is the average of each sampling time period, and n_1 is the sample quantity.

2.5. Feeder Structure Parameter Optimization

The movement of fertilizer within the feeder is crucial for the stability and uniformity of fertilizer discharge. Based on the initial feeder structure design and operational parameters, the structural parameters of the feeder's output rifling, inlet swirling guide vanes, fertilizer inlet section, and feeder chamber section were designed. Preliminary experiments using discrete element simulation were conducted to analyze the factors influencing the movement of fertilizer within the feeder. These experiments confirmed that the number of air inlet flow-guiding vanes directly affects the inlet air velocity and the rotational speed of the fertilizer particles. The fertilizer discharge contraction angle of the fertilizer discharge section directly influences the angle of collision and rebound of the fertilizer particles, thereby affecting their movement into the rifling output section. For the rifling output section, the number of ripper lines directly impacts the centrifugal force experienced by the fertilizer particles in the swirling acceleration output section, which in turn affects their movement. Therefore, the number of air inlet flow-guiding vanes, fertilizer discharge contraction angle, and the number of ripper lines were selected as the primary factors influencing feeder performance.

2.5.1. Box–Behnken Experiment Design

To further optimize the impact on fertilizer discharge performance, three key structural parameters are selected as experimental factors during the feeder structural design process. The experimental indicators are the gas–fertilizer mixing uniformity coefficient and the stability coefficient of variation. The Box–Behnken response surface optimization experiment is conducted. The factors and levels of the design are shown in Table 2.

Table 2. Encoding table of test factors.

Coding	Factors		
	Number of Air Inlet Flow-Guiding Vanes X_1 (pcs)	**Fertilizer Discharge Contraction Angle X_2 (°)**	**Number of Ripper Lines X_3 (Piece)**
1	2	22	4
0	4	24	6
−1	6	26	8

2.5.2. Results of the Orthogonal Simulation Experiments

Simulation experiments were conducted to analyze the feeder structural parameters that affect gas–fertilizer mixing uniformity and fertilizer transport stability. Based on the operational requirements, the optimal combinations of these parameters were determined. The software Design-Expert 13.0 is used to design the Box–Behnken orthogonal experiment. The experimental design and results are shown in Table 3.

Table 3. Results of orthogonal rotation combination tests.

No.	Number of Air Inlet Flow-Guiding Vanes X_1 (pcs)	Difertilizer Shrinkage Angle X_2/(°)	Number of Ribbed Chambers X_3/(piece)	Uniformity Coefficient Y_2/%	Stability Coefficient of Variation Y_2/%
1	−1	−1	0	79.86	20.15
2	1	−1	0	63.88	18.56
3	−1	1	0	87.64	20.31
4	1	1	0	85.99	17.73
5	−1	0	−1	91.32	26.31
6	1	0	−1	77.29	22.12
7	−1	0	1	85.63	14.09
8	1	0	1	83.92	15.33
9	0	−1	−1	81.41	25.61
10	0	1	−1	86.71	24.31
11	0	−1	1	71.13	14.77
12	0	1	1	90.57	16.52
13	0	0	0	93.21	20.87
14	0	0	0	96.34	22.35
15	0	0	0	94.67	19.86
16	0	0	0	93.52	20.18
17	0	0	0	94.15	21.22

2.6. Feeder Discharge Performance Tests

The feeder performance test for the pneumatic fertilizer applicator was conducted in the Laboratory of the College of Mechanical and Electrical Engineering at Shihezi University. The fertilizer particles used in the test were urea granules with 46.4% nitrogen content (China National Petroleum Corporation, Beijing, China). The urea granules have a sphericity of 93.6% and a particle size range of 1.2 to 3.0 mm. The feeder was 3D printed at a 1:1 scale using polylactic acid (PLA) material, as shown in Figure 6.

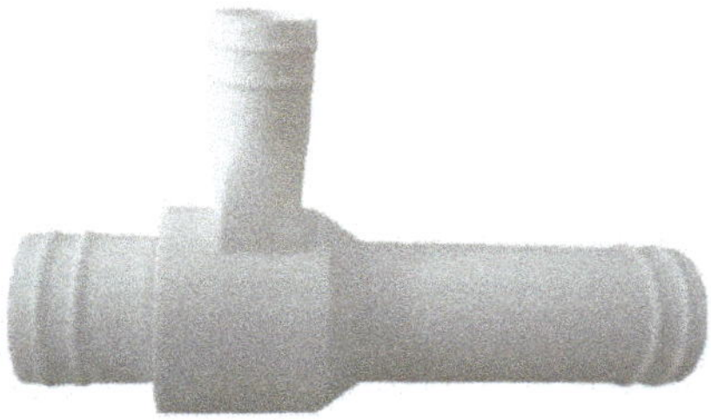

Figure 6. Feeder prototype.

A test prototype for the pneumatic fertilizer applicator feeder was constructed, and the structural schematic of the test prototype is shown in Figure 7. The test prototype consists of the test stand frame, stepper motor, variable fertilizer distributor, fan, fan inverter (model V8M4T2R2GB), feeder, fertilizer conveying pipeline, and fertilizer distributor. The fertilizer tank is made of stainless steel, with its bottom connected to an external slot wheel variable fertilizer distributor to ensure smooth fertilizer discharge. The discharge motor is a variable-speed motor to allow control of the fertilizer discharge rate. The frame is constructed with a high-strength angle iron, and the pneumatic fertilizer conveying the pipe is made from a smooth-walled steel wire hose.

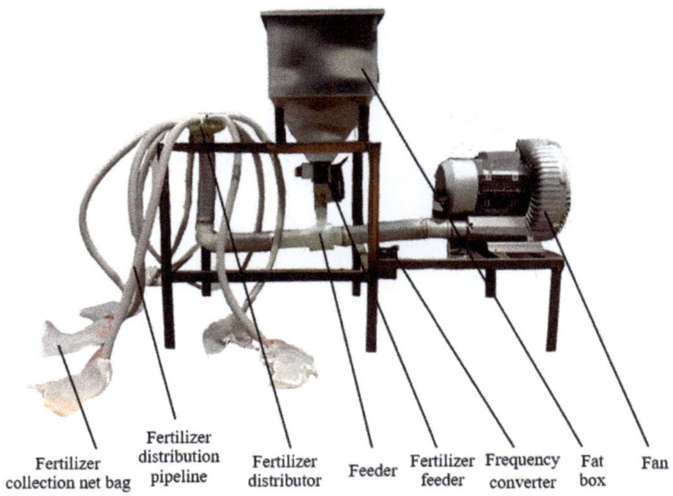

Figure 7. Feeder and fertilizer discharge performance bench.

Validation tests for the consistency coefficient of variation of the feeder were conducted. At the start of the tests, the fertilizer delivery pipeline at the rear of the feeder was removed, and the breathable mesh bag was installed at the outlet of the feeder. The mesh bag was placed directly above an electronic scale to measure the fertilizer discharge in real-time. The tests began by turning on the fan power switch, and once the airflow stabilized, the power to the fertilizer discharge motor was activated. Simultaneously, the video was recorded using a smartphone to capture the changes in weight on the electronic scale. The consistency coefficient of variation of the feeder's fertilizer delivery was then calculated based on the experimental data.

For the feeder uniformity index test, breathable mesh bags were installed at the outlet of each branch pipe of the fertilizer distributor, totaling nine bags. The fan power was turned on first, followed by the activation of the fertilizer discharge motor for the fertilization test. A stopwatch was used to begin timing the experiment. After a set period (3 min), the fan and fertilizer discharge motor were turned off to complete the test. The fertilizer discharge amounts from each branch of the distributor were then measured, and the variation coefficient of the fertilizer discharge across the branches of the pneumatic collection and distribution fertilization system was calculated. This coefficient was used to represent the feeder's uniformity index. The uniformity index was then evaluated to determine whether it met the fertilization requirements of the pneumatic collection and distribution fertilization machine.

3. Results

Based on the structural design of the flow-guiding chamber feeder as shown in Figure 3, which incorporates the selected flow-guiding rib structure, a comparative study was conducted to evaluate the effectiveness of the swirl-flow intake structure with guide vanes versus the outlet chamber rib structure. The comparison is made with single-phase swirl flow and steady flow conveying structures, as illustrated in Figure 8a,b.

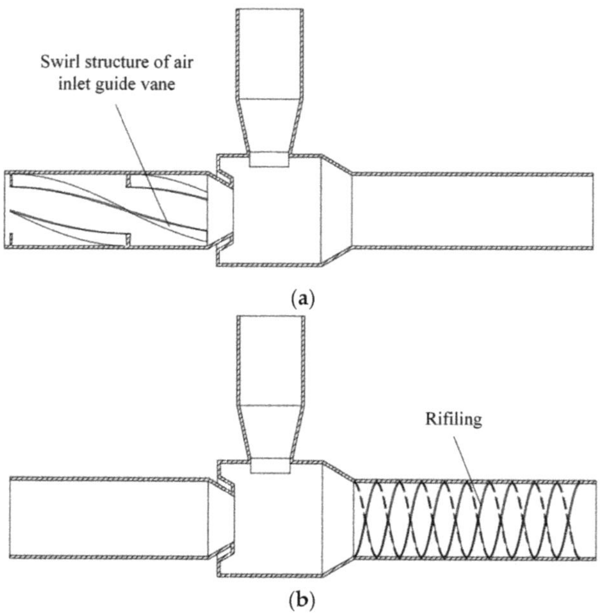

Figure 8. Motion Process of Ditching Plow Blade Cutting Soil. (**a**) Air Inlet Flow-Guiding Vane Structure; (**b**) Outlet Rifling Structure.

3.1. Simulation Results Analysis

The X-Y plane is selected as the sampling plane for the flow field, and the airflow velocity distribution contour maps for different air inlet swirl types of the feeder are calculated, as shown in Figure 9. The larger the area influenced by the airflow, the stronger the conveying performance. A comprehensive comparison of Figure 9b,c, as well as Figure 9a,c, reveals that both the air inlet flow-guiding vane structure and the outlet ribbed chamber structure increase the airflow influence area to some extent, thus improving the conveying capacity. Among them, the effect of the air inlet flow-guiding vane structure is more significant.

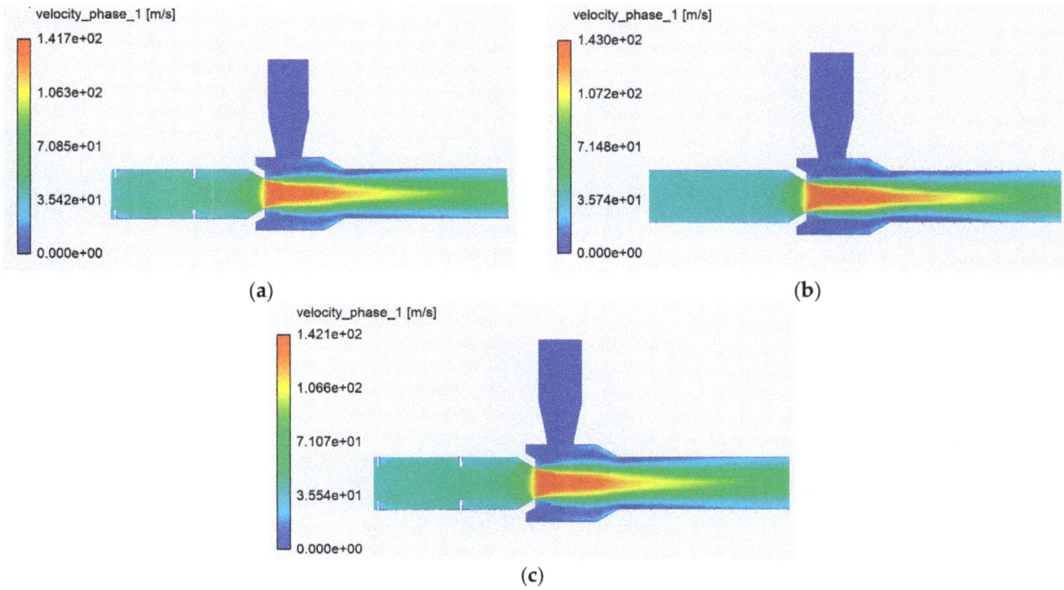

Figure 9. Analysis of airflow field velocity distribution of different structural schemes of the feeder. (**a**) Air Inlet Flow-Guiding Vane Structure; (**b**) Outlet Ribbed Chamber Structure; (**c**) Flow-Guiding Ribbed Chamber Structure (Combination of Air Inlet Flow-Guiding Vane and Outlet Ribbed Chamber).

Furthermore, the airflow velocity variation ranges in the airflow fields for the air inlet flow-guiding vane feeder, the outlet ribbed chamber feeder, and the flow-guiding ribbed chamber feeder are (35.4, 141.7) m/s, (35.7, 143.2) m/s, and (35.6, 142.1) m/s, respectively. A comprehensive comparison indicates that the application of the ribbed chamber structure reduces the range of airflow velocity fluctuations within the feeder to some extent. The ribbed chamber structure provides better stabilization of the airflow velocity inside the feeder, thereby improving the stability of fertilizer delivery to a certain extent.

The pressure distribution contour maps of the airflow field for feeders with different structures are shown in Figure 10. The pressure region and position obtained by the simulation under the three structures are not different. Comparing Figure 10a,c and Figure 10b,c, the increase of the outlet bore structure will make the high-pressure area larger and increase the maximum pressure. The increase of the inlet diversion structure reduces the pressure in the tube to a certain extent. By comparing the three, for the diversion bore (Figure 10c) the maximum pressure is moderate and in the minimum pressure range, and the airflow pressure stability is optimal.

The pressure variation ranges for the air inlet flow-guiding vane structure, outlet ribbed chamber structure, and flow-guiding ribbed chamber structure are (31.3, 125.2) kPa, (31.1, 124.2) kPa, and (29.8, 123.3) kPa, respectively. A comprehensive comparison reveals that the application of the ribbed chamber structure in the outlet section reduces the pressure variation range within the feeder, thereby improving airflow stability.

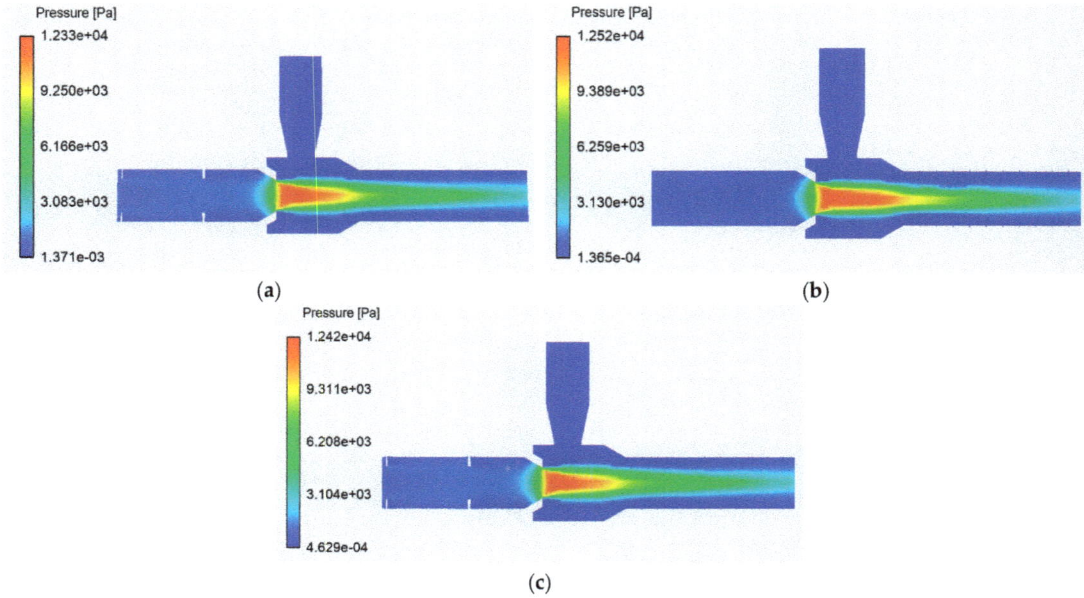

Figure 10. Analysis of airflow field pressure distribution in different structural schemes of the feeder. (**a**) Air Inlet Flow-Guiding Vane Structure; (**b**) Outlet Ribbed Chamber Structure; (**c**) Flow-Guiding Ribbed Chamber Structure (Combination of Air Inlet Flow-Guiding Vane and Outlet Ribbed Chamber).

3.2. Feeder Structure Parameter Optimization Results

3.2.1. Feeder Fertilizer Discharge Uniformity Index

The experimental results of the uniformity coefficient were analyzed using Design-Expert for variance analysis. The results are shown in Table 4. It can be seen that the model is highly significant ($p < 0.01$), and the lack of fit is not significant ($p > 0.05$). The coefficient of determination (R^2) for the experimental model is 0.9868, the adjusted R^2 is 0.9698, and the predicted R^2 is 0.8584. These values indicate that the model has a good fit, and it can effectively be used to analyze and predict the feeder's uniformity coefficient.

Additionally, based on the variance analysis results, non-significant factors were excluded, leading to the following regression equation for the optimized feeder fertilizer discharge uniformity index:

$$Y_1 = 94.38 - 4.17X_1 + 6.83X_2 + 3.58X_1X_2 + 3.08X_1X_3 + 3.54X_2X_3 - 6.48{X_1}^2 - 8.56{X_2}^2 - 3.36{X_3}^2 \quad (4)$$

From the significance analysis of the model, it can be observed that the linear terms X_1 (number of flow-guiding vanes), X_2 (fertilizer discharge contraction angle), and the quadratic interaction terms X_1X_2, X_1X_3, X_2X_3, ${X_1}^2$, ${X_2}^2$ have a highly significant effect on the feeder's uniformity index. The cubic term ${X_3}^2$ (number of ribbed chambers) has a significant effect, while X_3 (number of ribbed chambers) is a non-significant factor. By comparing the significance of each factor, the following order of impact on the feeder's uniformity index can be determined: X_2 (fertilizer discharge contraction angle) > X_1 (number of air inlet flow-guiding vanes) > X_3 (number of ribbed chambers).

Table 4. Variance analysis of uniformity coefficient.

Source of Variance	Qualified index					
	Quadratic Sum	Free Degree	Mean Square	F-Value	p-Value	Significance
Model	1240.98	9	137.89	58.07	<0.0001	**
X_1	139.19	1	139.19	58.63	0.0001	**
X_2	373.05	1	373.05	157.12	<0.0001	**
X_3	3.75	1	3.75	1.58	0.2489	-
X_1X_2	51.34	1	51.34	21.62	0.0023	**
X_1X_3	37.95	1	37.95	15.98	0.0052	**
X_2X_3	49.98	1	49.98	21.05	0.0025	**
$X_1{}^2$	176.54	1	176.54	74.36	<0.0001	**
$X_2{}^2$	308.54	1	308.54	129.95	<0.0001	**
$X_3{}^2$	47.61	1	47.61	20.05	0.0029	*
Total variation	16.62	7	2.37			
pure error	10.53	3	3.51	2.31	0.2183	
Unplanned item	6.09	4	1.52			
Residual	1257.60	16				

Note: $p \leq 0.001$ means highly significant, marked as "**"; $p \leq 0.005$, means significant, marked as "*". $0.01 < p < 0.05$, means not significant, marked as "-".

Furthermore, to provide a more intuitive and clear analysis of the interaction effects of the experimental factors on the uniformity index, the response surface plot of the regression model for the uniformity index is generated, as shown in Figure 11. The response values of the feeder, along with the interactions of the factors, form contour lines and response surfaces. The shape of the contour lines reflects the strength of the interaction between the two factors: circular contours indicate weak interactions, while the more elliptical the contours become, the stronger the interaction.

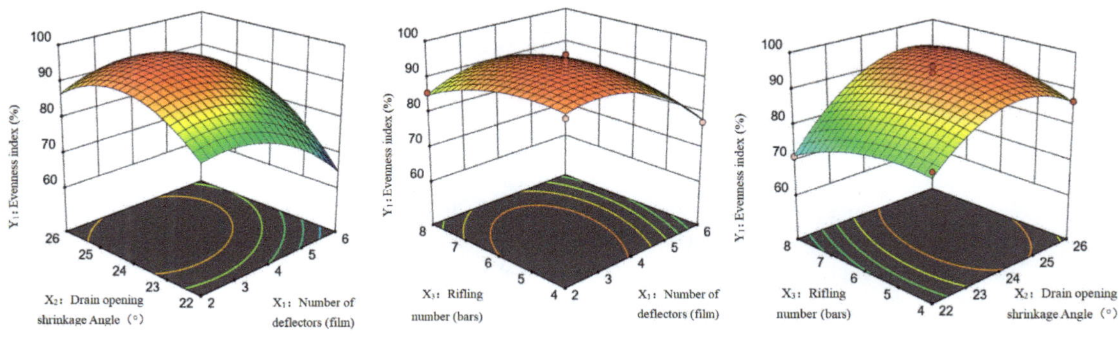

Figure 11. Each interaction effect on the uniformity index. (**a**) X_1 and X_2 interaction; (**b**) X_1 and X_3 interaction; (**c**) X_2 and X_3 interaction.

From the analysis of Figure 11a–c, it can be seen that the interaction surface of X_1X_2 shows a larger variation, followed by X_1X_3. Overall, the interactions between X_1X_2 and X_1X_3 have a greater impact on the feeder's uniformity index, while the X_2X_3 interaction has a relatively smaller effect. The fertilizer discharge contraction angle (X_2) has the largest influence on the uniformity index.

3.2.2. Feeder Stability Coefficient of Variation

The variance analysis of the stability coefficient of variation results is also conducted, as shown in Table 5. The results indicate that the model is highly significant ($p < 0.01$), and the lack of fit is not significant ($p > 0.05$). The coefficient of determination (R^2) for the experimental model is 0.9797, the adjusted R^2 is 0.9537, and the predicted R^2 is 0.9430, indicating that the model has a good fit and can effectively be used to analyze and predict the stability coefficient of variation of the feeder. Similarly, by excluding non-significant factors, the regression equation for the feeder's stability coefficient of variation is obtained:

$$Y_2 = 20.90 - 0.89X_1 - 4.71X_3 + 1.36X_1X_3 - 1.27{X_1}^2 \quad (5)$$

Table 5. Variance analysis of uniformity coefficient.

Source of Variance	Qualified Index					
	Quadratic Sum	Free Degree	Mean Square	F-Value	p-Value	Significance
Model	201.55	9	22.39	37.62	<0.0001	**
X_1	6.34	1	6.34	10.65	0.0138	*
X_2	0.01	1	0.01	0.01	0.9225	
X_3	177.10	1	177.10	297.52	<0.0001	**
X_1X_2	0.25	1	0.25	0.41	0.5416	
X_1X_3	7.37	1	7.37	12.38	0.0097	**
X_2X_3	2.33	1	2.33	3.91	0.0886	
${X_1}^2$	6.84	1	6.84	11.49	0.0116	*
${X_2}^2$	0.79	1	0.79	1.33	0.2860	
${X_3}^2$	0.11	1	0.11	0.18	0.6846	
Total variation	4.17	7	0.56			
pure error	0.36	3	0.12	0.1264	0.9396	
Unplanned item	3.81	4	0.95			
Residual	205.72	16				

Note: $p \leq 0.001$ means highly significant, marked as "**"; $p \leq 0.005$, means significant, marked as "*". $0.01 < p < 0.05$, means not significant, marked as "-".

From the significance analysis of the model, it can be observed that the linear term X_3 (number of ribbed chambers) and the quadratic interaction term X_1X_3 have a highly significant effect on the feeder's stability coefficient of variation ($p < 0.01$). The linear term X_1 (number of flow-guiding vanes) and the quadratic term ${X_1}^2$ have a significant effect on the stability coefficient of variation ($0.01 < p < 0.05$), while X_2 (fertilizer discharge contraction angle) is a non-significant factor. By comparing the significance of each factor, the following order of impact on the feeder's stability coefficient of variation can be determined: X_3 (number of ribbed chambers) > X_1 (number of flow-guiding vanes) > X_2 (fertilizer discharge contraction angle).

Furthermore, to provide a more intuitive and clear analysis of the interaction effects of the experimental factors on the stability coefficient of variation, the response surface plot of the regression model for the stability coefficient of variation is generated, as shown in Figure 12.

From the analysis of Figure 12, it can be observed that the interaction surface of X_1X_3 shows the largest variation, followed by X_2X_3. Overall, the interactions between X_1X_3 and X_2X_3 have a greater impact on the feeder's uniformity index, while the X_1X_2 interaction has a relatively smaller effect. Therefore, the number of ribbed chambers (X_3) is the main structural parameter influencing the fertilizer stability coefficient of variation.

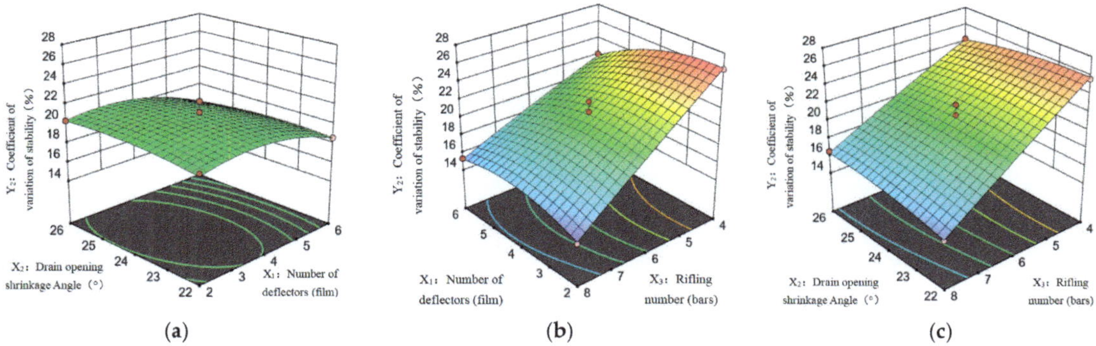

Figure 12. Effect of each interaction on the coefficient of variation of stability. (**a**) X_1 and X_2 interaction; (**b**) X_1 and X_3 interaction; (**c**) X_2 and X_3 interaction.

3.2.3. Feeder Structural Parameter Optimization Results

Based on the response surface experimental model, the parameter factor optimization module in Design-Expert software is used for multi-objective optimization of the feeder's structural parameters. The optimization objectives are to maximize the fertilizer uniformity index and minimize the fertilizer discharge stability coefficient of variation. The values of the structural parameters are within the specified levels. After setting the factor level ranges, the optimal solution is found to be (2.1 flow-guiding vanes, 23.2° discharge contraction angle, and 7.98 ribbed chambers). With these structural parameters, the predicted fertilizer discharge uniformity index is 81.84%, and the stability coefficient of variation is 14.08%.

Three experimental validations were conducted for the optimization results. Since the number of flow-guiding vanes and ribbed chambers must be integer values, the experimental validation was carried out with two flow-guiding vanes and eight ribbed chambers. The experimental validation results are shown in Table 6.

Table 6. Results of the optimized validation tests for structural parameters.

No.	Test Index	
	Uniformity Coefficient (%)	**Stability Coefficient of Variation (%)**
1	82.32	14.23
2	81.58	13.85
3	82.25	13.56
Mean value	82.02	13.88

The error between the test values and the predicted values is less than 5%, indicating that the test model is reliable. The error may be due to the rounding of the optimization results. Additionally, the main reason is that the simulation experiment simplifies and approximates the actual physical phenomena, such as the size of the fertilizer, to some extent. However, the overall error is within an acceptable range. In the above analysis, the optimal structural parameters of the feeder are two pieces of diversion tablets, 23.2 pieces of fertilizer shrinkage, and eight rifled lines.

3.3. Feeder Discharge Performance Tests Result

3.3.1. Analysis of Feeder Consistency Coefficient of Variation Validation Results

Based on the method described for measuring the stability coefficient of variation of the feeder, the following experiment was designed. The fertilizer weight change at the feeder's outlet was monitored in real-time using an electronic scale. During the total experimental duration, five time intervals were selected for testing. In each interval, fertilizer weight was recorded every 30 s, and a total of five measurements were taken. The average stability coefficient of variation was then calculated from these five sets of data. The results, shown in Table 7, provide an analysis of the feeder's consistency and a comparison with the simulation results.

Table 7. Test results of different coefficient of fertilizer stability of feeder.

No.	Mass of the Fertilizer Within 0–30 s/g	Mass of the Fertilizer Within 30–60 s/g	Mass of the Fertilizer Within 60–90 s/g	Mass of the Fertilizer Within 90–120 s/g	Mass of the Fertilizer Within 120–150 s/g	Average mass/g	Stability CoEfficient of Variation
1	235.00	320.00	284.00	297.00	236.00	274.40	13.78%
2	242.00	289.00	246.00	292.00	330.00	279.80	13.03%
3	284.00	250.00	234.00	300.00	309.00	275.40	11.67%
4	232.00	292.00	296.00	233.00	334.00	277.40	15.86%
5	229.00	269.00	303.00	296.00	249.00	269.20	12.17%
mean value						275.24	13.30%

The results of several bench tests show that the variation coefficient of the optimized feeder is 13.30%, and the difference between the optimized parameters and the simulation value of 13.88% is small, which further proves the accuracy of the simulation. In addition, compared with the results before and after optimization, the stability of the optimized feeder is significantly improved compared with the variation coefficient of 14.89% of the fertilizer discharge stability of the feeder before optimization of structural parameters.

3.3.2. Analysis of Feeder Delivery Uniformity Coefficient Validation

It is very difficult to measure the uniformity state of the pipe particles at the end of the feeder during the actual measurement process. Therefore, the measurement quality of each pipe is selected for statistics. The branch variation coefficient is then calculated to represent the uniformity index for evaluation. To evaluate the fertilizer delivery uniformity index of the feeder, as defined by the previously described method, the following experiment was conducted. The fan and fertilizer discharge motor were activated to initiate the fertilization process. Fertilizer discharged from each branch of the distributor was collected in individual mesh bags over a three-minute period. The mass of fertilizer collected in each bag was subsequently measured using an electronic balance. This procedure was replicated five times, and the results were recorded. The coefficient of variation of fertilizer discharge across the branches was then calculated from the collected data. The experimental results are shown in Table 8.

Table 8. Feeder feeding uniformity index test results.

No.	No. 1 Tube/g	No. 2 Tube/g	No. 3 Tube/g	No. 4 Tube/g	No. 5 Tube/g	No. 6 Tube/g	No. 7 Tube/g	No. 8 Tube/g	No. 9 Tube/g	Average Quality Production of/g	Sub-Branch Coefficient of Variation
1	1552.00	1740.00	1548.00	1568.00	1637.00	1690.00	1588.00	1764.00	1688.00	1641.6.00	3.61%
2	1588.00	1655.00	1687.00	1527.00	1757.00	1652.00	1668.00	1675.00	1668.00	1653.00	3.94%
3	1672.00	1534.00	1598.00	1648.00	1688.00	1587.00	1762.00	1652.00	1678.00	1646.50	4.03%
4	1542.00	1689.00	1576.00	1684.00	1765.00	1654.00	1648.00	1675.00	1598.00	1647.80	4.09%
5	1678.00	1596.00	1598.00	1694.00	1758.00	1643.00	1586.00	1546.00	1698.00	1644.10	4.13%
Mean value										1646.60	3.96%

Multiple bench test results indicated that the mean value variation coefficient of each branch of the pneumatic collection and distribution fertilization system, using the optimized feeder, was 3.96%, which is less than 5%. This confirms that the system meets the fertilization requirements for pneumatic collection and distribution. Comparisons before and after optimization also showed that the coefficient of variation of the branch coefficient of variation of the feeder decreased by 19.3%, from 4.91% before the optimization of the structural parameters. The stability of the feeder was significantly improved after optimization.

4. Discussion

The internal flow stability of the feeder can be improved by adopting the diversion rifling structure. It is believed that the basic mechanism of forming spiral flow at the air inlet is the influence of the guide vanes on the axial high pressure fluid. Through the shielding formed by the guide vanes, the airflow forms the relative air pressure difference on the windward and leeward sides of the guide vanes. When the air passes through two blade surfaces with pressure difference, the airflow deflects along the pressure gradient to form the gas rotation with a certain included angle. The analysis on the swirl principle shows that the bigger the swirl number is, the bigger the centrifugal force on the fertilizer particles is, and the fertilizer particles will attach to the pipe wall under the action of strong centrifugal force, resulting in a poor gas–fertilizer mixing effect. The output section of the rifling line is the key component for the feeder to form a stable swirl flow field. The stable swirl flow field formed by the rifling line can maintain the stability of the swirl flow field in the output section and reduce the problem that the fertilizer particles in the output pipeline are deposited at the bottom of the pipeline.

CFD-DEM simulation optimization played a crucial role in this study for enhancing the performance of the feeder in the pneumatic fertilizer distribution system. The established gas–solid coupling model using CFD-DEM methods enabled a detailed analysis of the airflow velocity and pressure distribution during the fertilizer discharge process. The simulation results showed that the air inlet flow-guiding vane structure increased the airflow influence area, thereby improving the conveying performance, while the ribbed chamber structure reduced the range of airflow velocity fluctuations and pressure variation, enhancing the stability of fertilizer delivery. This is in line with previous research that demonstrated the effectiveness of similar structural modifications in improving the performance of particulate material conveying systems [10,11]. The Box–Behnken response surface optimization experiment, based on the simulation results, was conducted to determine the optimal combinations of key structural parameters. The analysis of variance for the feeder fertilizer discharge uniformity index and stability coefficient of variation indicated that the model was highly significant, validating the reliability of the simulation-based optimization approach. The significant factors and their interactions identified through the simulation, such as the linear and quadratic terms of the number of flow-guiding vanes, fertilizer discharge contraction angle, and the number of ribbed chambers, provided valuable insights into the complex relationships between the structural parameters and the feeder's performance. However, the simulation and the establishment of the optimization model is based on the selection of a specific guide ripper feeder and specific parameters based on the actual requirements of cotton crops, so its applicability is limited, but the structure scheme selection, such as air inlet diversion and outlet ripper, provide ideas and reference for the design and optimization of cotton and other feeders.

The optimization results were verified by the bench test, and the variation coefficient of the fertilizer supply stability and the variation coefficient of the branch were significantly decreased before and after the optimization test. The coefficient of variation chosen to represent the uniformity index for evaluation is due to the actual measurement of the feeder end pipe particle uniformity state which is difficult to measure. We tried the image processing method, but to no success, so the selection of each pipe measurement quality, and calculation of branch coefficient of variation was undertaken to represent the uniformity index for evaluation. The comparative results prove the effectiveness of CFD-DEM simulation optimization in guiding the design and optimization of the feeder. However, in the verification, only the feeder performance bench test was conducted, and the feeder performance field test was not conducted. The field test may be affected by more factors, which requires the accuracy of simulation. In addition, the influence of tractor vibration on the feeding performance of the feeder device is simplified.

In conclusion, the CFD-DEM simulation optimization approach in this study provided a powerful tool for understanding and improving the performance of the feeder. It not only helped identify the optimal structural parameters but also enabled a comprehensive analysis of the underlying physical mechanisms. However, this paper, mainly for the strength of the collecting fertilizer feeder design optimization, did not involve the pneumatic set fertilizer system of conveying the elbow and fertilizer structure. Subsequent research will continue on the basis of the design of key components such as fertilizer research, better improving the operation performance of the pneumatic set fertilizer system.

5. Conclusions

This paper designs a diversion and bore feeder for the cotton pneumatic collector, carries out the gas–solid coupling simulation test for the cyclone feeder, and improves the mixing uniformity and stability of the fertilizer. Finally, the pneumatic collecting and discharging fertilization test bench is built to carry out the feeding performance test and verification. The feeder′s optimized design addresses issues of unstable fertilizer transport, uneven distribution, and damage, improving fertilizer application efficiency. This has significant technical value and broad prospects for promoting sustainable agricultural development. The following conclusions are drawn:

(1) Based on the feeding device's structural principle and swirling flow feeding theory, a guide-flow cavity-line feeding device is designed. It combines a wind-inlet vortex structure with guide vanes and an output cavity-line structure. A simulation model is established for the coupling of particle flow and airflow in the swirling cavity-line feeding device. The EDEM-Fluent fluid–solid coupling simulation is conducted to simulate the working process of the feeding device with fertilizer particles. The velocity and pressure distributions of the airflow field for different structural configurations of the feeding device are compared and analyzed; it proves that the design of the diversion rifled feeder is reasonable and effective.

(2) Using response surface optimization, we determine the best combination of structural parameters for the feeder. The performance is best when there are two guide vanes, a fertilizer discharge contraction angle of 23.2°, and eight cavity lines. Simulation experiments are conducted to verify the optimized structural parameters. The simulation results show that the optimized feeding device performs well, with all performance indicators meeting the predicted optimization values, proving that the response surface model is valid and reliable.

(3) The bench-mounted fertilizer extraction test was carried out for performance verification. The results of the bench test show that the variation coefficient of fertilizer discharge stability of the optimized feeder is 13.30%, which is consistent with the simulation test results. The accuracy of the simulation of the feeder is verified. The stability variation coefficient before optimization is 14.89%, and the stability of fertilizer discharge of the feeder is significantly improved after optimization. At the same time, the variation coefficient of fertilizer discharge of each branch of the pneumatic integrated discharge fertilizer system is 3.96%, less than 5%. It can better meet the operation

requirements of the pneumatic integrated discharge fertilizer system. After optimization, the variation coefficient of fertilizer discharge for each branch is reduced by 19.3%, and the stability of fertilizer discharge for the feeder is significantly improved.

Author Contributions: Conceptualization, L.Z.; methodology, W.D.; software, W.D. and X.Z.; data curation, X.Z. and Z.J.; validation, Y.G.; formal analysis, W.D.; investigation, X.H.; writing—original draft, W.D.; writing—review and editing, L.Z.; supervision, L.Z. All authors have read and agreed to the published version of the manuscript.

Funding: This research was funded by: 1. National Key R&D Program of China, grant number 2022ZD0115804; 2. The XPCC's science and technology plan project, grant number 2023112667; 3. The Silk Road Economic Belt Innovation-driven Development Pilot Zone, the Wuchangshi National Independent Innovation Demonstration Zone Science and Technology Development Plan project, grant number 2022BC004; 4. The XPCC's financial science and technology plan project, grant number 2020AB002.

Data Availability Statement: Due to restrictions on data ownership and technical limitations, we are currently unable to deposit the data in public repositories. Nevertheless, we remain committed to ensuring the availability and accessibility of the data. Therefore, we hereby declare that for researchers interested in the data, we will provide the data materials upon reasonable request. Researchers who wish to obtain the data are kindly requested to contact us via the email address provided, and we will promptly communicate with you and provide necessary arrangements for data access.

Conflicts of Interest: The authors declare no conflict of interest.

References

1. Li, J.W.; Yuan, W.S.; Zhai, H.L.; Zhu, S.S. Application of Pneumatic Conveying Technology in Agricultural Machinery. *Chin. Soc. Agric. Mec.* **2023**, *44*, 85–92.
2. He, Y.K.; Zhao, X.G.; Li, C.L.; Dou, H.J.; Li, S.; Wang, X. Research Status and Analysis of Mechanical Fertilization Technology for Topdressing Maize. *J. Agric. Mech. Res.* **2021**, *43*, 1–9.
3. Zhang, X.H.; Wang, Y.Z.; Zhang, L.; Peng, C.J.; Pan, G.J. Design and Experiment of Seeding Distribution System for Pneumatic Wheat Seeder-Collector. *Trans. Chin. Soc. Agric. Mac.* **2018**, *49*, 59–67.
4. Paul, H. *Foliar Feeding Good Supplement for Cotton Fertilization*; Southeast Farm Press: Auburn, AL, USA, 2011.
5. Zha, X.T.; Zhang, G.Z.; Zhang, S.J.; Hou, Q.X.; Wang, Y.; Zhou, Y. Design and Experiment of Centralized Pneumatic Deep Precision Fertilization Device for Rice Transplanter. *Int. J. Agric. Biol. Eng. (IJABE)* **2020**, *13*, 109–117. [CrossRef]
6. Andrii, Y.; Jean-Pierre, L. Establishing the Conveying Parameters Required for the Air-Seeders. *Biosyst. Eng.* **2018**, *166*, 1–12.
7. Niu, Y.; Zhang, J.; Qi, J.; Meng, H.; Peng, H.; Li, J. Design and Test of Soil–Fertilizer Collision Mixing and Mulching Device for Manure Deep Application Machine. *Agriculture* **2023**, *13*, 709. [CrossRef]
8. Ke, H.B.; Zhou, Y.; Zhang, G.Z.; Lü, W.; Liu, Y.; Huang, L. Design and Experiment of Pneumatic Fertilizer Collection and Distribution Device for Ratoon Rice. *J. Agric. Sci. Technol.* **2022**, *24*, 106–114.
9. Yang, Q.L.; Wang, Q.J.; Li, H.W.; He, J.; Lu, C.Y.; Wang, Y.B.; Yu, C.C. Structural Optimization and Experiment of Pneumatic Fertilizer Collection and Distribution System. *Trans. Chin. Soc. Agric. Eng.* **2020**, *36*, 1–10.
10. Al-Arkawazi, S.; Marie, C.; Benhabib, K.; Coorevits, P. Modeling the Hydrodynamic Forces between Fluid–Granular Medium by Coupling DEM–CFD. *Chem. Eng. Res. Des.* **2017**, *117*, 439–447. [CrossRef]
11. Yang, Q.L.; Li, Z.H.; Li, H.W.; He, J.; Wang, Q.J.; Lu, C.Y. Numerical Analysis of Particle Movement in Collective-Discharging Fertilizer Distributor Based on CFD-DEM. *Trans. Chin. Soc. Agric. Mac.* **2019**, *50*, 81–89.
12. Zhang, H.; Liu, M.L.; Li, T.J.; Huang, Z.Y.; Bo, H.L.; Dong, Y.J. Experimental Study on Plug Formation Characteristics of a Novel Draft Tube Type Feeder for Vertical Pneumatic Conveying of Coarse Particles. *Powder Technol.* **2016**, *301*, 730–736. [CrossRef]
13. Ma, H.L.; Zhao, C.J.; Yang, S.; Li, S.; Sun, X.H.; Zhai, C.Y. Research on a Newly Designed Venturi Ejector and Dazomet Particle Motion Characteristics for Dilute-Phase Pneumatic Conveying Systems Based on CFD-DEM. *Comput. Electr. Eng.* **2023**, *213*, 108240.
14. Kartal, V.; Emiroglu, M.E. Effect of Nozzle Type on Local Scour in Water Jets: An Experimental Study. *Ocean Eng.* **2023**, *277*, 114323. [CrossRef]

15. Ahad, J.; Ahmad, M.; Farooq, A.; Siddique, W.; Waheed, K.; Qureshi, K.R.; Shah, A.; Ahmed, A.; Bibi, A.; Irfan, N. Influence of Throat and Diverging Section on the Performance of Venturi Scrubber by Using Computational Fluid Dynamics. *Prog. Nucl. Energy* **2023**, *163*, 104800. [CrossRef]
16. Liao, Y.; You, Y.; Hui, Y.; Zhang, X.; Wang, D. Mixed Seeds of Oat and Vetch Based on DEM-Fluent Coupling Motion Simulation in a Venturi Tube. *Processes* **2023**, *11*, 1095. [CrossRef]
17. Gao, X.J.; Zhou, Z.Y.; Xu, Y.; Yu, Y.B.; Su, Y.; Cui, T. Numerical Simulation of Particle Motion Characteristics in Quantitative Seed Feeding System. *Powder Technol.* **2020**, *367*, 643–658. [CrossRef]
18. Chandramohan, P.; Balamurugan, P.; Dinesh, S. Effect of Fluid Flow Diversion and Venturi Size on Drip Irrigation Fertilizer Feeding System. *Flow Meas. Instrum.* **2019**, *65*, 250–256. [CrossRef]
19. Li, L.W.; Meng, Z.J.; Wang, X.O.; An, X.F.; Wang, P.; Wu, G.W. Simulation Analysis of Gas-Solid Two-Phase Flow in Fertilizer Conveying Device of Pneumatic Rice Fertilizer Applicator. *Trans. Chin. Soc. Agric. Mac.* **2018**, *49* (Suppl. S1), 171–180.
20. Guo, X.D. Study on Pneumatic System of Rice Side Deep Fertilizer Machine and Whole Machine Design. Master's Thesis, Zhejiang Sci-Tech University, Hangzhou, China, 2020.
21. Liang, J.; Pan, F.; Chen, J.; Zhang, H.; Ji, C. A Precise Simultaneous Sowed Control System for Maize Seed and Fertilizer. *Agriculture* **2024**, *14*, 192. [CrossRef]
22. Yuan, W.; Ji, C.; Liu, Z.; Jin, C.; Feng, Y. Influencing Factors of the Distribution Accuracy and the Optimal Parameters of a Pneumatic Fertilization Distributor in a Fertilizer Applicator. *Agronomy* **2022**, *12*, 2222. [CrossRef]
23. Gong, F.; Hu, M.; Bao, A.H.; Li, D.; Gao, T.; Wang, C.L. Parameter Calibration and Significance Analysis of Rice Straw Based on Hertz-Mindlin Model. *J. Southwest Univ. (Nat. Sci. Ed.)* **2022**, *44*, 186–196.
24. Jia, H.L.; Tan, H.W.; Wen, X.Y.; Wang, G.; Yuan, H.F.; Huang, D.Y. Design and Experiment of Pneumatic Precision Mixed Fertilizer Applicator with Concentrated Distribution. *Trans. Chin. Soc. Agric. Mac.* **2022**, *53* (Suppl. S2), 109–119.
25. Dong, W.C.; Bu, H.R.; Jiang, Z.; Yu, S.Y.; Zhang, L.X.; Xia, Y.Q. Design and Simulation Test of Cone Gear Pair for Troughed Belt Fertilizer Distributor. *J. Shihezi Univ. (Nat. Sci. Ed.)* **2023**, *41*, 529–536.
26. Bu, H.; Yu, S.; Dong, W.; Zhang, L.; Xia, Y. Analysis of the Effect of Bivariate Fertilizer Discharger Control Sequence on Fertilizer Discharge Performance. *Agriculture* **2022**, *12*, 1927. [CrossRef]
27. Bu, H.; Yu, S.; Dong, W.; Wang, Y.; Zhang, L.; Xia, Y. Calibration and Testing of Discrete Element Simulation Parameters for Urea Particles. *Processes* **2022**, *10*, 511. [CrossRef]
28. *T/SHZSAQS 00177-2023*; Technical Regulations for the Operation of Cotton Fertilizer-Seeding Combined Machines. Standardization Administration of China: Beijing, China, 2023.
29. *NY/T 1003-2006*; Technical Specification for the Quality Evaluation of Fertilizing Machinery. China Agricultural Machinery Test & Quality Supervision Station; Heilongjiang Provincial Water Field Mechanization Research Institute: Harbin, China, 2006.

agriculture

MDPI

Article

Smart Strip-Till One-Pass Machine: Winter Wheat Sowing Accuracy Assessment

Dariusz Jaskulski [1,*], Iwona Jaskulska [1], Emilian Różniak [2], Maja Radziemska [3], Barbara Klik [3] and Martin Brtnický [4]

[1] Department of Agronomy, Faculty of Agriculture and Biotechnology, Bydgoszcz University of Science and Technology, Prof. S. Kaliskiego 7, 85-796 Bydgoszcz, Poland; jaskulska@pbs.edu.pl
[2] Research & Development Centre Agro-Land Marek Różniak Śmielin, Kościelna 1, 89-110 Sadki, Poland; e.rozniak@mzuri-agro.eu
[3] Institute of Environmental Engineering, Warsaw University of Life Sciences, Nowoursynowska 159, 02-776 Warsaw, Poland; maja_radziemska@sggw.edu.pl (M.R.); barbara_klik@sggw.edu.pl (B.K.)
[4] Department of Agrochemistry, Soil Science, Microbiology and Plant Nutrition, Faculty of AgriSciences, Mendel University in Brno, Zemedelska 1, 61300 Brno, Czech Republic; martin.brtnicky@seznam.cz
* Correspondence: darekjas@pbs.edu.pl

Abstract: Modern agricultural machines are subject to requirements that result from developments in plant cultivation technology and environmental care. Agricultural practice demands multifunctional machines that perform several agrotechnical treatments in a single pass. Automated and digitalised management of machines and their working parts is also becoming standard. A strip-till one-pass machine was designed that automatically regulates and monitors sowing rate and depths and the application of fertiliser to loosened soil strips. Among other things, an electro-hydraulic depth regulator with a built-in linear potentiometer and an overload sensor was used. Laboratory and field tests assessed the accuracy of the rate and depth of sowing wheat grain and fertiliser application by the innovative machine. This study confirmed the machine's high quality of wheat sowing. The accuracy of the operating parameters was not less than 97% in laboratory tests and 92% in field conditions. The field emergence capacity of wheat was 88% and its sowing density can be considered good. The machine provides uniform operation of all 11 multifunctional assemblies (units, sections of loosening-applying tines and sowing coulters). The coefficient of variation (CV) of grain sowing and granular fertiliser application by individual assemblies was in the range of 4.27–7.29% and 3.74–6.90%, respectively. The sowing depth accuracy expressed as an accuracy coefficient (DA) was 87.33–93.67% with CV 4.62–9.65%. The machine's introduction onto the market can facilitate field cultivation of plants in accordance with the principles of conservation agriculture and Agriculture 4.0.

Keywords: agricultural machinery; depth regulator; strip tillage; fertiliser; sowing grain; sowing depth; emergence capacity

1. Introduction

Agriculture 3.0 and 4.0 integrate automation, robotisation and elements of IT to optimise agricultural production processes. It manifests in the implementation of modern machines and plant cultivation technologies that increase productivity, reduce human labour input, reduce costs and limit adverse environmental impacts [1,2]. Today, agriculture is evolving to meet the needs of contemporary society, where the food supply is secured through practices such as water and soil conservation and a reduction in agrochemicals, ensuring the production of high-quality crops. A necessary prerequisite for ensuring all of the above is the use of well-functioning agricultural machinery [3]. Digital technologies, automation and autonomisation are being adapted to agricultural machines and tools, including those for soil tillage, fertilisation and sowing. They enable the use of variable seed and fertiliser doses, automatic control of the operations of machine parts responsible for key cultivation processes, and performance monitoring [4,5]. Progress in techniques and technologies is being made in all aspects of modern field crop cultivation, including soil tillage, fertilisation and sowing, which are increasingly being combined into a single process [6]. One trend in contemporary agricultural mechanisation is the construction of multifunctional machines. One machine can perform many agrotechnical operations, from soil tillage to seed harvesting [7].

The development of soil cultivation technologies and techniques includes innovative solutions relating to both the methods and machines of cultivation, covering construction, functionality, operating parameters and their monitoring, and decision-support systems [8,9]. Among the new reduced-till methods, strip-till occupies an important and growing position [10,11]. However, such tillage, especially when combined with the simultaneous application of fertilisers and sowing of seeds (i.e., strip-till one-pass) requires specialised machinery [12].

Agricultural mechanisation also includes seed drills and machines for simultaneous sowing and fertilising. Design work is mainly focused on creating innovative, multifunctional machines that can perform agrotechnical tasks in a variety of agricultural systems and socio-economic conditions [13–15]. Particular attention is paid to solutions that determine the precision, uniformity and accuracy of seed sowing or fertiliser application. Therefore, research is focused on various of the assemblies and working parts of these machines, such as tanks, distributors, seed-transfer systems, and multifunctional assemblies for soil tillage and sowing seeds [16,17].

In the era of development of smart agriculture, automation, digitalisation and real-time environmental conditions monitoring systems are increasingly present in agricultural machines. These technologies support or replace mechanical operations in the field of precise traction of machines, their control and regulation of operating parameters during soil tillage, sowing, fertilisation, plant protection, harvesting [18]. Smart monitoring systems based on the Internet of Things technology analyse soil and field conditions, such as: moisture, temperature, pressure, light, air composition, CO_2 content, pH, NPK. This technology also has the ability to collect and transmit large amounts of data used in real-time and for planning agricultural operations [19]. Newly designed, innovative agricultural machines and tools are equipped with an increasing number of sensors for real-time operation monitoring and systems for automatic adjustment of working elements, e.g., stereo vision cameras, LiDAR, radars, thermal cameras, RGB cameras and laser scanners [20]. A sensor system comprising a linear potentiometer, inclinometer and optical distance sensor can measure the depth to which working parts penetrate the soil [21]. Sensors and automation systems are also used in shallow surface-cultivation tools. Rotary harrows equipped with stereo cameras that assess the levelling of the soil surface, and the size of soil aggregates make it possible to control the intensity and accuracy of seedbed cultivation required for precise seed sowing [22].

Ultrasonic sensors and linear differential sensors are used to precisely assess seeding depth in field conditions. Infrared sensors or photoelectric sensors are suitable for controlling seed rate and flow in seed transport ducts. It is essential to monitor these seeder parameters to ensure high-quality sowing and, consequently, uniform plant emergence [23]. Systems for monitoring sowing depth are also being developed, especially in high-speed seed drills. This is performed using combined data from laser, ultrasonic and angle sensors [24]. Environmental-condition sensors and work-monitoring systems based on the

Internet of Things are also used in reduced-tillage machines, e.g., no-till seed drills [25]. In modern seed drills, operating parameters such as depth, quantity and distribution of seeds in the soil are monitored and corrected in real time [26]. Sensors, as well as advanced control, data collection and processing systems enable the construction of next-generation machines. These include drones, robots, and autonomous machine [27,28].

In recent decades, there have been rapid advances in strip-till machines and their working parts [29,30], including in one-pass strip-till technology, which involves the tilling of soil strips, application of fertilisers and sowing of seeds to be performed simultaneously [31]. New and improved machine designs can operate even in difficult soil conditions [32]. Wang et al. [33] presented a strip-till machine with a system to automate the regulation of cultivation depth and the width of soil strips cleared of plant residues. Particular attention is being paid to constructing hybrid multifunctional strip-till machines enabling the sowing of various plant species while maintaining high-precision seed sowing [34].

An important sowing parameter that determines emergence and plant population is its accuracy. The distribution of seeds in the soil is also important for the growth of plants [35]. Seed-sowing precision depends on, among other things, seedbed preparation, sowing depth, seeder speed, and the number of seeds in the hopper [36,37]. No less important are the design solutions for the parts that convey seeds and determine dose sizes in seed drills [38]. The horizontal and vertical distribution of seeds in the soil depends on the adopted inter-row spacing and in-row plant spacing, sowing depth, pre-sowing soil preparation, technical parameters of the seeder, and operator skills. The indicators of sowing quality include the arithmetic mean of the seed spacing, the standard deviation of the plant spacing, and the coefficient of variation [39]. The density and uniformity of plants in the field resulting from different sowing techniques and methods are assessed by counting in the field or by image analysis [40]. The assessment results are expressed as indices created using statistical analysis and mathematical modelling [41,42], often using the coefficient of variation [43].

In line with the global trend of agricultural technological development, a consortium consisting of an agricultural machinery manufacturer, a research and development facility, and Poznań and Bydgoszcz universities of technology undertook to develop an automatically managed, regulated and monitored strip-till one-pass machine. The research hypothesis was that the introduced design and functional changes will allow the creation of a strip-till one-pass machine with automated regulation of basic operating parameters of working elements and the possibility of their real-time adjustment to changing conditions of crop fields. It will therefore be possible to abandon manual settings and rely on the subjective assessment of the operator during these activities. During industrial research and development work, the accuracy of the rate and depth of wheat grain sowing and the amount of granular fertiliser applied were tested, assuming that the field emergence capacity would be above 80% to ensure good plant density after emergence.

2. Materials and Methods

2.1. Subject, Scope and Location of Research

The subject of the research was a new model of Mzuri hybrid machine adapted to the strip-till one-pass technique used in the field cultivation of various crop species. The machine is equipped with sensors monitoring soil properties necessary for automatic control of operating parameters, i.e., depth of soil strip loosening, seed sowing depth and pressure force of the ground-following soil-consolidating wheel. The research and development work also results in new mechanical and electro-hydraulic solutions enabling automatic change in machine operating parameters in real time. The design and construction challenge was not only to determine the accuracy and range of operation of sensors and control elements, but also their mechanical implementation with the machine. Due to the difficult working environment—soil, stones, moisture, dust, and plant residues—the first prototypes of soil property assessment sensors were damaged and/or their operation was disrupted. Ultimately, the soil property sensors were mounted directly behind one of the loosening tines. The measurement of soil volumetric moisture and variation in relative soil compaction measured in real time or read from the field map are the basic elements of the algorithms for automatic management of the working depth of the loosening tines and sowing coulters, as well as the pressure force of the soil-consolidating wheels. Currently, seeds and fertiliser are evenly distributed from the central split tank to the working assemblies by two independent pneumatic heads. A central block of electro-magnetic and hydraulic valves (Figure 1A) manages the automatic regulation and monitoring systems of the operation of 11 multifunctional assemblies—sections of loosening-applying tines and sowing coulters. Each of these assemblies features four key elements. The first is a disc that cuts through the surface and clears residue, which is an existing solution, and so was not the subject of analysis herein. The second is a novel integrated soil-loosening tine with fertiliser-placement applicator. The final ground-contacting element is an innovative seed-sowing coulter (Figure 1B) combined with the ground-following soil-consolidating wheel. In earlier models of the machine, this connection was made by screws. The operator manually adjusted their length, and thus the sowing depth and wheel pressure force. These parameters were selected intuitively and were not changed during work in the field, despite differences in soil properties in different parts of the field. In the innovative solution, the screws were replaced with a cylinder and an electro-hydraulic regulator (Figure 1C). Its extension and wheel contact force depend on the soil conditions (structure, moisture) included in the device control algorithms. In dry, light soil, the coulter is sunk and the wheel pressure force increases so that the seeds have more favourable water conditions for germination. In heavy, moist soil, the coulter works shallowly and the wheel has less pressure force, reducing the risk of soil crusting after sowing. The sowing depth regulator installed in the machine was developed as a result of many critical studies of previous commercial electric and electro-hydraulic solutions. These devices (Figure 2A,B) were not sufficiently resistant to unfavourable weather conditions, did not have an appropriate response and controllability in real time, as well as durability in previously uncultivated soil and with a large amount of crop residue on the field surface. Only a controller with a built-in encoder and overload sensor (Figure 2C) could be used in the machine, meeting the design and functional assumptions in field conditions. The controller has a built-in linear potentiometer enabling automatic control and monitoring of the working depth. A solution similar in principle of operation also enables automatic regulation of the working depth of the loosening tines and its ongoing monitoring. For this purpose, a hydraulic cylinder with a built-in potentiometer was developed (Figure 3). The diagram of the hydraulic system of the valve block responsible for the smooth regulation of the working depth of the tine section and seed-sowing coulters is shown in Figure 4. For example, the algorithm of the tine working depth taking into account the soil properties determines the position of the potentiometer by opening the valve (4/3 flow direction control valve with control—Figure 4A). In turn, the pressure force of the compacting wheel behind the seed-sowing coulter depends on the oil pressure supplied to the cylinder, which is regulated by a valve placed in the system (System pressure regulating valve—Figure 4B). The quantities of dosed material (seeds, fertiliser) are monitored in real time by flow optical sensors (Figure 5). All sensors, control and regulation systems are managed by a central computer.

(**A**)
(**B**)
(**C**)

Figure 1. Central block of electro-magnetic and hydraulic valves (**A**), seed-sowing coulters (**B**), and electro-hydraulic regulator of soil-working depth (**C**) of strip-till one-pass machine.

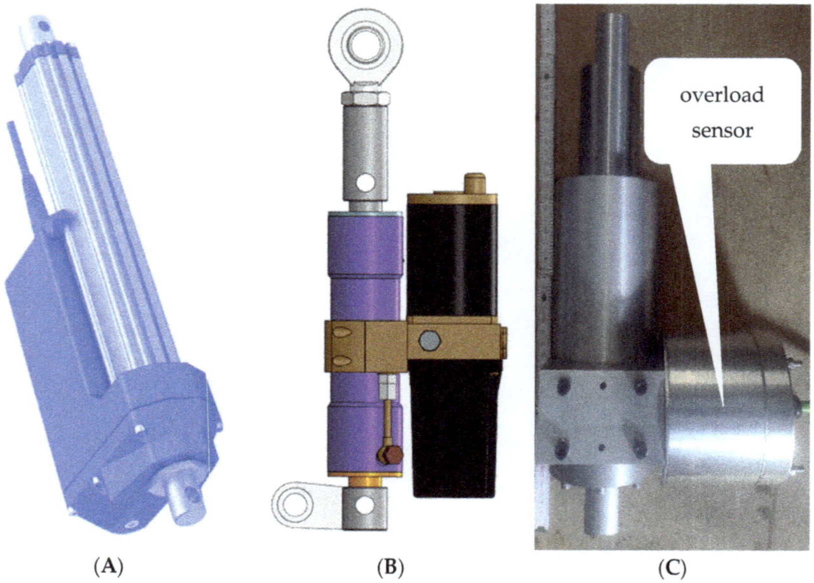

Figure 2. Coulter and tine depth regulators; electric (**A**), electro-hydraulic (**B**), with a built-in encoder and overload sensor (**C**).

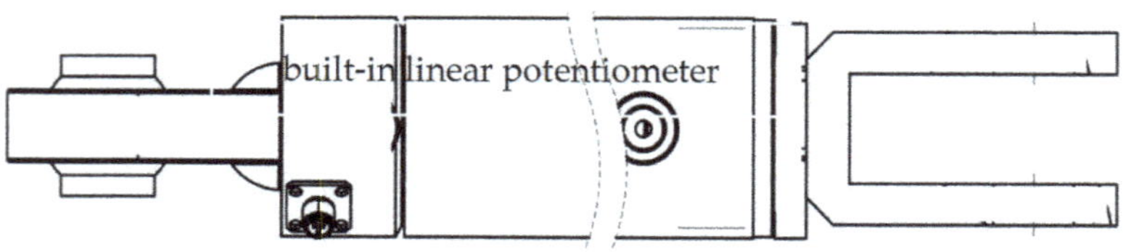

Figure 3. Working depth control cylinder with built-in linear potentiometer.

The design works and machine operation tests were performed at Mzuri World Sp. z o.o. and the Research and Development Center Agro-Means-Technique-Technology in Śmielin (53°09′04.0″ N; 17°29′10.7″ E), Kuyavian-Pomeranian Voivodeship, Poland with the participation of employees of the Departments of Agronomy of the Bydgoszcz University of Technology and Poznań University of Technology, Poland. The work was carried out in 2021–23 as part of the AGROTECH project co-financed by the European Union.

2.2. *Laboratory Tests and Field Studies*

In the first stage of industrial research, laboratory tests were carried out to determine the accuracy of the sowing amount of winter wheat grain and granulated fertiliser by, respectively, each coulter and each loosening-applying tine—ten repetitions. The research used wheat seed with a thousand-grain weight of 44 g and a utility value of 88%, assuming a sowing rate of 150 kg/ha. The fertiliser used for the tests was Polidap [NP (S), 18–46 (5)]. Ninety-two percent of the product is in the form of granules with dimensions 2.0–5.0 mm and bulk density 0.85–0.95 kg/dm^3. The fertiliser was applied at a dose of 150 kg/ha.

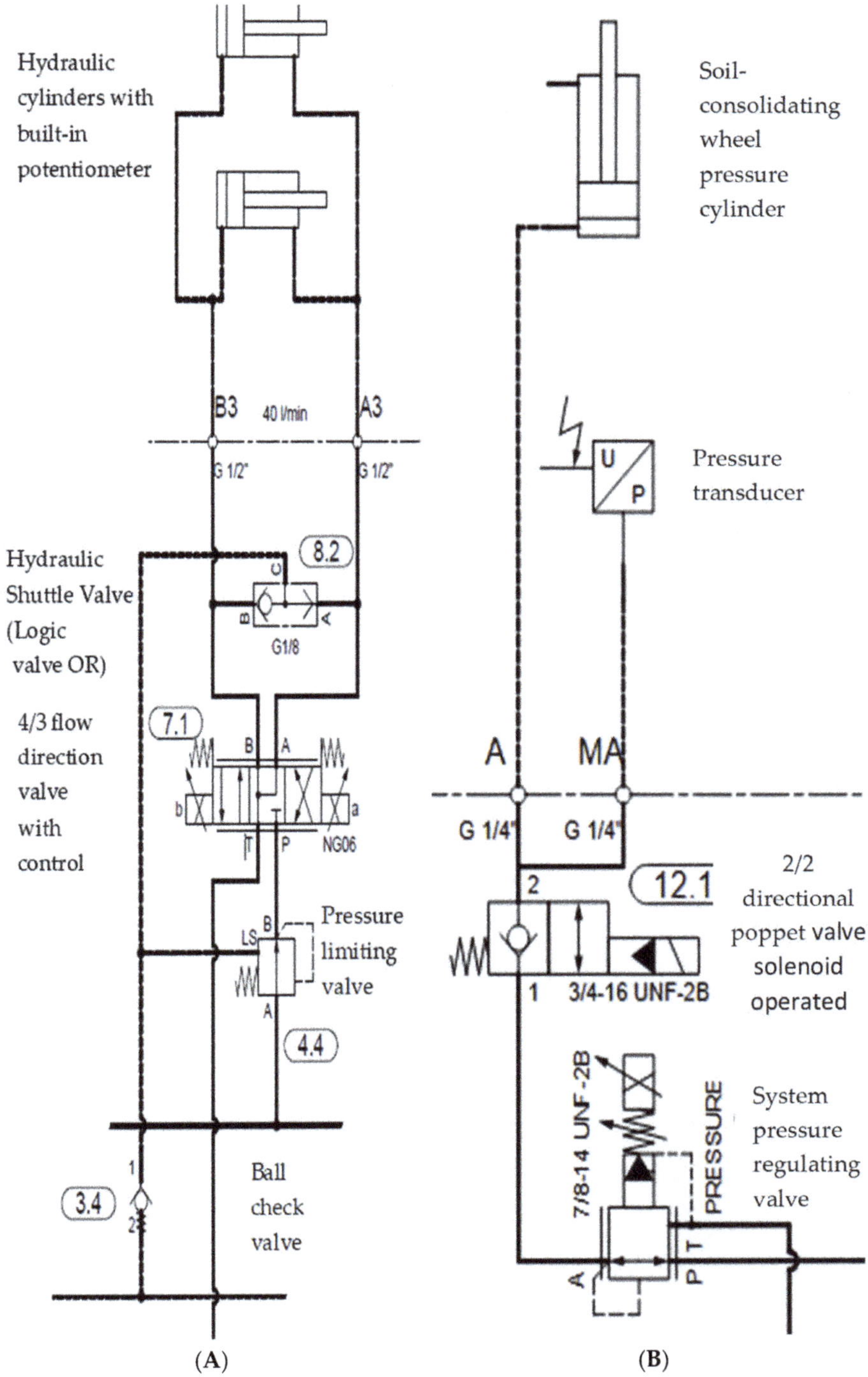

Figure 4. The diagram of the hydraulic system of the valve block responsible for the regulation of the working depth of the tine section (**A**) and seed-sowing coulters (**B**).

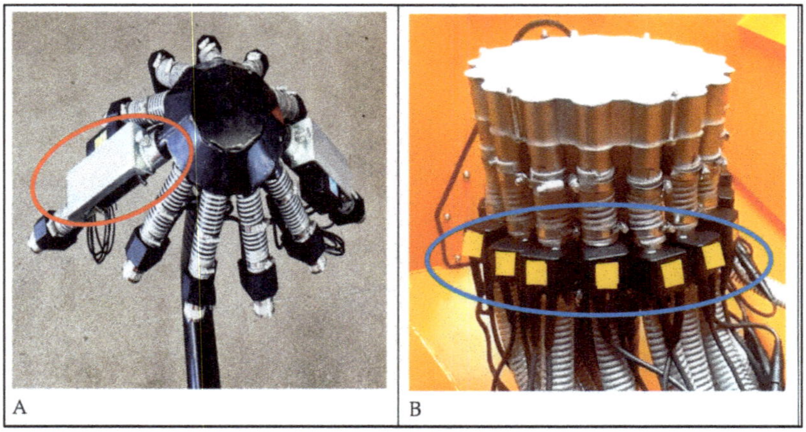

Figure 5. Flow optical sensors: seed head (**A**) and fertiliser head (**B**).

The indicators used as a measure of the accuracy of grain sowing and fertiliser application were those consistent with the sowing accuracy (SA) formula presented in the work of Liu et al. [44]. The accuracy of grain sowing and fertiliser application by individual coulters and loosening-applying tines was assessed on the basis of a test operation of the machine corresponding to work on an area of 0.1 ha. The test was repeated ten times. The data collected on the amount of grain and fertiliser from each of the 11 assemblies were converted into the values of the accuracy of grain sowing index (GA) and fertiliser application index (FA) according to the formulas:

$$GA = (1 - |(WGm/WGt) - 1|) \times 100\% \quad (1)$$

where WGm—grain mass measured during the test; WGt—theoretical grain mass resulting from the assumed sowing quantity (13.64 kg/coulter), and:

$$FA = (1 - |(WFm/WFt) - 1|) \times 100\% \quad (2)$$

where WFm—fertiliser mass measured during the test; WFt—theoretical fertiliser mass resulting from the assumed dose (13.64 kg/placement tine).

In the second stage, field tests were performed on Cambisol (loamy sand), with a soil volumetric moisture content of 11–17% after harvesting winter rape with crop residues on the field surface. The research site is located in the zone of influence of humid continental climate (cold, without dry season, warm summer). The average annual air temperature is 8.1 °C, and the total rainfall is 485 mm. The machine's travel speed was 8 km/h. Similarly to the first stage, the accuracy of seeding and fertiliser placement was tested. In addition, the accuracy of the working depth of the seeding coulters in the soil was assessed. The sowing depth accuracy index (DA) was calculated based on ten measurements of seed placement depth for each coulter relative to the assumed optimal depth of 3 cm, as follows:

$$DA = (1 - |(Dm/Dt) - 1|) \times 100\% \quad (3)$$

where Dm—depth of seed placement during the test; Dt—assumed optimal sowing depth.

In a field experiment, after sowing winter wheat on 2.5 ha at a density of 250 grains/m^2 and a field emergence capacity of 87%, the uniformity of plant density was assessed in a strip corresponding to the working width of the machine. The uniformity of planting in each of the 11 rows spaced 36.4 cm apart was determined by calculating the accuracy of the number of plants (PA) based on the number of plants per metre and the optimal number of plants according to the assumed seed material and sowing parameters.

$$PA = (1 - |(Pm/Pt) - 1|) \times 100\% \quad (4)$$

where Pm—number of plants detected after emergence; Pt—optimal number of plants according to the assumed sowing parameters—79.1 plants per 1 m of row.

2.3. Analysis of Results

The results of laboratory and field tests were statistically processed. The actual measurement data of individual sowing and plant emergence parameters and fertiliser placement are given in tables. Basic statistics (average, minimum, maximum, standard deviation, coefficient of variation, standard error) are presented for each coulter assembly and on average for the working width of the machine. The measurement data and theoretical data (those assumed as the standard) were used to calculate average accuracy indices values for each assembly, as well as error values and standard deviations. These values for each index are presented as box-and-whisker graphs illustrating each of the 11 assemblies. Based on all the measurement data determining the value of the accuracy index of a specific sowing parameter (from all assemblies and repetitions, n = 110), its coefficient of variation (CV) was calculated.

$$CV = (\text{standard deviation}/\text{mean value of measurements}) \times 100\% \quad (5)$$

Statistical analysis was performed using the Statistica.PL 12 program [45].

3. Results

In a static laboratory test, assuming the placement of 13.64 kg of wheat grain and 13.64 kg of fertiliser by each assembly, it was found that an average 0.18 kg more grain was sown than assumed and 0.09 kg less fertiliser was applied. The weight of grain sown by the machine ranged between 13.30 kg and 14.06 kg, while the weight of fertiliser was between 13.38 and 13.73 kg. In the field test, the corresponding difference between optimal and actual placed amounts of grain and fertiliser was 0.03 kg and 0.41 kg, respectively. The difference between the maximum and minimum size is 1.27 kg of grain and 1.06 kg of fertiliser. The standard deviation, coefficient of variation and standard error of the amount of grain sown and fertiliser applied in the field tests were higher than in the laboratory tests (Table 1). The sowing depth and the number of plants after emergence obtained by the Mzuri machine in field tests did not differ significantly from the assumed optimal values. These differences were 0.24 cm and 1.3 plants/m of row, respectively. The CV of none of the sowing parameters in the laboratory and field tests exceeded 3.50%.

There was a greater variation in sowing parameters among the 11 assemblies than in the average work of the entire machine (Table 2). The CV of seed sowing rate and fertiliser application in the laboratory was 2.19–3.46% and 2.49–5.36%, respectively, while in field conditions it was 4.27–7.29% and 3.74–6.90%. An even higher CV (7.98–11.71) was found for the variability of sowing depth. The variability in sowing depth was greatest in the outer coulter assemblies and smallest in those in the middle of the machine. The average CV for the working depth of the four central assemblies is 8.59%, and that of the four outermost, (two left and two right outermost) was 10.00%. The variability (CV) of number of plants after emergence in a 1-m section of row ranged from 5.63% to 10.43%.

Table 1. Laboratory and field tests of the strip-till machine operation—average results of all sowing coulters and loosening-applying tines.

Parameter	Unit	Average	Minimum	Maximum	Standard Deviation	Coefficient of Variation (%)	Standard Error
Laboratory tests							
Sowing grain	kg	13.82	13.30	14.06	0.31	2.24	0.10
Fertiliser application	kg	13.55	13.38	13.73	0.11	0.84	0.04
Field tests							
Sowing grain	kg	13.61	12.98	14.25	0.48	3.49	0.15
Fertiliser application	kg	14.05	13.52	14.58	0.35	2.52	0.11
Sowing depth	cm	2.76	2.64	2.89	0.07	2.58	0.02
Number of plants	pcs./m	80.38	78.36	83.18	1.50	1.87	0.48

Table 2. Coefficient of variation (CV (%)) of the working parameters of the individual machine assemblies.

Parameter	Assembly										
	1	2	3	4	5	6	7	8	9	10	11
Laboratory tests											
Sowing grain	2.19	2.40	2.59	3.06	2.43	2.57	2.25	3.22	2.61	3.46	2.59
Fertiliser application	2.49	4.67	4.26	5.05	4.83	5.36	3.11	3.04	3.91	4.98	3.49
Field tests											
Sowing grain	5.00	5.98	7.27	5.77	4.27	6.59	6.31	5.88	6.06	7.29	6.09
Fertiliser application	5.45	6.71	5.06	6.43	3.74	6.90	3.92	3.92	4.77	4.87	4.92
Sowing depth	9.76	9.51	10.07	8.55	9.51	7.98	8.32	11.71	11.20	10.06	10.66
Number of plants	8.45	5.63	8.74	7.24	10.43	9.79	8.23	6.76	6.00	8.50	8.33

In laboratory tests, the wheat grain sowing accuracy of individual assemblies was generally greater in the central part of the machine and lower at the sides. The GA index value ranged from 96.47% to 97.87% (Figure 6). Fertiliser application accuracy did not depend on loosening-applying tine position relative to the machine's centre line. The highest FA index (97.57%) described the work of assembly 7, and the lowest (95.37%) was for assemblies 4 and 6 (Figure 7).

In field tests, the accuracy of grain sowing (GA) and fertiliser application (FA) by individual assemblies did not depend to a large extent on their position on the machine. The grain was sown most accurately by coulter 5 and the least accurately by coulter 10. The maximum difference in GA index was 3.67 percentage points (Figure 8). The analogous difference in the FA index value was 2.72 percentage points and was the result of the most and least accurate applications (respectively, loosening-applying tines 3 and 4) (Figure 9).

The accuracy of working depth of the four central coulters was greater than that of the outer ones. The average DA index was 92.75% for the four central coulters, 88.67% for the three left outermost, and 88.42% for the four right outermost. The least stable in the soil was coulter 8 (DA = 87.33%), while the most stable was coulter 4 (DA = 93.67%) (Figure 10).

The uniformity of winter wheat density in the 11 individual rows, as expressed by the precision index of the number of plants (PA) per 1 m of row ranged from 91.00% for row 5 to 93.86% for row 2 (Figure 11).

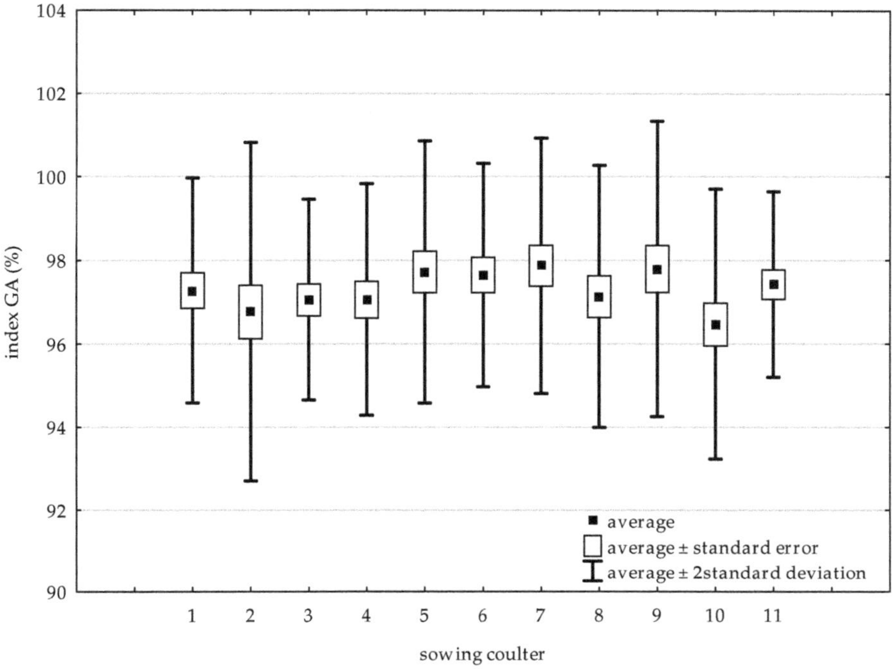

Figure 6. Grain sowing accuracy index (GA) of individual sowing coulters in laboratory tests.

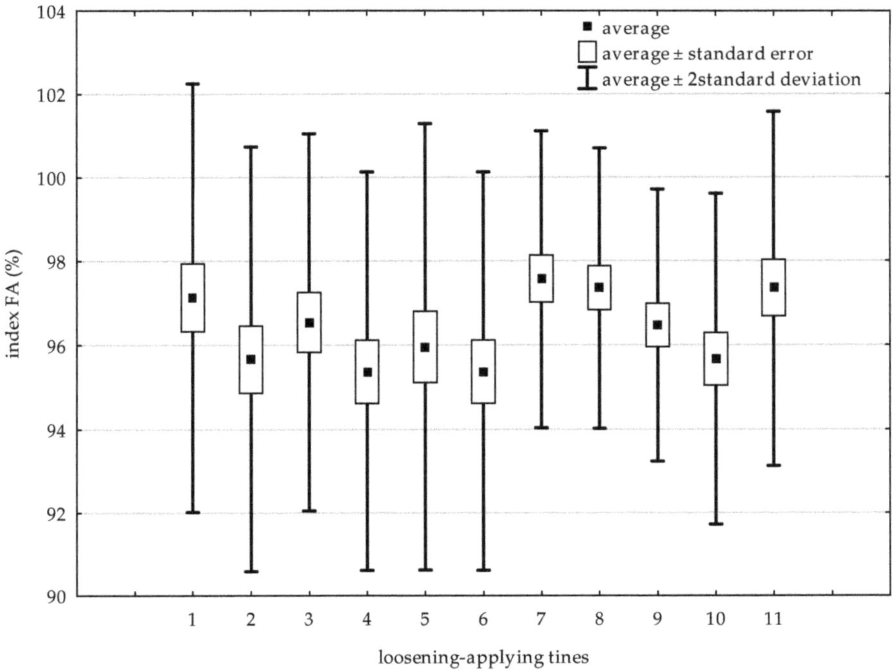

Figure 7. Fertiliser application accuracy index (FA) of individual loosening-applying tines in laboratory tests.

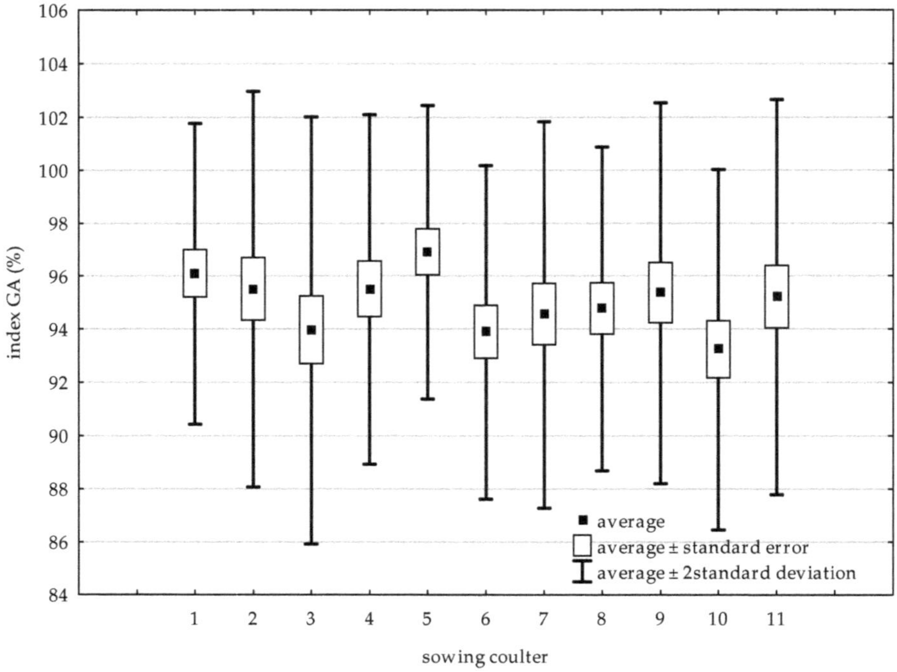

Figure 8. Grain sowing accuracy index (GA) of individual sowing coulters in field tests.

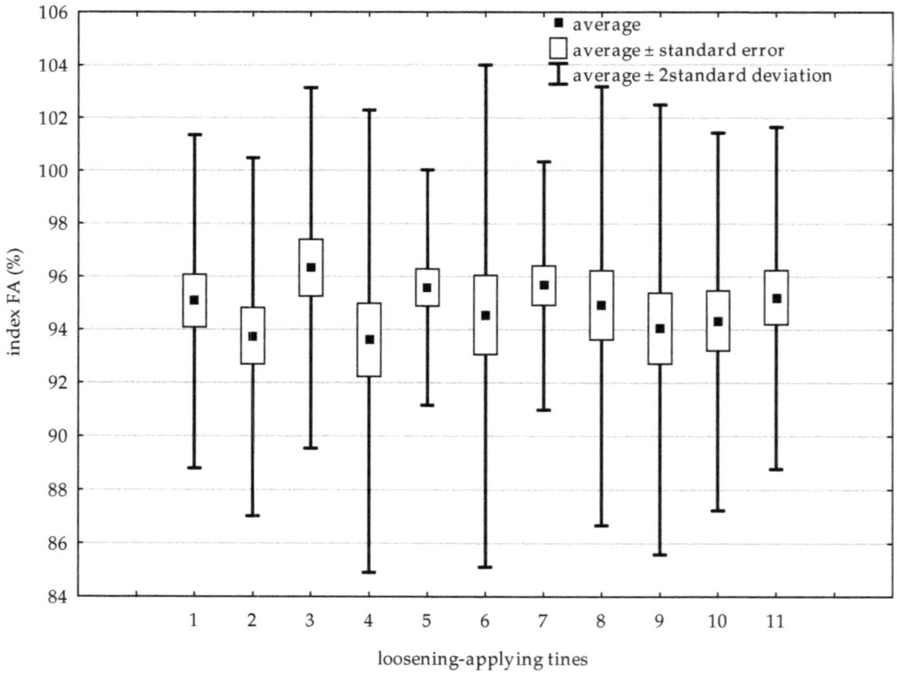

Figure 9. Fertiliser application accuracy index (FA) of individual loosening-applying tines in field tests.

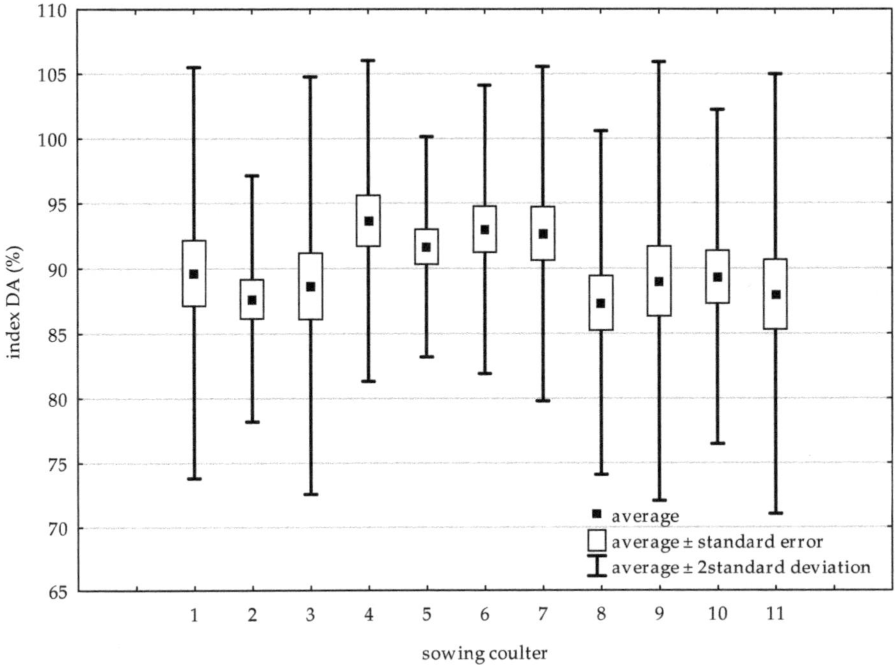

Figure 10. Sowing depth accuracy index (DA) of individual sowing coulters in field tests.

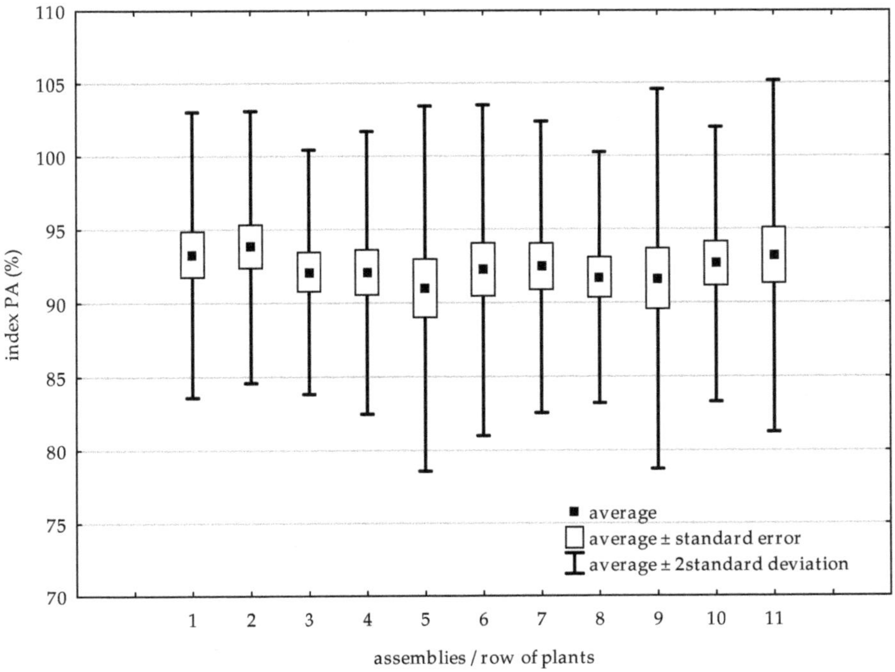

Figure 11. Number of plants accuracy index (PA) in the 11 individual rows.

Comparison of the average accuracy indices of each operating parameter of the entire machine indicates that, in field conditions, they differed only slightly from laboratory measurements. The difference for sowing accuracy (GA) and fertiliser application (FA) was 0.53 pp (percentage points) and 2.69 pp, respectively (Table 3). The least accurate operating parameter of all assemblies within the working width of the machine was sowing depth (DA = 92.00%). The CV coefficient of precision indices of sowing parameters assessed in the laboratory was below 1.00, whereas in field tests it ranged from 1.30% (sowing grain—GA) to 2.58% (sowing depth—DA).

Table 3. Accuracy indices (%) of the strip-till machine operation—average results of all seeding coulters and loosening-applying tines.

Index	Average	Minimum	Maximum	Standard Deviation	Coefficient of Variation (%)	Standard Error
Laboratory tests						
Sowing grain—GA	97.44	96.59	99.53	0.83	0.86	0.26
Fertiliser Application—FA	99.21	98.40	99.67	0.43	0.43	0.14
Field tests						
Sowing grain—GA	96.91	95.19	98.80	1.25	1.30	0.40
Fertiliser application—FA	96.52	92.78	99.67	2.32	2.40	0.73
Sowing depth—DA	92.00	87.88	96.36	2.37	2.58	0.75
Number of plants—PA	98.05	94.84	99.90	1.52	1.55	0.48

The analysis of the variability of all measurements of a given machine operation parameter (n = 110, 11 assemblies × 10 repetitions) indicates that, in laboratory tests, the rate of grain sown was more stable (i.e., a lower CV value) than the rate of fertiliser applied (Figure 12). The variability of fertiliser application rates was similar in the static laboratory tests as in the field studies. The difference in CV was 0.92 pp. In field conditions, the least stable parameter (CV = 9.83%) was sowing depth.

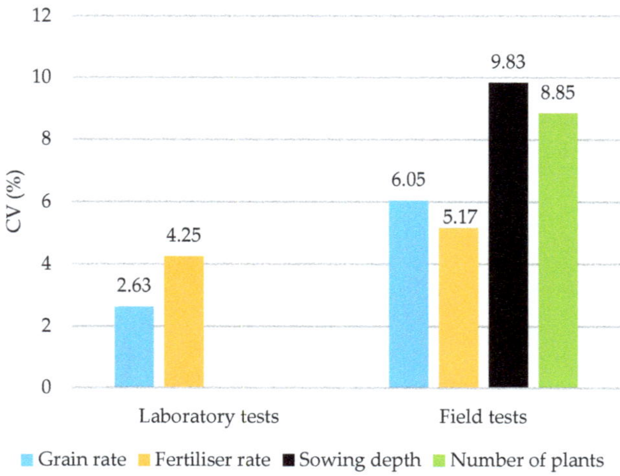

Figure 12. Coefficient of variation (CV) of the results of the machine operation parameters (n = 110, 11 assemblies × 10 measurements).

4. Discussion

Well-planned and well-executed sowing results in precise and uniform distribution of seeds in rows with the optimal spacing and appropriate depth for the plant species [46]. Sowing is a basic feature of agricultural technology that shapes the structure and architecture of a crop canopy. The construction of the seed drill, its operating parameters and their monitoring determine the uniformity of the horizontal and vertical distribution of seeds in the soil, which determines the living area of the plants [47,48]. In turn, the size of living space afforded to each plant influences the microclimate in the canopy, the efficiency of photosynthesis and, consequently, the yield [49]. This is equally important in the cultivation of crops sown in wide or narrow row spacing, e.g., cereals [50,51]. Zhai et al. [52] used a broadcast seeder to sow wheat with the theoretically most uniform grain distribution on the field surface and obtained more stalks per unit area without reducing the number of grains per ear and the thousand-grain weight than when sowing in rows. Greater uniformity of plant distribution in the canopy resulted in higher LAI, dry matter accumulation and grain yield.

The improvement of seed drills, often with the possibility of simultaneously tilling the soil and applying fertilisers, is an example of technological progress in agriculture. The assumed and expected result of new solutions is to improve the accuracy and uniformity of seed and fertiliser distribution in the soil, even in difficult field conditions. Automating and computerising these machines' management processes increases their efficiency and improves the quality of work [53,54]. Such activities also concern the construction of strip-till machines and their testing in field conditions [33]. These goals were adopted when designing the Mzuri one-pass strip-till machine. It ensured high process quality through its: pneumatic grain- and fertiliser-distribution heads for the coulter and loosening-applying tine elements with flow sensors in the transport ducts; electro-hydraulic smooth adjustment of working depth of the loosening/fertiliser-placement tines and sowing coulters with real-time parameter monitoring; ground-following, soil-consolidating wheels in the seed-sowing zone whose pressure force depends on soil properties (texture, moisture). In field conditions, the amount of wheat grain sown using the machine did not differ significantly from the theoretical amount assumed in accordance with the sowing standard, and the fertiliser dose was about 5.5 kg/ha higher than the assumed dose of 150 kg/ha. Cujbescu et al. [55] defined the sowing accuracy as the ratio of correctly sown seeds to the total expected number of seeds to be sown and obtained an index of 95.4% in a stationary test without vibration of the device and a CV coefficient of 9.1%. In field conditions, this rate was lower and the CV was higher, at 92.5% and 12.9%, respectively. The accuracy of sowing in the field deteriorated with increasing machine operating speed. Xiong et al. [56], referring to the GB/T 6973-2005 standard [57] assumed the "qualified index" as a measure of sowing accuracy. The standard assumes that the failure to place a seed where it should be sown is called a "missed sowing" (assessed using the "missed index"). It is also considered to be a missed sowing when the distance between successive seeds is greater than 1.5 times the theoretical (assumed) distance. The placement of two or more seeds instead of the intended one is called a "replay sowing" (assessed using the "replay index"). It is also considered to be a replay sowing when successive seeds are placed at a distance of less than or equal to 0.5 times the optimal distance. Therefore, sowing is "qualified" when individual seeds are spaced from each other within 0.5–1.5 times the theoretical distance. The authors, testing a newly designed mechanical-pneumatic sowing device in laboratory conditions, found that, for seeds of various sizes and properties (soybean, corn, rapeseed), the certified sowing rate exceeded 85%, and the missed sowing rate was below 10%. In our own research, the sowing accuracy (GA) index referring to the proportion of sown and unsown seeds used for laboratory testing of the designed Mzuri multifunctional hybrid machine was 97.4%, and in field conditions it was 96.9%.

Modernly designed pneumatic seed drills for precise grain sowing, which are limited by their low operating speed, nevertheless make it possible to achieve high-quality sowing. In the study by Fang et al. [58], this quality was described by a qualified sowing index of 91.66%, a replay sowing index of 5.98% and a missed sowing index of 2.36%. Sowing precision assessment indices are similarly created and calculated in accordance with the ISO 7256/1 [59] standard. The qualified seeding index and its coefficient of variation were used by Kamgar et al. [60] to evaluate the quality of work of a newly designed seed metering unit driven by a DC motor and equipped with digital encoders to monitor the rotational speed of the drive wheel. The used sowing unit improved the uniformity of wheat grain distribution in the soil compared to a traditional one. The coefficient of variation decreased from 4.40% to 1.84% at a working speed of 8 km/h and 50% plant residues in the field. The coefficient of variation in the sowing accuracy index of the Mzuri machine simultaneously tilling soil strips, applying fertilisers and sowing seeds in the laboratory and field tests was 0.86% and 1.30%, respectively. It is particularly difficult to achieve sowing precision (including accuracy of intended sowing rate and depth) in reduced-till conditions with a large amount of plant residues on the field surface. Xi et al. [61] improved the accuracy of wheat sowing by changing the design of the seeder by eliminating the tubular grain transport duct. While testing the prototype solution in the laboratory, a sowing accuracy of 83.84% was achieved. The coefficient of variation in seeding uniformity was 14.68%. In field conditions, these indicators were 83.13% and 15.12%, respectively, and the accuracy of the assumed sowing depth was 83.13%. The amount of grain sown by each coulter also depends on the position of the outlet relative to the grain tank [62]. In our own research, on average, each of the three outer coulters of the Mzuri machine (the furthest from the grain tank) tested in field conditions sowed 13.48 kg of grain. Meanwhile, each of the five middle coulters sowed 13.76 kg of grain.

The optimal and uniform sowing depth determines the uniform development of plants in the field, starting with germination and emergence. This sowing parameter is difficult to achieve, however, due to spatial variation in soil conditions within the field. Manually adjusted coulter pressure does not take into account variations in soil resistance during work. That is why modern solutions, including the developed Mzuri machine, have assemblies that dynamically adjust coulter load and depth of work. Such designs are used in modern sowing units of agricultural machine. They allow automatic adjustment of operating parameters depending on changing conditions within the field or between different fields. Sowing depth and the pressure force of the compaction wheel adapt to the field shape or soil moisture [63]. In studies by Nielsen et al. [64] and Nielsen et al. [65], it has been shown that such solutions improve the accuracy of seeding depth. Assuming a wheat sowing depth of 30 mm and a sowing speed of 8 km/h, the average deviation from this value was 5.3 mm—compared to 9.1 mm with traditional control. However, significant differences were found in the working depth of each of the eleven assemblies across the working width of the machine. The working depth range of the assemblies was from 14.2 mm to 25.9 mm. Sowing depth can also be better equalised by sowing assemblies equipped with a magnetorheological damper. A machine designed for ploughless sowing and equipped with a damper placed wheat grain at a depth of 39.8 mm, compared to the assumed 40 mm. The standard deviation of this sowing parameter was 5.8 mm and the coefficient of variation was 14.6% [66]. Field tests of the developed Mzuri machine have shown that the strip-till one-pass technology can sow wheat with a depth accuracy (DA) of 92.0% at a CV of 2.58%. At the same time, the variation in the working depth of the 11 assemblies across the working width of the machine did not exceed the limits from a minimum of 22 mm to a maximum of 33 mm with the assumed optimum size of 30 mm. The emergence capacity of wheat sown in strip-till technology using the newly designed machine was 88%, which is a very good result in conditions of a large amount of plant residue on the field surface and the lack of prior seedbed preparation. Research Carman et al. [67] indicate that the ability of plants to emerge depends on the design of the elements loosening the soil strips and sowing the seeds. In the cited studies, the ability of corn to emerge, depending on the machine used, ranged from 80.0 to 89.5%.

Equally important in modern agricultural technology equipped with real-time monitoring and control systems for working elements are seed drills that can apply fertilisers [68,69]. Yu et al. [70] indicate that, in a seed-fertiliser drill machine with fluted rollers, the dosing elements and systems adjusting their operation to environmental conditions and their degree of wear can maintain high precision of both seed sowing and fertilisation. According to the authors, the average relative deviations between actual and assumed (theoretical) application rates were 0.97% for wheat grain and 1.22% for fertiliser. The corresponding standard deviations were 0.69% and 0.62%. In turn, Yu et al. [71] assessed the precision of sowing and fertilisation based on real-time online measurements of grain and fertiliser mass and found that, in an area of 2.5 ha, the relative error of the cumulative application rate was 4.88%, and the mean error was 2.62% with a standard deviation of 1.7%. The average error in measuring the dynamic application rate on an area of ~2.2 ha was 3.15% with a standard deviation of 1.64%.

The key structural and functional element responsible for the uniformity of seed sowing and fertiliser application by individual assemblies of the machine is the dosing device. Adilet et al. [72] showed that replacing grooved devices with a pin-roller device increased the uniformity of sowing wheat grain and granular fertiliser. For wheat grain sowing, the CV was 96% and 37% for a twelve-grooved and roller-pin type, respectively, and for fertiliser distribution it was 80.74% and 37%. The authors of the paper, examining the accuracy of fertilisation by the 11 assemblies of the Mzuri machine in ten places on a 2.5 ha field, found that the average CV of the applied fertiliser dose was 2.52%, which was 1.68 pp higher than the static laboratory test results.

5. Conclusions

The expectations placed on the modern agricultural machine and equipment market are being met by industry and science. Cooperation among these areas of economic and social activity is particularly valuable. New technical solutions in agriculture should fit into the generally accepted direction of development of societies, e.g., Industry 4.0, Agriculture 4.0. In the field of plant cultivation, the strip-till one-pass technology is developing dynamically. It relates to the principles of conservation agriculture, which is environmentally friendly while maintaining its socio-economic role. The designed machine performs simultaneous tillage of soil strips, fertilisation and sowing and is equipped with a central computer that, through working assemblies, manages seed transport, sowing quantities, sowing depth, the pressure force of the ground-following/soil-consolidating wheels, and fertiliser dosing to implement its improvements in agricultural technology. The machine provides high-quality sowing and fertilising. The results of static laboratory tests were confirmed by tests under field conditions. The accuracy of the sowing amount, sowing depth and fertiliser dose of not less than 92% enabled uniform plant emergence—number of plants in a row (CV 1.55). The machines provision of an 88% field wheat emergence capacity resulted in a very good (expected) plant density. In turn, the uniform operation of all eleven assemblies in terms of rate of seed sown, sowing depth and fertiliser application results in the crop being uniform in the field space after emergence. The multifunctional Mzuri machine with automatic regulation and real-time monitoring of operating parameters, although it requires further research and technical improvement, is a significantly addition to the innovative agricultural machines on the global market. Further research should focus on improving sensors for assessing environmental properties in order to use the data for real-time machine control.

Author Contributions: Conceptualisation, D.J. and I.J.; methodology, D.J., I.J. and E.R.; investigation, D.J., I.J., E.R., M.R. and M.B.; resources, I.J., E.R., M.R. and B.K.; data curation, D.J. and I.J.; formal analysis, D.J., I.J., M.R., B.K. and M.B.; writing—original draft preparation, I.J., D.J., E.R., M.R., B.K. and M.B.; writing—review and editing, I.J., D.J., E.R., M.R., B.K. and M.B.; project administration, I.J. and D.J. All authors have read and agreed to the published version of the manuscript.

Funding: This research was supported by the National Centre for Research and Development (PL)—the industrial research and development work of the project Agrotech POIR.01.01.01-00-2279/20 entitled "A new generation of machines dedicated to the innovative strip-till one-pass technology adapted to smart fields agriculture and agriculture 4.0".

Data Availability Statement: Laboratory and field test data are included in this paper. Other questions regarding the method and results of the research work can be sent to the corresponding author.

Acknowledgments: The authors would like to thank the Agro-Land Marek Różniak Research and Development Centre in Śmielin for the opportunity to perform laboratory tests and field experiments.

Conflicts of Interest: The authors declare no conflicts of interest.

References

1. Zhai, Z.; Martínez, J.F.; Beltran, V.; Martínez, N.L. Decision support systems for agriculture 4.0: Survey and challenges. *Comput. Electron. Agric.* **2020**, *170*, 105256. [CrossRef]
2. Vlăduț, N.-V.; Ungureanu, N. Beyond Agriculture 4.0: Design and Development of Modern Agricultural Machines and Production Systems. *Agriculture* **2024**, *14*, 991. [CrossRef]
3. Asonja, A.; Desnica, E. Reliability of agriculture universal joint shafts based on temperature measuring in universal joint bearing assemblies. *Span. J. Agric. Res.* **2015**, *13*, e0201. [CrossRef]
4. Baillie, C.P.; Thomasson, J.A.; Lobsey, C.R.; McCarthy, C.L.; Antille, D.L. A Review of the State of the Art in Agricultural Automation. Part I: Sensing Technologies for Optimization of Machine Operation and Farm Inputs. In Proceedings of the 2018 ASABE Annual International Meeting, Detroit, MI, USA, 29 July–1 August 2018; American Society of Agricultural and Biological Engineers: St. Joseph, MI, USA, 2018.
5. Jha, K.; Doshi, A.; Patel, P.; Shah, M. A comprehensive review on automation in agriculture using artificial intelligence. *Artif. Intell. Agric.* **2019**, *2*, 1–12. [CrossRef]
6. Romaneckas, K. Sustainable Tillage and Sowing Technologies. *Agronomy* **2022**, *12*, 2467. [CrossRef]
7. Nandi, N.S.; Mallikarjuna, K.B.; Ashok, K.R.; Arunkumar, K.H.; Bale, A.S.; Vinay, N. Design and Fabrication of a Solar-Powered Bluetooth-Controlled Multipurpose Agromachine. In *Precision Agriculture for Sustainability*; Apple Academic Press: New York, NY, USA, 2024; pp. 221–240.
8. Fawzia, H.; Ahmeda, D.; Mostafac, S.A.; Fudzeec, M.F.M.; Mahmoodd, M.A.; Zeebareee, S.R.; Ibrahimf, D.A. A review of automated decision support techniques for improving tillage operations. *REVISTA AUS* **2019**, *26*, 219–240.
9. Zhang, Z.; Peng, X. Bio-tillage: A new perspective for sustainable agriculture. *Soil Tillage Res.* **2021**, *206*, 104844. [CrossRef]
10. Peixoto, D.S.; da Silva, L.D.C.M.; de Melo, L.B.B.; Azevedo, R.P.; Araújo, B.C.L.; Carvalho, T.S.; de Moreira, S.G.; Curi, N.; Silva, B.M. Occasional tillage in notillage systems: A global meta-analysis. *Sci. Total Environ.* **2020**, *745*, 140887. [CrossRef] [PubMed]
11. Różewicz, M. Review of current knowledge on strip-till cultivation and possibilities of its popularization in Poland. *Pol. J. Agron.* **2022**, *49*, 20–30.
12. Jaskulska, I.; Jaskulski, D. Strip-Till One-Pass Technology in Central and Eastern Europe: A MZURI Pro-Til Hybrid Machine Case Study. *Agronomy* **2020**, *10*, 925. [CrossRef]
13. Sims, B.; Kienzle, J. Sustainable Agricultural Mechanization for Smallholders: What Is It and How Can We Implement It? *Agriculture* **2017**, *7*, 50. [CrossRef]
14. Agrawal, A. Automated seed sowing machine: A review. *Asian J. Res. Soc. Sci. Humanit.* **2021**, *11*, 251–258. [CrossRef]
15. Zhang, B.; Liu, D.; Xi, X.; Zhang, Y.; Chen, C.; Qu, J.; Dong, H.; Zhang, R. The Analysis of the Applications of Crop Seed Tape Sowing Technology and Equipment: A Review. *Appl. Sci.* **2021**, *11*, 11228. [CrossRef]

16. Gierz, Ł.; Markowski, P. The Effect of the Distribution Head Tilt and Diffuser Variants on the Evenness of Sowing Rye and Oat Seeds with a Pneumatic Seed Drill. *Materials* **2020**, *13*, 3000. [CrossRef] [PubMed]
17. Li, Y.; Zhou, W.; Ma, C.; Feng, Z.; Wang, J.; Yi, S.; Wang, S. Design and optimization of the seed conveying system for belt-type high-speed corn seed guiding device. *Int. J. Agric. Biol. Eng.* **2024**, *17*, 123–131.
18. Reis, Â.V.; Medeiros, F.A.; Ferreira, M.F.; Machado, R.L.T.; Romano, L.N.; Marini, V.K.; Francetto, T.R.; Machado, A.L.T. Technological Trends in Digital Agriculture and Their Impact on Agricultural Machinery Development Practices. *Rev. Cienc. Agron.* **2020**, *51*, 2–12. [CrossRef]
19. Kim, W.S.; Lee, W.S.; Kim, Y.J. A review of the applications of the Internet of Things (IoT) for agricultural automation. *J. Biosyst. Eng.* **2020**, *45*, 385–400. [CrossRef]
20. Aby, G.R.; Issa, S.F. Safety of Automated Agricultural Machineries: A Systematic Literature Review. *Safety* **2023**, *9*, 13. [CrossRef]
21. Kim, Y.S.; Kim, T.J.; Kim, Y.J.; Lee, S.D.; Park, S.U.; Kim, W.S. Development of a real-time tillage depth measurement system for agricultural tractors: Application to the effect analysis of tillage depth on draft force during plow tillage. *Sensors* **2020**, *20*, 912. [CrossRef] [PubMed]
22. Riegler-Nurscher, P.; Moitzi, G.; Prankl, J.; Huber, J.; Karner, J.; Wagentristl, H.; Vincze, M. Machine vision for soil roughness measurement and control of tillage machines during seedbed preparation. *Soil Tillage Res.* **2019**, *196*, 104351. [CrossRef]
23. Kumar, R.; Kumar, S.; Kumar, N.; Kumari, S.; Danish, M. Sensor-based sowing machinery for precision farming. *Int. J. Chem. Stud.* **2020**, *8*, 13–15. [CrossRef]
24. Wang, S.; Yi, S.; Zhao, B.; Li, Y.; Li, S.; Tao, G.; Mao, X.; Sun, W. Sowing Depth Monitoring System for High-Speed Precision Planters Based on Multi-Sensor Data Fusion. *Sensors* **2024**, *24*, 6331. [CrossRef] [PubMed]
25. Rajesh, P.V.; Ashvin, J.; Harish, M.; Kishore, G.; Nidheesh, S. Fabrication of Equipment and IoT-Assisted Monitoring for Enhanced Zero-Till Farming. In *Emerging Technologies and Marketing Strategies for Sustainable Agriculture*; IGI Global Scientific Publishing: Hershey, PA, USA, 2024; pp. 209–233.
26. Kumar, R.; Kumar, A. Worldwide approach of sensorbased sowing machinery: A Review. *Curr. Adv. Agric. Sci.* **2023**, *15*, 1–7. [CrossRef]
27. Khan, N.; Ray, R.L.; Sargani, G.R.; Ihtisham, M.; Khayyam, M.; Ismail, S. Current Progress and Future Prospects of Agriculture Technology: Gateway to Sustainable Agriculture. *Sustainability* **2021**, *13*, 4883. [CrossRef]
28. Bai, Y.; Zhang, B.; Xu, N.; Zhou, J.; Shi, J.; Diao, Z. Vision-based navigation and guidance for agricultural autonomous vehicles and robots: A review. *Comput. Electron. Agric.* **2023**, *205*, 107584. [CrossRef]
29. Matin, M.A.; Desbiolles, J.M.A.; Fielke, J.M. Strip-tillage using rotating straight blades: Effect of cutting edge geometry on furrow parameters. *Soil Tillage Res.* **2016**, *155*, 271–279. [CrossRef]
30. Nadykto, V.; Domeika, R.; Golub, G.; Kukharets, S.; Chorna, T.; Čėsna, J.; Hutsol, T. Research on a Machine–Tractor Unit for Strip-Till Technology. *AgriEngineering* **2023**, *5*, 2184–2195. [CrossRef]
31. Barut, Z.B.; Ozdemir, S. Design Approaches of One-Pass Strip-Till Machines. In Proceedings of the 15th International Congress on Agricultural Mechanization and Energy in Agriculture, ANKAgEng 2023, Antalya, Turkey, 29 October–1 November 2023; Cavallo, E., Auat Cheein, F., Marinello, F., Saçılık, K., Muthukumarappan, K., Abhilash, P.C., Eds.; Lecture Notes in Civil Engineering 458. Springer: Cham, Switzerland, 2023.
32. Li, C.S.; Tang, Y.L.; McHugh, A.D.; Wu, X.L.; Liu, M.; Li, M.; Xiong, T.; Ling, D.; Tang, Q.; Liao, M.; et al. Development and performance evaluation of a wet-resistant strip-till seeder for sowing wheat following rice. *Biosyst. Eng.* **2022**, *220*, 146–158. [CrossRef]
33. Wang, Q.; Wang, B.; Sun, M.; Sun, X.; Zhou, W.; Tang, H.; Wang, J. Design and Testing of an Automatic Strip-Till Machine for Conservation Tillage of Corn. *Agronomy* **2023**, *13*, 2357. [CrossRef]
34. Benincasa, P.; Zorzi, A.; Panella, F.; Tosti, G.; Trevini, M. Strip tillage and sowing: Is precision planting indispensable in silage maize? *Int. J. Plant Prod.* **2017**, *11*, 577–588.
35. Zhao, S.; Chen, Y. Seed distribution in soil and its effects on early plant growth. *Appl. Eng. Agric.* **2015**, *31*, 415–423.
36. Ahmad, F.; Adeel, M.; Qiu, B.; Ma, J.; Shoaib, M.; Shakoor, A.; Chandio, F.A. Sowing uniformity of bed-type pneumatic maize planter at various seedbed preparation levels and machine travel speeds. *Int. J. Agric. Biol. Eng.* **2021**, *14*, 165–171. [CrossRef]
37. Ahmed, K.; Saleh, A.; Musa, A.D.; Kalu, G. Determination of selected parameters that influence planting space uniformity on maize seed. *Fudma J. Sci.* **2023**, *7*, 30–33. [CrossRef]
38. Jiang, X.; Zhang, J.; Liu, S.; Fu, L.; Huang, D. Development and Evaluation of a Universal Seed Discharger for Precision Planting in Corn-Soybean Cropping System. *Processes* **2024**, *12*, 1595. [CrossRef]
39. Turan, J.; Višacki, V.; Mehandžić, S.; Findura, P.; Burg, P.; Sedlar, A. Sowing quality indicators for a seed drill with overpressure. *Acta Univ. Agric. Silvic. Mendelianae Brun.* **2014**, *62*, 1487–1492. [CrossRef]
40. Liu, S.; Baret, F.; Andrieu, B.; Burger, P.; Hemmerlé, M. Estimation of Wheat Plant Density at Early Stages Using High Resolution Imagery. *Front. Plant Sci.* **2017**, *8*, 739. [CrossRef] [PubMed]

41. Liu, T.; Li, R.; Jin, X.; Ding, J.; Zhu, X.; Sun, C.; Guo, W. Evaluation of Seed Emergence Uniformity of Mechanically Sown Wheat with UAV RGB Imagery. *Remote Sens.* **2017**, *9*, 1241. [CrossRef]
42. Tomantschger, K.; Tadić, V. Partial differential equations for cereal seeds distribution. *Teh. Vjesn.* **2021**, *28*, 624–628.
43. Wang, B.; Gu, F.; Hu, Z.; Wu, F.; Chen, X.; Luo, W. Analysis and Evaluation of Influencing Factors on Uniform Sowing of Wheat with Wide Seed Belt after Sowing and Soil Throwing Device. *Agriculture* **2022**, *12*, 1455. [CrossRef]
44. Liu, W.; Hu, J.; Zhao, X.; Yao, M.; Lakhiar, I.A.; Zhao, J.; Liu, J.; Wang, W. An Adaptive Roller Speed Control Method Based on Monitoring Value of Real-Time Seed Flow Rate for Flute-Roller Type Seed-Metering Device. *Sensors* **2021**, *21*, 80. [CrossRef]
45. Statistica. *Data Analysis Software System*, Version 12; TIBCO Software Inc.: Palo Alto, CA, USA, 2017. Available online: http://sta-517tistica.io (accessed on 15 January 2019).
46. Petrović, V.D.; Radojević, L.R.; Tomantschger, K.; Golubović, Z. The Uniformity of Wheat Seeding Over an Area and Depth. In Proceedings of the DAS-29, 29th Danubia-Adria Symposium, University of Belgrade, Belgrade, Serbia, 26–29 September 2012; pp. 166–169.
47. Noor, H.; Khan, S.; Min, S.; Yu, S. Effect of different sowing methods and nitrogen rates on yield and quality of winter wheat in Loess Plateau of China. *Appl. Ecol. Environ. Res.* **2020**, *18*, 5701–5726. [CrossRef]
48. Šarauskis, E.; Kazlauskas, M.; Naujokienė, V.; Bručienė, I.; Steponavičius, D.; Romaneckas, K.; Jasinskas, A. Variable Rate Seeding in Precision Agriculture: Recent Advances and Future Perspectives. *Agriculture* **2022**, *12*, 305. [CrossRef]
49. He, S.; Li, X.; Chen, M.; Xu, X.; Zhang, W.; Chi, H.; Shao, P.; Tang, F.; Gong, T.; Guo, M.; et al. Excellent Canopy Structure in Soybeans Can Improve Their Photosynthetic Performance and Increase Yield. *Agriculture* **2024**, *14*, 1783. [CrossRef]
50. Lu, Y.; Zhang, X.; Chen, S.; Shao, L.; Sun, H.; Niu, J. Increasing the planting uniformity improves the yield of summer maize. *Agron. J.* **2017**, *109*, 1463–1475. [CrossRef]
51. Zhang, F.; Zhang, D.; Li, L.; Zhang, Z.; Liang, X.; Wen, Q.; Chen, G.; Wu, Q.; Zhai, Y. Effect of Planting Density on Canopy Structure, Microenvironment, and Yields of Uniformly Sown Winter Wheat. *Agronomy* **2023**, *13*, 870. [CrossRef]
52. Zhai, Y.; Wu, Q.; Chen, G.; Zhang, H.; Yin, X.; Chen, F. Broadcasting Winter Wheat Can Increase Grain Yield without Reducing the Kernels per Spike and the Kernel Weight. *Sustainability* **2018**, *10*, 4858. [CrossRef]
53. Bochtis, D.D.; Sørensen, C.G.C.; Busato, P. Advances in Agricultural Machinery Management: A Review. *Biosyst. Eng.* **2014**, *126*, 69–81. [CrossRef]
54. Zhang, L.; Liu, Y.; Wu, M.; Wu, Z. Sowing Performance of the Seeder Drill for the 2BYG-220 Type Combined Rapeseed Planter under Vibration Conditions. *Appl. Sci.* **2024**, *14*, 5091. [CrossRef]
55. Cujbescu, D.; Găgeanu, I.; Persu, C.; Matache, M.; Vlăduț, V.; Voicea, I.; Paraschiv, G.; Biriș, S.Ș.; Ungureanu, N.; Voicu, G.; et al. Simulation of Sowing Precision in Laboratory Conditions. *Appl. Sci.* **2021**, *11*, 6264. [CrossRef]
56. Xiong, D.; Wu, M.; Xie, W.; Liu, R.; Luo, H. Design and Experimental Study of the General Mechanical Pneumatic Combined Seed Metering Device. *Appl. Sci.* **2021**, *11*, 7223. [CrossRef]
57. *GB/T 6973-2005*; Testing Methods of Single Seed Drills. Standardization Administration of the P.R.C.: Beijing, China, 2005.
58. Fang, Z.; Zhang, J.; Chen, J.; Pan, F.; Wang, B.; Ji, C. Design and Experimental Analysis of an Air-Suction Wheat Precision Hill-Seed Metering Device. *Agriculture* **2024**, *14*, 1931. [CrossRef]
59. *ISO 7256-1:1984*; Sowing Equipment-Test Methods. Part 1. Single Seed Drills (Precision Drills). International Organization for Standardization: Geneva, Switzerland, 1984.
60. Kamgar, S.; Noei-Khodabadi, F.; Shafaei, S.M. Design, development and field assessment of a controlled seed metering unit to be used in grain drills for direct seeding of wheat. *Inf. Process. Agric.* **2015**, *2*, 169–176. [CrossRef]
61. Xi, X.; Gu, C.; Shi, Y.; Zhao, Y.; Zhang, Y.; Zhang, Q.; Jin, Y.; Zhang, R. Design and experiment of no-tube seeder for wheat sowing. *Soil Tillage Res.* **2020**, *204*, 104724. [CrossRef]
62. Jafari, M.; Hemmat, A.; Sadeghi, M. Comparison of coefficient of variation with non-uniformity coefficient in evaluation of grain drills. *J. Agric. Sci. Technol.* **2011**, *13*, 643–654.
63. Poncet, A.M.; Fulton, J.P.; McDonald, T.P.; Knappenberger, T.; Shaw, J.N.; Bridges, R.W. Effect of heterogeneous field conditions on corn seeding depth accuracy and uniformity. *Appl. Eng. Agric.* **2018**, *34*, 819–830. [CrossRef]
64. Nielsen, S.K.; Munkholm, L.J.; Lamandé, M.; Nørremark, M.; Skou-Nielsen, N.; Edwards, G.T.; Green, O. Seed drill instrumentation for spatial coulter depth measurements. *Comput. Electron. Agric.* **2017**, *141*, 207–214. [CrossRef]
65. Nielsen, S.K.; Munkholm, L.J.; Lamandé, M.; Nørremark, M.; Edwards, G.T.; Green, O. Seed drill depth control system for precision seeding. *Comput. Electron. Agric.* **2018**, *144*, 174–180. [CrossRef]
66. Galibjon, M.; Dimitris, S.; Hans, W. Implementation of a magnetorheological damper on a no-till seeding assembly for optimising sowing depth. *Comput. Electron. Agric.* **2018**, *150*, 465–475.
67. Carman, K.; Çıtıl, E.; Marakoğlu, T.; Gür, K.; Kırılmaz, H. Comparison of strip tillage systems for sillage maize production in middle anatolia. *Selcuk J. Agric. Food Sci.* **2018**, *32*, 430–435.

68. Lv, H.; Ren, J.; Liu, X.; Li, S.; Wu, G.; Zheng, Y. Review of the Monitoring Systems of the Machine for Precision Sowing and Fertilization of Wheat. In Proceedings of the 2018 ASABE Annual International Meeting, Detroit, MI, USA, 29 July–1 August 2018; American Society of Agricultural and Biological Engineers: St. Joseph, MI, USA, 2018; p. 1800736.
69. Jin, M.F.; Ding, Y.Q.; Yu, H.F.; Liu, H.T.; Jiang, Y.Z.; Fu, X.Q. Optimal Structure Design and Performance Tests of Seed metering Device with Fluted Rollers for Precision Wheat Seeding Machine. *IFAC-PapersOnLine* **2018**, *51*, 509–514.
70. Yu, H.; Ding, Y.; Liu, Z.; Fu, X.; Dou, X.; Yang, C. Development and Evaluation of a Calibrating System for the Application Rate Control of a Seed-Fertilizer Drill Machine with Fluted Rollers. *Appl. Sci.* **2019**, *9*, 5434. [CrossRef]
71. Yu, H.; Ding, Y.; Fu, X.; Liu, H.; Jin, M.; Yang, C.; Liu, Z.; Sun, G.; Dou, X. A solid fertilizer and seed application rate measuring system for a seed-fertilizer drill machine. *Comput. Electron. Agric.* **2019**, *162*, 836–844. [CrossRef]
72. Adilet, S.; Zhao, K.; Liu, G.; Sayakhat, N.; Jun, C.; Hu, G.; Bu, L.; Chen, Y.; Jin, H.; Zhang, S.; et al. Investigation of the pin-roller metering device and tube effect for wheat seeds and granular fertilizers based on DEM. *Int. J. Agric. Biol. Eng.* **2023**, *16*, 103–114. [CrossRef]

agriculture

MDPI

Article

Design and Realization of an Orchard Operation-Aid Platform: Based on Planting Patterns and Topography

Zhao Li [1,2], Can Li [1,2], Ye Zeng [1,2], Chaodong Mai [1,2], Runpeng Jiang [1,2] and Jun Li [1,2,3,*]

[1] College of Engineering, South China Agricultural University, Guangzhou 510642, China
[2] Guangdong Laboratory for Lingnan Modern Agriculture, Guangzhou 510642, China
[3] State Key Laboratory of Agricultural Equipment Technology, Beijing 100083, China
[*] Correspondence: autojunli@scau.edu.cn

Abstract: To address the lack of mechanical orchard operation-aid platforms that assist in the horticultural management tasks such as pruning, spraying, thinning flowers and fruits, and harvesting in litchi and longan orchards, this paper proposes an orchard operation-aid platform specifically tailored for hilly and mountainous orchards. The platform is optimized for orchards with tree and row spacing not exceeding 6 m and slopes not exceeding 15°. By considering the planting patterns and operational topography parameters of litchi and longan, the key components were meticulously designed, including the chassis, lifting device, extension device, and slope operation support device. The driving stability, slope operation stability, and the reachable workspace of the orchard operation-aid platform were analyzed, followed by a prototype experiment. The results demonstrate that the platform achieves an in situ turning radius of 1.2 m with no deviation in the turning path. It satisfies the passability and operational slope requirements of hilly terrains with both driving and operational slopes exceeding 15°. Additionally, the platform features a working height of 4.0 m and an operating radius of 3.7 m, meeting the operational requirements for multiple tasks. This research provides a practical and effective solution for enhancing operational efficiency in multiple stages of fruit cultivation, demonstrating significant practical value and potential for widespread application.

Keywords: agricultural machinery; hilly and mountainous terrains; orchard operation-aid platform; rollover analysis; workspace

1. Introduction

Orchard operation has the characteristics of difficulty, high labor intensity, and seasonal concentration. At present, in addition to spraying, fertilizing, and a few other aspects of orchard operations, pruning, flower and fruit thinning, and harvesting still mainly rely on manual labor. Furthermore, existing orchard machinery used for pruning, flower and fruit thinning, harvesting, and other operations makes it difficult to match the operational efficiency and effectiveness of manual labor, making commercialization challenging at this stage. Taking litchi and longan as examples, data from the China Litchi and Longan Industry Technology Research System show that the labor costs of litchi and longan currently account for 54.5% and 62.4% of the total production costs, respectively [1]. Excessive labor costs have limited the rapid development of China's fruit industry.

Mechanical operation-aid platform balances the advantages of full mechanization and manual operation, not only can effectively reduce worker up-tree-down time, but the orchard operation-aid platform's manned workbench can store operating tools, auxiliary workers to complete the pruning, spraying, flower thinning, fruit thinning, harvesting and other aspects of the operational tasks, effectively improve operational efficiency [2–4], reduce equipment idle time, still has a great advantage at this stage. According to the installation of the manned workbench, the form of movement, mechanical orchard operation-aid platform can be broadly divided into fixed type (manned workbench immobilized), vertical lifting type (manned workbench can be lifted vertically), boom type (manned workbench moves through space via cantilevered arms) several kinds.

Fixed-type orchard operation-aid platforms [5,6] generally provide standing space for workers around the fruit box loading and unloading devices and are mainly suitable for short rootstock densely planned fusiform tree shape of the plain orchard. Due to the differences in the distribution of fruit tree canopies in the vertical direction [7], researchers have divided the standing space into independent spaces with different vertical heights, i.e., multi-station orchard operation-aid platform, on the basis of fixed type orchard operation-aid platform [8,9]. After equipping each independent space with a lifting function [10], the worker at the back of the traveling direction can check and complement the work in front of them. The researchers also automatically adjusted the vertical height of each worker [11] and the forward speed of the operation-aid platform [12] according to the distribution estimation of the fruits with picking and the dynamic harvesting efficiency of the workers [13], thus effectively improving the picking efficiency.

When fruit trees exceed a certain height, it is difficult for the fixed orchard operation-aid platform to cover the full operating height. Unlike fixed orchard operation assistance platforms, vertical lifting type, especially scissor-type orchard operation assistance platforms [14], which vertically lift the entire manned workbench by means of a scissor-type mechanism, has a simple structure, and it is also possible to increase the lifting height by using a multistage scissor-type structure [15,16], which is the lifting mechanism with the highest market share at present. As more than 60% of China's orchards are distributed in provinces dominated by hilly and mountainous terrains [17], researchers have also conducted a lot of research on the rollover risk [18,19] of the scissor-type orchard operation-aid platform, which keeps the manned workbench level by adjusting the pitch angle and side inclination angle [20,21].

However, the two types of orchard operation-aid platforms, fixed and vertical lift, have a small working radius and poor adaptability to orchards with large crown widths and row spacing. The boom type of orchard operation-aid platform generally has a large lateral and vertical operating range to cover a wider portion of the fruit tree crown. The 3-DOF boom lifting platform designed by Duan [22] and others and the hilly operation-aid platform designed by Fan et al. [23] have expanded the working radius to some extent, but the expansion distance is relatively limited. The squirrel series elevating platform [24] from CRENDON, Australia, has a long boom that allows it to have a larger operating radius and height, but the long boom also limits the elevating platform from entering orchards with small row spacing.

In summary, at this stage, there is still a need for an orchard operation-aid platform suitable for hilly and mountainous terrains. This operation-aid platform needs to have the following features: a lift function to allow the equipment to cover a sufficient height range; a long lateral boom to allow the equipment to cover a sufficient crown of fruit tree range; a retractable lateral boom to enhance the equipment's trafficability; and the safety of driving and operating in most hilly mountain orchards. Based on the planting mode and agronomic requirements of orchards, this article designs an operation-aid platform with a lifting device and a telescopic boom. The key components were designed and selected, and the driving stability, slope operation stability, and reachable workspace of the whole machine were analyzed. A prototype field validation experiment was conducted to verify the performance indicators and the rationality of the overall design.

2. Materials and Methods

2.1. Overall Structure and Working Principle

The planting patterns and agronomic requirements of the target crops are crucial for determining the target workspace of the orchard operation-aid platform and its parameter design. Taking litchi and longan as examples, litchi and longan orchards in China [25–27] are mainly distributed in hilly and mountainous terrains (Figure 1), adopting wide-row narrow-planting or near-square planting methods, with planting distances $l_{row} \times l_{column}$ generally being (4 m–6 m) × (5 m–6 m), where l_{row} and $l_{columns}$, respectively, represent the distance between each row and each column of fruit trees. Depending on planting age, fruit trees can be classified into young trees, fruit-bearing trees, and aging trees. For fruit-bearing trees, the tree height h_{tree} is generally controlled at 3.5 m, and the distance from the ground to the retained thick upright branches and ascending branches is 1.2 m. The crowns of adjacent trees are 1.5 m apart, meaning the crown width l_{crown} is controlled within 3–4 m.

(a) (b)

Figure 1. Schematic diagram of planting patterns in litchi and longan orchards: (**a**) litchi orchard; (**b**) longan orchard.

Based on the orchard operation-aid platform's design requirement, the prototype and three-dimensional model of the orchard operation-aid platform are shown in Figure 2. The system mainly comprises the following components: walking mechanism, workspace expansion device, manned workbench, slope operations support device, hydraulic system, generator, and control system. The workspace expansion device consists of the slewing mechanism, lifting mechanism, and extension mechanism, which collaboratively enable precise positioning and movement of the manned workbench in reachable workspace. The manned workbench stores operational tools for workers to perform various tasks. The slope operation support device ensures system stability on uneven ground, enhancing operational safety and reliability. The hydraulic system provides power support for hydraulic components, ensuring smooth operation of each device. The control system coordinates and controls the system, ensuring efficient and stable operation of the orchard operation-aid platform.

Figure 2. Structural diagram of the orchard operation-aid platform: (**a**) prototype of the orchard operation-aid platform; (**b**) 3D model of the orchard operation-aid platform. (1) Walking mechanism; (2) slope operation support device; (3) lifting mechanism; (4) manned workbench; (5) extension mechanism; (6) hydraulic system; (7) 12 kW generator; (8) fruit basket; (9) slewing bearing; and (10) control box.

The working principle of the orchard operation-aid platform is as follows: workers control the platform to move between tree rows and approach the target fruit tree. Based on the orchard operation-aid platform's posture and the terrain, the working mode of the slope operation support device is determined, and the supporting legs are extended to adjust the orchard operation-aid platform to a horizontal state. The slewing mechanism rotates the manned workbench around the lifting mechanism axis, enabling left and right swinging of the platform between tree rows. The lifting mechanism and extension mechanism extend or retract through serial hydraulic cylinders, achieving vertical and horizontal translation of the manned workbench. Through the coordinated operation of the lifting mechanism, extension mechanism, and slewing mechanism, the manned workbench approaches the target operational area of the tree crown. Workers on the platform use tools stored on the manned workbench to perform tasks such as pruning, spraying, thinning flowers and fruit, or fruit harvesting. After completing tasks in the area, the lifting and extension mechanisms retract to their shortest lengths, and workers control the orchard operation-aid platform to move to the next target tree, repeating the above process.

2.2. Key Structural Design

2.2.1. Walking Mechanism

As mentioned earlier, the distance between the crowns of adjacent litchi and longan trees is less than 1.5 m due to lush branches resulting from infrequent tree management. To minimize collision damage to fruit trees during operation, the chassis width should be less than the distance between the crowns. The orchard operation-aid platform typically operates on soft roads or slopes in hilly areas; thus, the speed should not be too fast, and good steering and braking capabilities are required. The chassis uses brushless DC motors for drive and worm gears for speed reduction. The differential steering of the chassis is achieved by controlling the speed difference between the drive motors on either side. When the left and right tracks rotate in opposite directions at the same speed, the chassis can achieve in-place rotation, with the diagonal length of the rectangle projected onto the ground forming the chassis' in-place turning diameter. Due to significant friction between the tracks and the ground, deviations may occur during turning; hence, an actual turning radius of the chassis ≤1.5 m is considered acceptable. The design parameters of the chassis are summarized in Table 1.

Table 1. Chassis design parameters.

Parameter	Value
Length × Width × Height (mm)	1900 × 1100 × 600
Curb Weight (kg)	300
Minimum Turning Radius (mm)	1500
Climbing Angle (°)	15
Minimum Clearance (mm)	150
Motor Power (kW)	3
Reducer Type	Worm gear and worm type
Reduction Ratio	1:80

2.2.2. Slope Operation Support Device

The operational topography of the orchard operation-aid platform is generally in hilly and mountainous terrains, often requiring operations on slopes, which poses challenges to worker safety. Therefore, a slope operation support device is designed to increase the ground contact length of the chassis, as illustrated in Figure 3a.

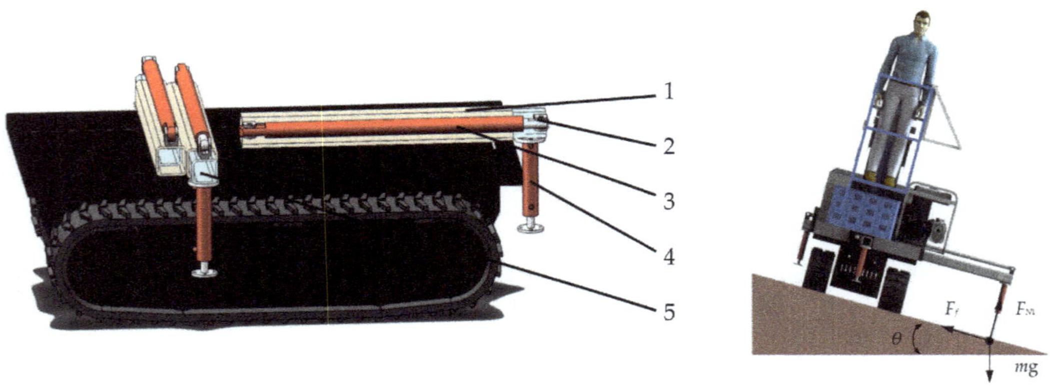

(a) (b)

Figure 3. Schematic diagram of the slope operation support device: (**a**) structural composition; (**b**) force analysis. (1) Fixed square tube; (2) cylinder fixing pin; (3) extension hydraulic cylinder; (4) supporting hydraulic cylinder; and (5) movable square tube.

The dimensions of the fixed and movable square tubes in the slope operation support device need to be calculated based on the bending moments and shear forces they endure. Considering the extreme loading scenario, the overall force analysis of the orchard operation-aid platform under this condition is shown in Figure 3b. In this scenario, the orchard operation-aid platform is about to roll over, but the tracks have not left the ground, and only one side's supporting leg is bearing the load. The supporting cylinder can be regarded as a two-force rod that is subjected to a support force F_{N1} from the ground and a pressure F_{N2} exerted by the movable square tube. Then, the force exerted by the supporting cylinder on the movable square tube is F_{N2}, and F_{N2} is equal and opposite to F'_{N2}. The maximum shear force $F_{Te\max}$ and maximum bending moment $M_{Te\max}$ on the movable square tubes are calculated as:

$$F_{Te\max} = F'_{N2} = -F_{N2} = F_{N1} \tag{1}$$

$$M_{Te\max} = F'_{N2}\, l_{ABTe} \tag{2}$$

In the formula, $F_{Te\max}$ is the maximum shear force of the movable square tube; $M_{Te\max}$ is the maximum bending moment in the movable square tube; l_{ABTe} is the overhanging length of the movable square tube when fully extended.

Rectangular tubes have complex dimensions, and individually checking the strength of the fixed and movable square tubes is labor-intensive. The Monte Carlo method (MCM) [28], based on the law of large numbers and the central limit theorem, is a numerical method utilizing random sampling to solve "non-formulaic" problems efficiently. Therefore, the MCM is employed to simplify the calculation. The steps for determining the dimensions of the fixed and movable square tubes are as follows:

(1) Define variable names and known parameter values.
(2) Define the value ranges for the parameters to be solved, assigning values based on a uniform distribution within these ranges.
(3) Establish relationships between variables, such as dimensions of the fixed and movable square tubes, shear forces, and bending moments.
(4) Filter parameter combinations based on constraints like shear force and bending moment limits.
(5) Select the optimal parameter combination.

By programming the MATLAB program, solving it, and comparing the results of the solution with the dimensions of the commercially available square tubes, we obtain the dimensions of the fixed and movable square tubes, which are presented in Table 2.

Table 2. Slope operation support device design parameters.

Component	Length (mm)	Width (mm)	Height (mm)	Thickness (mm)
Fixed tube	1250	120	120	8
Activity tube	1200	100	100	8

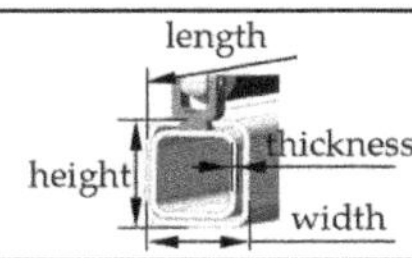

2.2.3. Workspace Expansion Device

Taking the intersection point of the slewing bearing's rotational axis and the ground as the origin O_0 and setting the forward direction of the orchard operation-aid platform as the positive X-axis, the upward direction of the lifting mechanism as the positive Z-axis, and determining the positive Y-axis using the right-hand rule, a basic coordinate system $O_0X_0Y_0Z_0$ is established, with units in millimeters. The coordinate system $O_0X_0Y_0Z_0$ can be positioned by the following methods: the X-axis is along the forward direction of the orchard operation-aid platform, and the chassis is symmetrical about the X-axis; the Z-axis is co-located with the axis of rotation of the swing bearing, and the center of gravity of the lifting mechanism is on the Z-axis, and the lifting mechanism is symmetrical about the Z-axis; the Y-axis can be determined by the right-hand rule; finally, the origin of the coordinate system is the intersection of the three coordinate axes. Thus, the X and Y coordinate values of any component in the coordinate system $O_0X_0Y_0Z_0$ can be determined by measuring the horizontal distance between the component and the lifting mechanism, and the Z coordinate value of the component can be determined by measuring the vertical distance between the component and the ground. The operational schematic of the orchard operation-aid platform is depicted in Figure 4.

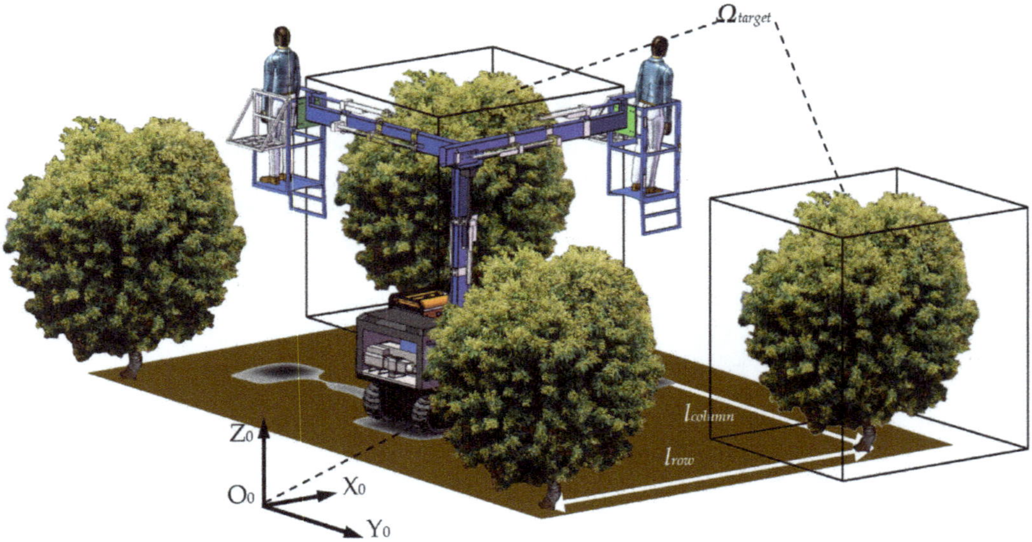

Figure 4. Operational schematic of the orchard operation-aid platform.

Considering the crown width and tree height of litchi and longan trees, the target workspace of the worker's hand, denoted as the orchard operation-aid platform's target workspace $\mathbf{\Omega}_{target1}$, can be defined as:

$$\mathbf{\Omega}_{target1} = \begin{bmatrix} \mathbf{D}_{tx1}, & \mathbf{D}_{ty1}, & \mathbf{D}_{tz1} \end{bmatrix}, \quad \begin{cases} \mathbf{D}_{tx1} = [0, l_{row}/2] \\ \mathbf{D}_{ty1} = [-l_{column}/2, l_{column}/2] \\ \mathbf{D}_{tz1} = [h_{trank}, h_{tree}] \end{cases} \tag{3}$$

In the formula, $\mathbf{D}_{tx1}$, $\mathbf{D}_{ty1}$, $\mathbf{D}_{tz1}$ represent the ranges of the target workspace $\mathbf{\Omega}_{target1}$ along the X, Y, and Z axis in the coordinate system $O_0X_0Y_0Z_0$, respectively.

Using Formulas (3), the maximum and minimum values of the orchard operation-aid platform's target workspace along the X, Y, and Z axis are calculated and summarized in Table 3.

Table 3. Orchard operation-aid platform target workspace dimensions.

Axis	X	Y	Z
Minimum (mm)	3000	3000	3500
Maximum (mm)	0	−3000	1200

The target workspace $\mathbf{\Omega}_{target1}$ is the union of the equipment racks' target workspace $\mathbf{\Omega}_{target2}$ and the worker's reachable workspace. According to the "GB/T 10000-2023 Human dimensions of Chinese adults" [29] the median values for adult males aged 18–70 are as follows: functional overhead reach height l_{p1} is 1993 mm, functional forward reach length of upper limbs l_{p2} is 710 mm, and shoulder height l_{p3} is 1373 mm.

Therefore, the equipment racks' target workspace $\mathbf{\Omega}_{target2}$ can be expressed as:

$$\begin{gathered}\mathbf{\Omega}_{target2} = \begin{bmatrix} \mathbf{D}_{tx2}, & \mathbf{D}_{ty2}, & \mathbf{D}_{tz2} \end{bmatrix}, \\ \begin{cases} \mathbf{D}_{tx2} = [0, l_{row}/2 - l_{p2}] \\ \mathbf{D}_{ty2} = [-(l_{column}/2 - l_{p2}), l_{column}/2 - l_{p2}] \\ \mathbf{D}_{tz2} = [h_{trank}, h_{tree} - l_{p1}] \end{cases}\end{gathered} \tag{4}$$

In the formula, $\mathbf{D}_{tx2}$, $\mathbf{D}_{ty2}$, $\mathbf{D}_{tz2}$ represent the ranges of the equipment racks' target workspace along the X, Y, and Z axis in the coordinate system $O_0X_0Y_0Z_0$.

The maximum lengths $l_{Stretching}$ and $l_{Lifting}$ of the extension and lifting mechanisms are then designed to satisfy:

$$\begin{cases} l_{Stretching} + l_{p2} + l_{gu2} \geq l_{row}/2 + l_{accuracy} \\ l_{Stretching} + l_{p2} + l_{gu2} \geq l_{column}/2 + l_{accuracy} \\ l_{Lifting} + l_{p1} + h_{Classic} + h_{Slewing} - h_{gu} \geq h_{tree} + l_{accuracy} \end{cases} \tag{5}$$

In the formula, h_{gu} is the manned workbench height; l_{gu1} and l_{gu2} is the manned workbench length and width, respectively.

To maximize operational requirements, parameters are set as follows: Considering that workers often lean their upper bodies out over the manned workbench guardrail during operations, and adhering to the size regulations for guardrails in high-altitude work, the manned workbench height h_{gu} is set at 1.1 m to balance safety and convenience, both the length l_{gu1} and width l_{gu2} of the manned workbench are designed to be 0.6 m. Planting distances l_{row} and l_{column} both take the maximum value of 6 m. Precision compensation $l_{accuracy}$ is set at 0.5 m, chassis height $h_{Classic}$ is 0.6 m, and slewing bearing height $h_{Slewing}$ is 0.2 m.

Using Formulas (5), the maximum design lengths for the extension and lifting mechanisms are calculated as $l_{Stretching}$ = 2190 mm and $l_{Lifting}$ = 2307 mm, respectively.

The extension mechanism needs to reach the limits of the equipment racks' target workspace when fully extended and minimize crown contact when retracted and rotating, requiring a high telescopic ratio. Therefore, a telescopic arm structure is adopted. Based on the endpoint values of the equipment racks' target workspace in the X and Y directions, a three-section telescopic arm is designed with a fixed arm length of 1 m, a movable arm length of 1 m, a fixed extension distance of 0.1 m, an extendable distance of 0.5 m, and a fully extended length of 2.8 m, exceeding the design requirement of 2190 mm.

Due to the significant offset load imposed by the extension mechanism and similar size requirements, the lifting mechanism also adopts a telescopic arm structure. With a fixed chassis height of 0.8 m and considering the endpoint values of the equipment racks' target workspace in the Z-axis direction, a two-section telescopic arm is selected with a fixed arm length of 1.1 m, a movable arm length of 1 m, a fixed extension distance of 0.1 m, an extendable distance of 0.5 m, and a fully raised length of 2.3 m, closely matching the design requirement of 2307 mm.

The extension and lifting mechanisms constitute a vertical cross-section of the equipment racks' reachable workspace. Combined with the coordination of the slewing bearing, they form the complete reachable workspace of the equipment racks. The three components work in harmony to expand the worker's reachable workspace.

2.3. Experimental Methods of Driving Ability

2.3.1. Analysis of Driving Stability

The roll angle [30] refers to the angle between the support plane of the orchard operation-aid platform chassis and the horizontal plane. The maximum stable roll angle is the angle at which the orchard operation-aid platform is about to roll over but has not yet done so; it is a crucial indicator of the safe operation of the orchard operation-aid platform. Given the high center of gravity of the orchard operation-aid platform and its frequent operation in hilly terrains, it is necessary to determine its maximum stable roll angle to define safe operating limits and reduce the risk of rollover during travel. Considering the operational scenarios of the orchard operation-aid platform, there are three primary attitudes: longitudinal downhill, longitudinal uphill, and lateral slope driving. The force analysis for these three attitudes is shown in Figure 5 (with the illustrated slope angle $\theta = 15°$), and the maximum stable roll angle of these primary attitudes must be greater than 15° to ensure the safety of operators. The forces acting on the orchard operation-aid platform mainly include its own weight m_0g, driving resistance F_{total}, and the resultant normal force F_{Nc} exerted by the ground on the entire track support area. To simplify the analysis, the following assumptions are made:

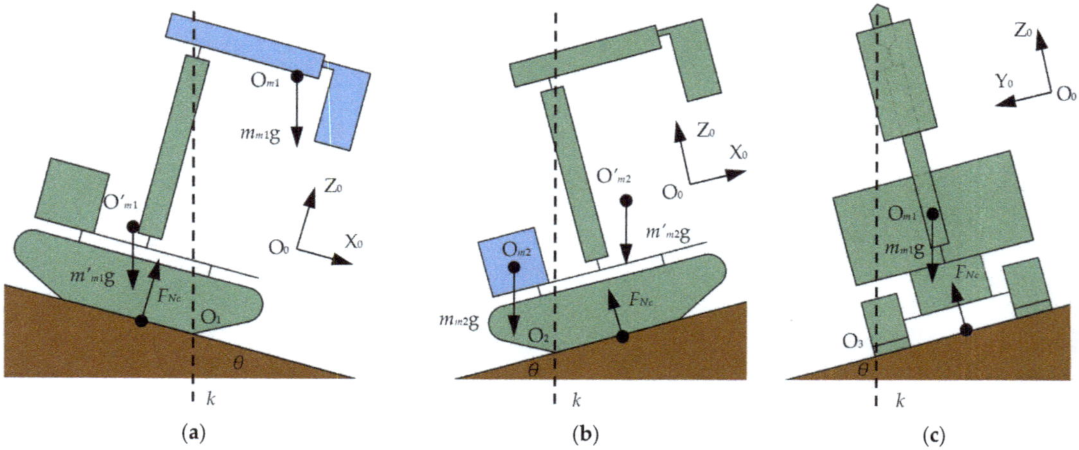

Figure 5. Force diagrams of the orchard operation-aid platform on slopes: (**a**) longitudinal downhill; (**b**) longitudinal uphill; and (**c**) lateral slope driving.

(1) When the orchard operation-aid platform is moving at a low constant speed on a slope, external factors can be neglected, and it can be assumed to be stationary on the slope.

(2) O_1, O_2 and O_3 represent the front, rear, and left support points of the track-ground contact area, respectively. Depending on the driving attitude, the rollover axis k passes vertically through the support point. The length of the track-ground contact area in the X-axis direction is $x_c = 1100$ mm, and in the Z-axis direction is $z_c = 1100$ mm.

(3) Since the operator and the manned workbench are basically kept at a relative standstill, the two are considered as one unit for calculating the center of mass, center of gravity, etc. As a rule of thumb, the operator is weighed to be 75 kg, while the manned workbench's gravity is weighed to be about 33 kg.

(4) Based on the mass and material parameters of each component of the orchard operation-aid platform, and using SolidWorks 2020 to define the properties of each part, the center of gravity O_{m0} of the orchard operation-aid platform in the base coordinate system $O_0X_0Y_0Z_0$ is determined to be (104, −3, 1031). Therefore, the distances between O_{m0} and the support points O_1, O_2 and O_3 along the X, Y, and Z axis are rounded to $x_{m01} = 446$ mm, $x_{m02} = 654$ mm, $x_{m03} = 553$ mm, $y_{m01} = y_{m02} = 0$, $y_{m03} = 553$ mm, $z_{m01} = z_{m02} = z_{m03} = 1031$ mm.

After simplification, the maximum stable roll angles for the three attitudes of longitudinal downhill, longitudinal uphill, and lateral slope driving of the orchard operation-aid platform are calculated separately.

- Maximum Stable Roll Angle of Longitudinal Downhill Driving

In Figure 5a, the orchard operation-aid platform has only one potential rollover scenario when moving downhill longitudinally: rollover clockwise about point O_1. As the slope angle θ changes, the masses on both sides of the rollover axis k change, and the moments generated by these masses change accordingly.

For ease of calculation, further simplifications are made for the longitudinal downhill attitude of the orchard operation-aid platform: the extension mechanism, worker, and manned workbench are all located on the same side of the rollover axis k, and their positions are relatively fixed. They can be considered as a whole with a mass m_{m1} = 366 kg and center of gravity O_{m1} at (751, −1, 2027) in the base coordinate system $O_0X_0Y_0Z_0$. The distances from O_{m1} to support point O_1 along the X and Z axis are x_{m11} = 201 mm and z_{m11} = 2027 mm, respectively. The remaining part of the orchard operation-aid platform is considered as another whole with mass m'_{m1} = 1216 kg and center of gravity O'_{m1} at (−91, 3, 730), with distances to O_1 of x'_{m11} = 641 mm and z'_{m11} = 730 mm along the X and Z axes, respectively.

Establishing the moment balance formula at point O_1:

$$M_{O1} = m'_{m1} \cdot g \cdot (x'_{m11} \cdot \cos\theta - z'_{m11} \cdot \sin\theta) - m_{m1} \cdot g \cdot (x_{m11} \cdot \cos\theta + z_{m11} \cdot \sin\theta) - F_{Nc} \cdot x_{i11}, \quad (6)$$

In the formula, M_{O1} is the moment acting at point O_1; when $M_{O1} < 0$, the orchard operation-aid platform will roll over. x_{i11} is the distance along the X-axis between the resultant normal force F_{Nc} and point O_1, which varies with the position of the center of gravity; when rollover or about to roll over, $x_{i11} = 0$.

From Formula (6), the maximum stable roll angle of the orchard operation-aid platform during longitudinal downhill is calculated to be approximately 23.4°, exceeding the design requirement of 15°.

- Maximum Stable Roll Angle of Longitudinal Uphill Driving

In Figure 5b, the orchard operation-aid platform has only one potential rollover scenario when moving uphill longitudinally: rollover counterclockwise about point O_2.

For ease of calculation, further simplifications are made for the longitudinal uphill attitude: the hydraulic station and the center of gravity of the orchard operation-aid platform are located on opposite sides of the rollover axis, so the hydraulic station is considered an independent unit with mass m_{m2} = 196 kg and center of gravity O_{m2} at (−614, −3, 784) in the base coordinate system. The distances from O_{m2} to support point O_2 are x_{m22} = 64 mm and z_{m22} = 784 mm along the X and Z axes, respectively. The remaining part of the orchard operation-aid platform is considered as another whole with mass m'_{m2} = 1385 kg and center of gravity O'_{m2} at (206, 1066, 3), with distances to O_2 of x'_{m22} = 756 mm and z'_{m22} = 1066 mm.

Establishing the moment balance formula at point O_2:

$$M_{O2} = m'_{m2} \cdot g \cdot (x'_{m22} \cdot \cos\theta - z'_{m22} \cdot \sin\theta) - m_{m2} \cdot g \cdot (x_{m22} \cdot \cos\theta + z_{m22} \cdot \sin\theta) - F_{Nc} \cdot x_{i22}, \quad (7)$$

In the formula, M_{O2} is the moment acting at point O_2; when $M_{O2} < 0$, the orchard operation-aid platform will roll over. x_{i22} is the distance along the X-axis between the resultant normal force F_{Nc} and point O_2, which varies with the position of the center of gravity; when rollover or about to roll over, $x_{i22} = 0$.

From Formula (7), the maximum stable roll angle of the orchard operation-aid platform during longitudinal uphill s calculated to be approximately 32.4° exceeding the design requirement of 15°.

- Maximum Stable Roll Angle of Lateral Slope Driving

In Figure 5c, the orchard operation-aid platform has only one potential rollover scenario when driving laterally on a slope: rollover counterclockwise about point O_3.

Establishing the moment balance formula at point O_3:

$$M_{O3} = mg\,(y_{m03}\cos\theta - z_{m03}\sin\theta) - F_{Nc}\,y_{i03}, \tag{8}$$

In the Formula, M_{O3} is the moment acting at point O_3; when $M_{O3} < 0$, the orchard operation-aid platform will roll over. y_{i03} is the distance along the Y-axis between the resultant normal force F_{Nc} and point O_3, which varies with the position of the center of gravity; when there is no rollover, $y_{i03} = 0.95$ m, and when rollover or about to roll over, $y_{i03} = 0$.

From Formula (8), the maximum stable roll angle of the orchard operation-aid platform during lateral slope driving is calculated to be approximately 28.2°, exceeding the design requirement of 15°.

In summary, the maximum stable roll angles of the orchard operation-aid platform during longitudinal downhill, longitudinal uphill, and lateral slope driving are 23.4°, 32.4°, and 28.2°, respectively, all exceeding the design requirement of 15°. Therefore, the orchard operation-aid platform meets the stability requirements for driving on slopes.

2.3.2. Experimental Methods

- Turning Radius

In October 2024, the experiments were conducted at the agricultural machinery depot, South China Agricultural University (40°08′11″ N, 116°10′45″ E). Measurement instruments included spray markers and tape measures. The orchard operation-aid platform was moved to an open area, with the manned workbench lowered to the lowest position. Spray markers were installed at the four corners of the vehicle. During turning, the remote control was used to operate the tracks on both sides to move in opposite circular motions, completing a full circle. The diameter of the central circle of the outer track was measured.

- Driving Stability

Due to limitations in experimental conditions, the contact pressures on the upper and lower sides of the chassis on slopes were measured using the following method to assess the stability of the orchard operation-aid platform under different roll angles. Experimental tools included an angle-measuring instrument, pressure sensors, and a laptop. The experiment site had three long slopes, with average slopes of 7.6°, 14.3°, and 13.5°, respectively. Experiments were conducted on flat ground, a 7.6° slope, and a 14.3° slope. During measurements, two pressure sensors were placed at the contact positions between the tracks and the ground on the upper and lower sides of the slope, collecting feedback data from the sensors.

2.4. Experimental Methods of Slope Operation Stability

2.4.1. Analysis of Slope Operation Stability

When workers prepare to perform tasks, such as pruning, spraying, thinning flowers and fruit, or fruit harvesting on the manned workbench, the extension and lifting mechanisms may extend or retract according to task requirements and rotate around the slewing bearing. At this time, the center of gravity of the orchard operation-aid platform shifts outward, changing the rollover moment and potentially causing a rollover. To ensure worker safety, the workers must deploy the slope operation support device before commencing work. Depending on the operational scenario, the slope operation support device has three working modes:

(1) When there is a higher slope on the left side of the orchard operation-aid platform, only the right and front support legs of the slope operation support device are extended.

(2) When there is a higher slope on the right side, only the left and front support legs are extended.

(3) When there are no high slopes on either side, all three sets of support legs are extended.

As the center of gravity of the orchard operation-aid platform changes and the slope operation support device acts, the rollover safety of the orchard operation-aid platform needs to be reassessed. It is evident that the most likely rollover scenario during operation is as shown in Figure 6, where the orchard operation-aid platform has the greatest potential to roll around support point S. Additionally, the following considerations are made:

(1) Due to the support from the support legs, the tracks on the side of the support legs may lose contact with the ground.

(2) The worker and manned workbench are located on the same side of the rollover axis k, with relatively fixed positions, considered as a whole with mass m_{w0} = 108 kg and center of gravity O_{w0} at (215, −2814, 27 14) in the base coordinate system. The distances from O_{w0} to support point S along the Y and Z axis are y_{w0} = 1514 mm and z_{w0} = 2714 mm.

(3) The extension mechanism is divided into two parts by the rollover axis; assuming the mass of the extension mechanism is uniformly distributed along its length, the masses of the two parts are m_{w1} and m_{w2}, with centers of gravity O_{w1} (122, 1572·tanθ–1825, 3144) and O_{w2} (122, 1572·tanθ–425, 3144) in the base coordinate system. The masses m_{w1} and m_{w2} are (96.4 + 288.6·tanθ) kg and (257 − m_{w1}) kg, respectively. The distances from O_{w1} and O_{w2} to support point S along the Y and Z axis are y_{w1} = (525–1572·tanθ) mm, z_{w1} = 3144 mm, and y_{w2} = (875 + 1572·tanθ) mm, z_{w2} = 3144 mm.

(4) The remaining part of the orchard operation-aid platform is considered as a whole with mass m_{w3} = 1216 kg and center of gravity O_{w3} at (−89, 3, 792). The distances from O_{w3} to support point S are y_{w3} = 1297 mm and z_{w3} = 792 mm along the Y and Z axes.

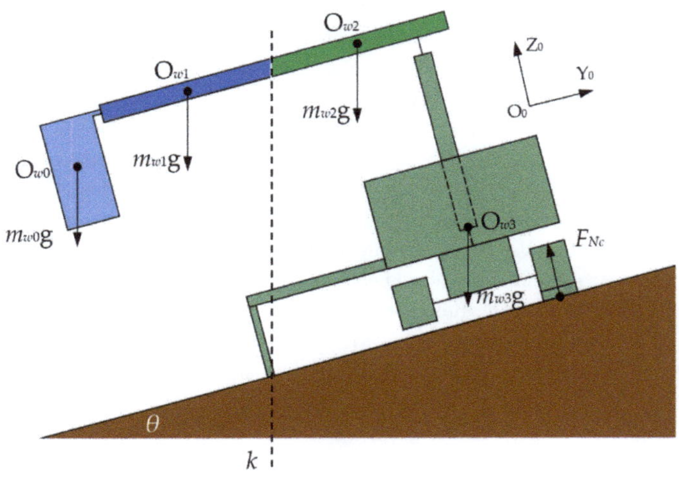

Figure 6. Force diagram of the orchard operation-aid platform during slope operations.

Establishing the moment balance formula at point S:

$$M_S = m_{w3} \cdot g \cdot (y_{w3} \cdot \cos\theta - z_{w3} \cdot \sin\theta) + m_{w2} \cdot g \cdot (y_{w2} \cdot \cos\theta - z_{w2} \cdot \sin\theta) - m_{w1} \cdot g \cdot (y_{w1} \cdot \cos\theta + z_{w1} \cdot \sin\theta) - m_{w0} \cdot g \cdot (y_{w0} \cdot \cos\theta + z_{w0} \cdot \sin\theta) - F_{Nc} \cdot y_{iS}, \quad (9)$$

In the formula, M_S is the moment acting at point S; when $M_S < 0$, the orchard operation-aid platform will roll over. y_{iS} is the distance along the Y-axis between the resultant normal force F_{Nc} and point S, which varies with the position of the center of gravity; when there is no rollover, y_{iS} = 1.6 m, and when rollover or about to roll over, y_{iS} = 0.

From Formula (9), the critical angle for slope operation of the orchard operation-aid platform is calculated to be approximately 36.07°, exceeding the design requirement of 15°, indicating safe operation in the preset scenarios.

2.4.2. Experimental Methods

The experimental site for slope operation stability is the same as the experimental site for driving stability. Experimental instruments such as an angle measuring instrument, pressure sensors, and a laptop were prepared. Experiments were conducted on flat ground, a 7.6° slope, and a 14.3° slope. According to the operational scenarios of the orchard operation-aid platform, the manned workbench was moved to a preset position, one side of the slope operation support device was extended, and three pressure sensors were placed at the contact positions between the support leg, the two tracks, and the ground. Feedback data from the sensors were collected.

2.5. Experimental Methods of Workspace

The reachable workspace of the orchard operation-aid platform refers to the set of all spatial points that can be reached by the worker's palm, representing the operational coverage of the orchard operation-aid platform. It is an important basis for evaluating the operational capability of the orchard operation-aid platform.

2.5.1. Establishment of Joint Coordinate Systems

According to the structure and working principle of the orchard operation-aid platform, the rotation of the slewing bearing and the rotation of the worker's arm about the shoulder constitutes 2 rotational joints. The worker's squatting and standing, lateral movement on the manned workbench, and the extension/retraction of the lifting and extension mechanisms constitute 4 prismatic joints. Therefore, during slope operations, the orchard operation-aid platform can be modeled as a 2R-4P serial open-chain mechanism. As shown in Figure 7, with each joint as the origin of a coordinate system, the Denavit–Hartenberg (D-H) method is used to establish the joint coordinate systems of the orchard operation-aid platform. Here, l_1 is the distance from the center of the upper surface of the slewing bearing to the ground, l_2 is the minimum distance along the lifting direction from the centroid of the extension mechanism to the center of the upper surface of the slewing bearing, l_3 is the minimum distance along the extension direction from the centroid of the lifting mechanism to the far end of the manned workbench, l_5 is the distance along the lifting direction from the centroid of the extension mechanism to the floor of the manned workbench, l_7 is the worker's functional forward reach length of the upper limbs, and l_4 and l_6 are 0. Measurements yield l_1 = 800 mm, l_2 = 1400 mm, l_3 = 900 mm, l_5 = 1100 mm, and l_7 = 710 mm.

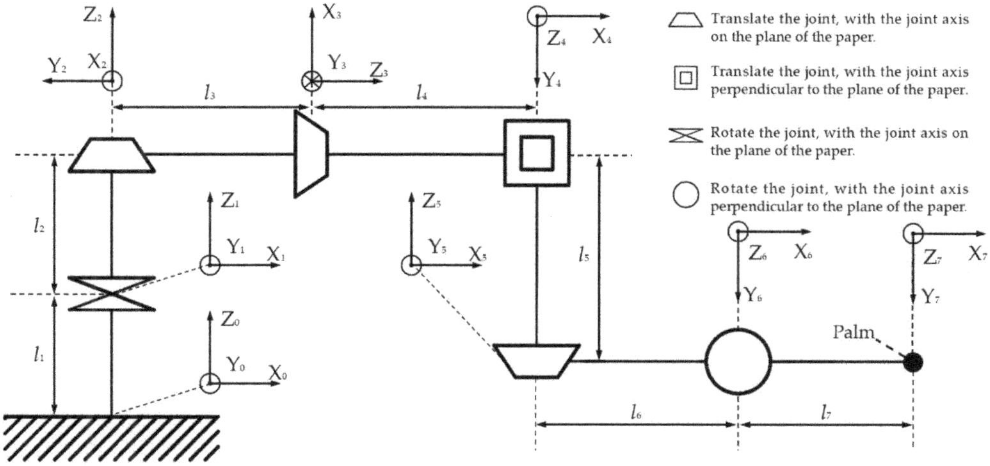

Figure 7. Joint coordinate systems of the orchard operation-aid platform.

Based on the joint coordinate systems, the structural parameters of the orchard operation-aid platform are shown in Table 4.

Table 4. Structural parameters of the orchard operation-aid platform.

Joint	α_i (°)	a_i (mm)	d_i (mm)	θ_i (°)	Range of Joint Variables
1	0	0	l_1	θ_1	$[-90, 90]$ (°)
2	0	0	d_2	−90	$[l_2, l_2 + 1000]$ (mm)
3	−90	0	d_3	−90	$[l_3, l_3 + 2100]$ (mm)
4	−90	0	d_4	−90	$[-l_{gu2}/2, l_{gu2}/2]$ (mm)
5	−90	0	d_5	0	$[-l_5, -l_5 + l_{p3}]$ (mm)
6	90	0	0	θ_6	$[-90, 90]$ (°)
7	0	l_7	0	0	–

In the table, α_i is the link twist angle, i.e., the angle through which Z_{i-1} must be rotated about X_{i-1} to align with Z_i; when the two links are parallel, $\alpha_i = 0°$; when perpendicular, $\alpha_i = 90°$. α_i is the link length, i.e., the distance along X_{i-1} from Z_{i-1} to Z_i. d_i is the link offset, i.e., the distance along Z_i from X_{i-1} to X_i; for rotational joints, d_i is constant; for prismatic joints, d_i is variable. θ_i is the joint angle, i.e., the angle through which X_{i-1} must be rotated about Z_i to align with X_i for rotational joints, θ_i is variable; for prismatic joints, θ_i is constant.

2.5.2. Kinematic Equation

The transformation matrix $\mathbf{T}_{i-1}^{i}$ between two adjacent coordinate systems can be determined using the homogeneous transformation formula:

$$\mathbf{T}_{i-1}^{i} = \begin{bmatrix} 1 & 0 & 0 & a_i \\ 0 & c\alpha_i & -s\alpha_i & 0 \\ 0 & s\alpha_i & c\alpha_i & 0 \\ 0 & 0 & 0 & 1 \end{bmatrix} \begin{bmatrix} c\theta_i & -s\theta_i & 0 & 0 \\ s\theta_i & c\theta_i & 0 & 0 \\ 0 & 0 & 1 & d_i \\ 0 & 0 & 0 & 1 \end{bmatrix} = \begin{bmatrix} c\theta_i & -s\theta_i & 0 & a_i \\ s\theta_i c\alpha_i & c\theta_i c\alpha_i & -s\alpha_i & -d_i s\alpha_i \\ s\theta_i s\alpha_i & c\theta_i s\alpha_i & c\alpha_i & d_i c\alpha_i \\ 0 & 0 & 0 & 1 \end{bmatrix}, \quad (10)$$

In the formula, $c\theta_i = \cos\theta_i$; $s\theta_i = \sin\theta_i$; $c\alpha_i = \cos\alpha_i$; $s\alpha_i = \sin\alpha_i$.

By substituting the structural parameters of each joint from Table 4 into the above formula, the transformation matrices between adjacent joints are obtained.

$$\mathbf{T}_0^1 = \begin{bmatrix} c_1 & -s_1 & 0 & 0 \\ s_1 & c_1 & 0 & 0 \\ 0 & 0 & 1 & l_1 \\ 0 & 0 & 0 & 1 \end{bmatrix} \quad \mathbf{T}_1^2 = \begin{bmatrix} 0 & 1 & 0 & 0 \\ -1 & 0 & 0 & 0 \\ 0 & 0 & 1 & d_2 \\ 0 & 0 & 0 & 1 \end{bmatrix} \quad \mathbf{T}_2^3 = \begin{bmatrix} 0 & 1 & 0 & 0 \\ 0 & 0 & 1 & d_3 \\ 1 & 0 & 0 & 0 \\ 0 & 0 & 0 & 1 \end{bmatrix} \quad \mathbf{T}_3^4 = \begin{bmatrix} 0 & 1 & 0 & 0 \\ 0 & 0 & 1 & d_4 \\ 1 & 0 & 0 & 0 \\ 0 & 0 & 0 & 1 \end{bmatrix},$$
$$\mathbf{T}_4^5 = \begin{bmatrix} 1 & 0 & 0 & 0 \\ 0 & 0 & 1 & d_5 \\ 0 & -1 & 0 & 0 \\ 0 & 0 & 0 & 1 \end{bmatrix} \quad \mathbf{T}_5^6 = \begin{bmatrix} c_6 & -s_6 & 0 & 0 \\ 0 & 0 & -1 & 0 \\ c_6 & s_6 & 0 & 0 \\ 0 & 0 & 0 & 1 \end{bmatrix} \quad \mathbf{T}_6^7 = \begin{bmatrix} 1 & 0 & 0 & l_7 \\ 0 & 1 & 0 & 0 \\ 0 & 0 & 1 & d_i \\ 0 & 0 & 0 & 1 \end{bmatrix} \tag{11}$$

For the orchard operation-aid platform, the end reference point, i.e., the palm center, relative to the base coordinate system $O_0X_0Y_0Z_0$ can be defined as:

$$\mathbf{T}_0^7 = \mathbf{T}_0^1\mathbf{T}_1^2\mathbf{T}_2^3\mathbf{T}_3^4\mathbf{T}_4^5\mathbf{T}_5^6\mathbf{T}_6^7, \tag{12}$$

By substituting the transformation matrices into the above formula, the transformation matrix of the end reference point relative to the base coordinate system, i.e., the kinematic model of the orchard operation-aid platform, is obtained.

$$\mathbf{T}_0^7 = \begin{bmatrix} n_x & o_x & a_x & p_x \\ n_y & o_y & a_y & p_y \\ n_z & o_z & a_z & p_z \\ 0 & 0 & 0 & 1 \end{bmatrix} = \begin{bmatrix} \mathbf{R}_0^7 & \mathbf{P}_0^7 \\ 0 & 1 \end{bmatrix}, \tag{13}$$

In the formula, $\mathbf{R}_0^7$ is the position vector of the end effector in the base coordinate system. $\mathbf{P}_0^7$ represents the kinematic relationship derived from the transformations.

Thus, the kinematic equations of the orchard operation-aid platform can be expressed as:

$$\mathbf{P}_0^7 = \begin{bmatrix} p_x, & p_y, & p_z \end{bmatrix}^{\mathrm{T}}, \\ \begin{cases} p_x = d_3\cos\theta_1 + d_4\sin\theta_1 + l_7\cos\theta_1\cos\theta_6 \\ p_y = d_3\sin\theta_1 - d_4\cos\theta_1 + l_7\sin\theta_1\cos\theta_6 \\ p_z = l_1 + d_2 + d_5 + l_7\sin\theta_6 \end{cases} \tag{14}$$

The reachable workspace of the orchard operation-aid platform can be obtained by mapping the joint space using the kinematic equations.

2.5.3. Reachable Workspace

Using the Monte Carlo method to construct a joint space following an inverse sine distribution, the joint space is mapped to the workspace using the kinematic equations, obtaining the reachable workspace of the orchard operation-aid platform, as shown in Figure 8.

From the figure, the maximum values of the reachable workspace of the orchard operation-aid platform along the X, Y, and Z axes can be obtained, rounded, and summarized in Table 5. The theoretical reachable workspace of the orchard operation-aid platform encompasses the target workspace.

Table 5. Reachable workspace dimensions of the orchard operation-aid platform.

Axis	X		Y		Z	
	Maximum	**Minimum**	**Maximum**	**Minimum**	**Maximum**	**Minimum**
Design value (mm)	3000	0	3000	−3000	3500	1200
Simulation value (mm)	3722	−300	3722	−3722	4183	392

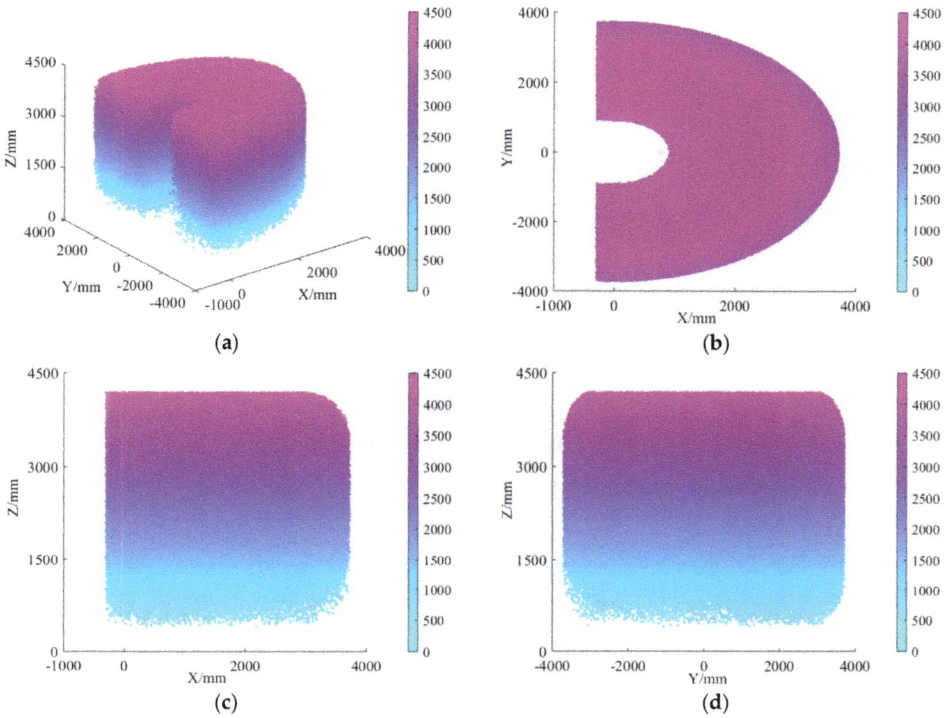

Figure 8. Reachable workspace of the orchard operation-aid platform: (**a**) three-dimensional view; (**b**) X-Y cross-sectional view; (**c**) X-Z cross-sectional view; and (**d**) Y-Z cross-sectional view.

2.5.4. Experimental Methods

A prototype validation experiment was conducted at the Wuhua Orchard in Maoming (110.969° E, latitude 21.752° N) to obtain the actual operational radius and height of the orchard operation-aid platform, evaluate its actual operational performance, and verify the orchard applicability, as well as the accuracy of the mechanism's dimension design. The experiment involved controlling the extension mechanism, lifting mechanism, and slewing bearing to their minimum and maximum values. The worker's position (standing or squatting) was adjusted as the lifting mechanism moved up or down. The distances from the worker's palm to the ground and the centroid of the lifting mechanism were measured, providing the boundary extremes of the actual reachable workspace of the orchard operation-aid platform. The worker's functional overhead reach height and functional forward reach length of the upper limbs are 2203 mm and 770 mm, respectively.

3. Results

In order to verify the key performances of the equipment, a prototype physical machine was constructed (Figure 2a), and field trials were conducted according to the previous experimental program. The experimental results and analyses are shown below.

3.1. Driving Ability

3.1.1. Turning Radius

The orchard operation-aid platform was moved to an open area, with the manned workbench lowered to the lowest position. Spray markers were installed at the four corners of the vehicle. The diameter of the central circle of the outer track was measured. The test scenario is shown in Figure 9.

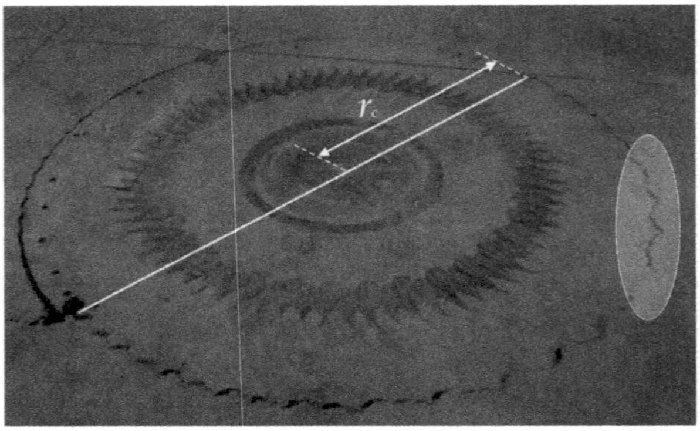

Figure 9. Field test of the orchard operation-aid platform's turning radius.

From Figure 9, it can be observed that the turning trajectory of the orchard operation-aid platform is approximately circular, indicating that the output power of the motors on both sides of the tracks is relatively consistent during turning. However, the outer trajectory drawn by the spray markers shows noticeable undulations, indicating periodic stuttering during turning, suggesting that the motor output logic needs to be optimized in future work. The test was repeated three times, and the results are recorded in Table 6.

Table 6. Results of the orchard operation-aid platform's turning radius test.

Test Item		Design Value (mm)	Permissible Value (mm)	Measured Value (mm)				Deviation Rate (%)
				1	2	3	Average Value	
Forward	Left	≤1500	1097.7	1180	1242	1085	1169	6.50
	Right			1200	1227	1237	1221	11.23
Back	Left			1165	1110	1172	1149	4.67
	Right			1207	1192	1180	1203	9.59

Calculating the average of the three measurements, the minimum turning radius for a single movement is obtained. The results indicate that the left and right turning radii during both forward and reverse operations are basically consistent, with an average minimum turning radius of 1186 mm, recorded as 1.2 m. Compared with the design value, the average deviation rate is ≤8.05%, demonstrating good steering flexibility of the orchard operation-aid platform.

3.1.2. Driving Stability

The experiment scenarios are shown in Figure 10.

From Figure 10e, it can be observed:

(1) Due to multiple contact points between the tracks and the ground, the data feedback from the two sensors is less than the actual value; however, the consistent trends in the feedback data reflect their reference value.

(2) As the tracks' contact with the ground is not continuous, and the tracks rolling over can cause displacement of the sensors, the feedback data from the sensors show fluctuations and delays. Therefore, the maximum value of the measured data in each experiment process is taken as the experiment result.

The above experiment steps were repeated multiple times, and the measurement results are recorded in Table 7.

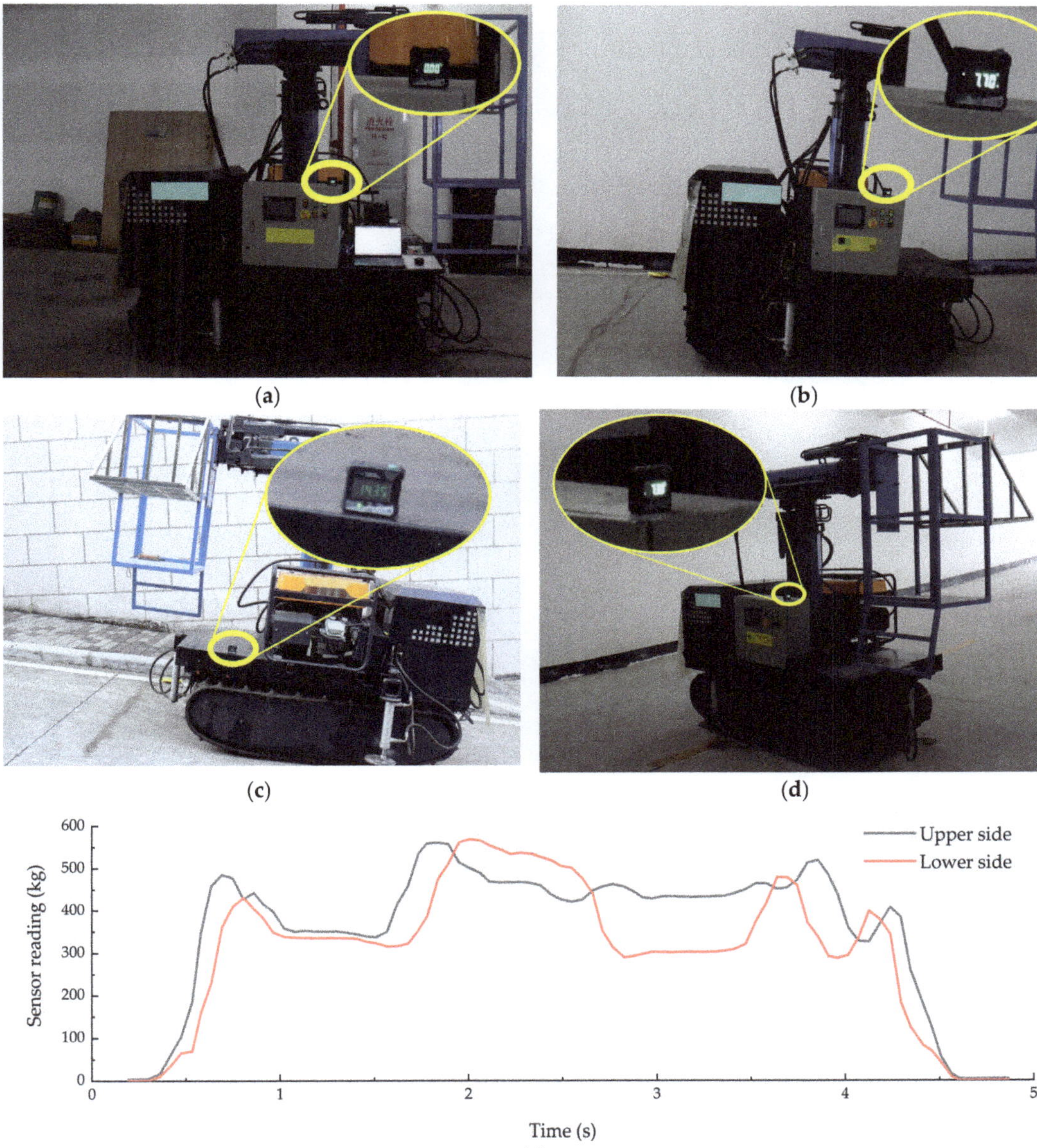

Figure 10. Field experiment of the orchard operation-aid platform's driving stability: (**a**) driving on 0° ground; (**b**) descending on a 7.6° slope; (**c**) ascending on a 14.3° slope; (**d**) lateral driving on a 7.6° slope; and (**e**) sensor data when the equipment is moving longitudinally downhill on a surface with a gradient of 0°.

Table 7. Results of the orchard operation-aid platform's driving stability experiment.

Experiment Item		Upper Side (kg)				Lower Side (kg)			
		1	2	3	Average Value	1	2	3	Average Value
Longitudinal downhill	0°	550	559	544	551	590	570	588	583
	7.6°	418	419	428	422	697	710	695	701
	14.3°	233	234	236	234	910	900	893	901
Longitudinal uphill	0°	613	616	606	612	521	507	523	517
	7.6°	484	496	482	487	655	626	657	646
	14.3°	388	389	373	383	733	755	759	749
Lateral slope driving	0°	588	589	595	591	529	542	544	538
	7.6°	509	498	495	501	616	622	624	621
	14.3°	300	308	299	302	827	831	829	829

From Table 7, it can be concluded:

(1) The mass is not symmetrically distributed along the X-axis; the side with the generator has a greater mass. When driving laterally, it is advisable to have the generator side on the upper side of the slope to reduce the risk of rollover.

(2) As the slope increases, the contact pressure between the chassis and the upper side of the slope gradually decreases, with the lowest pressure during downhill on a 14.3° slope, equivalent to the gravity of a 234 kg object, i.e., 2293 N.

(3) The contact pressure between the chassis and the lower side of the slope gradually increases with the slope, with the highest pressure during downhill on a 14.3° slope, equivalent to the gravity of a 901 kg object, i.e., 8830 N.

The experimental results are consistent with theoretical analysis, indicating that the orchard operation-aid platform can operate normally in typical hilly orchards.

3.2. Slope Operational Stability

The experiment scenario is shown in Figure 11.

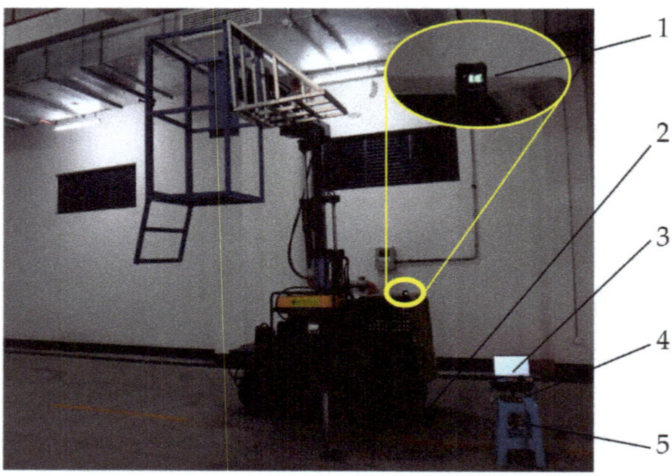

Figure 11. Field experiment of the orchard operation-aid platform's slope operation: (1) angle measuring instrument; (2) pressure sensor; (3) sensor monitoring software; (4) remote controller; and (5) force digital transmitter.

As shown in Figure 11, for the operational stability experiment conducted on flat ground, three sensors were placed in a straight line beneath the supporting leg and the two tracks. Sensor data were obtained and processed using sensor monitoring software (SEMT 1.27.3). Repeat the experimental steps, take multiple measurements, and record the results in Table 8.

Table 8. Results of the orchard operation-aid platform's slope operation experiment.

Experiment Item	Supporting Leg (kg)	Track 1 (kg)	Track 2 (kg)
0	550	0	589
7.6°	766	0	477
14.3°	957	0	363

From the table, it can be observed that as the slope increases, the contact pressure between the chassis and the upper side of the slope gradually decreases, with the lowest pressure at a slope of 14.3°, equivalent to the gravity of a 363 kg object, i.e., 3557 N. The contact pressure between the support leg and the ground gradually increases with the slope, reaching a maximum equivalent to the gravity of a 957 kg object, i.e., 9379 N, at a 14.3° slope. This indicates that the slope operation support device is effective and that rollover does not occur even at a slope of 14.3°, providing numerical evidence for standardizing the operational scenarios of the orchard operation-aid platform.

3.3. Workspace

The experiment scenario is shown in Figure 12.

Figure 12. Field experiment of the orchard operation-aid platform's workspace.

As shown in Figure 12, we conducted experiments on relatively flat ground in an orchard. We controlled the equipment according to the joint parameters corresponding to the limit values of the workspace obtained from the simulation and measured the operational radius and operational height using a tape measure. Repeat the experimental steps, take multiple measurements, and record the results in Table 9.

From Table 9, it can be seen that the operational radius and height of the orchard operation-aid platform, i.e., the maximum distances from the origin along the X, Y, and Z axis in the actual reachable workspace, are 3694 mm, 3730 mm, and 4067 mm, respectively.

Therefore, the operational radius and height of the orchard operation-aid platform are 3.7 m and 4.0 m, respectively, meeting the design requirements.

Table 9. Results of the orchard operation-aid platform's workspace experiment.

Experiment Item		Design Value (mm)	Measured Value (mm)			
			1	2	3	Average Value
X	Maximum	3000	3678	3694	3711	3694.3
	Minimum	0	−920	−895	−909	−908.0
Y	Maximum	3000	3676	3729	3703	3702.7
	Minimum	−3000	−3713	−3756	−3722	−3730.3
Z	Maximum	3500	4072	4086	4043	4067.0
	Minimum	1200	602	567	572	580.3

It is important to note that there are significant discrepancies between the measured values of the actual reachable workspace and the theoretical values. Of course, these errors do not affect the overall assessment of the experimental results. The main reasons for this discrepancy may be as follows:

(1) To minimize friction from the movement of the telescopic sleeves in the lifting and extension mechanisms, ultra-high-molecular-weight polyethylene (UHMW-PE) sliders are installed at the ends of each telescopic arm. The thickness of these sliders affects the gap between the adjacent telescopic arm, causing the axis of the adjacent telescopic arm not to be colinear when fully extended, which leads to "bending" of the lifting and extension mechanisms.

(2) The uneven terrain of the orchard results in variations in ground angles, contributing to errors in height measurements.

(3) The worker has six working postures: extending their arms forward, backward, to the left, to the right, and upward while standing, as well as extending their arms downward while squatting. To simplify the kinematic model, the previous discussion did not consider the backward, leftward, and rightward postures. These three postures have little impact on the final experimental results, specifically the working radius and working height.

4. Discussion

Overall, this research develops an orchard operation-aid platform based on planting patterns and topography, including the design of the telescopic arm and support devices, analysis of driving stability, analysis of operational stability, workspace analysis, and field experiments. This work contributes to the mechanization of orchards in hilly areas and has broad application potential.

The orchard operation-aid platform developed in this research features a large operational range, with its lifting and extension accomplished through a telescopic arm, enabling it to navigate narrow roads. In contrast, the orchard operation-aid platform referenced in the literature [31] has a frame length of 3437 mm and a frame width of 1650 mm, resulting in a small operational radius and a large turning radius, making it unsuitable for narrow roads in hilly areas. Furthermore, the center of gravity of the equipment designed changes continuously with the rotation of the slewing bearing. Although the works in the literature [32,33] can maintain the center of gravity of the equipment within a certain slope range, their operational range is limited, and the adjustment is slow, making them unsuitable for this equipment. The support device designed in this research effectively increases the operational slope range of the equipment, thereby enhancing the safety of workers during operation.

During the field experiments, some issues arose with the prototype, and summarizing these issues can provide important references for future improvements:

(1) The thickness of the sliders installed at the ends of each telescopic arm is too thin, and the reserved space is too large, leading to the bending of the telescopic arm during the lifting and extension processes. Future work needs to balance the thickness of the sliders of the telescopic arm, friction resistance, and the amount of deformation during operation.

(2) The operational speed of the hydraulic components is relatively slow and needs to be increased. However, a hydraulic component that operates too quickly can affect the stability of the equipment, potentially leading to tipping. Therefore, future work must balance the operating speed of hydraulic components with the safety of the equipment.

(3) The telescopic arm employs industrial products that, while meeting some functional requirements, have clearly excessive design parameters such as load-bearing capacity. This results in increased weight, raising the overall center of gravity of the equipment. In future work, the telescopic arm should be designed based on the operational scenario.

(4) When workers are on the manned workbench, transferring materials to the ground becomes challenging. Future improvements could include a fixed pulley system to facilitate this process.

5. Conclusions

In order to address the limited adaptability of the existing orchard operation-aid platforms to the orchard, this paper establishes the design requirements, overall structure, and main technical parameters based on planting patterns and topography. The operation-aid platform can assist the operator in completing the tasks of pruning, spraying, thinning flowers and fruits, harvesting, etc., in hilly and mountainous orchards with plant and row spacing not exceeding 6 m and slopes not exceeding 15°. This innovation enhances operational efficiency and reduces labor intensity.

(1) Combining the planting patterns and operational topography parameters of litchi and longan orchards, the key components of the orchard operation-aid platform, including the chassis, lifting mechanism, extension mechanism, and slope operation support device, were designed using parameters like driving speed, turning radius, lifting height, extension length, and support forces as indicators.

(2) The maximum stable roll angles of the orchard operation-aid platform were analyzed under three typical driving conditions—longitudinal downhill, longitudinal uphill, and lateral slope driving—as well as during extreme conditions of slope operation. In all cases, the roll angles exceeded the design requirement of 15°, demonstrating the platform's stability and adaptability in challenging terrains.

(3) The kinematic model of the orchard operation-aid platform was developed using the D-H parameter method, and its reachable workspace was obtained through the Monte Carlo method. The simulation experiment results show that the maximum values of the orchard operation-aid platform in the X, Y, and Z axes are 3722 (mm), 3722 (mm), and 4183 (mm), respectively, exceeding the design values. This indicates that the reachable workspace of the orchard operation-aid platform is larger than the target workspace, meeting the design requirements.

(4) By analyzing the operational safety and operational performance of the orchard operation-aid platform, the experiment results show that the in situ turning radius of the orchard operation-aid platform is 1.2 m, the driving slope and slope operation angles both exceed 15°, the operational height is 4.0 m, and the operational radius is 3.7 m. The orchard operation-aid platform meets the passability and operational requirements of hilly orchards.

Author Contributions: Conceptualization, Z.L.; methodology, Z.L.; software, C.M.; validation, C.M. and R.J.; formal analysis, C.L. and Y.Z.; investigation, Z.L. and J.L.; resources, J.L.; data curation, Z.L.; writing—original draft preparation, Z.L.; writing—review and editing, Z.L.; visualization, Z.L.; supervision, C.L. and J.L.; project administration, J.L.; funding acquisition, J.L. All authors have read and agreed to the published version of the manuscript.

Funding: This research was supported by the earmarked fund for the Guangdong Laboratory for Lingnan Modern Agriculture (Grant NZ2021040 NT2021009), the Guangdong Province Rural Revitalization Strategy Special Project under (Grant 2023-440000-60010000-9818), the Special Project of Rural Vitalization Strategy of Guangdong Academy of Agricultural Sciences (Grant No. TS-1-4), the China Agriculture Research System of MOF and MARA (Grant No. CARS-32-13), and the Specific University Discipline Construction Project of South China Agricultural University in 2023 (Grant 2023B10564002).

Institutional Review Board Statement: Not applicable.

Data Availability Statement: The raw data supporting the conclusions of this study are available from the corresponding author upon reasonable request.

Acknowledgments: The authors thank the helpful feedback from reviewers, which contributed to improving the quality of this manuscript.

Conflicts of Interest: The authors declare no conflicts of interest.

References

1. South Subtropical Crops Center, Ministry of Agriculture and Rural Affairs. *Report on the Development of China's Litchi and Longan Industries*; China Tropical Agriculture: Beijing, China, 2024; pp. 5–7.
2. Coppock, G.E.; Jutras, P.J. An Investigation of the Mobile Picker's Platform Approach to Partial Mechanization of Citrus Fruit Picking. *Proc. Fla. State Hortic. Soc.* **1960**, *73*, 258–262.
3. Baugher, T.; Schupp, J.; Lesser, K.; Harsh, R.M.; Seavert, C.; Lewis, K.; Auvil, T. Mobile Platforms Increase Orchard Management Efficiency and Profitability. In *Acta Horticulturae*; International Society for Horticultural Science (ISHS): Leuven, Belgium, 2009; pp. 361–364. [CrossRef]
4. Radicioni, L.; Bono, F.; Cinquemani, S. On the use of vibrations and temperatures for the monitoring of plastic chain conveyor systems. *Mech. Syst. Signal Process.* **2025**, *223*, 111935. [CrossRef]
5. Bandit Cub Info. Automated Ag. Available online: https://www.automatedag.com/bandit-cub-info (accessed on 4 December 2024).
6. Zhang, Z.; Heinemann, P.H.; Liu, J.; Schupp, J.R.; Baugher, T.A. Design and Field Test of a Low-Cost Apple Harvest-Assist Unit. *Trans. ASABE* **2016**, *59*, 1149–1156. [CrossRef]
7. Berlage, A.G.; Langmo, R.D.; Yost, G.E. Limitations of Single- and Multi-Man Platform Harvesting Aids. 1972. Available online: https://ir.library.oregonstate.edu/concern/administrative_report_or_publications/kk91fk93k (accessed on 4 December 2024).
8. Mechanization Is Rolling Along—Video—Good Fruit Grower. Available online: https://www.goodfruit.com/mechanization-is-rolling-along/ (accessed on 4 December 2024).
9. Yang, Z.; Wang, Y.; Han, B.; Jin, H.; Chen, J. Research and Design on Conveying System for Tractive Orchard Picking Platform. *J. Chin. Agric. Mech.* **2017**, *38*, 24–28.
10. Li, Q.; Fan, G.; Wang, X.; Zhang, X.; Wang, J. Development of a Fruit Conveying Device for Self-Propelled Orchard Picking Platform. *J. Agric. Mech. Res.* **2019**, *41*, 116–119+152. [CrossRef]
11. Fei, Z.; Vougioukas, S.G. Co-Robotic Harvest-Aid Platforms: Real-Time Control of Picker Lift Heights to Maximize Harvesting Efficiency. *Comput. Electron. Agric.* **2021**, *180*, 105894. [CrossRef]
12. Fei, Z.; Vougioukas, S.G. A Robotic Orchard Platform Increases Harvest Throughput by Controlling Worker Vertical Positioning and Platform Speed. *Comput. Electron. Agric.* **2024**, *218*, 108735. [CrossRef]
13. Peterson, D.L. Development of a Harvest Aid for Narrow-Inclined-Trellised Tree-Fruit Canopies. *Appl. Eng. Agric.* **2005**, *21*, 803–806. [CrossRef]
14. CUI, Z.; GUAN, C.; CHEN, Y.; GAO, Q.; YANG, Y. Design of Small Multi-Functional Electric Crawler Platform for Greenhouse. *Trans. Chin. Soc. Agric. Eng.* **2019**, *35*, 48–57.
15. Zhang, H. Design of Multifunctional Orchard Lifting Platform in Electric Crawler Type. Master's Thesis, South China Agricultural University, Guangzhou, China, 2017.

16. Liu, D.; Xie, F.; Li, X.; Wang, X. Design and experiment of small lifting platform in orchard. *Trans. Chin. Soc. Agric. Eng.* **2015**, *31*, 113–121.
17. Yongjun, Z.; Shijie, J.; Bingtai, C.; Haotun, L.; Chang, W.; Feng, K. Review on Technology and Equipment of Mechanization in Hilly Orchard. *Trans. Chin. Soc. Agric. Mach.* **2020**, *51*, 1–20.
18. Cutini, M.; Brambilla, M.; Bisaglia, C.; Melzi, S.; Sabbioni, E.; Vignati, M.; Cavallo, E.; Laurendi, V. A Study of the Lateral Stability of Self-Propelled Fruit Harvesters. *Agriculture* **2017**, *7*, 92. [CrossRef]
19. Melzi, S.; Sabbioni, E.; Vignati, M.; Cutini, M. Numerical Analysis of Rollover Risk of a Fruit-Harvesting Truck. *J. Adv. Veh. Eng.* **2017**, *3*, 112–120.
20. Yongchao, Y.; Feng, K.; Yongjun, Z.; Haotun, L.; Yaxiong, W. Design and simulation of the automatic-leveling high-position platform in orchards. *J. Beijing For. Univ.* **2021**, *43*, 150–159.
21. Liu, X.; Zhong, B.; Chu, Y.; Liu, X.; Qin, X.; Zhang, C. Design and analysis of lifting and leveling mechanism for new orchard picking platform. *J. Chin. Agric. Mech.* **2020**, *41*, 80–85+106. [CrossRef]
22. Duan, Z.; Qiu, W.; Ding, W.; Liu, Y.; Ouyang, Y.; Huang, L. Tilting Stability Analysis and Experiment of the 3-DOF Lifting Platform for Hilly Orchards. *Int. J. Agric. Biol. Eng.* **2018**, *11*, 73–80. [CrossRef]
23. Fan, G.; Wang, Y.; Wang, J.; Zhang, X. Development and Experiment of Platform for Orchards in Hill Area. *J. Agric. Mech. Res.* **2016**, *38*, 77–81.
24. Crendon Machinery. Available online: https://www.crendon.com.au/ (accessed on 4 December 2024).
25. *NY/T 5174-2002*; Pollution-Free Food Litchi Production Technical Regulations. Ministry of Agriculture of the People's Republic of China: Bejing, China, 2002.
26. Zhang, W. Interpretation of DB4403/T 295-2022, Code of Practice for Longan Production. *China Stand.* **2023**, *20*, 145–150.
27. Gan, Y.; Yang, B.; Li, Z.; Luo, J.; Gao, X.; Li, Y. Technical Points of Guigang Litchi High Grafting and Closed Orchard Transformation Technology. *South. Hortic.* **2024**, *35*, 57–60.
28. Li, Z.; Fan, G.; Liang, Z.; Niu, C. Workspace Analysis and Experiments of Orchard Platform Based on D-H Method. *Trans. Chin. Soc. Agric. Eng.* **2020**, *36*, 25–34.
29. *GB/T 10000-2023*; Human Dimensions of Chinese Adults. National Standards of the People's Republic of China: Bejing, China, 2023.
30. *GB/T 14172-2021*; Bench Test Methods of Static Roll Stability for Motor Vehicles, Trailers and Combination Vehicles. National Standards of the People's Republic of China: Beijing, China, 2021.
31. Liu, J. Frame Design and Optimization of an Orchard Picking Operation Platform. Master's Thesis, Shandong Jiaotong University, Jinan, China, 2023.
32. Lee, S.; Kim, J.; Park, W. Structural analysis for the development of a vertically raise type aerial work machinery. *J. Korea Inst. Inf. Electron. Commun. Technol.* **2017**, *10*, 225–231.
33. Jiang, Y.; Sun, Z.; Wang, R.; Xia, C.; Ye, Q.; Guo, Y. Design and performance test of the omnidirectional leveling system for crawler work machine in hilly areas. *Trans. Chin. Soc. Agric. Eng.* **2023**, *39*, 64–73.

agriculture

MDPI

Article

Severity Estimation of Inter-Turn Short-Circuit Fault in PMSM for Agricultural Machinery Using Bayesian Optimization and Enhanced Convolutional Neural Network Architecture

Mingsheng Wang [1,*], Wuxuan Lai [2], Peng Sun [2], Hong Li [1] and Qiang Song [2,*]

[1] College of Mechanical Electrification Engineering, Tarim University, Alar 843300, China; 120140003@taru.edu.cn
[2] National Engineering Laboratory for Electric Vehicles, Beijing Institute of Technology (BIT), Beijing 100081, China; 3120195251@bit.edu.cn (W.L.); sydney_peng@126.com (P.S.)
* Correspondence: wangmingsheng_08@163.com (M.W.); songqiang@bit.edu.cn (Q.S.)

Abstract: The permanent magnet synchronous motor (PMSM) is a key power component in agricultural machinery. The harsh and variable working environments encountered during the operation of agricultural machinery pose significant challenges to the safe operation of PMSMs. Early diagnosis of inter-turn short-circuit (ITSC) faults is crucial for improving the safety of the motor. In this study, a fault diagnosis method based on an improved convolutional neural network (CNN) architecture is proposed, featuring two main contributions. First, a dilated convolutional neural network is combined with residual structures, multi-scale structures, and channel attention mechanisms to enhance the training efficiency of the model and the quality of feature extraction. Second, Bayesian optimization algorithms are applied for the automatic tuning of architecture hyperparameters in deep learning models, achieving automatic optimization of the hyperparameters for the fault diagnosis model of ITSCs. To validate the effectiveness of the proposed algorithm, 17 simulated tests of ITSC fault severities were conducted under both constant conditions and dynamic conditions. The results show that the proposed model achieves the best performance regarding the validation accuracy (98.2%), standard deviation, *F*1 scores, and feature learning capability compared to four other models with different architectures, demonstrating the effectiveness and superiority of the algorithm.

Keywords: agricultural mechanization; fault diagnosis; permanent magnet synchronous motors (PMSMs); inter-turn short-circuit (ITSC) fault; Bayesian optimization

1. Introduction

Since the 21st century, the world has been experiencing rapid changes due to issues such as global, ecological, and climate shifts, along with population growth [1]. Environmental protection and food security have garnered increasing attention, creating an urgent need for safe, intelligent, and sustainable solutions [2]. Agricultural mechanization plays a vital role in advancing agricultural modernization and sustainable development, making intelligent fault diagnosis research of paramount importance [3]. Agricultural machinery is extensively utilized in all aspects of modern agricultural production, including tillage, fertilization, sowing, and harvesting. Given their vital role in the production process, the efficient operation of these machines directly impacts both the efficiency and yield of agricultural output [4]. Motors are key power components in agricultural machinery, being responsible for providing stable torque and achieving efficient energy conversion, thereby enhancing mechanical efficiency and reducing energy loss, which makes them more environmentally friendly [5]. PMSMs are widely used in agricultural mechanization due to their excellent torque control performance, high power density, and high efficiency, coupled with China's natural advantage in rare earth resources [5]. They are utilized in equipment such as electric tractors, seeders, harvesters, spraying equipment, tillage, and seeding machinery, significantly enhancing the intelligence and automation levels of agricultural production. However, the operating conditions and environment of agricultural machinery pose challenges to the safe operation of PMSMs [6]. The faults of PMSMs can generally be categorized into mechanical faults, permanent magnet faults, and electrical faults [7]. Mechanical faults primarily refer to the failures caused by the damage of mechanical components, such as bearings, rotors, and shafts, with common fault types including eccentricity and bearing failures. Permanent magnet faults refer to irreversible partial or total demagnetization of the permanent magnets fixed on the rotor, which is unique to PMSMs and can be caused by various factors. Electrical faults usually occur due to damage to the stator windings, and the main fault types include open-circuit winding faults, ITSC faults, phase-to-phase short-circuit faults, and winding ground faults. Due to the limited installation space and the high power density requirements of PMSMs in agricultural machinery, the winding design becomes highly compact, which poses significant challenges for the heat dissipation of the motor windings. Furthermore, the operating environment of agricultural machinery is particularly harsh and variable, including exposure to dust, high temperatures, high humidity, complex modal vibrations, as well as frequent instantaneous overloads and fluctuating loads [8]. These factors make ITSC faults one of the most common failures in PMSMs [9]. The occurrence of these faults generates significant fault currents within the short windings, which not only affects the distribution of the air gap's magnetic field and exacerbates motor vibrations but also causes excessive heat generation in the affected windings. If these issues are not detected and addressed in time, it can lead to a rapid increase in the stator winding temperature, damaging the insulation of nearby windings and further worsening the fault condition [10]. This may even result in a loss of control over the motor and agricultural machinery, leading to catastrophic accidents and significant economic losses. Therefore, it is crucial to diagnose and address ITSC faults in their early stage.

Traditional fault diagnosis methods that rely heavily on regular maintenance and experience not only fail to provide early warnings but also are inefficient and costly [11]. Thanks to the advancements in computer technology and sensor technology, intelligent fault diagnosis methods have received widespread attention and application in recent years [12,13]. Jiang et al. implemented fault diagnosis for the rolling bearings of a combine harvester using an improved variational modal decomposition (VMD) and machine learning method, with experimental results demonstrating the superiority of this approach [14]. Parvin proposed a transformer neural network (TNN) model for diagnosing the severity of ITSC faults [15]. By employing a multihead attention mechanism, this algorithm enables the model to concentrate on specific aspects of the input signals, achieving an experimental accuracy exceeding 96%. Li et al. used the correlation coefficient of permutation entropy as an evaluation index, combining random forest algorithms with support vector machines to identify the engine state of a tractor [16]. Their experiments show that this algorithm has good recognition accuracy under small sample conditions. Fan et al. implemented a sparse classification framework for the composite fault diagnosis of tractor bearings, utilizing adaptive feature dictionary learning to automatically extract fault features, which improved the accuracy of fault state identification under heavy noise conditions [17]. Lee et al. proposed an ITSC fault diagnosis model by combining an attention mechanism with a recurrent neural network (RNN) to realize the fault severity estimation [18]. Xu et al. used Time Generative Adversarial Networks (Time GANs) for data augmentation to overcome the issue of limited fault samples and combined it with transformers to perform fault diagnosis of tractor transmission systems [19].

Despite significant achievements in research on fault diagnosis using machine learning algorithms, there has been limited study on early fault diagnosis, let alone the early diagnosis of ITSC faults [20]. In ITSC faults, early fault diagnosis is crucial, as overcurrent and overheating can lead to more severe issues. The existing ITSC fault models inadequately consider the impact of the coil structure within the winding on the fault model, failing to accurately reflect the relationship between winding parameters and fault severity [21]. Moreover, the three-phase current signals utilized are generally lengthy one-dimensional signals that are highly susceptible to electromagnetic interference and can change with varying operating conditions [22]. Consequently, accurately diagnosing ITSC faults requires the extraction of more profound and higher-dimensional features from the collected current signals, particularly when dealing with signals under dynamic operating conditions. This necessitates that the deep learning models employed have sufficient network depth and complexity [23]. However, tests indicate that when the depth of the model increases to a certain extent, its performance tends to saturate and then rapidly decline, which is different from overfitting [24]. Therefore, as the network depth increases, some performance degradation issues will arise. Additionally, the automatic tuning of hyperparameters for the network model is another pressing problem that needs to be addressed. The hyperparameters for the network architecture and training in the aforementioned studies largely rely on manual tuning based on experience, which can consume a significant amount of time and computational resources, even for those with considerable experience.

To address the aforementioned issues, a novel Bayesian optimization-based improvement algorithm was proposed for the enhancement of the ITSC fault diagnosis model. The primary improvements of this paper are outlined as follows:

(1) By conducting a mechanism analysis of PMSMs, this study investigates the relationship between the parameters of different winding components and the severity of ITSC faults. It proposes a fault model for ITSCs that considers the winding coil structure, as well as indicators that can be used to guide the setting of the severity of ITSC faults.
(2) A well-crafted deep learning network is proposed, which incorporates residual structures, multi-scale structures, and channel attention mechanisms. This network utilizes dilated convolutions for signal feature extraction, employs residual structures to enhance learning efficiency, and leverages multi-scale structures to enrich the scale of extracted features. Finally, the channel attention mechanism is used to adjust the weight of effective features in fault recognition, thereby improving the accuracy of fault severity identification.
(3) The Bayesian optimization algorithm is employed to address the tuning of hyperparameters for the fault diagnosis, enabling the automatic optimization of the model's hyperparameters. Building upon the automatic optimization of model training hyperparameters using Bayesian optimization, the network's feature extraction layers are divided into a three-layer architecture, integrating three improved CNN structures to achieve automatic optimization of the model architecture hyperparameters.
(4) The effectiveness of the proposed fault diagnosis method was evaluated through simulated ITSC fault tests conducted under both constant and dynamic operating conditions. By comparing it with five other fault diagnosis models of different structures, the advantages of the proposed method were validated.

The remainder of this paper is structured as follows: Section 2 presents the ITSC model that considers the winding coil structure and derives an index that can be used to set ITSC fault parameters. Section 3 introduces the proposed algorithm model along with the structure and components of each part. Section 4 describes the experimental equipment used and the settings required for simulating fault tests, as well as detailing the generated dataset. In Section 5, the fault diagnosis model proposed in this paper is compared with five other models of different structures, with experimental results demonstrating the effectiveness and superiority of the proposed algorithm. Finally, Section 6 summarizes the work presented in this paper and discusses future improvements.

2. ITSC Fault in PMSMs

The estimation of ITSC faults is critically important for two main reasons. On one hand, these faults are very difficult to detect in their early stages [25]. On the other hand, an ITSC fault can lead to overcurrent and overheating, which can cause more severe issues [26]. In previous research, no index is particularly suitable for the estimation of an early-stage ITSC fault. In this paper, an equivalent circuit model is proposed, and an index is derived to guide the setting of the ITSC fault severity in experiments.

Currently, the winding structure of a PMSM mostly uses distributed winding arrangements. The coils are wound into appropriate shapes and distributed across two stator slots with a specific pitch. When an ITSC fault occurs in a few turns of the coil within a particular slot, the wires within the corresponding slot will also be shorted, as shown in Figure 1a. Figure 1a is a cross-sectional view of a PMSM with 8 poles and 36 slots. Every turn of the wire within the slot is labeled as Pc-t. For example, A1-3 denotes the 3rd turn wire of the first coil within winding phase A. The red section of the stator winding in the figure indicates the location where the ITSC fault happens, and the corresponding enlarged view shows the labels of the wires involved in the short circuit. Assuming an ITSC fault occurs in the first coil of winding phase A, the schematic diagram of the equivalent circuit model is shown in Figure 1b. From the figure, it can be seen that after the fault occurs, the faulty phase winding will be divided into two parts. One part is the shorted section, and the other is the remaining healthy section. Additionally, the winding of the shorted section will form a new closed loop at the point of the shorted wires. When the current of phase A winding flows through the newly formed closed loop, it divides into the current i_f passing through the fault resistance R_f and the current (i_a–i_f) passing through the shorted winding. Let N_c be the number of coils in each phase winding, N_t be the number of turns per coil, and N_s be the number of turns shorted in the case of an ITSC fault. The degree of winding shorted can be expressed as:

$$\mu = \frac{N_s}{N_c N_t} \tag{1}$$

where μ indicates the proportion of shorted turns in the fault phase winding relative to the total number of turns in that phase winding. Based on the above analysis, the description of the equivalent circuit model is as follows:

$$\boldsymbol{V}_{abcn} = \boldsymbol{R}_{abcf}\boldsymbol{I}_{abcf} + \frac{d}{dt}(\boldsymbol{L}_{abcf}\boldsymbol{I}_{abcf}) + \boldsymbol{e}_{abcf} \tag{2}$$

where

$$\begin{aligned}
\boldsymbol{V}_{abcn} &= \begin{bmatrix} v_{an} & v_{bn} & v_{cn} & 0 \end{bmatrix}^T \\
\boldsymbol{R}_{abcf} &= \begin{bmatrix} R_{ah} + R_{af} & & & R_{af} \\ & R_b & & \\ & & R_c & \\ R_{af} & & & R_{af} + R_f \end{bmatrix} \\
\boldsymbol{I}_{abcf} &= \begin{bmatrix} i_a & i_b & i_c & -i_f \end{bmatrix}^T \\
\boldsymbol{L}_{abcf} &= \begin{bmatrix} L_{ah} + L_{af} + 2M_{ahf} & M_{ahb} + M_{afb} & M_{ahc} + M_{afc} & L_{af} + M_{ahf} \\ M_{ahb} + M_{afb} & L_{bb} & M_{bc} & M_{afb} \\ M_{ahc} + M_{afc} & M_{bc} & L_{cc} & M_{afc} \\ L_{af} + M_{ahf} & M_{afb} & M_{afc} & L_{af} \end{bmatrix} \\
\boldsymbol{e}_{abcf} &= \begin{bmatrix} e_{fah} + e_{faf} & e_{fb} & e_{fc} & e_{faf} \end{bmatrix}^T = \tfrac{d}{dt}\begin{bmatrix} \Psi_{fah} + \Psi_{faf} & \Psi_{fb} & \Psi_{fc} & \Psi_{faf} \end{bmatrix}^T
\end{aligned}$$

In the formula, R_{ah}, R_{af}, and R_f represent the resistance of the remaining healthy portion, the resistance of the shorted portion, and the fault resistance at the shorted point in fault phase winding A, respectively. i_a, i_b, and i_c represent the current flowing through phase winding A, phase winding B, and phase winding C, respectively. v_{an}, v_{bn}, and v_{cn} represent the voltages of the three-phase windings with respect to the neutral point. i_f represents the current flowing through the fault resistance. L_{aa} and L_{bb} denote the self-inductance of phase windings B and C, respectively. L_{ah} and L_{af} denote the self-inductance of the remaining healthy portion and the shorted portion of fault phase winding A, respectively. M_{bc} indicates the mutual inductance between phase windings B and C. M_{ahf} indicates the mutual inductance between the two portions of fault phase winding A. M_{ahb} and M_{afb} represent the mutual inductance between the two portions of fault phase winding A and phase winding B, respectively. M_{ahc} and M_{afc} denote the mutual inductance between the two portions of fault phase winding A and phase winding C, respectively. e_{fah} and e_{faf}

represent the induced electromotive forces generated by the permanent magnet in the shorted part and the remaining healthy part of phase winding A, respectively. e_{fb} and e_{fc} represent the induced electromotive forces generated by the permanent magnet in phase winding B and phase winding C, respectively. Ψ_{fah} and Ψ_{faf} represent the flux linkage of the healthy portion and the shorted portion of fault phase winding A, respectively, associated with the permanent magnet. Ψ_{fb} and Ψ_{fc} represent the flux linkage of phase winding B and phase winding C induced by the permanent magnet, respectively.

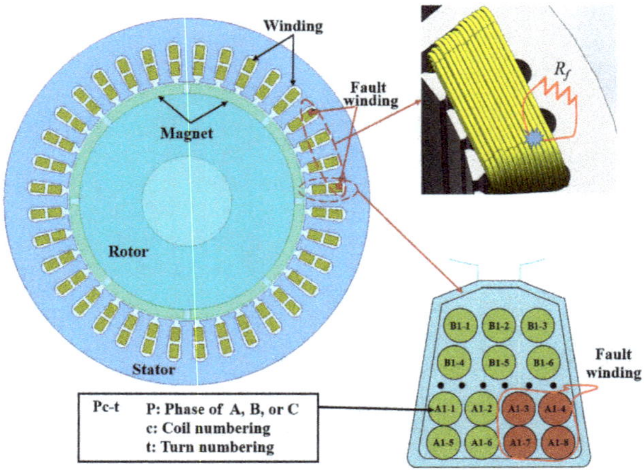

(**a**)

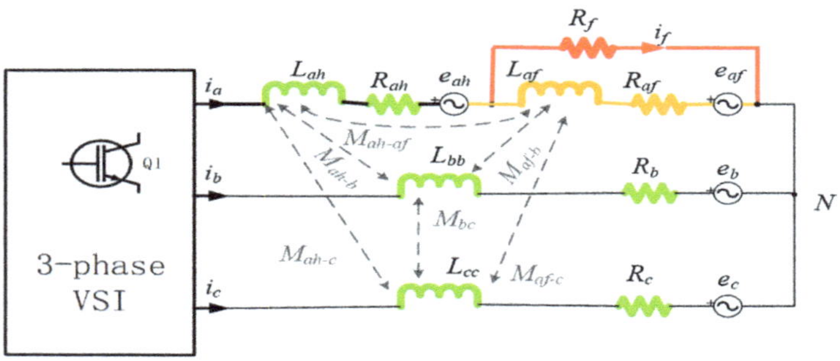

(**b**)

Figure 1. (**a**) Schematic representation of a motor cross-section with an ITSC fault. (**b**) Equivalent circuit diagram of the motor with an ITSC fault.

Determining the parameters in the fault model is a crucial step for modeling and studying different fault states of the motor. The resistances of the healthy portion and the shorted portion are proportional to the contribution of each part to the total number of turns in the faulted phase winding. The expressions are as follows:

$$\begin{aligned} R_{af} &= \mu R_a \\ R_{ah} &= (1-\mu) R_a \end{aligned} \quad (3)$$

where R_a stands for the resistance of phase winding A when there is no ITSC fault.

The flux linkage of the permanent magnet in the winding is proportional to the number of turns of the winding. The flux linkages of the healthy portion and the shorted portion of the fault phase winding are represented as follows:

$$\begin{aligned} \Psi_{faf} &= \mu \Psi_f \\ \Psi_{fah} &= (1-\mu)\Psi_f \end{aligned} \tag{4}$$

where Ψ_f stands for the flux linkage of the permanent magnet in phase winding A when there is no ITSC fault.

In the fault model of ITSCs in PMSMs, determining the parameters for the stator winding's self-inductance and mutual inductance is the most complex part. This complexity arises from the changes in the magnetic field caused by the presence of the ITSC fault. The stator winding of a motor is typically composed of multiple coils, as shown in Figure 2. For each coil within a given phase winding, it is necessary to separately discuss the coil's self-inductance, the mutual inductance between this coil and other coils within the same phase winding, and the mutual inductance between this coil and different coils in the rest of the phase windings.

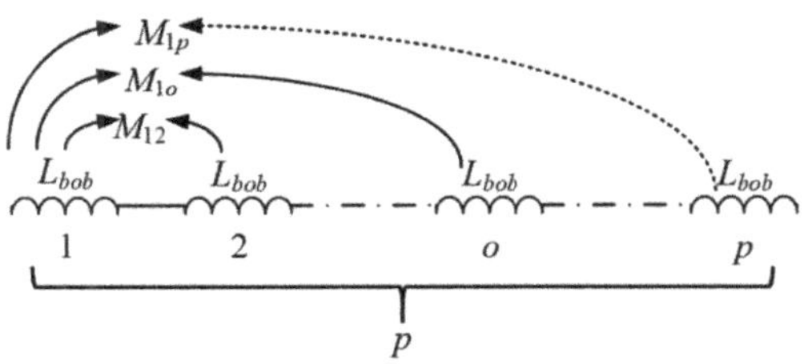

Figure 2. Schematic diagram of the coil composition within the phase winding and the mutual inductance relationships between the coils.

When studying the relationship between the mutual inductance of a coil within a given phase winding and another phase winding, and if the fault occurs only within a single coil, the mutual inductances between the two portions of the fault coil with another phase winding are described by the following equations:

$$\begin{aligned} M &= \sum_{i=1}^{N_c} M_{ip} = M_{afp} + M_{ahp} \\ M_{afp} &= \eta M_{1p} \\ M_{ahp} &= (1-\eta)M_{1p} + \sum_{i=2}^{N_c} M_{ip} \end{aligned} \tag{5}$$

where M stands for the mutual inductance between the given phase winding and another phase winding. M_{ip} stands for the mutual inductance between the i-th coil within the given phase winding and another phase winding. M_{afp} represents the mutual inductance between the shorted wires within the fault coil and another phase winding, while M_{ahp} represents the mutual inductance between the remaining unshortened wires of the fault phase winding and another phase winding.

When the fault occurs in more than one coil, assuming that the ITSC fault occurs in the first n coils, where $n \geq 2$, and the first $n-1$ coils are also shorted, the mutual inductances between the two portions of coils within the fault phase winding with another phase winding are described by Equation (6), The meanings of the parameters are consistent with those described earlier.

$$\begin{aligned} M &= \sum_{i=1}^{N_c} M_{ip} = M_{afp} + M_{ahp} \\ M_{afp} &= \sum_{i=1}^{n-1} M_{ip} + (\mu N_c - n + 1) M_{np} \\ M_{ahp} &= (n - \mu N_c) M_{np} + \sum_{i=n+1}^{N_c} M_{ip} \end{aligned} \tag{6}$$

When studying the self-inductance and mutual inductance relationships between coils in a phase winding, since each phase winding is composed of multiple coils connected in series, and assuming a symmetrical distribution of stator winding coils, the self-inductance of each coil is essentially consistent. However, the mutual inductance between coils is related to their relative positions. Let L_{bob} denote the self-inductance of a single coil in the phase winding, and M_{ij} denote the mutual inductance between two coils in the same winding, which depends on their relative positions, as described by Expression (7).

$$\begin{gathered} M_{ij} = M_{ji} \,\&\, M_{ij} = M_{kl},\ if |i-j| = |k-l| \ or\ |i-j| = |k-l-n| \ or\ |i-j| = |k-l+n| \\ i,j,k,l \in Z, 1 \leq i,j,k,l \leq N_c \,\&\, i \neq j, k \neq l \end{gathered} \tag{7}$$

Here, i, j, k, and l represent the positions of each coil in the A-phase winding.

Based on the above analysis, the inductance of each coil can be described as:

$$L = N_c\left(L_{bob} + \sum_{i=1}^{N_c-1} M_{ij}\right) \tag{8}$$

where $\sum_{i=1}^{N_c-1} M_{ij}$ represents the mutual inductance between the chosen coil and the remaining coils in the same winding. L denotes the self-inductance of the phase winding. Assuming an ITSC fault occurs on the first coil of phase winding A under the condition of no distinction and neglecting the leakage inductance between the wires within the coil, the inductance between the coils in the fault phase winding satisfies the following relationship:

$$\begin{aligned} & L_{bobf} = \eta^2 L_{bob} = (\mu N_c)^2 L_{bob} \\ & L_{bobh} = (1-\eta)^2 L_{bob} = (1 - N_c\mu)^2 L_{bob} \\ & M_{bobfh} = \eta(1-\eta) L_{bob} = N_c\mu(1 - N_c\mu) L_{bob} \\ & L_{bobf} + 2M_{bobfh} + L_{bobh} = L \end{aligned} \tag{9}$$

where L_{bobf} represents the self-inductance of the shorted wires within the fault coil. L_{bobh} represents the self-inductance of the unshortened wires within the fault coil. M_{bobf} stands for the mutual inductance between the shorted wires and unshortened wires within the fault coil. The mutual inductances between the two portions of the fault coil and the other remaining coils within the fault winding satisfies the relationship:

$$\begin{aligned} M_{bobf} &= \eta \sum_{i=1}^{N_c-1} M_{ij} = N_c\mu \sum_{i=1}^{N_c-1} M_{ij} \\ M_{bobh} &= (1-\eta) \sum_{i=1}^{N_c-1} M_{ij} = (1 - N_c\mu) \sum_{i=1}^{N_c-1} M_{ij} \end{aligned} \tag{10}$$

where M_{bobf} represents the mutual inductance between the shorted wires within the fault coil and the other remaining coils within the fault winding. M_{bobh} represents the mutual inductance between the unshortened wires within the fault coil and the other remaining coils within the fault winding.

Based on the above analysis, the inductance of each portion in the fault winding can be described as:

$$
\begin{aligned}
L_{af} &= (\mu N_c)^2 L_{bob} \\
L_{ah} &= (1 - N_c\mu)^2 L_{bob} + (n_p - 1)(L_{bob} + \sum_{i=1}^{N_c-2} M_{ij}) + 2(1 - N_c\mu)\sum_{i=1}^{N_c-1} M_{ij} \\
M_{afh} &= N_c\mu(1 - N_c\mu)L_{bob} + N_c\mu \sum_{i=1}^{N_c-1} M_{ij}
\end{aligned}
\tag{11}
$$

Substituting Equations (3)~(11) into Equation (2), the resistance, inductance, and back electromotive force in the voltage balance voltage equation under ITSC fault conditions can be described as:

$$
\begin{aligned}
\boldsymbol{R}_{abcf} &= \begin{bmatrix} R_a & & & \mu R_a \\ & R_b & & \\ & & R_c & \\ \mu R_a & & & \mu R_a + R_f \end{bmatrix} \\
\boldsymbol{L}_{abcf} &= \begin{bmatrix} L_{aa} & M_{ab} & M_{ac} & L_{af} + M_{ahf} \\ M_{ab} & L_{bb} & M_{bc} & \mu M_{ab} \\ M_{ac} & M_{bc} & L_{cc} & \mu M_{ac} \\ L_{af} + M_{ahf} & \mu M_{ab} & \mu M_{ac} & L_{af} \end{bmatrix} \\
\boldsymbol{e}_{abcf} &= \begin{bmatrix} e_{fa} & e_{fb} & e_{fc} & e_{ff} \end{bmatrix}^T = \frac{d}{dt}\begin{bmatrix} \Psi_{fa} & \Psi_{fb} & \Psi_{fc} & \mu\Psi_{fa} \end{bmatrix}^T
\end{aligned}
\tag{12}
$$

The e_{fa} represents the induced electromotive forces generated by the permanent magnet in phase winding A. Ψ_{fa} represents the flux linkage of phase winding A induced by the permanent magnet.

Since the analyzed stator winding is Y-connected, it follows, from Kirchhoff's Current Law, that:

$$
i_a + i_b + i_c = 0 \tag{13}
$$

From Equations (2), (12), and (13), the expression for the fault current can be derived as:

$$
i_f = \frac{\mu v_{an} + (\mu(L_{af} + M_{ahf}) - L_{af})\frac{di_f}{dt}}{\mu R_a + Rf - \mu^2 R_a} \tag{14}
$$

Let $d_1 = \mu R_a + R_f - \mu^2 R_a$, $d_2 = \mu(L_{af} + M_{ahf}) - L_{af}$, and $v_{an} = v_a - v_n$; then, the above equation can be rewritten as:

$$
\frac{di_f}{dt} = \frac{1}{d_2}(d_1 i_f) - \frac{1}{d_2}\mu(v_a - v_n) \tag{15}
$$

Since the focus of the study is on the early stage of ITSC faults, the amplitude of voltage v_n is much smaller than that of v_a, so $v_a \approx v_a - v_n$. Assuming $v_a = V_a \sin(\omega t)$, the analytical solution of Equation (15) can be described as:

$$
i_f = e^{\frac{d_1 t}{d_2}}\left(i_f(0) - \frac{\mu V_a \omega}{d_2(\omega^2 + \frac{d_1^2}{d_2^2})}\right) + \frac{\mu V_a(\omega\cos(\omega t) + \frac{d_1 \sin(\omega t)}{d_2})}{d_2(\omega^2 + \frac{d_1^2}{d_2^2})} \tag{16}
$$

At the early stage of an ITSC fault, the fault usually occurs in a single coil, there are fewer shorted turns of wires, and the fault resistance at the shorted point is relatively large. Therefore, $d_1 > 0$, $d_2 < 0$, and $| d_1 | >> | d_2 |$. As a result, d_1/d_2 tends towards $-\infty$, and d_2/d_1 tends towards 0. Thus, from Equation (16), the approximate expression for the current amplitude can be obtained:

$$
I_f \approx \frac{\mu V_a}{\mu R_a + R_f - \mu^2 R_a} \tag{17}
$$

According to reference [27], it is known that the amplitude of the three-phase voltage in the stator winding of the PMSM is positively correlated with the motor speed. Therefore, Equation (17) can be rewritten as:

$$I_f \approx K \frac{\mu \omega_r}{\mu R_a + R_f - \mu^2 R_a} \tag{18}$$

where ω_r represents the mechanical speed of the PMSM. K represents a known coefficient. By analyzing the above equation, it can be seen that the resistance of fault phase winding R_a can be regarded as a known quantity in the equation, and the remaining parameters μ, R_f, and ω_r can directly affect the amplitude of the fault current i_f. However, among these parameters, μ and R_f are related to the severity of the ITSC fault, while ω_r is not. If ω_r is excluded from Equation (18), an expression related only to the shorted degree μ and the fault resistance R_f will be derived:

$$\frac{I_f}{\omega_r} \approx K \frac{\mu}{\mu R_a + R_f - \mu^2 R_a} := FI \tag{19}$$

where FI stands for the severity index of the ITSC fault. When the tested motor is in a healthy state, this index is 0. When the winding of a certain phase of the motor is completely shorted and the fault resistance is 0, this index becomes infinite.

In the early stages of an ITSC fault, this index is essentially unaffected by speed and increases as the fault resistance R_f decreases or the degree of shorted turns μ increases, and vice versa. Each fault severity can be considered as a combination of different R_f and μ values. Of course, in actual motor operation, it is difficult to directly detect the fault resistance R_f and the degree of shorted turns μ, so this severity index is not suitable for estimating the severity of an ITSC fault. However, it can be used as an index for fault severity in experiments to guide the setting of ITSC fault severity.

3. Proposed Algorithm

3.1. Convolutional Neural Networks

A CNN is an evolved form of artificial neural networks, currently widely used in image processing and fault diagnosis. A typical convolutional neural network structure is generally a multi-layer feedforward neural network composed of an input layer, convolutional layer, and output layer [2]. The convolutional layer typically needs to be used in conjunction with various functional layers, such as pooling layers, activation layers, normalization layers, and dropout layers, to enhance the performance of the convolutional module [28]. A deep convolutional network structure is formed by stacking multiple convolutional modules, with its complete structure shown in Figure 3. Compared to traditional artificial neural networks, the convolutional layer of a CNN has the characteristics of weight sharing and local connections, which significantly reduces the number of model parameters and lowers the difficulty of training the model. A CNN model with multiple hidden layers can automatically extract multiple features from the input signal, where the lower hidden layers learn the generalized features of the input signal, and the higher hidden layers can obtain more abstract, high-dimensional features through the abstraction and extraction of lower-level features, enabling more precise classification tasks.

In this research, the dilated CNN is adopted due to its ability to enlarge the receptive field while minimizing the loss of resolution, which is an important quality for forming a deep network architecture [29]. The process of dilated convolution can be expressed as:

$$F(x) = (S *_d f)(x) = \sum_{i=0}^{k-1} f(i) \cdot S_{x\text{-}d\cdot i} \tag{20}$$

where S stands for the 1-dimensional series data of the input signal and $S \in R^n$, f represents the convolutional kernel and f: {0, 1,. . ., k − 1} → R, d denotes the size of dilation, k stands for the convolutional kernel size, and $x\text{-}d\cdot i$ represents the elements of segment x undergoing the i-th iteration of the operation. Conventional convolution is a special case where the dilation factor $d = 1$. As the network depth increases, the dilation factor also grows, and the receptive field of the final output layer expands accordingly.

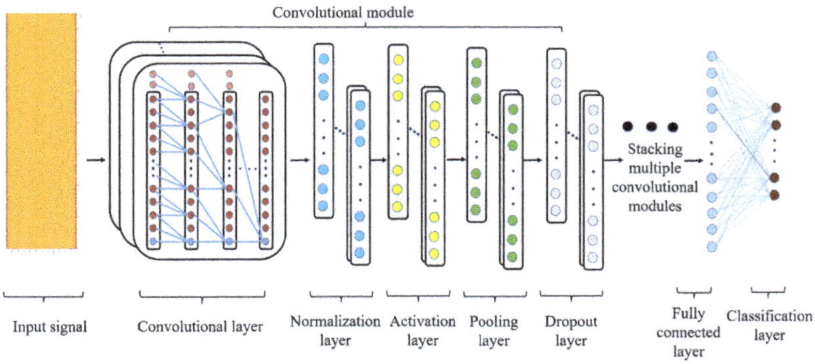

Figure 3. Schematic diagram of a conventional CNN model structure.

3.2. *Improvement Measures for Network Architecture*

In the early stages of CNN development, scholars generally believed that increasing the width and depth of the network could enhance the model's fitting ability. Therefore, when designing models, they aimed to improve the performance by continuously expanding the width and depth of the network. However, as the complexity of models increased, several common issues in deep learning began to emerge, such as overfitting, gradient explosion or vanishing problems, and network degradation [30]. To address these issues, improvement measures such as residual network structures, multi-scale network structures, and attention mechanisms have gained increasing attention in the design of deep learning models.

3.2.1. Residual Neural Network

Residual neural networks (Resnets) are mainly used to solve issues such as gradient vanishing or explosion and network degradation during the training process of deep learning models [28]. Theoretically, due to the convolution operation, as the number of network layers increases, the feature information extracted by the model is progressively compressed layer by layer. If the network depth becomes too large, the extracted features will be overly compressed, which will affect the final recognition accuracy, leading to the network degradation problem [24]. In theory, if the newly added layers simply repeat the features from the previous layer without learning new features (called identity mapping), the model's performance will neither improve nor decline. Inspired by this, He et al. introduced identity mapping between branches of different depths in the network, ensuring that the subsequent layers contain more enriched feature information than the previous ones, thereby addressing the degradation issue caused by an increasing network depth [31].

For a residual network structure, let the input be x, and after passing through the residual branch, the learned feature representation is $F(x)$. By directly connecting x and $F(x)$ through identity mapping and integrating them, y represents the total output of the residual network structure. This process can be expressed as:

$$y = x + F(x) \tag{21}$$

A schematic diagram of the residual network structure module is shown in Figure 4. This module consists of ReLU activation layers, batch normalization layers, dilated convolution layers, and dropout layers. In a standard residual network structure, the input x is directly added to the features extracted by the residual branch, $F(x)$, as shown in Figure 4a.

However, if there is a dimensional inconsistency between the two, a 1 × 1 convolution is required in the residual branch to match their dimensions, followed by feature addition, as shown in Figure 4b. If the feature information after the residual branch is not zero, the model's performance can be enhanced by adding more layers. Conversely, if the feature information after the residual branch is zero, the model's performance will neither improve nor degrade. Therefore, increasing the network depth through residual structures can avoid the network degradation problem.

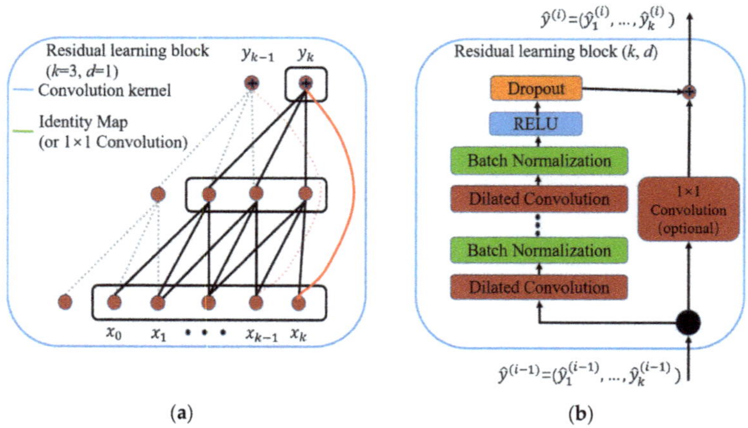

Figure 4. (**a**) An example of a residual neural network. The black line represents the convolutional kernel in the Resnet, and the green line represents the identity mapping in the Resnet. (**b**) Residual neural network. When the input and output of the Resnet have different dimensions, a 1 × 1 convolution is added.

3.2.2. Multi-Scale Kernel Network

The core of the multi-scale kernel network is the Inception module. Before its introduction, most CNN models could only enhance their performance by increasing the number of convolution kernels or the depth of convolutional layers. However, this not only leads to a significant increase in the computational burden but may also cause overfitting or even network degradation [32]. To address this issue, the Inception module was proposed, with its structure shown in Figure 5. As illustrated, when data flow into the Inception module, convolution operations are performed simultaneously in multiple convolution kernels of various sizes in parallel branches, extracting features at different scales from the input data [33]. These features are then adjusted to a consistent dimension for concatenation and integration. Thanks to this characteristic, the network structure enhances the model's perceptive ability while also using 1 × 1 convolution kernels to reduce the number model's parameters and overall computational burden.

There are a total of four parallel branches in the Inception module. The orange boxes represent the dilated CNN modules, with the convolution kernel sizes from right to left being 1 × 5, d = 5; 1 × 3, d = 3; and 1 × 3, d = 1. The other boxes represent conventional CNN modules, with convolution kernel sizes of 1 × 5, 1 × 3, and 1 × 1, respectively. Since the features extracted by the dilated CNN are discontinuous, the introduction of conventional CNNs in each branch can supplement the types of extracted features. After the input data pass through convolution kernels of different sizes and types, features at various scales can be extracted. Let X represent the input information to this module, and then the expressions of the feature vectors extracted at different scales by the three branches are as follows:

$$\begin{aligned} X_1 &= X * C_{1\times1} * D_{f=1} \\ X_2 &= X * C_{1\times1} * C_{1\times3} * D_{f=3} \\ X_3 &= X * C_{1\times1} * C_{1\times5} * D_{f=5} \end{aligned} \quad (22)$$

Here, X_1, X_2, and X_3 represent the feature vectors extracted by different branches, the symbol $*$ denotes the convolution operation, $C_{1\times1}$, $C_{1\times3}$, and $C_{1\times5}$ represent conventional convolutions, and $D_{f=1}$, $D_{f=3}$, and $D_{f=5}$ represent dilated convolutions.

In this parallel branch structure, features at different scales are fused through stacking, so the feature vector output by this module can be expressed as:

$$Y = \{X + X_1, X + X_1 + X_2, X + X_1 + X_2 + X_3\} \tag{23}$$

Here, the symbol + represents the element-wise addition of corresponding elements in the feature vectors, and the symbol · represents the concatenation operation of feature vectors from different branches.

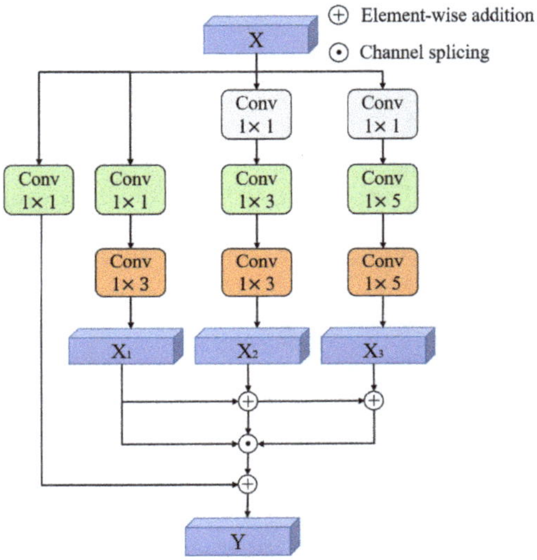

Figure 5. Schematic diagram of a multi-scale network structure.

From the structure of the Inception module, it can be seen that the residual network can serve as a branch of the multi-scale parallel structure connected to this module. Therefore, after the input information passes through this module, the output feature vector not only contains features extracted by dilated convolutions with different kernel sizes and dilation factors but also includes features extracted by conventional convolutions with different kernel sizes. This gives the model a richer ability to represent multi-scale features.

3.2.3. The Attention Mechanism

Recent research has shown that in addition to network structure parameters and training parameters, another factor affecting the performance of deep learning models, the attention mechanism, has also been receiving increasing attention [34]. The essence of the attention mechanism is to guide the model toward the task objective by adjusting the weights, thereby filtering out information irrelevant to the task and focusing the model's "attention" on the feature information that is more useful for achieving the task objective.

Commonly used attention mechanisms in CNN models mainly include channel attention, spatial attention, and mixed attention. In this paper, the channel attention mechanism is adopted to enhance the performance of the ITSC fault diagnosis model. By adjusting the weights of the input features from different channels, the model's attention is redistributed, thereby adjusting the contribution of different feature channels to the model and improving its recognition accuracy.

Figure 6 shows a schematic diagram of a typical network structure for the channel attention mechanism. This network structure consists of three main steps: squeezing, excitation, and scaling. The squeezing and excitation steps are the core parts of the module, which is why this module is called SENet (Squeeze and Excitation Net) [35]. During model training, SENet continuously adjusts the weight distribution of different feature channels through its internal squeeze and excitation modules. By increasing the weights of important feature channels, SENet highlights their contribution and suppresses the contribution of irrelevant and redundant feature channels.

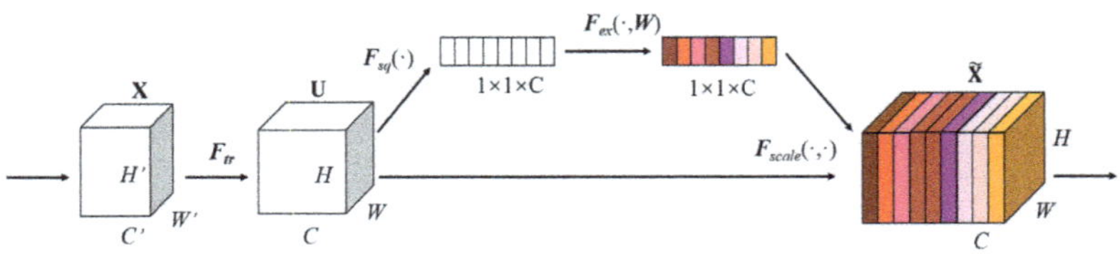

Figure 6. Schematic diagram of the channel attention mechanism.

In the channel attention mechanism, the squeezing operation is performed first. This involves compressing the feature information of the channels to adjust the weight relationships between channels. In this process, the input feature information with dimensions $W \times H \times C$ is converted into $1 \times 1 \times C$ features through global average pooling, transforming the entire spatial feature of all channels into C global features. This process can be expressed as:

$$z_c = F_{sq}(u_c) = \frac{1}{W \times H}\sum_{i=1}^{W}\sum_{j=1}^{H} u_c(i,j) \tag{24}$$

Here, Z_c represents the feature output by the squeezing operation, F_{sq} represents the squeezing operation, and u_c represents the input features. After obtaining global features through the squeezing operation, the next step is excitation to capture the relationships between different channels. In the excitation operation, the compressed global feature information is delivered to a fully connected layer, with the dimension of the fully connected layer being $C \div r \times C$, where r is the scaling factor of a channel, being mainly used to reduce the number of channels. As a result, the parameters and computational complexity of the entire module are correspondingly reduced. The output information of this fully connected layer is passed through a ReLU activation layer and then into a second fully connected layer, where the number of feature channels reduced in the previous fully connected layer is restored in this layer. Throughout this process, the two fully connected layers mainly function to organize the feature information obtained using different feature channels, and the Sigmoid activation layer maps the input feature information of each channel into the range of (0, 1), obtaining the weights corresponding to each channel, thus completing the weight adjustment of the channel attention. This process can be expressed as:

$$s = F_{ex}(z_c, W_i) = \delta(W_2\sigma(W_1 z_c)) \tag{25}$$

where F_{ex} represents the excitation operation, σ represents the ReLU activation function, W_1 and W_2 represent the first and second fully connected layers, respectively, δ represents the Sigmoid activation function, and s represents the weights of each channel.

Once the excitation operation is completed and the weights corresponding to each channel are obtained, the scaling operation is performed on the features of each channel. The features within the channel are multiplied element wise by the corresponding channel weights obtained in the previous excitation operation, thereby completing the recalibration of the original features in the channel dimension and achieving the readjustment of attention across all channel features. The entire scaling process can be expressed as:

$$\tilde{x}_c = F_{scale}(u_c, s_c) = s_c \cdot u_c \tag{26}$$

where $\widetilde{x}_c$ represents the features after channel attention adjustment, and F_{scale} represents the scaling operation.

3.3. Bayesian Optimization Algorithm

The CNN-based ITSC fault diagnosis model has a flexible structure and numerous hyperparameters. The impact of varying these hyperparameters on the model's convergence speed and validation accuracy is difficult to predict, which increases the complexity of model design and parameter tuning. To achieve a high-performance diagnosis model, it is necessary to optimize the combination of hyperparameters. Common optimization algorithms include grid search, random search, and Bayesian optimization. Grid search and random search are exhaustive methods that not only require significant computational resources but also can waste resources on ineffective hyperparameter combinations, especially when computational resources are limited. In contrast, Bayesian optimization is an efficient global optimization algorithm, named due to the use of the well-known Bayes' theorem in its optimization framework [36]. It can adjust the optimization strategy based on prior knowledge of existing data, making it more efficient. Therefore, this study uses Bayesian optimization to fine-tune the hyperparameters of the inter-turn short-circuit fault diagnosis model. The expression of Bayes' theorem is shown in Equation (27). The way Bayesian optimization is applied to the hyperparameter optimization of the ITSC fault diagnosis model is achieved by approximating the posterior distribution of the objective function based on Bayes' theorem. By constructing a mapping relationship between the observed values in the set and the maximum value of the objective function, it searches for the hyperparameter combination that maximizes the objective function.

$$p(f \mid D_{1:t}) = \frac{p(D_{1:t} \mid f)p(f)}{p(D_{1:t})} \tag{27}$$

Here, f stands for the unknown objective function, and in the hyperparameter optimization of the ITSC fault diagnosis model, it refers to the performance metric of the model. $D_{1:t}$ represents the set of observed points, where the number of observed points in the set is t. $p(f \mid D)$ and $p(f)$ denote the posterior and prior probability distributions of f, respectively. $p(D \mid f)$ represents the likelihood distribution of the observed points in the set, and $p(D)$ represents the marginal likelihood distribution of f.

The core of the Bayesian optimization algorithm mainly consists of two parts: the acquisition function and the surrogate model [37]. The surrogate model aims to model the distribution of the unknown objective function using a prior model, thereby finding the optimal parameter combination within the given search space. The acquisition function selects the next observation point to evaluate based on the results obtained from the surrogate model. The framework of the entire Bayesian optimization algorithm is shown in Algorithm 1.

Algorithm 1. Bayesian optimization algorithm framework process.

Input: Surrogate model f, acquisition function α, and two values, t and k.
Output: The optimal combination in the hyperparameter vector X.
1: Initialize D_t;
2: **for** $i = 1, 2, \ldots, k$ **do** //**Iterative search**;
3: Update the initial surrogate model *Surg*;
4: Maximize the acquisition function to select a new evaluation point:
$x_{t+1} = \mathbf{arg\ max}_{x \in X}\ \alpha(x \mid Surg, D_{1:t})$;
5: Evaluate the sample point $y_{t+1} = f(x_{t+1}) + \varepsilon_t$;
6: Update the dataset $D_{t+1} = D_t \cup \{x_{t+1}, y_{t+1}\}$, and update the surrogate model;
7: **end for**.

First, it is necessary to choose the type of surrogate model and acquisition function, determine the maximum number of iterations k for the entire optimization process, and generate t initial sample points as the initialization samples for the Bayesian optimization algorithm. Then, the iterative search process of Bayesian optimization begins. The surrogate model is updated based on the existing sample set D_t, and the potential point x_{t+1} of the next optimal value is calculated based on the optimized acquisition function. The objective function is evaluated at the sample point to obtain new observations, and the surrogate model is updated based on the new observations. This iterative process is repeated until the maximum number of iterations k is reached, at which point the optimal hyperparameter combination is output, completing the entire Bayesian optimization process.

3.4. *Bayesian Optimization-Based Improvement Algorithm for CNN Models*

The performance of the deep learning models used for fault diagnosis is greatly influenced by the model architecture hyperparameters and training hyperparameters. The process of building and training the model involves tuning different combinations of hyperparameters. Selecting the right combination of hyperparameters for model building and training can enhance its performance and improve the accuracy of fault diagnosis identification. However, there are numerous types and quantities of hyperparameters in the model. The architecture hyperparameters of the model include the depth of the CNN mode, the number of convolutional kernels in each layer, the size of the convolutional kernels, and the dropout probability for each dropout layer, among others. Additionally, the type of different layers and the connection methods between different layers will also have a significant impact on the fault diagnosis model. The training hyperparameters of the model include key factors such as the initial learning rate, the decay strategy of learning rate and decay factor, the type of optimizer used, the $L2$ regularization coefficient, and the size of the mini-batch, among others. In deep learning models, due to the large search space, numerous parameters, and the difficulty in representing the connections between layers, architecture hyperparameters are typically set based on experience and are rarely optimized using optimization algorithms, which will increase the uncertainty of the model. In contrast, training hyperparameters are typically determined based on experience to define the optimization range and then optimized using hyperparameter optimization methods.

In this study, to improve the efficiency of hyperparameter tuning and enhance the model's performance, Bayesian optimization is utilized to optimize the architecture hyperparameters and training hyperparameters of the model. To facilitate the optimization of architecture hyperparameters, the model's feature extraction layers are divided into a three-stage architecture based on the characteristics of the features extracted by convolutional networks. The number of convolutional blocks within each structure is an adjustable parameter, and each convolutional block employs the same number of convolutional kernels, which is also an optimizable parameter. The number of layers within each convolutional block is fixed. The optimization range for each architecture hyperparameter is determined by the experience, the characteristics of the model, and the available computer hardware resources. The maximum number of layers for the model is set as d and the maximum allowed width for each layer as w. The maximum number of layers for each stage of the three-stage structure is denoted as d_1, d_2, and d_3, with the maximum allowed width for each structure also being w. The number of combinations for training hyperparameter optimization is $\boldsymbol{n}$. The comparison of computational complexity between the two architectures is shown in Figure 7. It can be observed that the number of hyperparameters has been reduced from $(d + 1)$ dimensions to six dimensions, comparing the parameter amount of a conventional model with the three-stage structure with that of a conventional model structure, which makes the automated hyperparameter optimization possible for the model. In addition, the three-stage structure ensures that the model possesses sufficient flexibility and the ability to represent fault features.

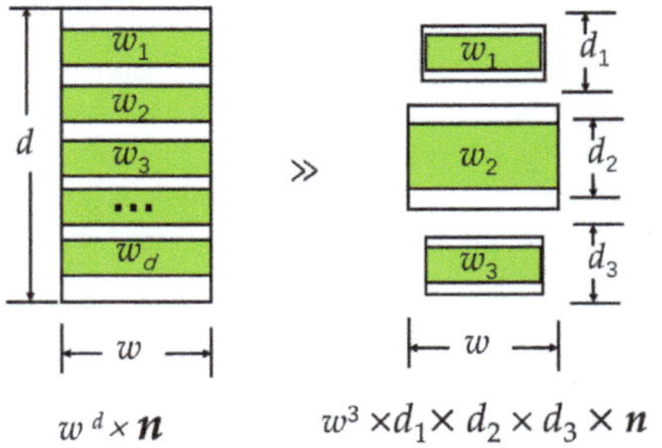

Figure 7. Comparison of the complexity of hyperparameters to be optimized under two architectures.

To achieve time efficiency while ensuring model performance, the previously mentioned Bayesian optimization algorithm with global optimization capabilities is used for automated hyperparameter tuning. The structural diagram of tuning the model's hyperparameters using the Bayesian optimization algorithm is shown in Figure 8, which is the core part of the Bayesian optimization-based improvement algorithm for CNN models. From the figure, it can be seen that the whole process is divided into two parts: one is the Bayesian optimization part (indicated by the blue box) and the other is the training part of the ITSC fault diagnosis model (indicated by the red box). The entire Bayesian optimization process involves the iterative exploration of different hyperparameter combinations until the set stopping condition is met. At that point, the optimization process concludes, selecting the optimal combination of hyperparameters, which corresponds to the best model for ITSC fault diagnosis under the current conditions.

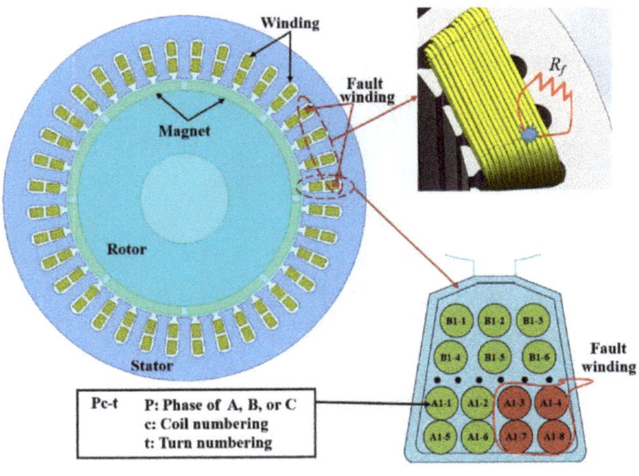

Figure 8. Schematic diagram of hyperparameter tuning for fault diagnosis models based on Bayesian optimization algorithms.

The Bayesian optimization section primarily focuses on two aspects: the initialization of hyperparameter combinations and the updating of model parameters based on the training results feedback from the ITSC fault diagnosis model. The hyperparameter combinations that need to be optimized include architecture hyperparameters and training hyperparameters. Architecture hyperparameters include the depth of convolution blocks in each stage (d_1, d_2, and d_3), the width of convolution blocks in each stage (w_1, w_2, and w_3), and the dropout layer probability (P). Training hyperparameters include the initial learning rate (L_{init}), regularization coefficient (L_{2R}), and gradient optimization coefficient (G_1).

The tasks required for training the ITSC fault diagnosis model include constructing a CNN model according to the architecture hyperparameters provided by Bayesian optimization and training the model based on the given training hyperparameters and dataset. Once the model training is completed, the training results are fed back into the Bayesian optimization algorithm, which then uses the newly obtained hyperparameter combinations to reconstruct and retrain the model. This process of Bayesian optimization and model training continues to iterate until the set termination conditions are met.

The flowchart of the Bayesian optimization-based improvement algorithm for CNN models is shown in Figure 9. The entire process can be divided into five steps:

(1) Data preparation: Set different combinations of fault resistance and shorted ratios to simulate varying severities of faults. Conduct different operating conditions for each fault severity and collect the three-phase currents during the experiment.
(2) Building of dataset: Perform preprocessing operations such as filtering, downsampling, normalization, segmentation, and label classification. Then, divide all data segments into two non-overlapping datasets, namely the training set and the testing set.
(3) Initialization: Determine the structure of the CNN blocks, the combinations of hyperparameters to be optimized, and the search ranges for each parameter. Use Bayesian optimization to select the initialization parameters for the model and construct the network structure of the fault diagnosis model according to the specified parameter set.
(4) Model training and optimization: Train the fault diagnosis model based on the given training hyperparameters, and then test the model's performance using the testing dataset. Record the results obtained along with the corresponding hyperparameter combinations and feed the testing results back into the Bayesian optimization algorithm to update the model's hyperparameters. Repeat the process until the stopping criteria are triggered.
(5) Output result: When the maximum optimization iterations are met, select the best testing accuracy of the ITSC fault diagnosis model and its relevant hyperparameter combination as the results and output them.

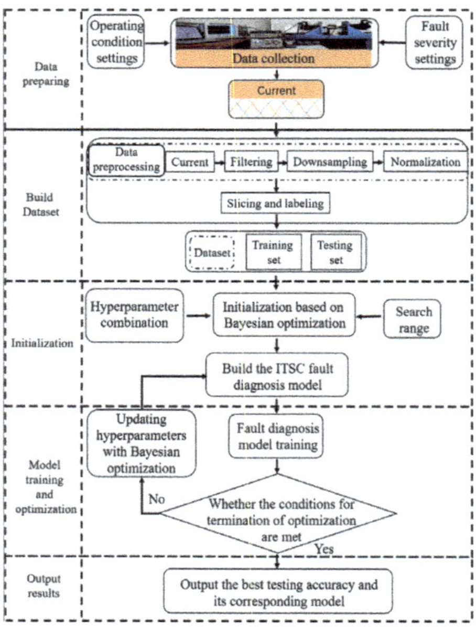

Figure 9. The flowchart of the Bayesian optimization-based improvement algorithm for CNN models.

4. Experimental Setup and Data Description

In order to verify the validity of the proposed Bayesian optimization-based improvement algorithm for the ITSC fault diagnosis model. Experiments are carried out on a PMSM. The setup consists of a simulated fault motor and its controller, an auxiliary test motor and its controller, data acquisition equipment, etc., as shown in Figure 10. The fault motor is an 8-pole, 36-slot PMSM, with the windings configured in a star connection, featuring 108 turns of wire per phase. The specific parameters of the faulty motor are shown in Table 1. The fault motor simulates different severities of ITSC faults by combining different fault resistances and shorted ratios. To prevent damage to the fault resistor, a cooling device is required for heat dissipation during the experiment. Temperature monitoring of the entire setup is conducted during the experiment to prevent overheating and damage. The fault resistor and its cooling device are shown in Figure 10c, the fault motor and its shorted winding point terminals are shown in Figure 10b, and the temperature measurement device is shown in Figure 10d.

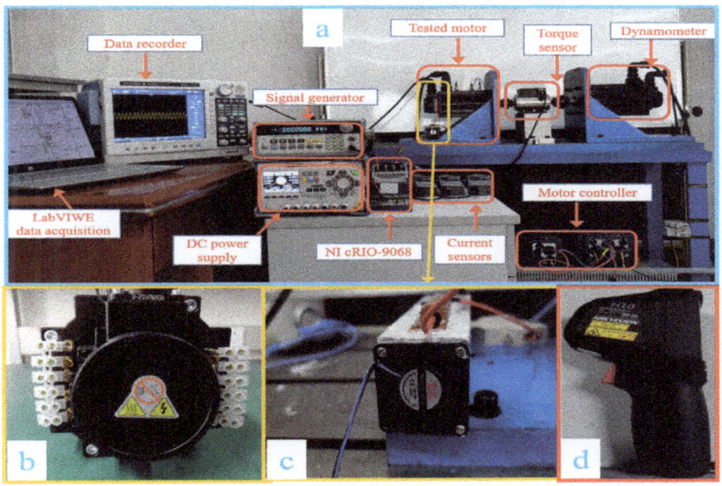

Figure 10. Composition diagram of equipment for ITSC simulation test bench. (**a**) Test bench and its testing equipment. (**b**) The faulty motor. (**c**) The fault resistance and its heat dissipation device. (**d**) The temperature measurement device.

Table 1. Specifications of the tested PMSM.

Parameters	Values	Parameters	Values
Rated power	2.3 kW	Line–line resistance	1.1 Ω
Rated torque	15 Nm	Line–line inductance	4.45 mH
Rated current	9.5 A	Number of turns per phase	108
Rated speed	1500 rpm	Number of coils per phase	12
Pole pairs	4	Voltage constant	114 V/1000 r/min

A fault motor simulation test was conducted using a test bench to replicate 17 different fault states of a PMSM exhibiting ITSC faults. This includes one healthy state and sixteen distinct fault conditions. The severity of the ITSC faults is determined by combinations of shorted degrees and fault resistances. The shorted degrees are defined as 5 turns, 9 turns, 11 turns, and 15 turns, totaling four categories. The fault resistances are set at 5 Ω, 1 Ω, 0.5 Ω, and 0.1 Ω, also totaling four categories, resulting in 16 fault levels. Considering the healthy state of the motor as having a fault level of 0, the experimental data encompass a total of 17 fault severities. To simulate the motor's operating conditions during agricultural machinery acceleration, deceleration, and constant speed driving, 8 constant speed scenarios and 2 variable speed scenarios were established during the bench test, as detailed in Table 2.

Table 2. Operating conditions of the tested PMSM.

Case	Constant				Dynamic
	8				2
Speed (rpm)	150	450	900	1350	850~1550~850
Torque (Nm)	3.0/7.5	3.0/7.5	3.0/7.5	3.0/7.5	3.0/7.5

It can be seen from Table 2 that there are 10 different operating conditions in the test process, each generated by combinations of five speeds and two torques. The two load torques are both constant, while among the five speeds, four are constant and one represents an acceleration and deceleration condition. The dynamic speed variation ranges from 850 rpm to 1550 rpm and then back to 850 rpm, as shown in Figure 11. For each distinct fault condition of the motor, ITSC fault tests are conducted under the aforementioned 10 conditions. The Yokogawa DL850EA oscilloscope is used to record the three-phase current, with a sampling frequency of 1 MHz. The data sampling duration for the fault motor under each operating condition is 10 s. The entire data collection process employs a field-oriented control (FOC) strategy using the VFD037C23A inverter, operating at the switching frequency of 15 kHz, with the auxiliary test motor using speed closed-loop control and the tested motor using current closed-loop control.

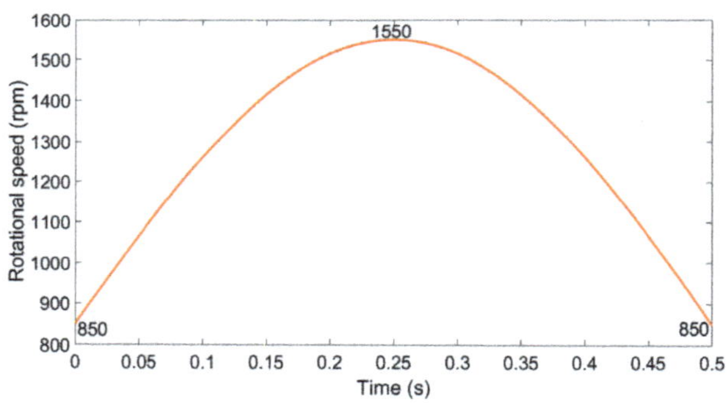

Figure 11. Schematic diagram of speed variation under variable operating conditions.

During the experiment, due to the absence of hardware filtering, a relatively high sampling frequency of 1 MHz was chosen to avoid signal aliasing caused by interference and other factors during data acquisition. If the raw data were directly used for dataset construction, it would impose a significant challenge on computer hardware resources and severely impact the training speed. The goal of this study is to use deep learning models to extract low-frequency features from the acquired experimental data that are useful for classifying the severity of ITSC faults. Therefore, during data preprocessing, the acquired data are first filtered and then down-sampled to retain low-frequency features while reducing the memory usage of the dataset. A zero-phase low-pass filter is applied to the data, and the down-sampled sampling frequency is set to 15 kHz, matching the switching frequency of the controller. To facilitate the comparison of data under different fault severities and operating conditions and to accelerate the convergence of the deep learning model, the acquired data are normalized to the range of $[-1, 1]$. To aid in training the deep learning model, the down-sampled three-phase current data are divided into equal-length data slices, each containing sufficient feature information. The length of each data slice is set to 3000 sampling points, which ensures that, at the lowest operating speed, the three-phase current signal collected over one cycle of the motor's rotation is captured in each slice.

The labels of the data slices correspond to their fault severity, as shown in Table 3. In the labels, "HL" denotes the data collected under healthy motor conditions, while "A*R*" stands for the data collected under different combinations of fault resistors and shorted ratios. "A2", "A4", "A5", and "A6" represent shorted turns of 5, 9, 11, and 15, respectively. "R5", "R1", "R0.5", and "R0.1" indicate fault resistances of 5 Ω, 1 Ω, 0.5 Ω, and 0.1 Ω, respectively. The fault severities in Table 3 are arranged in ascending order based on the severity calculated using Equation (18). The sampled data were organized into datasets according to different fault severities, ensuring that the amount of data for each condition under a specific fault severity was equal and the quantities of data corresponding to each fault severity were also equal. For each fault level, the number of data samples is set at 1200, with 360 samples randomly selected for testing, leaving 840 samples for training, resulting in a ratio of 3:7. Ultimately, all training samples form the training set, while all testing samples comprise the validation set.

Table 3. Dataset description.

Label	Fault Setting		Sample Size		
	Fault Resistance (Ω)	Shorted Turn Ratio (%)	Training	Testing	Total
HL	Inf	0	840	360	1200
A2R5	5	4.6	840	360	1200
A4R5	5	8.3	840	360	1200
A5R5	5	10.2	840	360	1200
A6R5	5	13.8	840	360	1200
A2R1	1	4.6	840	360	1200
A4R1	1	8.3	840	360	1200
A2R0.5	0.5	4.6	840	360	1200
A5R1	1	10.2	840	360	1200
A6R1	1	13.8	840	360	1200
A4R0.5	0.5	8.3	840	360	1200
A5R0.5	0.5	10.2	840	360	1200
A6R0.5	0.5	13.8	840	360	1200
A2R0.1	0.1	4.6	840	360	1200
A4R0.1	0.1	8.3	840	360	1200
A5R0.1	0.1	10.2	840	360	1200
A6R0.1	0.1	13.8	840	360	1200

The comparison of the data before and after preprocessing is shown in Figure 12. In each figure, the left side displays the original three-phase current signal, while the right side shows the three-phase current after data preprocessing. Figure 12a illustrates the three-phase current under healthy conditions at a speed of 150 rpm and a torque of 3.0 Nm. Figure 12b depicts the three-phase current of a faulty motor with the fault label "A5R0.1", collected under dynamic speed conditions at a torque of 3.0 Nm.

(a)

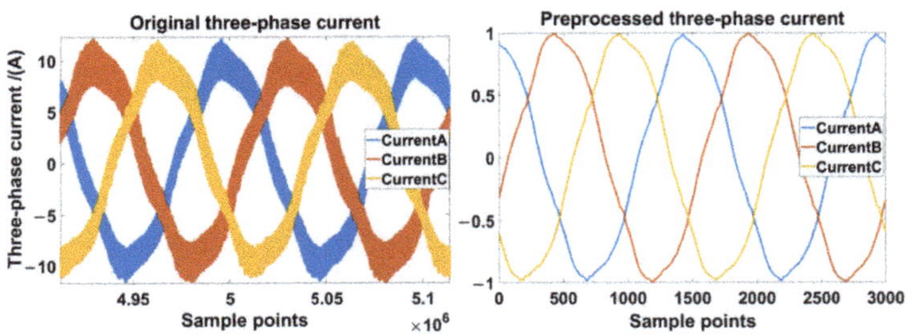

Figure 12. *Cont.*

(**b**)

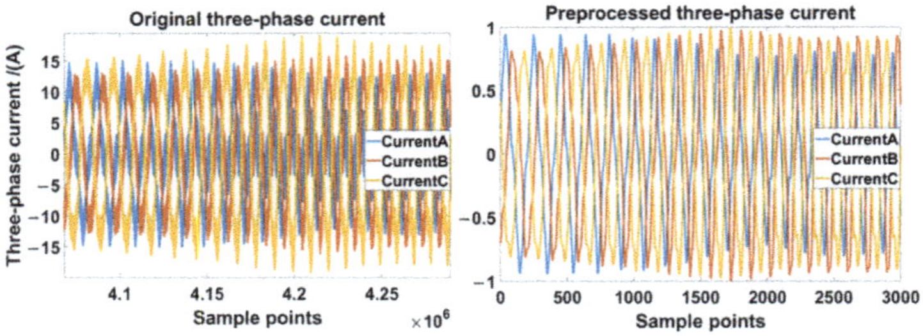

Figure 12. Comparison of three-phase current signals before and after data preprocessing. The left side of the figure shows the original signal, while the right side shows the preprocessed signal. (**a**) The three-phase current signals are collected under constant operating conditions. (**b**) The three-phase current signals are collected under dynamic operating conditions.

5. Results and Comparisons

After completing the data preprocessing and dataset construction, the proposed Bayesian optimization-based ITSC fault diagnosis model is used to analyze the three-phase current signals. The whole procedure is carried out offline. The hyperparameter combinations to be optimized and their search space are shown in Table 4.

Table 4. Hyperparameters to be optimized.

Hyperparameters	Search Intervals	Data Types	Transform	Best Result
L_{init}	$[1 \times 10^{-5}\ 1]$	real	log	1.6227×10^{-4}
G_1	[0.5 1]	real	log	0.8747
L_{2R}	$[1 \times 10^{-10}\ 1 \times 10^{-2}]$	real	log	7.4777×10^{-8}
d_1	[2 8]	integer	none	5
d_2	[4 16]	integer	none	9
d_3	[2 8]	integer	none	6
w_1	[2 60]	integer	none	18
w_2	[40 160]	integer	none	66
w_3	[2 60]	integer	none	38
P	$[1 \times 10^{-5}\ 1]$	real	log	5.1585×10^{-4}

Among them, L_{init} represents the initial learning rate of the entire model, $G1$ represents the gradient optimization coefficient of the Adam optimizer, L_{2R} represents the $L2$ regularization coefficient, P represents the probability of dropout, and the data type for the above hyperparameters is all real numbers. The depths of the three convolutional layers are denoted by d_1, d_2, and d_3, and the numbers of convolutional kernels for each layer are represented by w_1, w_2, and w_3. Both the number of convolutional kernels and the depth of the convolutional layers are integer types. The size of the convolutional kernels is set to a fixed value of 1×3, the dilation rate is set to 2, the learning rate decay step size is set to 20, and the decay factor is set to 0.1. "Transform" indicates whether the hyperparameters are searched on a logarithmic scale during the search process in the set space. Based on experience, the maximum number of iterations for Bayesian optimization is set to 60, with 40 training epochs per iteration. The values of the hyperparameters for the optimal combination obtained are shown in Table 4, and the corresponding schematic diagram of the optimal model architecture is presented in Figure 13.

To verify the advantages of the proposed improved CNN architecture and to compare the performance improvements brought by different combinations of enhancements, several models are constructed: a conventional CNN model without any architecture enhancement (CNN); a conventional CNN model with the residual network structure (Res); a CNN model that shares both residual and multi-scale networks (MK-Res); and a CNN model that shares both residual and attention mechanisms (SE-Res). The architecture hyperparameters of the feature layers for these four models are set consistently with the proposed improved CNN model. The training hyperparameters for the four models were obtained through hyperparameter tuning using Bayesian optimization. The error loss and validation accuracy of the five models throughout the training process were recorded as they varied with the number of training epochs, and the results are compared in Figure 14.

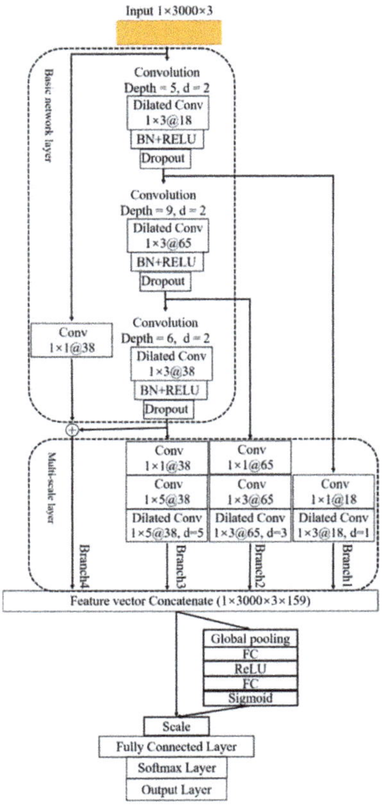

Figure 13. Schematic diagram of the optimal model structure obtained using a Bayesian optimization algorithm.

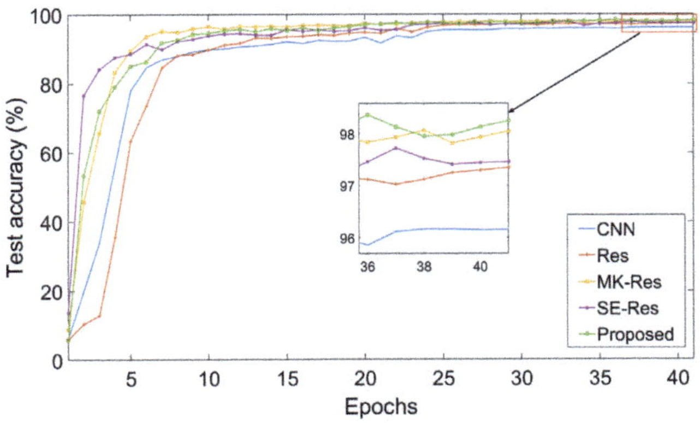

(**a**)

Figure 14. *Cont.*

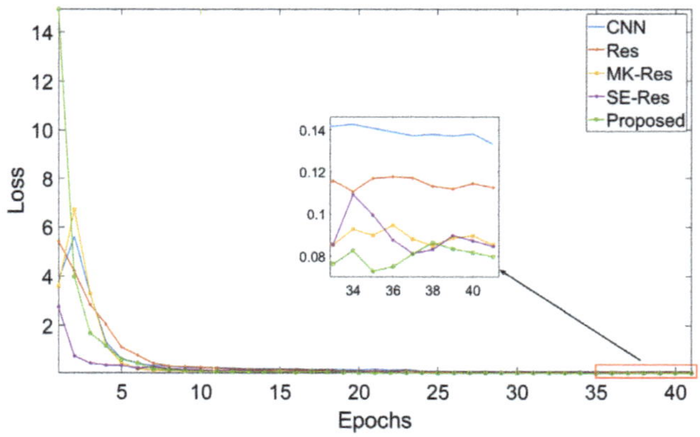

(b)

Figure 14. Comparison diagram of ITSC fault diagnosis results of the five models. (**a**) Overall testing accuracy trends of the five algorithms. (**b**) The trend of loss function changes in each algorithm.

Figure 14a and Table 5 compare the test accuracy trends of the five models as training epochs progress. It can be seen that, compared to the CNN model, all four improved models exhibit varying extends of enhancement in the final test accuracy. The final test accuracy of the CNN model is 96.16%. The final test accuracy of the Res model is 97.35%, which represents an improvement of 1.19% over the CNN model. The MK-Res model achieves a final test accuracy of 98.06%, improving by 1.90% compared to the CNN model. The SE-Res model has a final test accuracy of 97.47%, an increase of 1.31% over the CNN model. The proposed model reaches a final validation accuracy of 98.25%, marking an improvement of 2.09% compared to the CNN model.

Table 5. Result comparison of different methods.

Method	Test Accuracy	Loss
CNN	96.16%	0.1333
Res	97.35%	0.1125
MK-Res	98.06%	0.0854
SE-Res	97.47%	0.0847
Proposed	98.25%	0.0799

It is equally important to note that the feature extraction layers of all five models are consistent, with the differences between the models lying in the use of various improved architectures within the feature extraction layers. From the final results, it is evident that the residual network structure, multi-scale network structure, and channel attention mechanism all contribute to varying degrees of performance improvement, with the combination of all three achieving the most significant enhancement. Based on the principles of these improved architectures, the channel attention mechanism is able to discard irrelevant parameters during training, thus not only improving the model's performance but also accelerating the overall convergence speed. The residual network structure helps the model train more effectively and improves recognition accuracy. The multi-scale network architecture enriches the scale of the extracted fault features, enhancing the diversity of the fault feature space, which, in turn, boosts the model's final recognition accuracy. From the final results, it can be seen that for complex tasks such as ITSC fault severity recognition, the multi-scale architecture has the greatest impact on the model's performance, followed by the channel attention mechanism, with the combined use of all three yielding the best results. Figure 14b and Table 5 show the comparison of the loss trends of the five models as training epochs progress. From the figure, it is evident that the final error losses of the four improved models are all better than those of the CNN model. Among them, the proposed model has the smallest error loss and exhibits the best generalization capability, followed by the MK-Res and SE-Res models. The Res model has the highest error loss among the four improved models.

To accurately assess the performance of the proposed model in different severity labels, three metrics are introduced for comprehensive evaluation: recall (r), precision (p), and $F1$ score. In large datasets, there exists a tradeoff between recall and precision. The $F1$ score takes into account both recall and precision, thereby providing a more holistic representation of the algorithm's performance. The specific definitions of these evaluation metrics are presented in Equation (28):

$$\begin{aligned} p &= \frac{TP}{TP+FP} \\ r &= \frac{TP}{TP+FN} \\ F1 &= \frac{2p\times r}{p+r} \end{aligned} \tag{28}$$

To comprehensively compare the performance of the proposed model with the four other models, the confusion matrices of the five models on the test dataset are compared. The confusion matrices for the five models are shown in Figures 15–19. The leftmost labels of the confusion matrix represent the actual severity of ITSC faults contained in the test dataset, categorized into 17 types, arranged in ascending order according to the fault severity calculated using Equation (18). According to the definitions of precision and recall, the precision for each label is derived from the ratio of the number of samples in the diagonal to the total number of samples in that column, as shown in the row vector at the bottom of the confusion matrix. The recall for each label is determined by the ratio of the number of samples in the diagonal to the total number of samples in that row, as indicated by the column vector on the right side of the confusion matrix. The classification accuracy of the models is calculated as the ratio of the number of correctly classified samples on the diagonal to the total number of samples in the test dataset.

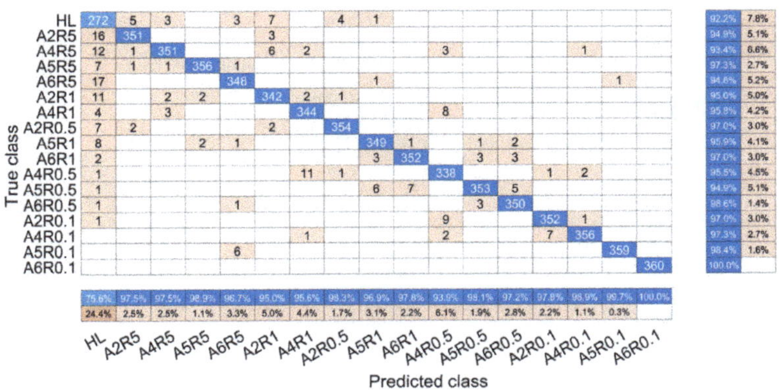

Figure 15. The confusion matrix of the CNN model.

From the figures, it can be observed that, compared to the confusion matrix of the CNN model, the four improved ITSC fault diagnosis models show a significant reduction in the number of misclassified samples. All four improved models exhibit varying degrees of improvement in terms of "false alarms" and "missed detections", although there remains room for further enhancement.

To further compare the performance of the five models across different fault severity labels, the $F1$ scores and overall test accuracy for each model based on the precision and recall from the confusion matrices in the test dataset are calculated. The comparison results are shown in Table 6. From the table, it can be seen that while the four improved ITSC fault diagnosis models show varying degrees of improvement in the overall test accuracy compared to the CNN model, the *F*1 scores for different fault severity labels reveal mixed performances among the five models. The four improved models exhibit clear advantages in labels associated with lighter fault degrees, showing significant increases in *F*1 scores. Among the 17 different fault classifications, the proposed models achieved the highest scores in 12 of the fault categories, demonstrating the best performance.

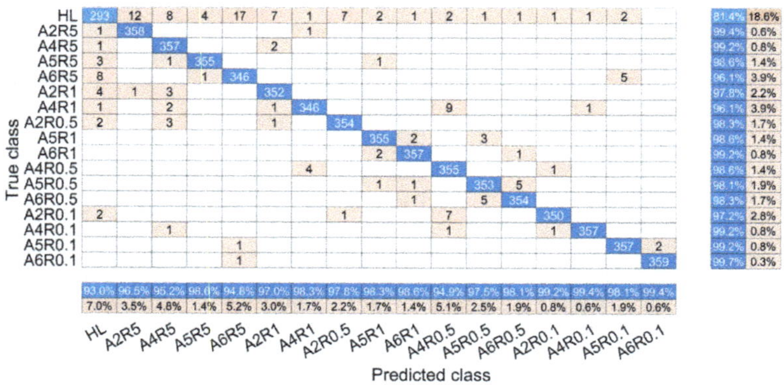

Figure 16. The confusion matrix of the Res model.

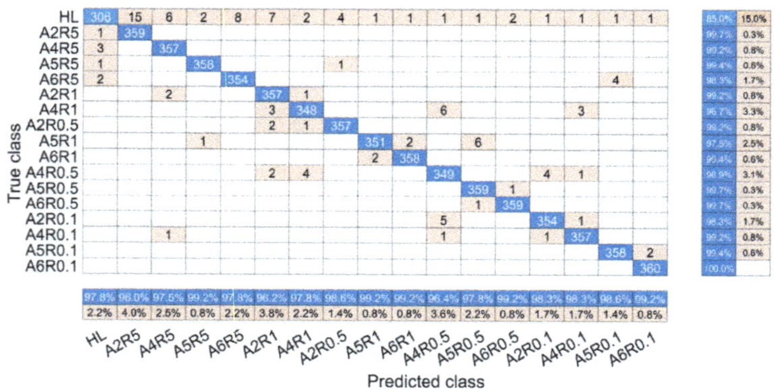

Figure 17. The confusion matrix of the MK-Res model.

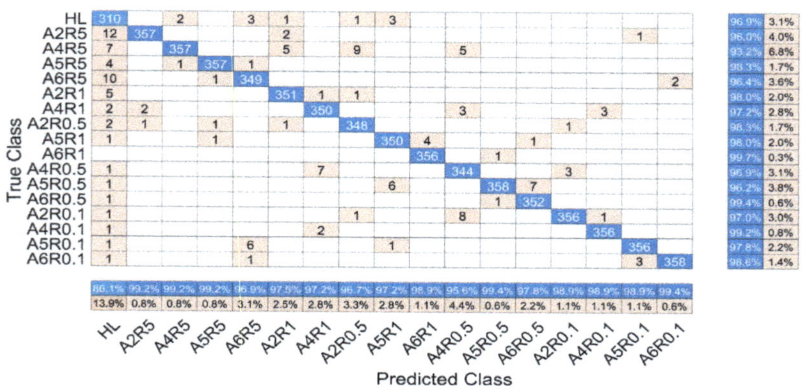

Figure 18. The confusion matrix of the SE-Res model.

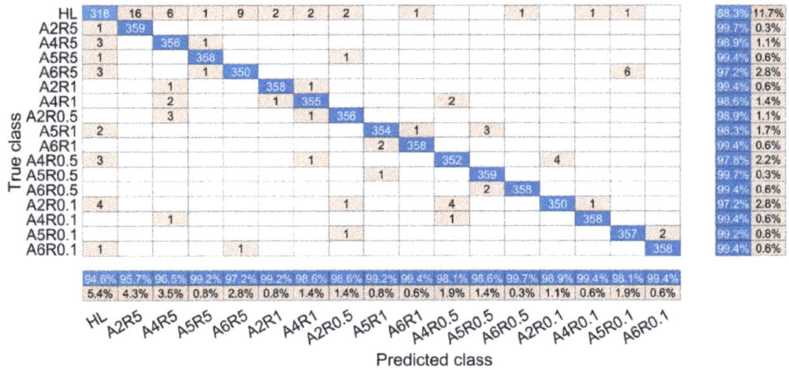

Figure 19. The confusion matrix of the proposed model.

Table 6. *F*1 score comparison of five methods under different fault labels.

Label	CNN (%)	Res (%)	MK-Res (%)	SE-Res (%)	Proposed (%)
Acc	96.16	97.35	98.06	97.47	98.25
HL	83.05	86.81	90.94	91.18	91.38
A2R5	96.16	97.95	97.82	97.54	97.69
A4R5	95.38	97.14	98.35	96.10	97.67
A5R5	98.07	98.61	99.31	98.76	99.31
A6R5	95.74	95.45	98.06	96.68	97.22
A2R1	95.00	97.37	97.67	97.77	99.31
A4R1	95.69	97.19	97.21	97.22	98.61
A2R0.5	97.66	98.06	98.89	97.48	98.75
A5R1	96.41	98.47	98.32	97.63	98.74
A6R1	97.37	98.89	99.31	99.30	99.44
A4R0.5	94.68	96.73	96.68	96.22	97.91
A5R0.5	96.45	97.78	98.76	97.81	99.17
A6R0.5	97.90	98.20	99.45	98.60	99.58
A2R0.1	97.37	98.18	98.33	97.94	98.04
A4R0.1	98.07	99.30	98.76	99.03	99.44
A5R0.1	99.03	98.62	99.03	98.34	98.62
A6R0.1	100	99.58	98.06	99.03	99.44

To reduce the impact of randomness, the diagnostic results from five repeated experiments were averaged, and the standard deviation of the results for each experiment was calculated. Additionally, the time taken by the model to recognize the test set in each experiment was recorded, and the average recognition time per data slice was computed, as shown in Table 7. From Table 7, it can be seen that the proposed model not only achieves the highest average accuracy of 98.20% but also has the smallest standard deviation of 0.105%, indicating both good accuracy and stability. The complexity of the deep learning model is represented by the total number of adjustable parameters, including weights and biases, as shown in Table 7. It is evident from the table that each improvement measure added to the model increases its complexity. The proposed model, incorporating the most improvements, has the highest complexity. The average recognition time of the model reflects its data processing speed. From the table, it can be seen that the proposed model has the longest average recognition time, reaching 1.14 ms, but this is still much smaller than the 0.2 s sampling time per data slice, meaning the model can meet the required time for data processing.

Table 7. Stability comparison of different methods and the average computation time for each slice of test data.

Method	Average Accuracy (%)	Average Computation Time (ms)	Model Complexity
CNN	96.19 ± 0.133	0.946 ± 0.0195	5,552,359
Res	97.19 ± 0.137	0.652 ± 0.0039	5,952,918
MK-Res	97.91 ± 0.112	0.852 ± 0.0693	24,510,130
SE-Res	97.59 ± 0.147	0.689 ± 0.0431	5,954,419
Proposed	98.20 ± 0.105	1.140 ± 0.0052	24,689,824

Through a comprehensive analysis of the five ITSC fault diagnosis models, it is evident that the proposed model exhibits the best performance in terms of the final test accuracy and stability. In the *F*1 scores across 17 different fault severity labels, the proposed model demonstrates overall superior performance, making it the best-performing model among the five. To validate the feature learning capability of the proposed model, the t-distribution stochastic neighbor-embedding algorithm (T-SNE) was used to visualize the features from the final output layer of the ITSC fault diagnosis model, and the results were compared with the other four models. The two-dimensional visualization results are shown in Figure 20. From the figure, it can be observed that the feature map contains 17 colors, each corresponding to a specific fault severity label, with each point representing a data sample. Figure 20a shows the feature distribution of the input layers of each model. It is apparent that the feature distribution of the input data is chaotic, with significant overlap among samples of different colors, making it difficult to discern the fault severity of the corresponding samples in the dataset based solely on the input data. Figure 20b–f display the feature distribution maps of the classification layers for the CNN model, Res model, MK-Res model, SE-Res model, and the proposed model, respectively. From these figures, it can be seen that after feature extraction by the model, the samples within the same ITSC fault severity labels exhibit good intra-class clustering characteristics. The proposed ITSC fault diagnosis model has the fewest misclassified sample points compared to the other four models. Additionally, the boundaries between different ITSC fault labels are clear and more distant, resulting in better separation characteristics among different categories. Thus, the proposed model demonstrates superior feature learning and discrimination capabilities.

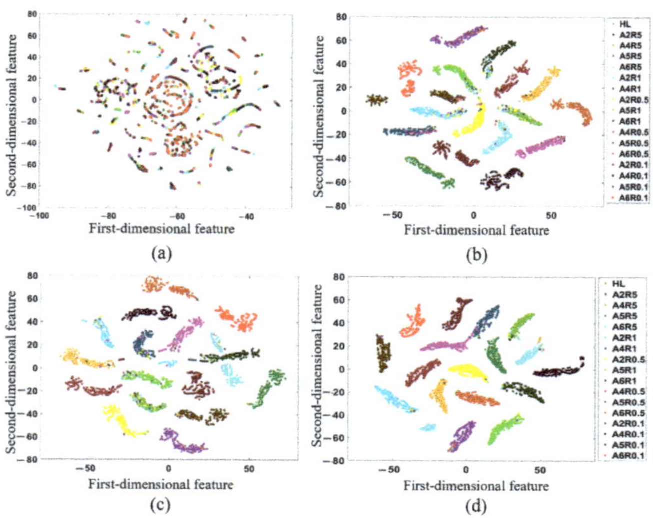

Figure 20. *Cont.*

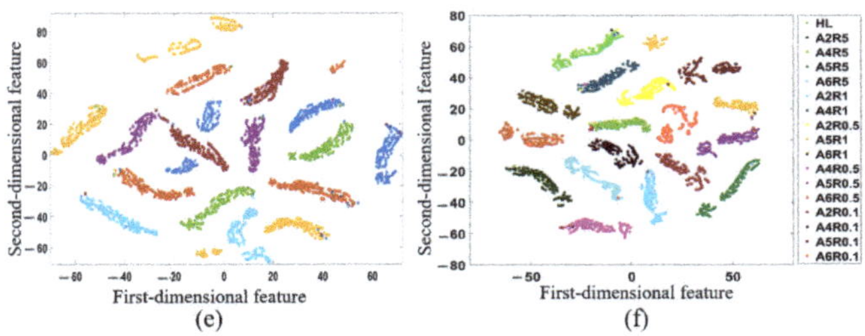

Figure 20. Comparison of visualized features extracted by different algorithms. (**a**) Feature map of the input data. (**b**) Feature map of the CNN model. (**c**) Feature map of the Res model. (**d**) Feature map of the MK-Res model. (**e**) Feature map of the SE-Res model. (**f**) Feature map of the proposed model.

6. Conclusions

In this research, a novel Bayesian optimization-based improvement algorithm was proposed for the enhancement of an ITSC fault diagnosis model. The results indicate that the proposed method is applicable under both dynamic and steady-state operating conditions. Firstly, a fault model was proposed for the analysis of ITSC faults, and a severity index was derived for the guiding of ITSC fault severity settings. Secondly, a residual network, multi-scale network, and attention mechanism were applied to prevent network functionality degradation, increase the richness of the extracted features, and enhance the proportion of useful features in the model, ultimately enhancing the network's performance. Then, to facilitate the optimization of architecture hyperparameters, the model's feature extraction layers were divided into a three-stage architecture based on the characteristics of the features extracted by convolutional networks. Furthermore, the ITSC fault motor test was carried out with the fault severity set to 17 different levels. The proposed algorithm was conducted to analyze the three-phase current signals that were collected in the motor test. Conventional CNN, Res, MK-Res, and SE-Res models were also employed in the same dataset for comparison. The results illustrated that the proposed algorithm not only achieved the best final test accuracy but also provided the best feature extraction capability.

This study aims to improve the accuracy of ITSC fault diagnosis using deep learning methods. The approach relies on supervised training with a sufficient and balanced sample size. However, in practical applications, challenges such as an insufficient sample size, imbalanced sample distribution, or lack of labeled samples often arise. Under these conditions, the application of the proposed method would be significantly limited. Future research will focus on addressing these challenges and improving the accuracy of ITSC fault diagnosis under such adverse conditions.

Author Contributions: Conceptualization, M.W., Q.S. and W.L.; methodology, M.W. and W.L.; software, M.W. and W.L.; validation, M.W.; formal analysis, M.W. and W.L.; investigation, M.W., Q.S. and W.L.; resources, Q.S.; data curation, M.W. and W.L.; writing—original draft preparation, M.W.; writing—review and editing, M.W., H.L., P.S. and W.L.; visualization, M.W.; supervision, Q.S.; project administration, M.W. and Q.S.; funding acquisition, Q.S. All authors have read and agreed to the published version of the manuscript.

Funding: This research received no external funding.

Institutional Review Board Statement: Not applicable.

Data Availability Statement: The raw data supporting the conclusions of this study are available from the corresponding author upon reasonable request (wangmingsheng_08@163.com).

Conflicts of Interest: The authors declare no conflicts of interest.

References

1. Xie, F.; Li, G.; Liu, H.; Sun, E.; Wang, Y. Advancing Early Fault Diagnosis for Multi-Domain Agricultural Machinery Rolling Bearings through Data Enhancement. *Agriculture* **2024**, *14*, 112. [CrossRef]
2. Xie, F.; Sun, E.; Wang, L.; Wang, G.; Xiao, Q. Rolling Bearing Fault Diagnosis in Agricultural Machinery Based on Multi-Source Locally Adaptive Graph Convolution. *Agriculture* **2024**, *14*, 1333. [CrossRef]
3. Xie, F.; Wang, Y.; Wang, G.; Sun, E.; Fan, Q.; Song, M. Fault Diagnosis of Rolling Bearings in Agricultural Machines Using SVD-EDS-GST and ResViT. *Agriculture* **2024**, *14*, 1286. [CrossRef]
4. Wang, H. Comparative investigation and evaluation of electric-drive seed-metering systems across diverse speed ranges for enhanced high-precision seeding applications. *Comput. Electron. Agric.* **2024**, *222*, 108976. [CrossRef]
5. Scolaro, E.; Estevez, M.P.; Renzi, M.; Mattetti, M. Electrification of Agricultural Machinery: A Review. *IEEE Access* **2021**, *9*, 164520–164541. [CrossRef]
6. Ni, H.; Lu, L.; Sun, M.; Bai, X.; Yin, Y. Research on Fault Diagnosis of PST Electro-Hydraulic Control System of Heavy Tractor Based on Support Vector Machine. *Processes* **2022**, *10*, 791. [CrossRef]
7. Riera-Guasp, M.; Antonino-Daviu, J.A.; Capolino, G.A. Advances in electrical machine, power electronic, and drive condition monitoring and fault detection: State of the art. *IEEE Trans. Ind. Electron.* **2015**, *62*, 1746–1759. [CrossRef]
8. Li, Y.; Liu, Y.; Ji, K.; Zhu, R. A Fault Diagnosis Method for a Differential Inverse Gearbox of a Crawler Combine Harvester Based on Order Analysis. *Agriculture* **2022**, *12*, 1300. [CrossRef]
9. Husari, F.; Seshadrinath, J. Stator Turn Fault Diagnosis and Severity Assessment in Converter-Fed Induction Motor Using Flat Diagnosis Structure Based on Deep Learning Approach. *IEEE J. Emerg. Sel. Top. Power Electron.* **2023**, *11*, 5649–5657. [CrossRef]
10. Mahmoudi, A.; Jlassi, I.; Cardoso, A.J.M.; Yahia, K.; Sahraoui, M. Inter-Turn Short-Circuit Faults Diagnosis in Synchronous Reluctance Machines, Using the Luenberger State Observer and Current's Second-Order Harmonic. *IEEE Trans. Ind. Electron.* **2022**, *69*, 8420–8429. [CrossRef]
11. Choe, H.O.; Lee, M.-H. Artificial Intelligence-Based Fault Diagnosis and Prediction for Smart Farm Information and Communication Technology Equipment. *Agriculture* **2023**, *13*, 2124. [CrossRef]
12. Yang, G.; Cheng, Y.; Xi, C.; Liu, L.; Gan, X. Combine Harvester Bearing Fault-Diagnosis Method Basedon SDAE-RCmvMSE. *Entropy* **2022**, *24*, 1139. [CrossRef] [PubMed]
13. Liu, Z.; Sun, W.; Chang, S.; Zhang, K. Corn Harvester Bearing Fault Diagnosis Based on ABC-VMD and Optimized EfficientNet. *Entropy* **2023**, *25*, 1273. [CrossRef] [PubMed]
14. Jiang, W.; Shan, Y.; Xue, X.; Ma, J.; Chen, Z.; Zhang, N. Fault Diagnosis for Rolling Bearing of Combine Harvester Based on Composite-Scale-Variable Dispersion Entropy and Self-Optimization Variational Mode Decomposition Algorithm. *Entropy* **2023**, *25*, 1111. [CrossRef] [PubMed]
15. Parvin, F.; Faiz, J.; Qi, Y.; Kalhor, A.; Akin, B. A Comprehensive Interturn Fault Severity Diagnosis Method for Permanent Magnet Synchronous Motors Based on Transformer Neural Networks. *IEEE Trans. Ind. Inform.* **2023**, *19*, 10923–10933. [CrossRef]
16. Li, J.; Li, X.; Li, Y.; Zhang, Y.; Yang, X.; Xu, P. A New Method of Tractor Engine State Identification Based on Vibration Characteristics. *Processes* **2023**, *11*, 303. [CrossRef]
17. Fan, W.; Yang, C.; Chen, C.; He, C.; Yuan, Y.; Li, Y. Adaptive Feature-Oriented Dictionary Learning and Sparse Classification Framework for Bearing Compound Fault Diagnosis. *IEEE Trans. Instrum. Meas.* **2024**, *73*, 3518010. [CrossRef]
18. Lee, H.; Jeong, H.; Koo, G.; Ban, J.; Kim, S.W. Attention Recurrent Neural Network-Based Severity Estimation Method for Interturn Short-Circuit Fault in Permanent Magnet Synchronous Machines. *IEEE Trans. Ind. Electron.* **2021**, *68*, 3445–3453. [CrossRef]
19. Xu, L.; Zhang, G.; Zhao, S.; Wu, Y.; Xi, Z. Fault Diagnosis of Tractor Transmission System Based on Time GAN and Transformer. *IEEE Access* **2024**, *12*, 107153–107169. [CrossRef]
20. Husari, F.; Seshadrinath, J. Early Stator Fault Detection and Condition Identification in Induction Motor Using Novel Deep Network. *IEEE Trans. Artif. Intell.* **2022**, *3*, 809–818. [CrossRef]
21. Zhao, J.; Guan, X.; Li, C.; Mou, Q.; Chen, Z. Comprehensive Evaluation of Inter-Turn Short Circuit Faults in PMSM Used for Electric Vehicles. *IEEE Trans. Intell. Transp. Syst.* **2021**, *22*, 611–621. [CrossRef]
22. Zhang, Y.; Liu, G.; Zhao, W.; Zhou, H.; Chen, Q.; Wei, M. Online Diagnosis of Slight Interturn Short-Circuit Fault for a Low-Speed Permanent Magnet Synchronous Motor. *IEEE Trans. Transp. Electrif.* **2021**, *7*, 104–113. [CrossRef]
23. Ehya, H.; Skreien, T.N.; Nysveen, A. Intelligent Data-Driven Diagnosis of Incipient Interturn Short Circuit Fault in Field Winding of Salient Pole Synchronous Generators. *IEEE Trans. Ind. Inform.* **2022**, *18*, 3286–3294. [CrossRef]
24. Zhao, M.; Zhong, S.; Fu, X.; Tang, B.; Pecht, M. Deep Residual Shrinkage Networks for Fault Diagnosis. *IEEE Trans. Ind. Inform.* **2020**, *16*, 4681–4690. [CrossRef]
25. Jafari, A.; Faiz, J.; Jarrahi, M.A. A Simple and Efficient Current-Based Method for Interturn Fault Detection in BLDC Motors. *IEEE Trans. Ind. Inform.* **2021**, *17*, 2707–2715. [CrossRef]
26. Hu, R.; Wang, J.; Mills, A.R.; Chong, E.; Sun, Z. Current-Residual-Based Stator Interturn Fault Detection in Permanent Magnet Machines. *IEEE Trans. Ind. Electron.* **2021**, *68*, 59–69. [CrossRef]
27. Hang, J.; Zhang, J.; Cheng, M.; Huang, J. Online Interturn Fault Diagnosis of Permanent Magnet Synchronous Machine Using Zero-Sequence Components. *IEEE Trans. Power Electron.* **2015**, *30*, 6731–6741. [CrossRef]

28. Yao, Y.; Wang, J.; Xie, M. Adaptive residual CNN-based fault detection and diagnosis system of small modular reactors. *Appl. Soft Comput.* **2022**, *114*, 108064. [CrossRef]
29. Bai, S.; Kolter, J.Z.; Koltun, V. An Empirical Evaluation of Generic Convolutional and Recurrent Networks for Sequence Modeling. *arXiv* **2018**, arXiv:1803.01271.
30. Shao, S.; Yan, R.; Lu, Y.; Wang, P.; Gao, R.X. DCNN-Based Multi-Signal Induction Motor Fault Diagnosis. *IEEE Trans. Instrum. Meas.* **2020**, *69*, 2658–2669. [CrossRef]
31. He, K.; Zhang, X.; Ren, S.; Sun, J. Deep Residual Learning for Image Recognition. In Proceedings of the 2016 IEEE Conference on Computer Vision and Pattern Recognition (CVPR), Las Vegas, NV, USA, 27–30 June 2016; pp. 770–778.
32. Xu, Y.; Yan, X.; Sun, B.; Liu, Z. Hierarchical Multiscale Dense Networks for Intelligent Fault Diagnosis of Electromechanical Systems. *IEEE Trans. Instrum. Meas.* **2022**, *71*, 3505312. [CrossRef]
33. Li, Y.; Du, L.; Wei, D. Multiscale CNN Based on Component Analysis for SAR ATR. *IEEE Trans. Geosci. Remote Sens.* **2022**, *60*, 5211212. [CrossRef]
34. Weng, C.; Lu, B.; Gu, Q. A multi-scale kernel-based network with improved attention mechanism for rotating machinery fault diagnosis under noisy environments. *Meas. Sci. Technol.* **2022**, *33*, 055108. [CrossRef]
35. Hu, J.; Shen, L.; Albanie, S.; Sun, G.; Wu, E. Squeeze-and-Excitation Networks. In Proceedings of the IEEE Conference on Computer Vision and Pattern Recognition, Long Beach, CA, USA, 15–20 June 2019; Volume 42, pp. 2011–2023.
36. Gelbart, M.A.; Snoek, J.; Adams, R.P. Bayesian Optimization with Unknown Constraints. *arXiv* **2014**, arXiv:1403.5607.
37. Han, S.; Eom, H.; Kim, J.; Park, C. Optimal DNN architecture search using Bayesian Optimization Hyperband for arrhythmia detection. In Proceedings of the 2020 IEEE Wireless Power Transfer Conference (WPTC), Seoul, Republic of Korea, 15–19 November 2020; pp. 357–360.

agriculture

MDPI

Article

Design and Implementation of an AI-Based Robotic Arm for Strawberry Harvesting

Chung-Liang Chang * and Cheng-Chieh Huang

Department of Biomechatronics Engineering, National Pingtung University of Science and Technology, Neipu, Pingtung 91201, Taiwan
* Correspondence: chungliang@mail.npust.edu.tw; Tel.: +886-8-7703202 (ext. 7586)

Abstract: This study presents the design and implementation of a wire-driven, multi-joint robotic arm equipped with a cutting and gripping mechanism for harvesting delicate strawberries, with the goal of reducing labor and costs. The arm is mounted on a lifting mechanism and linked to a laterally movable module, which is affixed to the tube cultivation shelf. The trained deep learning model can instantly detect strawberries, identify optimal picking points, and estimate the contour area of fruit while the mobile platform is in motion. A two-stage fuzzy logic control (2s-FLC) method is employed to adjust the length of the arm and bending angle, enabling the end of the arm to approach the fruit picking position. The experimental results indicate a 90% accuracy in fruit detection, an 82% success rate in harvesting, and an average picking time of 6.5 s per strawberry, reduced to 5 s without arm recovery time. The performance of the proposed system in harvesting strawberries of different sizes under varying lighting conditions is also statistically analyzed and evaluated in this paper.

Keywords: fruit picking robot; deep learning; visual servoing; fuzzy logic control; hydroponic greenhouse

1. Introduction

Farmers are increasingly adopting automated agricultural machinery to improve production efficiency in response to challenges such as unstable crop yields caused by extreme weather, labor shortages, and shrinking available farmland. In recent years, advancements in high-speed computing have matured, and the application of artificial intelligence (AI) in agriculture is now seen as a key future trend. AI has the potential to reduce crop production costs, minimize labor requirements, and even replace manual labor. Notably, mobile automated handling machines and robotic arms integrated with computer vision systems are widely used for highly repetitive tasks like grasping, moving, and placing objects [1]. Unlike industrial robots that function with minimal external interference, agricultural robots must adapt to dynamic outdoor environments to perform tasks such as leaf pruning [2,3], sowing [4], pollination [5], fruit picking [6–10], and transplanting [11]. These rigid robots can withstand higher loads, and their arms are typically designed with multiple axes to enhance workspace flexibility [5]. Some even use dual-arm setups to perform more complex tasks [12]. However, such robots require numerous sensors and actuators, often relying on a ground vehicle for movement to execute tasks. Due to the complexity of positioning control, operations such as pruning, pollination, picking, stem cutting, or fruit suction for individual plants tend to be time-consuming. Although two-degree-of-freedom four-bar robots have been proposed to quickly grasp and place seedling pots on conveyor belts, completing each cycle in approximately 2.25 s on average [11], low-degree-of-freedom mechanisms are more suited for simple transplanting operations. For more complex tasks, like fruit picking, an appropriate control method is essential. Additionally, pneumatic cylinders, suction cups, or soft materials are often used for handling or transporting seedlings. However, for delicate seedlings or mature fruits, improper contact forces can easily cause damage.

A soft arm composed of bio-inspired materials uses fewer actuators to achieve a high degree of freedom [13]. It provides a high degree of flexibility and can be controlled through various methods, including hydro-pneumatic pressure, wire drives, temperature variations, air pressure, and electrorheological or magnetorheological techniques [14–17]. These control methods help reduce operation time and enhance positioning accuracy [10,18,19].

Hofer et al. [20] introduced a spherical soft robotic arm made of fabric, incorporating a spherical pneumatic joint between two arms. The joint is actuated by three actuators, providing two degrees of freedom, and the study analyzes how pressure adjustments in the actuators affect positioning accuracy under different stiffness conditions. Sparrman et al. [16] developed a pneumatic soft robotic arm using silicone, with molds created by 3D printing. The arm comprises hollow silicone wave tubes, where the silicone composition was adjusted to balance rigidity and flexibility, producing an optimal arm that is strong enough to maintain shape, flexible enough to expand and contract, and durable enough to withstand repeated inflation cycles. However, due to the arm's complex structure and highly nonlinear motion, precise positioning and control remain challenging.

On the other hand, for the robotic arm to operate effectively in environments with heavier loads, its material structure must be strengthened. This can be achieved by connecting multiple rigid joints in series to increase its load capacity. When combined with the wire-driven method, the system mimics the contraction of natural tendons. Adding springs or other damping components to support the arm and isolating the drive mechanism from the arm itself can effectively reduce both the size and weight, offering the advantages of a quick response and low latency [21]. Li et al. [22] presented a wire-driven, bio-inspired, multi-jointed robot modeled after the structure and movement of snake-like and continuum robots. Multiple joints replicated the skeleton of a snake, while wire-driven systems imitated the muscles of an octopus's foot. Flexible hoses were inserted into the joints to distribute stress evenly during bending, resulting in smoother movements. This wire-driven approach requires only two to three motors to enable bending in a snake-like robotic arm. Compared to link-based arms, this structure offers significantly higher fault tolerance; even with minor component failures, the system can maintain correct operation [23]. Integrating cable-driven robotic arms of varying sizes can further expand the overall workspace [24]. However, due to the complex movement structure of the arm, accurately estimating and controlling its spatial position remains a considerable challenge [20].

The design of end effectors used in fruit picking directly affects the control methods that follow [25,26]. Common types include gripper claw picking [27], scissor-type cutting [28], and suction [10]. Fan et al. [19] developed a three-fingered rubber gripper, investigating various apple picking methods and comparing the resulting surface damage on the apples. Zhang et al. [10] proposed a vacuum-based apple picking system, where a soft silicone structure conforms closely to the fruit surface to maximize suction efficiency. However, excessive separation force can exert undue strain on the branches, increasing the risk of detaching neighboring fruits. These methods are unsuitable for harvesting delicate fruits like strawberries. Although de Preter et al. [18] improved gripper materials to fully enclose strawberries and reduce surface damage, the picking mechanism still applies a gentle pull and rotational force to separate the calyx, stem, and fruit. In practice, farmers prefer to retain the calyx to minimize the risk of contamination during washing, which remains challenging for current robotic harvesting solutions.

For more fragile fruits like strawberries, a scissor mechanism is designed to cut the fruit stem, while a clamp holds the fruit until the arm reaches the target container, preventing damage to the surface of the strawberry. However, accurately identifying the pedicel cutting point and achieving precise visual positioning control for the robotic arm remains a significant challenge [29].

Machine vision systems are frequently used to assess the state of objects, including their appearance, location, presence of disease, or ripeness. Typically, one or more cameras capture multiple images [30–32], which are then fed into recognition systems to detect target objects and their positions within the image.

Among the diverse object detection methods, image processing and learning-based techniques are the most widely used. Hayashi et al. [33] proposed an image processing method that converts the RGB color space to the hue–saturation–intensity (HSI) color space, allowing the system to determine strawberry ripeness and locate the stem position in the image, achieving a 60% success rate in detecting the target stem. Ge et al. [34] utilized Mask R-CNN along with a refinement method to identify strawberries of varying ripeness, achieving an F1-score of 0.94 for mature strawberries. Sun et al. [35] further integrated a deep learning model with active sensing to detect and locate strawberries. However, as the number of target categories increases, the recognition time also rises, requiring substantial computational resources for real-time picking applications. Although Perez-Borrero et al. [36] developed a strawberry instance segmentation method based on a fully convolutional neural network to improve real-time performance, low accuracy remains a limiting factor, reducing the overall effectiveness of robotic picking.

Utilizing the real-time capabilities and reduced reliance on large datasets of the YOLO (You Only Look Once) architecture, several enhanced methods have proven effective in detecting fruits and their 2D spatial positions [37,38] or in locating stem positions [39]. Furthermore, technologies such as LiDAR, stereo vision, and depth cameras are employed to measure the depth distance of fruits from specific viewpoints [9,40,41], enabling a precise determination of their 3D spatial coordinates. By incorporating additional information on fruit size, scale, and orientation, the 3D position of the stem can be accurately estimated. These data, combined with the robotic arm's kinematic parameters, allow the system to calculate the optimal approach angle and joint movements for effective fruit grasping [40]. Finally, the drive system directs the robotic arm to the target position, enabling it to perform tasks such as spraying or fruit harvesting.

Manually annotating large amounts of data to train recognition models has proven effective in handling complex orchard environments, including variable lighting and fruit diversity (in terms of color, size, and shape). However, most of the existing literature focuses on fruit object recognition in changing environments [42], with little discussion on the performance of robotic arm positioning control. In the application of robotic gripping and control, the arm often requires assistance from contact-based sensors, such as tactile sensors, force sensors, and bending sensors, to perform the operations effectively [43, 44]. Fuzzy logic is commonly used to address control issues in uncertain and dynamic conditions within complex scenarios. Originally proposed by Zadeh [45], fuzzy logic is widely applied for precise motor control in agricultural functions such as plowing, sowing, watering, pesticide spraying, and temperature monitoring [46,47]. In 2015, Dimeas et al. [27] developed a fuzzy controller-based gripper with a pressure sensor to adjust grasping force and detect uneven force distribution. This method is often combined with other intelligent control approaches, such as neural networks [48,49]. Subsequently, joint parameters, sensor values, and joint angle variations in the robotic arm are used to build a knowledge base or train the network model. The end effector is driven to the target position through a vision-based servo system and the aforementioned intelligent control methods. However, the picking process still requires considerable time, leaving room for improvement in real-time fruit picking efficiency.

This study presents a precise and stable strawberry picking system designed specifically for handling delicate fruits, overcoming the limitations of existing methods. Unlike the robotic arm joint design in [50], the arm in this study employs multiple ball joints in series, mimicking muscle contractions observed in snake-like movements. This configuration offers enhanced operational flexibility in irregular or variable agricultural environments, surpassing traditional designs that rely solely on rigid structures. Additionally, a two-stage fuzzy logic control system (2s-FLC) is used to precisely control the length of arm and bending angle, significantly improving picking accuracy and simplifying the control complexity compared to conventional systems. Furthermore, unlike traditional gripping or suction-based approaches, the end effector in this system is designed to cut and hold the strawberry stem by clamping only at the stem and base, avoiding direct contact with the fruit surface. This design effectively minimizes the risk of damage to fragile fruits like strawberries and is rarely seen in the current literature.

This paper is organized as follows: Section 2 outlines the methodology for establishing the fruit picking platform and the process of picking fruits, covering the kinematics of the robot arm, the design of the multi-joint arms and grippers, the fruit picking platform, and the vision-based position control method for the robotic arm. Sections 3 and 4 present the experimental results and discussion. The Section 5 provides the conclusion of the paper.

2. Methodology

This chapter discusses the kinematics of the multi-joint arm, the design and development of the arm and clamps, an outline of the parts of the picking platform, and the position control method for fruit picking. It also details the 2s-FLC method for regulating the motion of robotic arm.

2.1. Motion Model

The robotic arm proposed in this study is composed of multiple joint elements connected in series. The Y-axis swinging motion of the three-joint arm in the 2D plane is illustrated in Figure 1. Assuming the length of each joint is represented as l_n ($\mathrm{n} = 1, 2, \ldots, N$), N represents the number of joints, and the link length between each joint is equal, with $l_1 = l_2 = \ldots = l$ cm, and the total length of the joint arm is lN. The initial posture of the arm is vertical to the ground. When the arm bends, assuming the bending angle between each joint is equal, $\theta_1 = \theta_2 = \ldots = \theta_N = \theta$, the total bending angle is $\theta_{\mathrm{Total}} = N\theta$. When the displacement at each joint is equal, $\Delta_1 = \Delta_2 = \ldots = \Delta_N = \Delta$, the total displacement is $\ell = N\Delta$. In the 2D plane, assuming the origin of the arm is at coordinate $\mathrm{p}_0(y_0, z_0)$, the coordinates of the n-th joint point $\mathrm{p}_\mathrm{n}(y_\mathrm{n}, z_\mathrm{n})$ when the arm bends are as follows.

$$y_\mathrm{n} = y_0 + \sum_{\mathrm{k}=1}^{\mathrm{n}} l \sin(\mathrm{k}-1)\theta \tag{1}$$

$$z_\mathrm{n} = z_0 - l - \sum_{\mathrm{k}=1}^{\mathrm{n}} l \cos(\mathrm{k}-1)\theta \tag{2}$$

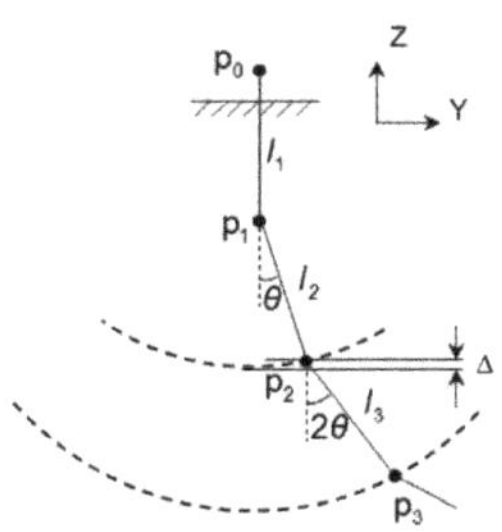

Figure 1. Schematic of joint arm swing (the black dotted line indicates the trajectory of the arm swing).

In this study, the robotic arm can be considered a multi-link structure utilizing spherical joints. The total degrees of freedom of the arm can be calculated using the Chebychev–Grübler–Kutzbach criterion as follows:

$$F = 6(N-1-b) - \sum_{j=1}^{b} f_j \tag{3}$$

where b represents the number of joints, and f_j is the number of constrained degrees of freedom for the j-th joint. For spherical joints, each joint constrains 3 degrees of freedom, allowing 3 rotational degrees of freedom. For example, in a three-link mechanism (with $N = 4$ links, including the base) with $b = 3$ spherical joints, the total degrees of freedom is 9, where $f_j = 3$ for each joint $j = 1, 2, 3$.

2.2. Mechanism and Harvesting Platform

2.2.1. Robot Arm

Several joint components are connected in series to form an arm (Figure 2), which is made up of four tapered spherical components printed using a 3D printer. The sphere of each joint is inserted into the spherical recess on the neighboring joint, allowing it to rotate freely. The root joint of the arm is a level base (upper left in Figure 2), designed for attachment to the motor bracket. Small holes are drilled at the four outer corners of each tapered component, through which four thin wires are threaded. These wires are fastened at the four corners of the terminal joint (bottom left in Figure 2). When the wires are pulled, the arm swings.

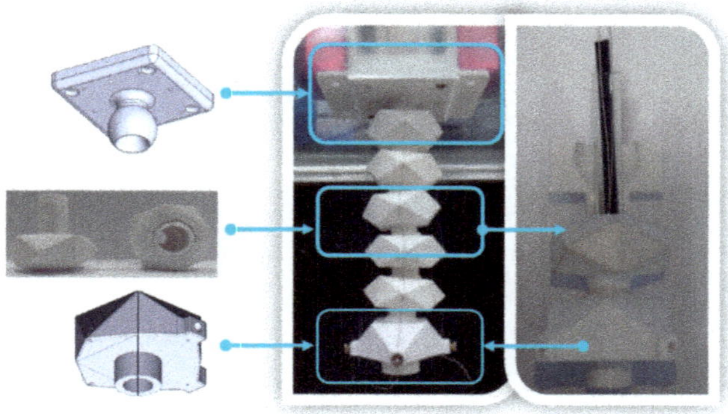

Figure 2. Structure of the multi-jointed robotic arm (**center**); base of the arm (**top left**) and end joint (**bottom left**); internal hoses and thin wires within the arm (**right**).

Additionally, after the joint components are connected in series, a plastic hose is inserted inside the arm (right in Figure 2), ensuring smoother motion during swinging. A thin wire is threaded through the hose, with its end secured to the gripping mechanism at the end of the arm. The driving mechanism adjusts the tension on the wire to open or close the gripping mechanism.

2.2.2. Clamping and Cutting

Two sets of clamping and cutting tools are designed and mounted on the upper and lower sides of the endpoint of arm. This design allows the arm to pick strawberries on the left and right cultivation tubes without needing to rotate, enabling rapid fruit picking. The design concept of the cutter is inspired by the structure of scissors and consists of two blades. Each blade consists of a three-layer structure—top, middle, and bottom (see Figure 3a)—all made of acrylic material. The middle layer contains a trapezoidal groove to hold the blade for cutting strawberry stems, while the top and bottom layers are securely fitted to enclose the blade within the central layer. The ends of the two blades are gear-shaped and meshed together. One end of a thin wire is connected to a round hole at the end of the middle layer of the blade on one side, while the other end is attached to the disk. The center of the disk is coupled to the motor bearing. When the disk is idle, the springs attached to both sides of the clamp keep it in an open position (as shown in Figure 3b). As the disk rotates, the wire is pulled, generating sufficient force to overcome the tension of the springs and bring the two blades together toward the center (as shown in Figure 3c). Figure 3d,e show the physical appearance of the clamping tool and its integration with the joint arm. When the blades close, the stem is cut and securely gripped, ensuring that the strawberry does not fall (Figure 3f). Additionally, a nozzle is fixed at the center of the joint of arm (bottom left in Figure 3f), connected to a plastic tube, and is used for spraying.

Figure 3. Design of clamp and cutting tool. (**a**) The structure of clamping and cutting tools; (**b**) clamp in the open state; (**c**) clamp in the closed state; (**d**) prototype of the two sets of clamps; (**e**) mounting of the clamp on the joint arm (with the upper clamp in the open state and the lower clamp in the closed state); (**f**) the clamp in action for picking strawberries (the nozzle is installed inside the tube, bottom left).

It is worth noting that in the experiment, the gripper is designed to simultaneously clamp and cut the fruit stem. It securely holds the stem attached to the fruit while avoiding the main plant stem above the blade, ensuring efficient harvesting (see Figure 4). Soft foam lining provides a gentle yet firm grip on the fruit stem.

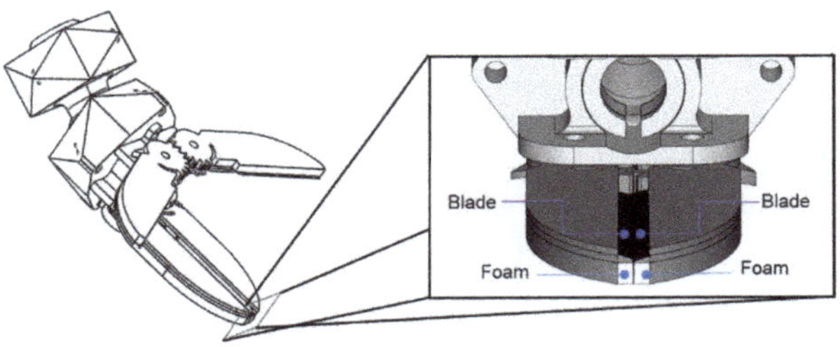

Figure 4. Clamp cutting part with two blades and gripping part with two foam pads.

2.2.3. Platform for Harvesting Strawberries

The robot arm module is mounted on a modified pulley system [51], which moves back and forth along an aluminum extrusion track (Figure 5) on the cultivation support. Two cameras (Model: Brio Ultra 4K HD, Logitech Inc., San Jose, CA, USA) are mounted on either side of the module at a fixed height above the ground to capture images of strawberries on water pipes positioned to the left and right. A microcontroller (Model: Arduino UNO R3, Arduino, Turin, Italy) combined with a sensor expansion board (Model: Sensor Shield V5.0) is used to drive the lifting mechanism within the module (Figure 6). This mechanism includes four built-in servo motors (Model: LD-27MG, Hiwonder Inc., Shenzhen, China): the two lower, symmetrically placed motors are used to regulate the bending angle of the arm, while the two upper motors control vertical arm movement and gripper operation, each connected to a disk wound with thin wire. As the motors rotate, the wire length is adjusted, thereby controlling the arm length, bending angle, and the opening and closing of the clamp.

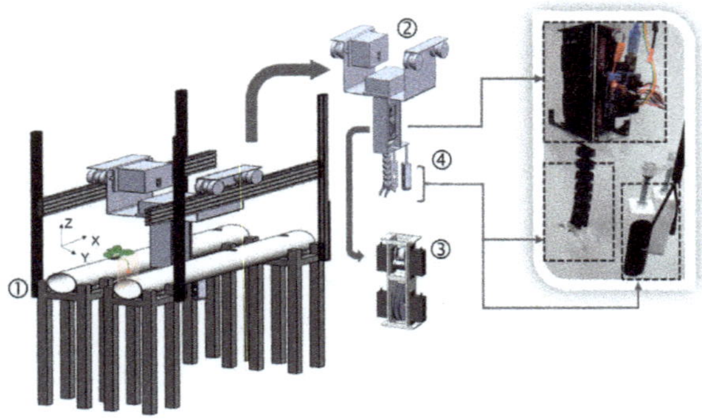

Figure 5. Hydroponic fruit picking platform: ① hydroponic PVC pipe and aluminum extrusion track; ② pulley module; ③ module for raising and lowering the arm; ④ arm with two camera units.

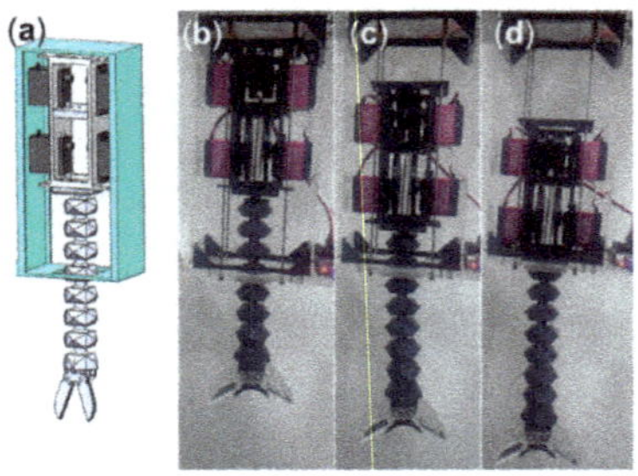

Figure 6. Robotic arm and lifting module: (**a**) prototype of lifting module and mechanism; (**b–d**) show the actions of extending the robotic arm.

2.3. *Vision-Based Control for Harvesting*

This chapter presents the development of the fruit recognition model, the visual positioning control for strawberries, and the lifting and bending mechanisms of the robotic arm.

2.3.1. Fruit Detection and Positioning

The strawberry recognition model is based on YOLOv4, with CSPDarknet53 as its main framework. SPPNet and PANNet architectures are used for feature extraction, and the extracted features are then passed to the Head layer for object detection, loss function calculation, and intersection over union (IoU) evaluation [52].

The process of establishing the strawberry detection model is illustrated in Figure 7. Replica strawberry models, commonly used for landscaping, were selected as picking targets in three sizes: 3.5 cm, 4 cm, and 6 cm (Figure 8). Handheld cameras and experimental cameras were used to capture strawberry objects at different viewing angles. Image augmentation techniques, including scaling, color transformation, affine transformation, and blur processing, were applied to increase the number of images. The dataset was divided into test and validation sets in predetermined ratios, which were used for model training, testing, and verification. Once the model is constructed, it is deployed on the required devices, and its object recognition accuracy is evaluated under different IoU conditions. Once the strawberry within the image is detected and bounded, the inrange() function from the OpenCV library is used to specify a target color to identify the desired object within the bounding box. The cv.contourArea() function is then applied to estimate the total pixel count of the object, representing its contour area, denoted as a. Finally, the circle() function is used to determine the center point of object, indicating the location of the fruit.

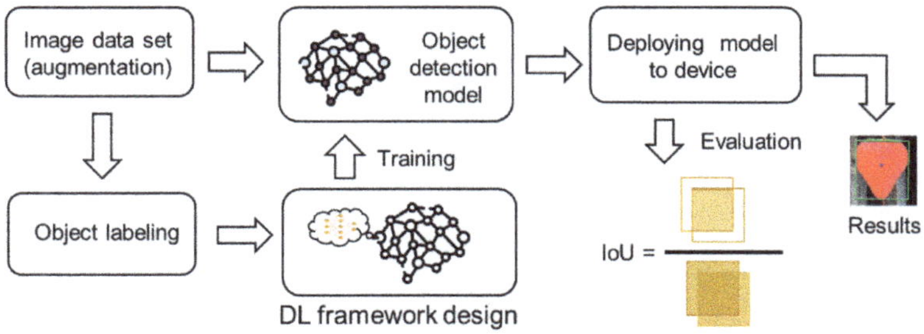

Figure 7. Process of creating the object model.

Figure 8. Side view of fruit models in three different sizes, labeled Size 1 (**a**), Size 2 (**b**), and Size 3 (**c**).

The spatial configuration of the joint arm and strawberry is illustrated in Figure 9. The point p_0 represents the arm's origin in the X-Y-Z coordinate frame. Let $o'(x', y', z')$ be the center point of fruit in the X-Y-Z frame, and let its pixel coordinates in the U-V frame, as viewed by the camera, be $\tilde{o}'(u', v')$, where u' and v' indicates the position of fruit along the U-axis and V-axis in pixels, respectively. A movable base, equipped with a vision-based positioning system, adjusts the position of arm along the X-axis.

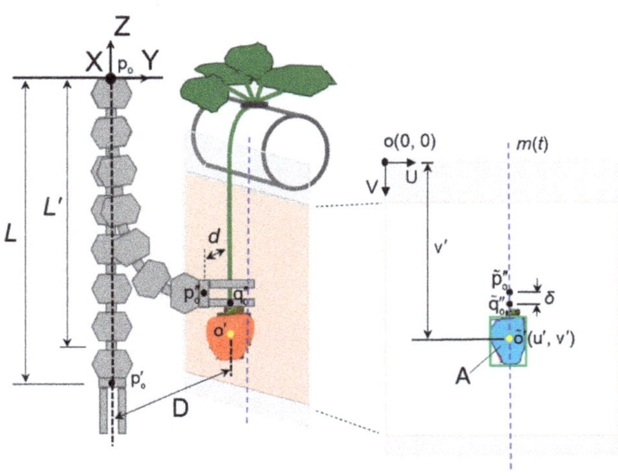

Figure 9. Coordinate configuration of arm and strawberry.

Next, assuming the total length of the joint arm is L and the distance D between the center of fruit and the arm is unknown, the arm must first adjust its length to $|L - L'|$ for fruit picking operations. L' represents the adjusted length of the arm. The end of the arm should be moved as close as possible to the strawberry while ensuring the initial position of the upper and lower cutting tools at the end of arm remains open to prevent the arm from making contact with the strawberry while bending.

Based on the above conditions and the known strawberry outline area A in the image, a 2s-FLC system is employed to adjust the position $p'_o(x'_o, y'_o, z'_o)$ of the endpoint of arm, guiding it to position p''_o, effectively controlling the joint displacement of arm along the *Y*-axis and *Z*-axis. When the arm bends to position p''_o, the lower shearing tool closes, cutting the strawberry stem at point q''_o. The variable δ represents the pixel distance between the cutting point $\tilde{q}''_o$ in the image and the endpoint of arm $\tilde{p}''_o$, which is negligible as its value approaches 0. Therefore, the projected distance between point p''_o and the stem can be approximated as $d = \overline{p''_o \tilde{p}''_o}$. The depth error can be defined as $\eta = \left| d - \hat{d} \right|$, where $\hat{d}$ represents the estimated depth distance.

2.3.2. Lateral (*X*-Axis) Position Control

The vision-based drive control system moves the platform along the *X*-axis. Initially, when the platform is stationary ($t = 0$), it is assumed that a line parallel to the image coordinate system and the V-axis in the image captured by the camera is denoted as $m(t) = u$, where the u is a fixed constant corresponding to the position of arm in space. When the center point of frame aligns with this line, the platform will stop, positioning the arm directly with the center of fruit. Let the U-axis value of the center point of object $\tilde{o}'$ in the image be represented as $u'(t)$. The speed control condition for the platform is as follows:

$$v(t+\Delta t) = \begin{cases} 0 & u = u'(t) \\ v_{max} & u \neq u'(t) \end{cases} \tag{4}$$

$v(t)$ and v_{max} stand for the speed and maximum speed of the moving system at time t. Δt is the sampling time, which includes the time for image processing and updating the speed control commands, and $p_o(t + \Delta t) = p_o(t) + v(t)\Delta t$ is the position of the end point of arm. The speed control goal is to reduce the side displacement error $\varepsilon \to \min|u - u'|$.

2.3.3. *Y*- and *Z*-Axis Position Control

The bending angle of the joints is controlled by four thin wires that pass through holes at opposite corners on both sides of each joint and are secured at the four corners of the end joint. Two of the wires are fixed at the base joint, ensuring the arm swings in a specific direction only. The other two wires are attached to two symmetrical points on a disk driven by a servo motor. As the servo motor rotates, the disk turns, indirectly enabling the winding or unwinding of these two wires, which causes the arm to bend. Assuming the radius of the disk is r, θ_s represents the rotation angle of the servo motor.

$$\theta_s / 180 = \pi r / \ell \tag{5}$$

The picking position of the arm is affected by the fruit's height and size. For strawberries of varying dimensions, even when their center points are at the same height, the arm must adjust both its extension and bending angle to suit each fruit's size. To ensure the end effector reaches close to the target for effective gripping, a 2s-FLC system is implemented to independently control the arm's length and bending angle. The architecture of proposed system is presented in Figure 10.

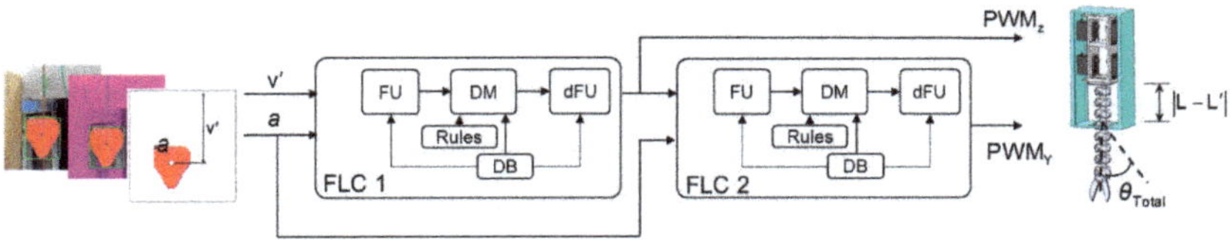

Figure 10. The block diagram of 2s-FLC system.

First, the position $\widetilde{o}'(u', v')$ of the center point of strawberry in the image, its v' value, and the area of strawberry a (in pixels) serve as the input variables for FLC 1, while the pulse width modulation signal PWM_Z is the output variable of FLC 1. This signal drives the servo motor, which rotates the disk and indirectly adjusts the length of the arm, enabling vertical movement along the Z-axis. The PWM_Z value and a are the input variables for FLC 2, with the output variable of FLC 2 being the pulse width modulation signal PWM_Y. This signal drives one servo motor to rotate the disk by an angle θ_s, indirectly adjusting the bending angle of arm θ_{Total}. Since the arm only swings in the Y-axis direction, the other servo motor remains stationary. The internal structure of both controllers includes fuzzification (FU), rules (Rules), decision-making (DM), defuzzification (dFU), and a knowledge base (DB). The design parameters of the 2s-FLC system are listed in Table 1.

Table 1. Illustration of the parameters for 2s-FLC.

Parameters	**Description/Value**	
	FLC 1	**FLC 2**
Input variables	• v' The length between center point $\widetilde{o}'$ and upper bound of captured image • a: The area of strawberry	• PWM_Z: Pulse width modulation signal (Z-axis) • a: The area of strawberry
Output variables	• PWM_Z: Pulse width modulation signal (Z-axis)	• PWM_Y: Pulse width modulation signal (Y-axis)
Membership functions	Triangular and trapezoidal	
Number of rules	15	15
Inference method	Mamdani	
Defuzzification method	Centroid method	
Performance metrics	• ε: Displacement error • η: Depth error	

A. Fuzzification and knowledge database

Fuzzification is defined as the process of converting the precise numerical values of input or output variables in the controlled system into corresponding fuzzy sets. The input or output variables are defined as the observed values or the control actions applied to the system. During the fuzzification of input variables, fuzzy terms must first be defined, such as 'Big' (B), 'Normal' (N), 'Small' (S), 'Low' (L), 'Medium' (M), 'High' (H), 'Medium Low' (ML), 'Medium Medium' (MM), 'Medium High' (MH), 'High High' (HH), 'High Medium' (HM), 'High Low' (HL), 'Low High' (LH), 'Low Medium' (LM), and 'Low Low' (LL). Each term represents a fuzzy set. Four variables, including v', a, PWM_Z, and PWM_Y, have corresponding fuzzy sets, which may consist of multiple membership functions. Figure 11 shows the triangular and trapezoidal membership functions, the number of functions, and their distribution used for fuzzifying the input and output variables in FLC 1 and FLC 2. The membership function distributions for input variables v' and a of the FLC 1 are shown in Figure 11a and Figure 11b, respectively. Input variable a also serves as the input for FLC 2. The distributions of membership functions of output variable PWM_Z of FLC 1 is also shown in Figure 11c. Figure 11d,e illustrate the distribution of the five and nine membership functions for input variable PWM_Z of FLC 2 and output variable PWM_Y of FLC 2, respectively. During the fuzzification process, the variable corresponds to multiple fuzzy sets and the degree of membership to each set. For example, if the input variable v' in FLC 1 is 800 pixels, its membership degrees for 'H', 'MH', 'M', 'ML', and 'L' are 0, 0, 0.28, 0.72, and 0, respectively (Figure 11a). On the other hand, if the fruit area is 1×10^5 pixels, its membership degrees for 'S', 'N', and 'B' are 0, 0.43, and 0.58, respectively (Figure 11b). Another example, if $PWM_Z = 2000$, the membership degrees for 'LH', 'LM', 'LL', 'ML', 'MM', 'MH', 'HH', 'HM', and 'HL' are 0, 0, 0, 0, 0, 0.2, 0.8, 0, and 0, respectively (Figure 11c).

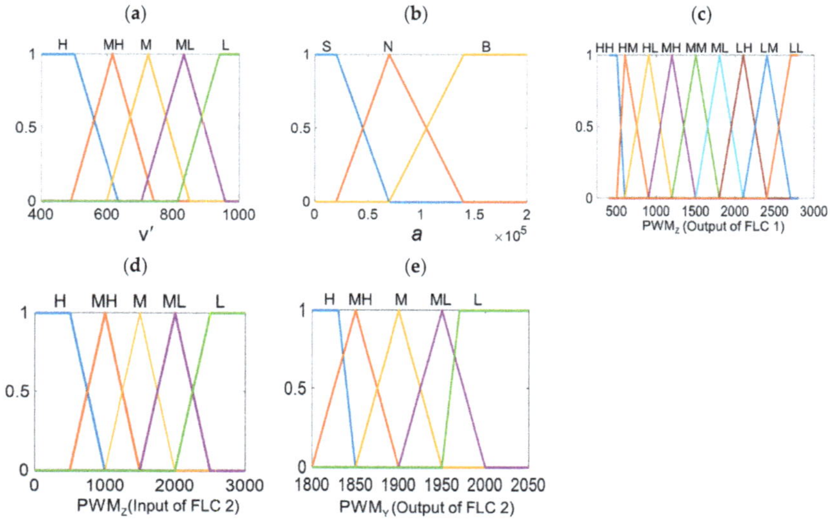

Figure 11. Input and output variable fuzzification for FLC 1 and FLC 2. (**a**) v' for input of FLC 1; (**b**) a for input of FLC 1 and FLC 2; (**c**) PWM_Z for output of FLC 1; (**d**) PWM_Z for input of FLC 2; (**e**) PWM_Y for output of FLC 2.

B. Rules, decision-making, and defuzzification

Rules are formulated based on expert knowledge and are used for inference and decision-making. In FLC 1, a total of 15 rules are established, with each rule expressed in the 'IF-THEN' format.

(1) IF (v' is H) AND (a is B) THEN (PWM_Z is HH)
(2) IF (v' is H) AND (a is N) THEN (PWM_Z is HH)
(3) IF (v' is H) AND (a is S) THEN (PWM_Z is HM)
(4) IF (v' is MH) AND (a is B) THEN (PWM_Z is HL)
(5) IF (v' is MH) AND (a is N) THEN (PWM_Z is HL)
(6) IF (v' s MH) AND (a is S) THEN (PWM_Z is HL)
(7) IF (v' is M) AND (a is B) THEN (PWM_Z is MH)
(8) IF (v' is M) AND (a is N) THEN (PWM_Z is MM)
(9) IF (v' is M) AND (a is S) THEN (PWM_Z is MM)
(10) IF (v' is ML) AND (a is B) THEN (PWM_Z is ML)
(11) IF (v' is ML) AND (a is N) THEN (PWM_Z is LH)
(12) IF (v' is ML) AND (a is S) THEN (PWM_Z is LH)
(13) IF (v' is L) AND (a is B) THEN (PWM_Z is LM)
(14) IF (v' is L) AND (a is N) THEN (PWM_Z is LM)
(15) IF (v' is L) AND (a is S) THEN (PWM_Z is LL)

Similarly, in FLC 2, a total of 15 rules are established, as detailed below.

(1) IF (PWM_Z is H) AND (a is B) THEN (PWM_Y is H)
(2) IF (PWM_Z is H) AND (a is N) THEN (PWM_Y is H)
(3) IF (PWM_Z is H) AND (a is S) THEN (PWM_Y is H)
(4) IF (PWM_Z is MH) AND (a is B) THEN (PWM_Y is MH)
(5) IF (PWM_Z is MH) AND (a is N) THEN (PWM_Y is MH)
(6) IF (PWM_Z is MH) AND (a is S) THEN (PWM_Y is MH)
(7) IF (PWM_Z is M) AND (a is B) THEN (PWM_Y is M)
(8) IF (PWM_Z is M) AND (a is N) THEN (PWM_Y is M)
(9) IF (PWM_Z is M) AND (a is S) THEN (PWM_Y is M)
(10) IF (PWM_Z is ML) AND (a is B) THEN (PWM_Y is ML)
(11) IF (PWM_Z is ML) AND (a is N) THEN (PWM_Y is ML)
(12) IF (PWM_Z is ML) AND (a is S) THEN (PWM_Y is ML)
(13) IF (PWM_Z is L) AND (a is B) THEN (PWM_Y is L)
(14) IF (PWM_Z is L) AND (a is N) THEN (PWM_Y is L)
(15) IF (PWM_Z is L) AND (a is S) THEN (PWM_Y is L)

The Mamdani method is used for fuzzy inference. It involves converting two sets of input values into corresponding fuzzy sets. Each rule is then applied to the inputs, and the membership degree of each fuzzy set is obtained. These sets are then intersected, and the result corresponds to an output fuzzy set (as shown in the blue area on the right side of Figure 12). The output fuzzy sets of all rules are then aggregated using a union operation, forming a single geometric shape (as shown in the blue area at the bottom right). Finally, the centroid method is used to convert the fuzzy output set into a specific value (represented by the thick red line). Additionally, by observing the pattern of the fuzzy inference surface, the suitability of the inference rules can be evaluated, and the rules can be adjusted if necessary to modify the initial inference surface (Figure 13).

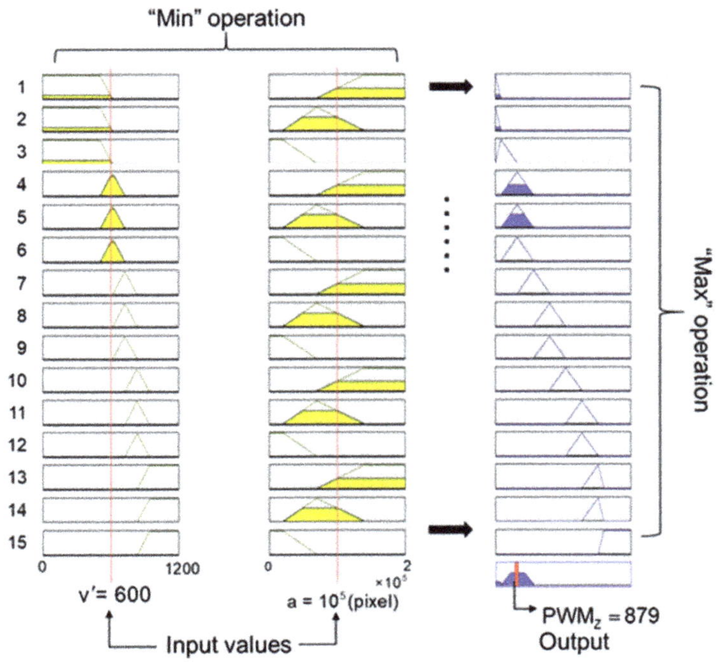

Figure 12. Example of fuzzy inference and defuzzification; fuzzy inference results when v′ = 600 and $a = 10^5$(pixel) (FLC 1).

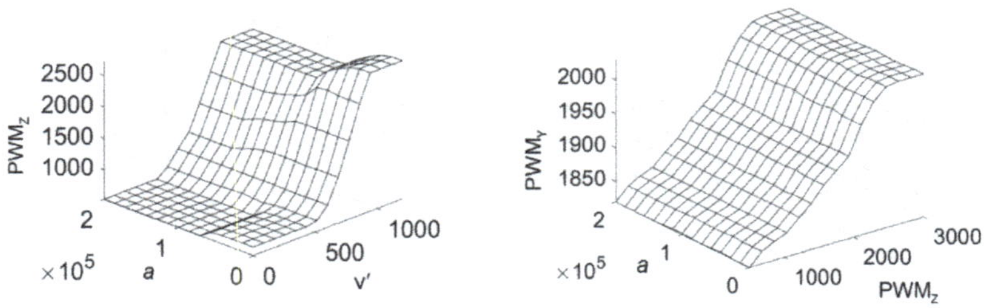

Figure 13. Fuzzy inference surfaces of FLC 1 (**left**) and FLC 2 (**right**).

3. Performance Analysis and Results

This section presents the performance results of the system used to control the joint arm for fruit grasping, focusing on lateral movement control and the positioning accuracy achieved through a 2s-FLC system. Finally, the fruit-grasping experiments were conducted in a greenhouse, where the picking performance of the robotic arm was evaluated under different weather conditions.

3.1. Object Detection

The image dataset consists of 2000 images, with 1400 used for training, 400 for validation, and 100 for testing. The required image sizes for the model during training are 416 × 416 pixels. The training parameters are as follows: batch size set to 16, 200 epochs, and the model is optimized using SGD with a learning rate of 0.0001. The number of iterations is set to 6000, and the loss function, comprising classification loss, localization loss, confidence loss, and object classification loss, is evaluated after each iteration [52]. Next, tissues were twisted into thin strips, dyed green to resemble stems, and tied to the tops of fruit models of different sizes (to the left of Figure 14). These fruit replicas were then randomly suspended along the side of the hydroponic pipe. The identification results are shown in Figure 14 (right), where, with an F1-score of 0.94, a precision of over 88% was achieved.

Figure 14. Strawberry identification results.

3.2. Positioning Accuracy

3.2.1. Range of Motion

The Simulink R2022a software tool was used to develop a bending model of the joint arm and simulate its bending behavior [53], resulting in a smooth bending curve (Figure 15a). However, in the actual arm bending test, if the plastic tube is not placed in the intra-arm channel, the bending is irregular (Figure 15b). When the tube is inserted, the joint arm's bending closely matches the simulation results (Figure 15c). The swing trajectory of the arm reveals a clear difference between the cases where the tube is absent and when it is inserted into the joint channel (Figure 16a). The red line represents the arm's movement without the tube, while the blue line shows the trajectory with the tube. Swing trajectories for arms of different lengths are shown in Figure 16b, indicating that the maximum swing range reaches approximately 20 cm.

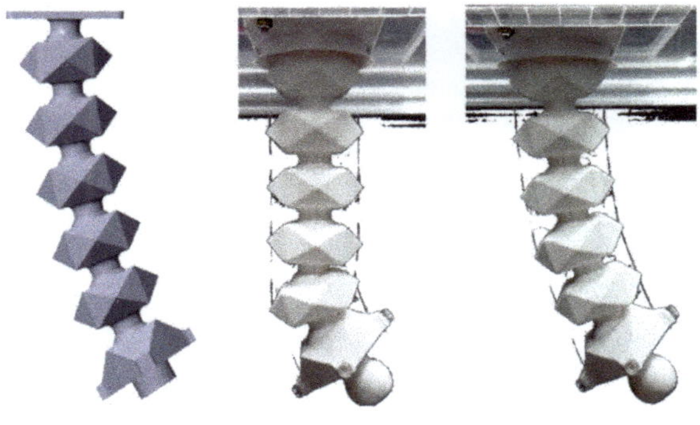

(**a**) (**b**) (**c**)

Figure 15. Bending test of the jointed arm. (**a**) Simulated joint arm bending using the Simulink tool, (**b**) arm bending without the plastic tube inserted, and (**c**) arm bending with the plastic tube inserted.

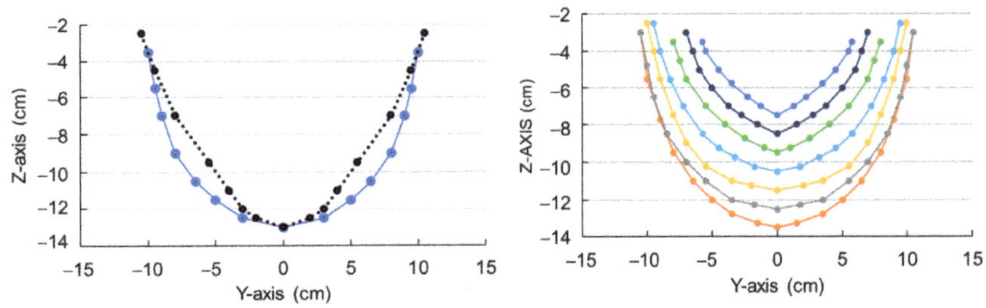

(**a**) (**b**)

Figure 16. Swing trajectory of the joint arm. (**a**) Bending trajectories of PVC plastic pipes with insertion (blue line) and without insertion (black dashed line); (**b**) Relationship between joint arm lengths and swing trajectories (each color represents the swing trajectory for a different joint arm length).

3.2.2. Lateral and Depth Error

In the fruit picking experiment, metrics such as displacement deviations ε, η, FPR, and FDR were used to evaluate the object grasping performance of the fruit picking platform. The Scikit-fuzzy toolkit was used to implement control of the Z-axis and Y-axis displacement and the bending angle of the robotic arm. The operation process was divided into three stages: platform stopping, arm bending, and gripper picking. First, we observed whether the platform could align the robotic arm with the strawberry replica in the grasping direction. Each size of the strawberry replica was tested five times. The results showed that the average displacement deviations for Size 1, Size 2, and Size 3 objects were 2 mm, 1.4 mm, and 0.8 mm, respectively.

A tissue was rolled into thin strips, dyed green to resemble stems, and then attached to the tops of fruit models of various sizes. Next, using three different fruit sizes, each fruit model was suspended at five random positions. Under the assumed conditions, we observed the distance d between the end position p_o'' of the robotic arm after bending the fruit stem. Additionally, the expected values for this distance were set based on the volumes of Size 1, Size 2, and Size 3, assumed to be 10 mm, 15 mm, and 30 mm, respectively. Ideally, the value of d matches the expected value. Manual measurement is used to obtain d. Table 2 presents a comparison of the average error values obtained using linear control and 2s-FLC. The results also show that using the 2s-FLC method achieves lower error values. In addition, the process of successfully picking a single fruit was observed, and the time taken for each step was recorded, including arm lifting, bending and adjustment, clamping and cutting, lowering the arm, opening the clamp, and returning to the arm origin. Bending and adjusting the arm took the longest, averaging about 1.9 s (Figure 17), followed by arm lifting, which averaged 1.5 s. The clamping and cutting, along with opening the clamp, required the shortest time, averaging about 0.5 s. The overall average time to pick a single fruit was 6.5 s. Excluding the arm's reset time, the average picking time per fruit is reduced to 5 s.

Table 2. Depth error η between the gripper and the target using a linear control method and a 2s-FLC method, respectively.

Fruit	Test 1	Test 2	Test 3	Test 4	Test 5	Average Error (mm)
Size 1	28.0/24.0 [z]	25.0/20.0	31.0/30.0	29.0/27.0	28.0/29.0	28.2/26.0
Size 2	19.0/19.0	20.0/14.0	19.0/16.0	17.0/13.0	15.0/11.0	18.0/14.6
Size 3	9.0/9.0	6.0/4.0	11.0/13.0	12.0/6.0	10.0/7.0	9.6/7.8

[z]: Linear control/2s-FLC.

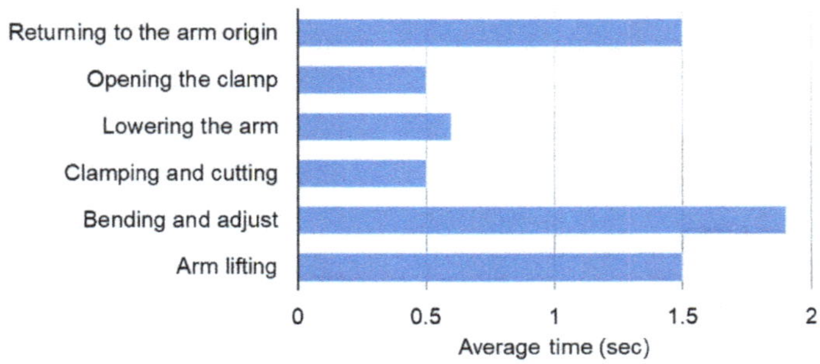

Figure 17. Average time per fruit for single fruit picking operation.

Using three different fruit sizes randomly placed at five locations, the performance of the gripper in grasping the fruit pedicel was tested. The results were categorized as 'successful picking', 'dropped after picking', or 'failed picking'. Table 3 illustrates the findings. Importantly, in Test 4 of the treatments, successful picking results were achieved for all three fruit sizes, with the observed positions of the fruits consistently located toward the lower central area of the image frame.

Table 3. Test results of fruit picking by grippers.

Size of Fruit	Treatments				
	Test 1	Test 2	Test 3	Test 4	Test 5
Size 1	△	•	×	•	×
Size 2	△	•	•	•	•
Size 3	•	×	•	•	△

•: successful picking. △: dropped after picking. ×: failed picking.

3.3. *Environmental Condition and Results*

The trained model was then deployed to the embedded development board (Model: Jetson NANO, NVIDIA Co., Santa Clara, CA, USA) located on the side of the lifting module. Five fruits of Size 1, Size 2, and Size 3 were randomly suspended on one side of the hydroponic pipe. The platform moved at a speed of 10 cm/s, as shown in Figure 18, with the hydroponic pipe measuring 3.5 m in length. The experiment lasted for three days under mostly cloudy weather. The test period ran from 10:00 AM to 4:00 PM, with the mobile module moving from one end of the pipe to the other once every hour. During the movement, a single camera continuously captured images at 10 frames per second (fps) while the identification system detected strawberries in the images. A total of 2450 images were collected per day. Three evaluation metrics, including Precision, Recall, and Average Precision (AP), were used to assess the identification performance of deep learning model.

Figure 18. Snapshot of the experimental site (strawberry models of different sizes hung on one side).

The average identification performance of the detection model during each time period over three days is shown in Figure 19. The average precision, recall, AP@0.5, and AP@0.5:0.05:0.95 were 0.891, 0.880, 0.891, and 0.718, respectively. AP@0.5:0.05:0.95 represents the average AP value calculated at IoU thresholds ranging from 0.5, 0.55, . . . , up to 0.95.

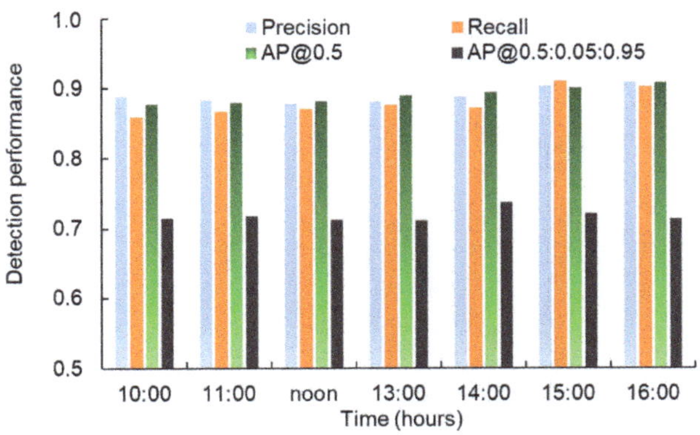

Figure 19. Performance comparison of the detection model at various times.

The fruit picking experiment for this platform was conducted in an outdoor greenhouse at various time intervals and under different weather conditions (longitude: 120°60′61.30″ E; latitude: 22°64′65.96″ N) during the summer of 2023. First, on sunny days, three periods were selected for testing: early morning (5:00–7:00), mid-morning (9:00–11:00), and late afternoon (16:00–18:00). Strawberry replicas were positioned along the sides of the hydroponic pipes (Figure 20). During the early morning and mid-morning sessions, picking was carried out on one side only, while both sides were tested in the late afternoon. For the single-sided experiment, 10 strawberry replicas were hung on one side of the pipe, and for the double-sided experiment, 9 replicas were hung on each side, totaling 18. The stems of the strawberry replicas were tied with thin plastic strings that break easily. The results of fruit picking and positioning are shown in Table 4. The camera images were evenly divided into upper, middle, and lower sections, representing the regions where the fruit was successfully picked. Figure 21 shows a side view of the operator continuously picking two fruits using the system.

Figure 20. Strawberry picking experiment site (with strawberry models of different sizes hanging on both sides).

Table 4. Test results of fruit picking (single day).

Time Slots	Fruit Location	Number of Fruits	Number of Picked Fruits	Size of Fruits			Location of Fruit		
				Size 1	Size 2	Size 3	Upper	Medium	Lower
05–07	One side	10	6	1	3	2	1	2	3
09–11	One side	10	7	0	4	3	0	4	3
16–18	Double side	18	6	1	2	3	1	4	1

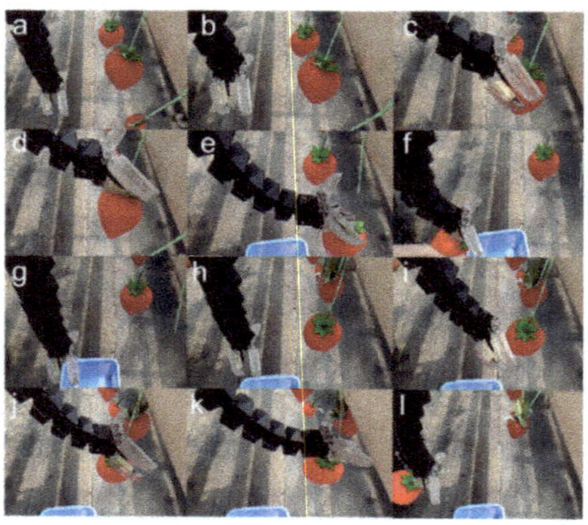

Figure 21. Snapshots of the joint arm grasping a strawberry. (**a**) The joint arm is lowered and aligned with the target; (**b**) the joint arm rises; (**c**) the joint arm bends; (**d**) the gripper cuts the stem; (**e**) the gripper clamps the stem; (**f**) the arm is lowered; (**g**) the gripper releases the stem; (**h**) the mobile platform moves to the next target. Images (**i–l**) respectively illustrate the lifting and bending of arm toward the strawberry stem (**i**,**j**), the gripping action (**k**), and finally the arm in a lowered position (**l**).

Finally, the experiment was repeated three times, and both the fruit picking success rate (FPR) and the fruit drop rate (FDR) after grasping were measured. The standard for successful picking is that the arm bends and the gripper securely grasps the fruit until the arm returns to its original position. In contrast, the fruit drop rates show that during the process, the arm either fails to grasp the fruit or the fruit drops mid-operation. As shown in

Table 5, the FPR reached up to 82%. The lowest success rate, only 32%, occurred between 5 and 7 AM under cloudy conditions. Interestingly, under cloudy conditions and near evening, there was no fruit drop at all.

Table 5. Fruit picking success rate.

Time / Weather	Early Morning (05:00–07:00)		Morning (09:00–11:00)		Late Afternoon (16:00–18:00)	
	FPR (%)	FDR (%)	FPR (%)	FDR (%)	FPR (%)	FDR (%)
Sunny	82	18	64	27	55	9
Partly cloudy	55	36	73	18	55	27
Cloudy	37	9	46	18	37	0

4. Discussion

The success rate of fruit picking depends on the accurate identification of the center of the fruit. We observed that the highest success rate occurs during bright early mornings (5–7 AM). In contrast, during overcast mornings or in the late afternoon (4–6 PM) under cloudy skies, the success rate significantly decreases. The light intensity during these periods is relatively low, leading us to infer that weak lighting prevents precise object detection. This result is consistent with earlier research findings [54].

In the experiment, the object detection system operated at 10 fps to perform strawberry identification and estimate the fruit area. However, signal transmission delays between the microcontroller and peripheral devices, at times, led to excessive lateral positioning errors in the mobile platform. This misalignment caused the strawberry stem to deviate significantly from the intended picking position, preventing the cutting tool from effectively cutting the stem. Higher-performance graphic processing units (GPUs) and controllers can execute complex deep learning and image processing algorithms more efficiently, which would improve the performance of the object detection system [55]. The most time-consuming operation of the proposed system is the adjustment of the bending angle of the arm and the picking position, which takes approximately 1.9 s. Using a more advanced processor would help reduce picking time, enhance the system's tolerance for positional errors, and reduce the chance of significant lateral misalignment.

On the other hand, once the harvesting system detects the center point of the target, the platform halts, adjusting the length of the arm and bending angle. However, slight shaking or instability during operation can easily cause the fruit to drop. Furthermore, variations in the placement and height of the water pipe shelf relative to the ground may prevent the robotic arm from precisely aligning with the fruit stem. The flexibility of the wire also influences harvesting accuracy. During experiments, continuous wire tensioning altered its length, causing the end of the arm to miss the intended position. Furthermore, as the arm bends, the wire lengths on the inner and outer sides of the joints vary. This slack can cause the wire to slip out of the rotating shaft groove, leading to it coming off the shaft. Finally, when the gripper at the end of the arm tries to grab the fruit, it sometimes fails to cut the stem. This happens when the stem is near the edge of the gripper, where the cutting force is too weak. The experimental results demonstrate that the wire-driven joint arm offers excellent flexibility in precise position control. However, maintaining consistent tension on the wires can lead to issues such as the wire snapping or slipping off the rotating shaft. Additionally, while the 3D-printed components are cost-effective, they tend to become fragile when exposed to temperature fluctuations and sunlight in outdoor greenhouse environments. This issue could be addressed with more careful material selection.

According to our observations, the successfully picked fruits were all located in the lower central area of the image frame. In this region, the 2s-FLC appeared particularly effective, with the picking mechanism displaying minimal sway. Adjusting the distribution surface of the fuzzy rules within the 2s-FLC system might yield different results, which warrants further investigation. This harvesting system is optimized for movement along a mounted rail, and the current gripper is specifically designed for handling herbaceous plants. Most harvesting robot prototypes achieve a picking success rate of around 66%, with an average picking cycle of 33 s per fruit [56]. In contrast, our experimental results show an average picking time of 6.5 s per fruit, with a maximum success rate of 82%, demonstrating the high efficiency of this design. Excluding the arm reset time, the picking time can be further reduced to 5 s, comparable to the single fruit picking time for kiwifruit achieved by Mu et al. [57]. However, due to constantly changing crop backgrounds and lighting conditions, the position of the end effector during picking can vary, affecting the picking success rate [8,42,58]. This variability impacts the overall performance of the picking mechanism. In addition, to improving harvest rates, the non-contact gripping method used in this study minimizes fruit surface damage, enhancing the commercial value of the harvester and the quality of the fruit.

5. Conclusions

The fruit picking mechanism and precision control strategy proposed in this study were initially applied to strawberry harvesting in a hydroponic greenhouse. The cutting gripper, designed to grasp the fruit stem, minimizes surface damage to the strawberries. Three models of ripe strawberries of different sizes were used to validate the effectiveness of the proposed system. The harvesting system achieved a detection accuracy of 90% and a harvesting success rate of 82%, with an average picking time of 6.5 s per strawberry, reduced to 5 s when arm recovery was omitted, enabling the system to potentially pick around 553 strawberries per hour. However, performance is affected by lighting conditions: success rates drop to 32% in early cloudy mornings, while higher success rates are recorded in bright early mornings. Light intensity notably influences picking accuracy, with increased errors in low-light conditions. Future work will include assessing the system with real fruit under natural conditions, broadening image recognition types (e.g., fruit quality inspection), and integrating a spraying system to enhance flexibility.

Author Contributions: Conceptualization, C.-L.C. and C.-C.H.; methodology, C.-L.C.; software, C.-C.H.; verification, C.-C.H.; data management, C.-C.H.; writing—manuscript preparation, C.-L.C. and C.-C.H.; writing—review and editing, C.-L.C.; visualization, C.-L.C.; supervision, C.-L.C.; project management, C.-L.C.; fund acquisition, C.-L.C. All authors have read and agreed to the published version of the manuscript.

Funding: This research was funded by the National Science and Technology Council, grant number NSTC 112-2221-E-020-013; NSTC 113-2221-E-020-018.

Institutional Review Board Statement: Not applicable.

Informed Consent Statement: Not applicable.

Data Availability Statement: Data are contained within the article.

Conflicts of Interest: The authors declare no conflicts of interest.

References

1. Montoya-Cavero, L.E.; de León Torres, R.D.; Gómez-Espinosa, A.; Cabello, J.A.E. Vision systems for harvesting robots: Produce detection and localization. *Comput. Electron. Agric.* **2022**, *192*, 106562. [CrossRef]
2. Botterill, T.; Paulin, S.; Green, R.; Williams, S.; Lin, J.; Saxton, V.; Mills, S.; Chen, X.; Corbett-Davies, S. A robot system for pruning grape vines. *J. Field Robot.* **2017**, *34*, 1100–1122. [CrossRef]
3. Zahid, A.; Mahmud, M.S.; He, L.; Heinemann, P.; Choi, D.; Schupp, J. Technological advancements towards developing a robotic pruner for apple trees: A review. *Comput. Electron. Agric.* **2021**, *189*, 106383. [CrossRef]
4. Bechar, A.; Vigneault, C. Agricultural robots for field operations. Part 2: Operations and systems. *Biosyst. Eng.* **2017**, *153*, 110–128. [CrossRef]
5. Li, K.; Huo, Y.; Liu, Y.; Shi, Y.; He, Z.; Cui, Y. Design of a lightweight robotic arm for kiwifruit pollination. *Comput. Electron. Agric.* **2022**, *198*, 107114. [CrossRef]
6. Zujevs, A.; Osadcuks, V.; Ahrendt, P. Trends in robotic sensor technologies for fruit harvesting: 2010–2015. *Procedia Comput. Sci.* **2015**, *77*, 227–233. [CrossRef]

7. Xiong, Y.; From, P.J.; Isler, V. Design and evaluation of a novel cable-driven gripper with perception capabilities for strawberry picking robots. In Proceedings of the 2018 IEEE International Conference on Robotics and Automation (ICRA), Brisbane, Australia, 21–25 May 2018; pp. 7384–7391. [CrossRef]
8. Williams, H.A.; Jones, M.H.; Nejati, M.; Seabright, M.J.; Bell, J.; Penhall, N.D.; MacDonald, B.A. Robotic kiwifruit harvesting using machine vision, convolutional neural networks, and robotic arms. *Biosyst. Eng.* **2019**, *181*, 140–156. [CrossRef]
9. Ling, X.; Zhao, Y.; Gong, L.; Liu, C.; Wang, T. Dual-arm cooperation and implementation for robotic harvesting of to-matoes using binocular vision. *Robot. Auton. Syst.* **2019**, *114*, 134–143. [CrossRef]
10. Zhang, K.; Lammers, K.; Chu, P.; Li, Z.; Lu, R. System design and control of an apple harvesting robot. *Mechatronics* **2021**, *79*, 102644. [CrossRef]
11. Rahul, K.; Raheman, H.; Paradkar, V. Design and development of a 5R 2DOF parallel robot arm for handling paper pot seedlings in a vegetable transplanter. *Comput. Electron. Agric.* **2019**, *166*, 105014. [CrossRef]
12. Kamandar, M.R.; Massah, J.; Jamzad, M. Design and evaluation of hedge trimmer robot. *Comput. Electron. Agric.* **2022**, *199*, 107065. [CrossRef]
13. Chubb, K.; Berry, D.; Burke, T. Towards an ontology for soft robots: What is soft? *Bioinspir. Biomim.* **2019**, *14*, 063001. [CrossRef]
14. Huang, C.C.; Chang, C.L. Design and implementation of bio-inspired snake bone-armed robot for agricultural irrigation application. *IFAC-PapersOnLine* **2019**, *52*, 98–101. [CrossRef]
15. Chang, C.L.; Shie, J.L. Design and implementation of an actuator for the swing mechanism of a bio-mimosa robot. *J. Chin. Soc. Mech. Eng.* **2013**, *34*, 137–142.
16. Sparrman, B.; du Pasquier, C.; Thomsen, C.; Darbari, S.; Rustom, R.; Laucks, J. Printed silicone pneumatic actuators for soft robotics. *Addit. Manuf.* **2021**, *40*, 101860. [CrossRef]
17. Gariya, N.; Kumar, P. A review on soft materials utilized for the manufacturing of soft robots. *Mater. Today Proc.* **2021**, *46*, 11177–11181. [CrossRef]
18. de Preter, A.; Anthonis, J.; De Baerdemaeker, J. Development of a robot for harvesting strawberries. *IFAC-PapersOnLine* **2018**, *51*, 14–19. [CrossRef]
19. Fan, P.; Yan, B.; Wang, M.; Lei, X.; Liu, Z.; Yang, F. Three-finger grasp planning and experimental analysis of picking patterns for robotic apple harvesting. *Comput. Electron. Agric.* **2021**, *188*, 106353. [CrossRef]
20. Hofer, M.; D'Andrea, R. Design, fabrication, modeling, and control of a fabric-based spherical robotic arm. *Mechatronics* **2020**, *68*, 102369. [CrossRef]
21. Salman, F.; Cui, Y.; Imran, Z.; Liu, F.; Wang, L.; Wu, W. A wireless-controlled 3D printed robotic hand motion system with flex force sensors. *Sens. Actuators A Phys.* **2020**, *309*, 112004. [CrossRef]
22. Li, Z.; Du, R.; Lei, M.C.; Yuan, S.M. Design and analysis of a biomimetic wire-driven robot arm. In Proceedings of the ASME International Mechanical Engineering Congress and Exposition, Denver, CO, USA, 11–17 November 2011; pp. 191–198.
23. Dong, X.; Raffles, M.; Guzman, S.C.; Axinte, D.; Kell, J. Design and analysis of a family of snake arm robots connected by compliant joints. *Mech. Mach. Theory* **2014**, *77*, 73–91. [CrossRef]
24. Li, Z.; Ren, H.; Chiu, P.W.Y.; Du, R.; Yu, H. A novel constrained wire-driven flexible mechanism and its kinematic analysis. *Mech. Mach. Theory* **2016**, *95*, 59–75. [CrossRef]
25. Tang, Y.; Qiu, J.; Zhang, Y.; Wu, D.; Cao, Y.; Zhao, K.; Zhu, L. Optimization strategies of fruit detection to overcome the challenge of unstructured background in field orchard environment: A review. *Precis. Agric.* **2023**, *24*, 1183–1219. [CrossRef]
26. Zhang, Y.; Zhang, K.; Yang, L.; Zhang, D.; Cui, T.; Yu, Y.; Liu, H. Design and simulation experiment of ridge planting strawberry picking manipulator. *Comput. Electron. Agric.* **2023**, *208*, 107690. [CrossRef]
27. Dimeas, F.; Sako, D.V.; Moulianitis, V.C.; Aspragathos, N.A. Design and fuzzy control of a robotic gripper for efficient strawberry harvesting. *Robotica* **2015**, *33*, 1085–1098. [CrossRef]
28. van Herck, L.; Kurtser, P.; Wittemans, L.; Edan, Y. Crop design for improved robotic harvesting: A case study of sweet pepper harvesting. *Biosyst. Eng.* **2020**, *192*, 294–308. [CrossRef]
29. Kaleem, A.; Hussain, S.; Aqib, M.; Cheema, M.J.M.; Saleem, S.R.; Farooq, U. Development challenges of fruit-harvesting robotic arms: A critical review. *AgriEngineering* **2023**, *5*, 2216–2237. [CrossRef]
30. Zhao, Y.; Gong, L.; Liu, C.; Huang, Y. Dual-arm robot design and testing for harvesting tomatoes in a greenhouse. *IFAC-PapersOnLine* **2016**, *49*, 161–165. [CrossRef]
31. Shaw, J.; Cheng, K.Y. Object identification and 3-D position calculation using eye-in-hand single camera for robot gripper. In Proceedings of the International Conference on Industrial Technology, Taipei, Taiwan, 14–17 March 2016; pp. 1622–1625. [CrossRef]
32. Ge, Y.; From, P.J.; Xiong, Y. Multi-view gripper internal sensing for the regression of strawberry ripeness using a mini-convolutional neural network for robotic harvesting. *Comput. Electron. Agric.* **2024**, *216*, 108474. [CrossRef]
33. Hayashi, S.; Shigematsu, K.; Yamamoto, S.; Kobayashi, K.; Kohno, Y.; Kamata, J.; Kurita, M. Evaluation of a strawberry-harvesting robot in a field test. *Biosyst. Eng.* **2010**, *105*, 160–171. [CrossRef]
34. Ge, Y.; Xiong, Y.; From, P.J. Instance segmentation and localization of strawberries in farm conditions for automatic fruit harvesting. *IFAC-PapersOnLine* **2019**, *52*, 294–299. [CrossRef]
35. Sun, T.; Zhang, W.; Miao, Z.; Zhang, Z.; Li, N. Object localization methodology in occluded agricultural environments through deep learning and active sensing. *Comput. Electron. Agric.* **2023**, *212*, 108141. [CrossRef]

36. Perez-Borrero, I.; Marin-Santos, D.; Vasallo-Vazquez, M.J.; Gegundez-Arias, M.E. A new deep-learning strawberry instance segmentation methodology based on a fully convolutional neural network. *Neural Comput. Appl.* **2021**, *33*, 15059–15071. [CrossRef]
37. Tian, Y.; Yang, G.; Wang, Z.; Wang, H.; Li, E.; Liang, Z. Apple detection during different growth stages in orchards using the improved YOLO-V3 model. *Comput. Electron. Agric.* **2019**, *157*, 417–426. [CrossRef]
38. Zhou, X.; Ampatzidis, Y.; Lee, W.S.; Zhou, C.; Agehara, S.; Schueller, J.K. Deep learning-based postharvest strawberry bruise detection under UV and incandescent light. *Comput. Electron. Agric.* **2022**, *202*, 107389. [CrossRef]
39. He, Z.; Karkee, M.; Zhang, Q. Detecting and localizing strawberry centers for robotic harvesting in field environment. *IFAC-PapersOnLine* **2022**, *55*, 30–35. [CrossRef]
40. Eizentals, P.; Oka, K. 3D pose estimation of green pepper fruit for automated harvesting. *Comput. Electron. Agric.* **2016**, *128*, 127–140. [CrossRef]
41. Li, Y.; Wang, W.; Guo, X.; Wang, X.; Liu, Y.; Wang, D. Recognition and positioning of strawberries based on improved YOLOv7 and RGB-D sensing. *Agriculture* **2024**, *14*, 624. [CrossRef]
42. Tang, Y.; Chen, M.; Wang, C.; Luo, L.; Li, J.; Lian, G.; Zou, X. Recognition and localization methods for vision-based fruit picking robots: A review. *Front. Plant Sci.* **2020**, *11*, 510. [CrossRef]
43. Zhang, B.; Xie, Y.; Zhou, J.; Wang, K.; Zhang, Z. State-of-the-art robotic grippers, grasping and control strategies, as well as their applications in agricultural robots: A review. *Comput. Electron. Agric.* **2020**, *177*, 105694. [CrossRef]
44. Han, C.; Lv, J.; Dong, C.; Li, J.; Luo, Y.; Wu, W.; Abdeen, M.A. Classification, advanced technologies, and typical applications of end-effector for fruit and vegetable picking robots. *Agriculture* **2024**, *14*, 1310. [CrossRef]
45. Zadeh, L.A. Fuzzy sets. *Inf. Control* **1965**, *8*, 338–353. [CrossRef]
46. Narendran, V.; Edberg, C.P.L.; Gandhi, G.M. Autonomous robot for e-farming based on fuzzy logic reasoning. *Int. J. Pure Appl. Math.* **2018**, *118*, 3811–3821.
47. Chang, C.L.; Chen, H.W.; Ke, J.Y. Robust guidance and selective spraying based on deep learning for an advanced four-wheeled farming robot. *Agriculture* **2023**, *14*, 57. [CrossRef]
48. Tian, G.; Zhou, J.; Gu, B. Slipping detection and control in gripping fruits and vegetables for agricultural robot. *Int. J. Agric. Biol. Eng.* **2018**, *11*, 45–51. [CrossRef]
49. Liang, X.; Peng, M.; Lu, J.; Qin, C. A visual servo control method for tomato cluster-picking manipulators based on a TS fuzzy neural network. *Trans. ASABE* **2021**, *64*, 529–543. [CrossRef]
50. Li, Z.; Du, R. Design and analysis of a bio-inspired wire-driven multi-section flexible robot. *Int. J. Adv. Robot. Syst.* **2013**, *10*, 209. [CrossRef]
51. Chang, C.L.; Chung, S.C.; Fu, W.L.; Huang, C.C. Artificial intelligence approaches to predict growth, harvest day, and quality of lettuce (*Lactuca sativa* L.) in an IoT-enabled greenhouse system. *Biosyst. Eng.* **2021**, *212*, 77–105. [CrossRef]
52. Bochkovskiy, A.; Wang, C.; Liao, H. YOLOv4: Optimal speed and accuracy of object detection. *arXiv* **2020**, arXiv:2004.10934.
53. Huang, C.C.; Chang, C.L. Design and implementation of wire-driven multi-joint robotic arm. In Proceedings of the 2022 International Conference on Advanced Robotics and Intelligent Systems, Taipei, Taiwan, 24–27 August 2022. [CrossRef]
54. Chang, C.L.; Xie, B.X.; Chung, S.C. Mechanical control with a deep learning method for precise weeding on a farm. *Agriculture* **2021**, *11*, 1049. [CrossRef]
55. Hua, X.; Li, H.; Zeng, J.; Han, C.; Chen, T.; Tang, L.; Luo, Y. A review of target recognition technology for fruit picking robots: From digital image processing to deep learning. *Appl. Sci.* **2023**, *13*, 4160. [CrossRef]
56. Arad, B.; Balendonck, J.; Barth, R.; Ben-Shahar, O.; Edan, Y.; Hellström, T.; Hemming, J.; Kurtser, P.; Ringdahl, O.; Tielen, T.; et al. Development of a sweet pepper harvesting robot. *J. Field Robot.* **2020**, *37*, 1027–1039. [CrossRef]
57. Mu, L.; Cui, G.; Liu, Y.; Cui, Y.; Fu, L.; Gejima, Y. Design and simulation of an integrated end-effector for picking kiwifruit by robot. *Inf. Process. Agric.* **2020**, *7*, 58–71. [CrossRef]
58. Hemming, J.; Ruizendaal, J.; Hofstee, J.W.; van Henten, E.J. Fruit detectability analysis for different camera positions in sweet-pepper. *Sensors* **2014**, *14*, 6032–6044. [CrossRef]

agriculture

MDPI

Article

The Effect of Varying Compaction Levels on Soil Dynamic Properties and the Growth of Canola (*Brassica napus* L.)

Ernest Owusu-Sekyere and Ying Chen *

Department of Biosystems Engineering, University of Manitoba, Winnipeg, MB R3T 2N2, Canada; owususee@myumanitoba.ca
* Correspondence: ying.chen@umanitoba.ca

Abstract: Extremely low field emergence rates for canola are primarily attributed to soil compaction from field traffic during and after planting. This study aimed to determine the critical compaction level for canola emergence across different soil types. A laboratory experiment was conducted using sandy loam, silt clay, and clay soils, compacted to five levels (zero to four) using Proctor hammer drops after sowing canola (*Brassica napus* L.). The lab results were validated through two years of field experiments in sandy loam, applying four compaction levels (zero to three) using a tractor. Soil properties (bulk density and surface resistance) and canola growth parameters (plant emergence rate, count, height, and above-ground biomass) were measured. Zero compaction resulted in lower bulk density and surface resistance across all soil types. Laboratory results showed maximum emergence rates of 95% for sandy loam, 100% for silt clay, and 60% for clay, while field emergence rates were 63% and 87.59% in the first and second years, respectively, both at zero compaction. Recommendations include light or no compaction for sandy loam, and zero compaction for silt clay, while clay soil did not achieve the 80% emergence target at any compaction level. These results can assist agricultural producers in optimizing their seeding equipment setup and managing field traffic for canola production.

Keywords: soil; compaction; surface resistance; canola; emergence rate; biomass

1. Introduction

Canola (*Brassica napus* L.) is a commercially significant crop in the Canadian Prairies, with uses ranging from consumable oil to biofuels. Annual revenue doubled from CAD 15.4 billion to CAD 26.7 billion between 2007 and 2015 [1], driving an expansion of canola cultivation areas across Canada. However, emergence rates of canola as low as 50% have been reported [2]. This implies a 50% loss in revenue. Low emergence rates may reduce the yield and cause an economic loss due to the high cost of canola seeds. Factors contributing to low crop emergence rate include inadequate soil moisture, inappropriate seeding depth [3], seed variety and size [4], and soil compaction. Among these, soil compaction is considered a major cause [5]. High soil surface resistance of compacted soil prevents small seedlings from emerging [6]. Soil compaction also increases soil bulk density and surface resistance, which is detrimental to crop emergence and early growth of small seeds, like canola [5].

Soil structural properties significantly influenced by compaction include bulk density, surface resistance, and related characteristics [7]. These properties are associated with soil deformation resulting from induced stresses that vary according to the magnitude of the applied pressure [8] (such as tractor traffic) and other influencing factors [9]. Previous research has shown that an increase in soil bulk density and penetration resistance with the number of tractor passes is accompanied by a notable reduction in soil porosity [7]. Soil properties relevant to this study were bulk density and surface resistance, as compaction was found to notably increase both [10,11], with high values observed at high compaction levels.

The effects of soil compaction extend beyond changes in soil properties, impacting root and plant growth and reducing crop yield [12,13]. Varying conclusions regarding the effects of compaction on crop growth and yield have been reported from different academic studies. For example, Gelder et al. (2007) [14] reported no significant effect of traffic-induced compaction on corn yield and plant emergence. In contrast, other investigations have demonstrated a pronounced influence of compaction on crop growth and yield [7]. Ishaq et al. (2001) [15] documented subsoil compaction as a factor contributing to reduced grain and straw yield, diminished fertile tillers, and compromised water and nutrient use efficiency in wheat. Excessive soil compaction increases soil resistance to emerging seedlings and compromises root development, leading to poor crop performance [2,11,16]. The decrease in seedling emergence with an increase in soil compaction has been reported for many crops, such as corn [11], sorghum, wheat [10], and soybean [16]. A study performed by Buttery et al. (1998) [16] showed that a decrease in overall soybean and common bean growth was correlated with an increasing compaction level. The work conducted by Chan et al. (2006) [17] revealed a 34% reduction in yield and inhibited root growth in canola due to the compaction induced by tractor wheel traffic. These findings highlight the need for further research on how compaction affects soil properties and canola growth.

Existing studies have not specifically examined the effects of post-seeding soil compaction on the emergence of small seeds, such as canola, which are sensitive to soil resistance. Therefore, studying post-seeding compaction is crucial, as the results can guide the down pressure design of press wheels of seeders and the timing management of field traffic. The ideal soil compaction level favorable for canola emergence is currently unknown, making it essential to determine the compaction level that allows seedlings to emerge at an acceptable rate. Soil properties, such as surface resistance, vary with soil type and composition [16,18]. Therefore, it is important to include different soil types when investigating the effects of soil compaction levels on the emergence and growth of canola. Given that canola seeds are small relative to soil aggregates, controlled lab tests may help mitigate errors associated with highly non-homogeneous field soil conditions. However, to draw valid conclusions on the effect of compaction on soil properties and canola growth, field tests are necessary to confirm whether the lab results apply in real-world settings. This will be highly valuable to agricultural producers.

Therefore, the specific objectives of this study were (1) to evaluate in a laboratory the effect of different soil compaction levels on soil properties and canola (*Brassica napus* L.) emergence and growth under three distinct soil types, and (2) to validate the laboratory results of soil and plant responses in a field over two years. The results helped to determine the critical compaction level needed for achieving a target canola emergence rate for each soil type.

2. Materials and Methods

2.1. Laboratory Experiment

2.1.1. Description of Soil and Seed

The canola emergence test was first conducted under controlled laboratory conditions in 2020. The canola (*Brassica napus* L.) seed variety was Westar Canola, from a commercial seed supplier in Manitoba, Canada. Three soils, namely sandy loam, silt clay, and clay soils, were sourced from three different farms (Piney, the Ian N. Morrison research station at Carman, and a research station at the University of Manitoba Fort Garry campus, respectively) in southern Manitoba, Canada. The average moisture level for each soil type was adjusted to near its field capacity. Table 1 shows the compositions of these soils along with their gravimetric moisture content (d.b.), determined by oven-drying the soil samples for 24 h at 105 °C [19].

Table 1. Soil composition and gravimetric moisture contents for the laboratory experiment.

Soil Type	Clay (%)	Silt (%)	Sand (%)	Moisture Content (d.b.)%
Sandy Loam	14	16	70	24
Silt Clay	54	42	4	33
Clay	77	19	4	36

2.1.2. Laboratory Experimental Design and Test Procedure

The laboratory experiment used a completely randomized design. The treatments were five soil compaction levels: zero compaction level/no compaction ($C0_L$), compaction level one ($C1_L$), level two ($C2_L$), level three ($C3_L$), and level four ($C4_L$). The creation of these compaction levels is described below. Each treatment was replicated four times. Thus, 20 sample units (5 compaction levels × 4 replicates) were carried out for each soil type, totaling 60 tests for the 3 soil types in the experiment. The experimental design is illustrated in Figure 1a.

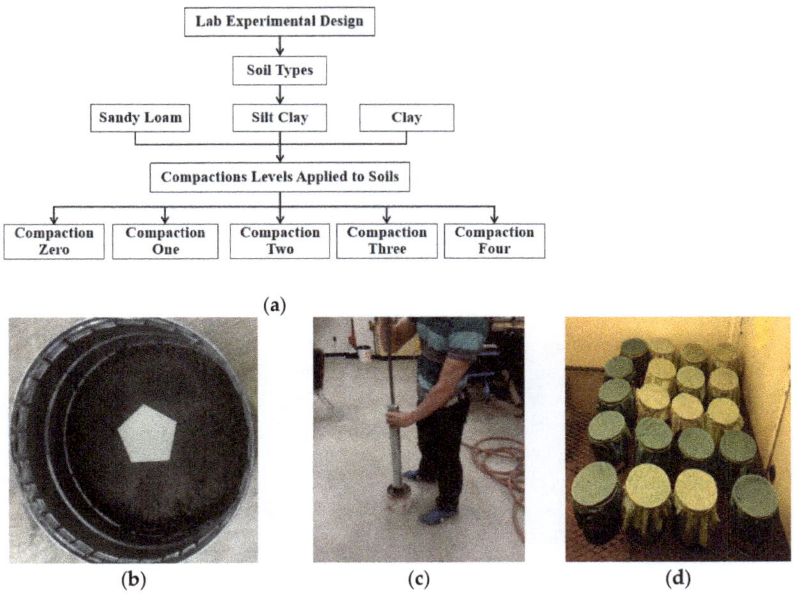

Figure 1. Laboratory experimental procedure: (**a**) flow chart of the experimental design, (**b**) seed positioning template, (**c**) soil compaction with a proctor, and (**d**) covered seeded containers in the environmental chamber.

For each soil type, 20 cylinders (166 mm in height and 154 mm in diameter) were employed as seeding containers and filled with soil up to a height of 20 mm below the container's rim. Subsequently, five canola seeds were positioned on the seedbed following a pentagon-shaped template as a guide, with one seed placed at each corner of the pentagon (Figure 1b). The distance between adjacent seeds was 30 mm. A predetermined amount of soil was then used to cover the seeds in each container, ensuring an appropriate depth of soil cover. This entire procedure, from soil preparation to seeding, was consistently applied across all the three soil types.

Following seeding, the 20 containers within each soil type were subdivided into 5 groups for compaction of the seed cover. A soil proctor (ASTM/Modified Proctor Compaction Rammer, ELE International, Loveland, CO, USA) with a hammer of mass 4.5 kg and a drop height of 0.457 m, was utilized for this purpose (Figure 1c). The group of containers, denoted by treatment $C0_L$, underwent no compaction. The remaining groups of containers received one, two, three, and four drops of the hammer of the proctor, corresponding to treatments $C1_L$, $C2_L$, $C3_L$, and $C4_L$, respectively. Given the hammer mass (m) of 4.5 kg, the gravitational acceleration (g), and the drop height (h) of 0.457 m, the values of compaction energy (potential energy = mgh) for the compaction levels $C0_L$, $C1_L$, $C2_L$, $C3_L$, and $C4_L$ were 0, 20.2, 40.4, 60.6, and 80.8 J, respectively. After seeding, the containers were transferred to an environmental chamber set at predetermined conditions of 19 °C, 10% RH, and an 8–16 h dark-photo-period (Figure 1d). To maintain soil moisture content, the containers were covered with wet clothes that were rewetted twice daily during the initial five days to mitigate soil water evaporation.

2.1.3. Laboratory Measurements

Soil compaction levels in the containers were quantified for bulk density and surface resistance. For this, separate containers of the same soil compaction levels as used in the seeding test were prepared without seeding. A Dail Pocket Penetrometer (Model: HM-502, Gilson Company Inc., OH 43035, USA) with a specified plunger surface area of 314 mm^2 was used to measure the surface resistance at three random locations in each container. For each container, a soil core was taken using a 50 × 50 mm cylindrical soil sampler. The gravimetric moisture content and soil dry bulk density were calculated after oven-drying at 105 °C for 24 h [19].

The number of plants that emerged in each container and the height of those plants were recorded daily for a total of 14 days after seeding. At the end of the 14 days, the plants were cut off at the soil surface, and their above-ground biomass was determined by oven-drying for 72 h at 60 °C [19].

2.2. *Field Experiment*

2.2.1. Description of the Study Site, Soil, and Seed

The field experiment was conducted during the summers of 2023 and 2024, from May to August. The experimental site was in Piney, Manitoba, Canada (49°07′15″ N, 96°07′56″ W). During the study months, the site recorded an average maximum temperature of 23.4 °C and a minimum of 10.5 °C, with 67% relative humidity, 0.8 mm/day of rainfall, and wind speeds averaging 13.31 km/h in 2023 [20]. For 2024, the site recorded an average maximum temperature of 21.9 °C and a minimum of 10 °C, with 74% relative humidity, 2.9 mm/day of rainfall, and wind speeds averaging 14.16 km/h [20]. The field had sandy loam soil with the same texture as that employed in the laboratory experiment, and the canola variety was also the same. At the time of measurements, the field soil moisture content was 19.67% and 35.58% (d.b.) for the first- and second-year tests, respectively.

2.2.2. Field Experimental Design and Test Procedure

In the field experiment, different compaction levels were created using tractor wheels on the field. The experiment was a completely random design with four compaction levels, namely zero compaction level/no compaction ($C0_F$), compaction level one ($C1_F$), level two ($C2_F$), and level three ($C3_F$), as treatments. Each treatment was randomly replicated 5 times, giving a total of 20 plots (4 compaction levels × 5 replications). The field plot layout is illustrated in Figure 2a.

Before compacting the soil, canola was seeded in the field using a disc seeder (4-row Plotter choice, Shelbyville, IN, USA). The seeding rate was 1 kg/ha and 3.8 kg/ha for the first- and second-year tests, respectively. As in the laboratory experiment, post-seeding compaction was applied to the soil by driving a tractor (John Deere 1023E tractor, weighing 656 kg, equipped with ballasts) at 2 km/h in the seeded field in the direction perpendicular to the seeding direction (Figure 2a). The tractor made one pass for the $C1_F$ treatment, two passes along the same track for the $C2_F$ treatment, and three passes along the same track for the $C3_F$, creating tracks (Figure 2b) for each compaction level.

On the wheel tracks, measurement areas were marked (Figure 2b) with each area measuring 1.8 × 0.22 m. The width of the measured area was the active tire width illustrated in Figure 2c. Six measurement areas were randomly selected in each plot, with three areas for soil property measurements and three areas for canola growth measurements. Additionally, in each plot, six random areas of the same dimensions (1.8 × 0.22 m), that did not have wheel tracks, were used for measurements of the zero compaction level ($C0_F$).

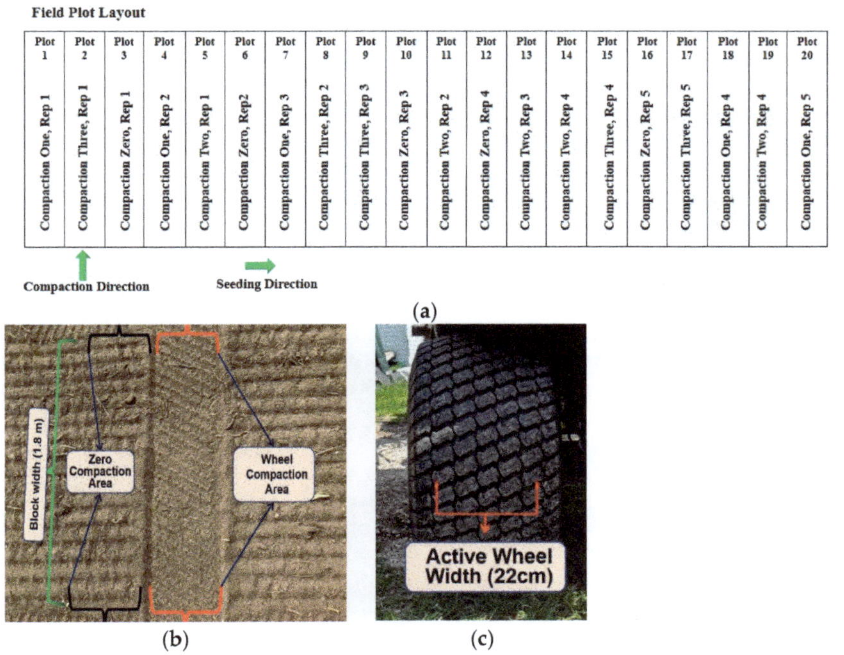

Figure 2. Field experimental process: (**a**) plot layout, (**b**) tractor wheel tracks showing compacted and non-compacted areas, and (**c**) active wheel width contributing to wheel tracks.

2.3. Field Measurements

As in the laboratory experiment, the soil compaction levels in the field were quantified for bulk density and surface resistance. Three soil cores (diameter 50 mm and height 100 mm) were taken from each measurement area per plot to measure bulk density. The soil surface resistance was measured using the same instrument as in the laboratory test at three measurement areas in each plot.

After 48 h of seeding, plant counting was conducted in the measurement areas of each plot every 48 h for five consecutive counts. After allowing the crops to fully emerge, a final count was conducted 27 days after seeding. Subsequently, the emergence rate was determined. Fully matured above-ground plants were collected from all plots three months post-seeding. The biomass of the plants was determined using the same oven-drying method performed in the laboratory experiment.

2.4. Data Analysis

One-way analysis of variance (ANOVA) was performed using R programming software version R4.2.2 and the 'car' package to analyze the effects of the treatments (soil compaction levels) on the measured parameters (soil properties and plant responses). The Tukey HSD test was utilized to determine the significant differences in the measured variables between all treatments at a confident level of 95%, using the same software.

3. Results

3.1. Soil Physical Properties

3.1.1. Bulk Density

In the laboratory test, the initial soil bulk density without compaction ($C0_L$) was the highest for the sandy loam soil (Figure 3a), followed by the silt clay (Figure 3b), and finally the clay soil (Figure 3c). The silt clay and clay soils had slight changes in magnitude, with no significant difference between $C3_L$ and $C4_L$ (Figure 3b,c). Similar results were reported by Li et al. (2016) [21]. However, the $C3_L$ of the sandy loam resulted in a lower bulk density than $C4_L$ (Figure 3a). Overall, soil bulk density was the highest for the sandy loam soil and lowest for the clay soil at all compaction levels. At compaction levels of $C2_L$ and lower, all the soils had a similar response to the compaction efforts, in terms of soil bulk density. The highest compaction in the sandy loam was 35.35% greater than in the silt clay and 59.05% greater than in the clay. Field soil bulk density results did not show significant differences between the compaction levels. In the first-year test, soil bulk density also increased when the compaction level was increased from $C0_F$ to $C2_F$, and there was no further increase in soil bulk density for $C3_F$ (Figure 3d). Although this trend was not statistically significant, it did not contradict the explained patterns in the lab experiment. However, the second-year test followed an increasing trend from $C0_F$ to $C3_F$ with small changes between compaction levels (Figure 3e). It registered lower bulk densities than the first-year test at all levels.

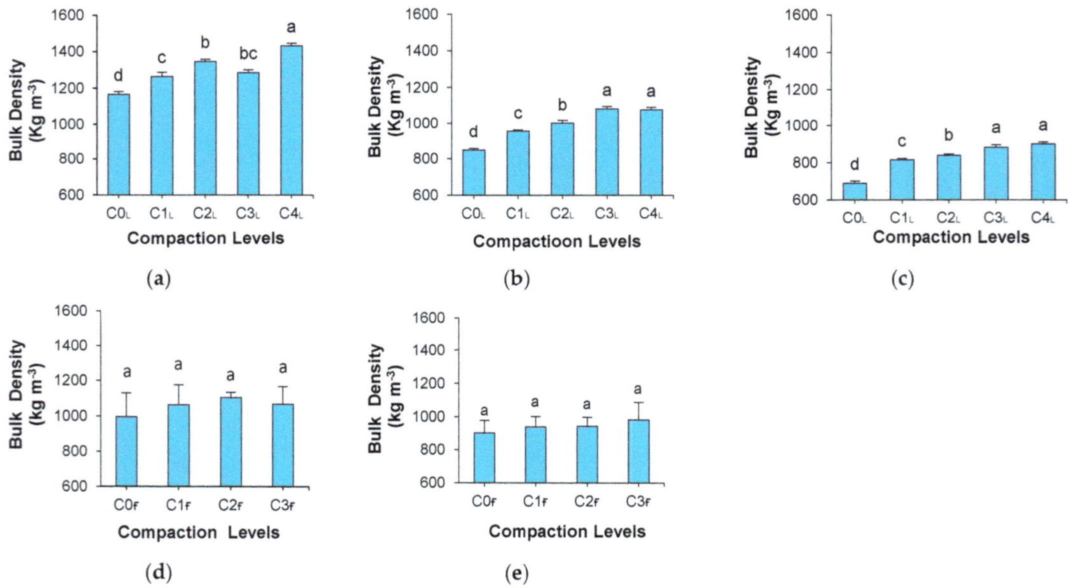

Figure 3. Bulk densities at varying compaction levels; (**a**–**c**) are lab measurements taken for the sandy loam, silt clay, and clay soils, respectively ($C0_L$, $C1_L$, $C2_L$, $C3_L$, and $C4_L$ represent compaction levels zero, one, two, three, and four, respectively); (**d**,**e**) are field measurements of the sandy loam soil for first- and second-year tests, respectively ($C0_F$, $C1_F$, $C2_F$, and $C3_F$, represent compaction levels zero, one, two, and three, respectively). Means labeled with different letters in a graph were significantly different at $p < 0.05$.

3.1.2. Soil Surface Resistance

The laboratory test revealed an increasing trend in surface resistance with increasing compaction levels for all the three soil types (Figure 4a–c). All three soils had a very low resistance at $C0_L$. However, the rate of increase was different among the different soil types. When the compaction level was increased from $C0_L$ to $C3_L$, the soil resistance increased

29 times for the sandy loam soil (Figure 4a), 19 times for the silt clay soil (Figure 4b), and only 5 times for the clay soil (Figure 4c). For the sandy loam and silt clay soils, there was no further increase in resistance for $C4_L$. The change in the surface resistance of the clay soil between $C3_L$ and $C4_L$ was significant. Comparing the magnitudes of surface resistance at $C4_L$, the resistance of the sandy loam was 54% higher than that of silt clay, and 76% higher than that of clay.

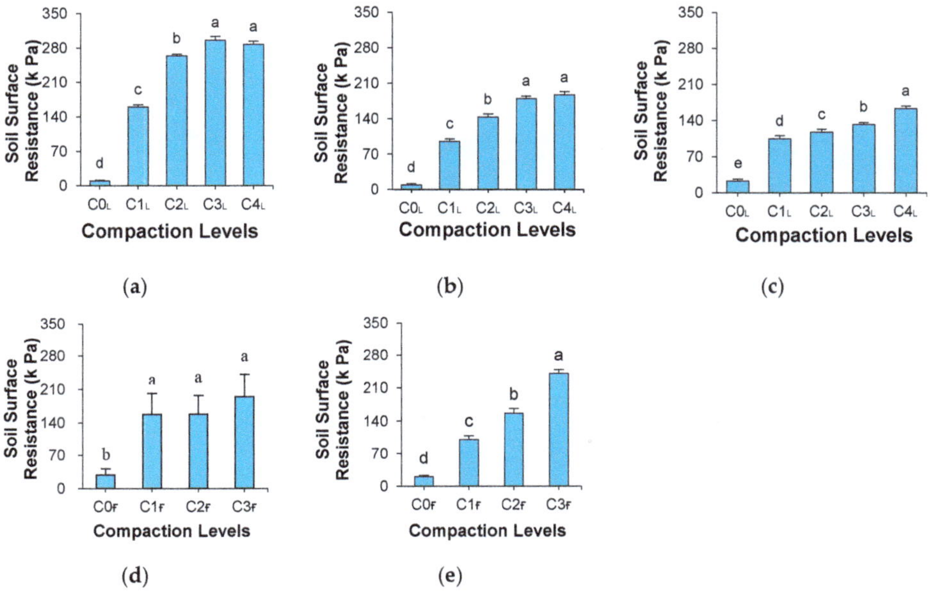

Figure 4. Soil surface resistances at varying compaction levels, where (**a–c**) are lab measurements taken for the sandy loam, silt clay, and clay soils, respectively ($C0_L$, $C1_L$, $C2_L$, $C3_L$, and $C4_L$ represent compaction levels zero, one, two, three, and four, respectively); (**d**,**e**) are field measurements on the sandy loam soil for first- and second-year tests, respectively ($C0_F$, $C1_F$, $C2_F$, and $C3_F$ represent compaction levels zero, one, two, and three, respectively). Means labeled with different letters in a graph were significantly different at $p < 0.05$.

The first-year field test exhibited significantly lower soil surface resistance at the $C0_F$. There were significant disparities between compaction levels zero and one (Figure 4d). However, the differences in soil resistance between $C1_F$, $C2_F$, and $C3_F$ were not statistically significant. Also, there was a general increasing trend of surface resistance with the soil compaction level (Figure 4d). This trend from the field was consistent with those obtained from the laboratory test [5]. However, the second-year test showed an increasing pattern with significant differences among all compaction levels (Figure 4e). This trend is consistent with the general trend in bulk density.

3.2. *Plant Emergence and Growth*

3.2.1. Qualitative Assessment of Canola Emergence

In the lab experiment, canola emergence and growth can be visualized from the photographs taken at the end of the two-week study (Figure 5a–c). These photos show 3 groups of 20 containers, with 1 group for each soil type. Within a group, the five containers in a row represent the five compaction levels ($C0_L$, $C1_L$, $C2_L$, $C3_L$, and $C4_L$), while the four containers in a column represent the four replicates (R1, R2, R3, and R4) of the treatment. Three visual comparisons were made from these photos. Row-wise comparison within each group showed that better canola emergence was observed at a lower compaction level for all three soils. Column-wise comparison within each group showed that the four replications had consistent emergence, indicating low experimental errors. Comparison between the groups showed that the sandy loam soil had the worst plant growth, while clay soil had the best growth.

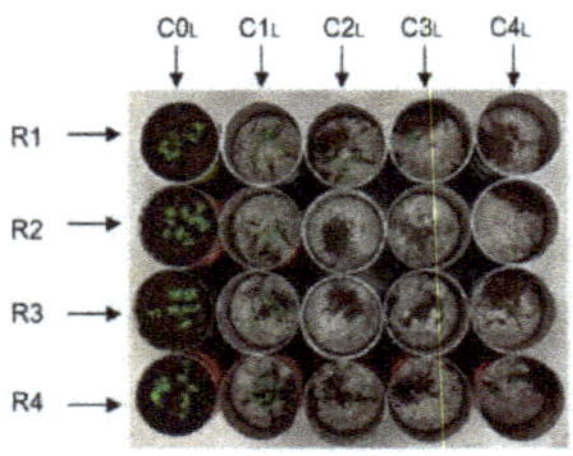

(**a**)

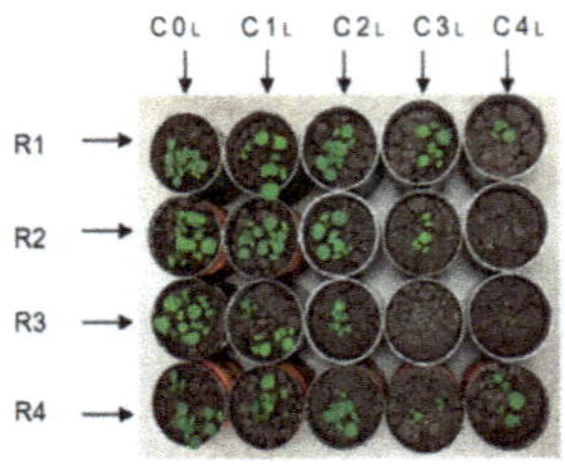

(**b**)

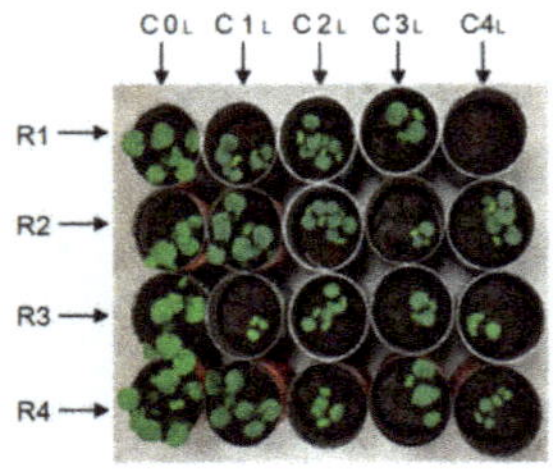

(**c**)

(**d**)

Figure 5. Canola emergence at varying compaction levels, where (**a**–**c**) are lab measurements taken 14 days post-seeding for the sandy loam, silt clay, and clay soils, respectively ($C0_L$, $C1_L$, $C2_L$, $C3_L$, and $C4_L$ represent compaction levels zero, one, two, three, and four, respectively; R1, R2, R3, and R4 represent replicates one, two, three, and four, respectively); (**d**) field results showing emerged canola plants at compacted and non-compacted sandy loam soil areas taken 27 days post-seeding.

Analogous results to those shown in the lab test were noticed in the field test for the same sandy loam soil (Figure 5a,d). Field plots on tractor wheel tracks (distinguished by popsicle sticks) exhibited a significant reduction in canola plants, while a greater number of plants were evident in the no compaction areas for same reasons as stated above.

3.2.2. Plant Count

In the lab test, the numbers of canola plants that emerged in each pot were counted daily over the entire observation period. The daily total number of plants of all the replicates are presented here. For the sandy loam soil, the first emergence was observed for $C0_L$. For this compaction level, canola seedlings started to emerge on the fifth day after seeding (Figure 6a). Increasing the compaction level to $C1_L$ resulted in a one-day delay in emergence. Further increase in compaction level to $C2_L$, $C3_L$, and $C4_L$ resulted in a delay by another day. At the early stages of observation, the two lower compaction levels ($C0_L$ and $C1_L$) demonstrated a steep increase in seedling emergence, and seedlings emerged over a 10-day period. There was no increase in plant count afterwards. The emergence rate for treatments $C2_L$, $C3_L$, and $C4_L$ increased much slower over time than the rates for the lower compaction levels, and then emergence ceased on the 8th day. At the end of the observation period, the plant count was reduced by 5% for $C0_L$, 20% for $C1_L$, 80% for $C2_L$ and $C4_L$, and 95% for $C3_L$.

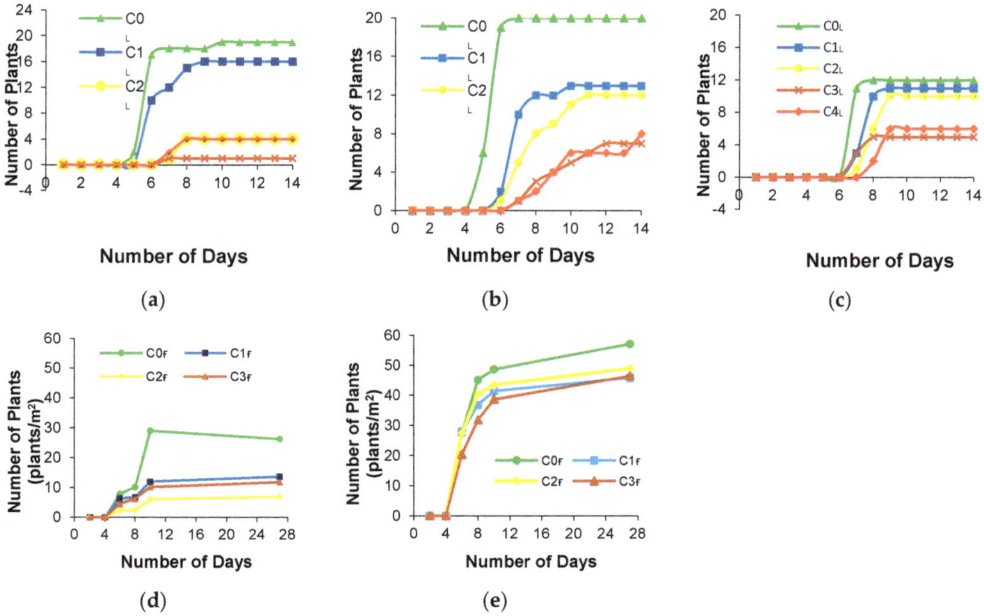

Figure 6. Plant population counts at varying compaction levels, where (**a–c**) are lab measurements for the sandy loam, silt clay, and clay soils, respectively ($C0_L$, $C1_L$, $C2_L$, $C3_L$, and $C4_L$ represent compaction levels zero, one, two, three, and four, respectively); (**d**,**e**) are field measurements using the sandy loam soil for first- and second-year tests, respectively ($C0_F$, $C1_F$, $C2_F$, and $C3_F$ represent compaction levels zero, one, two, and three, respectively).

For the silt clay soil, the first emergence was also observed on the fifth day, which was for $C0_L$, followed by $C1_L$ and $C2_L$ on the sixth day, and $C3_L$ and $C4_L$ on the seventh day (Figure 6b). The plants of $C3_L$ and $C4_L$ emerged until the end of the observation period. $C0_L$ had the highest plant count, with all the seedlings emerging by the seventh day. The plant count for $C1_L$ also demonstrated a rapid increase initially and slowly reached its maximum count of 12 plants on the 10th day. The count for $C2_L$ was lower, and new plants did not emerge after the 11th day. $C0_L$ had a 0% reduction in plant count, while the maximum reduction was 65% for $C3_L$.

For the clay soil, the first emergence for $C0_L$, $C1_L$, $C2_L$, and $C3_L$ occurred two days later than it did for the other two soil types and occurred on the eighth day for $C4_L$ (Figure 6c). No new seedlings emerged after 9 days of seeding for all the treatments. The number of plants was reduced by 40% for $C0_L$, with a maximum reduction of 75% for $C3_L$.

In the field test, canola emergence was assessed by the number of plants per unit field area over the observation period. Results from the two-year test showed that seedling emergence started six days after seeding for all compaction levels (Figure 6d,e). Both tests revealed a general increase in the number of emerging plants over time. A higher number of emerged plants were observed at the zero compaction level compared to the other compaction levels in both years. Towards the end of the twenty-seven-day period, a slight reduction in plant population was observed for $C0_F$ in the first-year test due to external environmental factors, such as nocturnal field grazing by deer. A higher number of plants were observed in the second year than the first year, due to the increased seeding rate in the second year.

3.2.3. Plant Emergence Rate

The emergence rate of the laboratory experiment was determined as the ratio of the total number of emerged plants to the total number of seeds sown. The results showed a decrease in seedling emergence rate with an increase in soil compaction level.

Emergence for the sandy loam soil showed two distinct groups of statistical differences (Figure 7a). The group with a high rate consisted of the $C0_L$ and $C1_L$ compaction levels, and the group with a much lower rate consisted of $C2_L$, $C3_L$, and $C4_L$. The average emergence rate for $C0_L$ was quite high (95%), while that for $C1_L$ was 80%. However, the average emergence rate for the latter groups averaged to 15%, much lower than the literature-reported range of 50–60% [22].

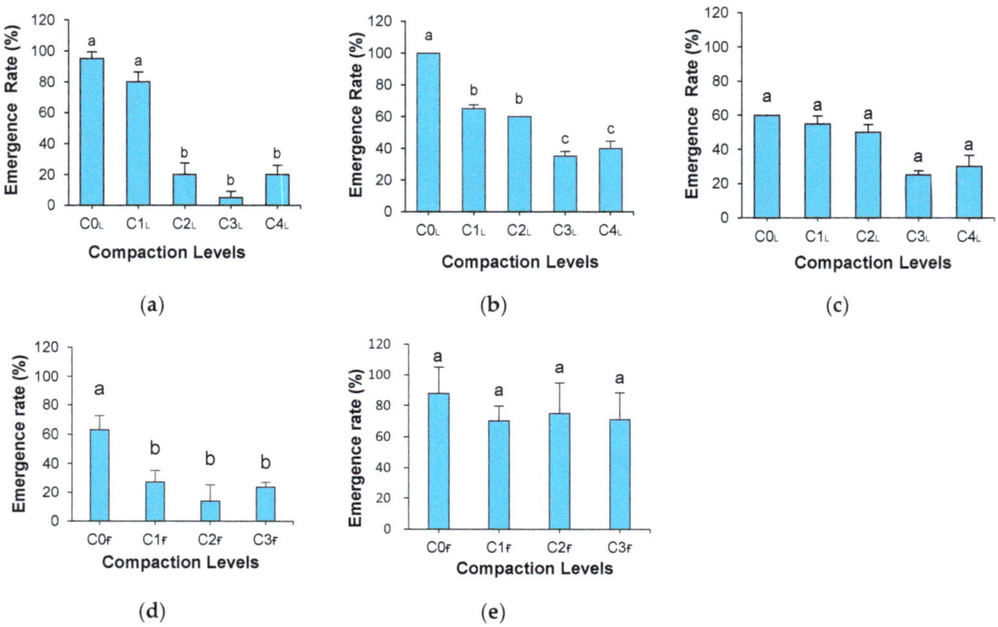

Figure 7. Canola emergence rate at varying compaction levels, where (**a**–**c**) are lab measurements taken for the sandy loam, silt clay, and clay soils, respectively ($C0_L$, $C1_L$, $C2_L$, $C3_L$, and $C4_L$ represent compaction levels zero, one, two, three, and four, respectively); (**d**,**e**) are field measurements on the sandy loam soil for first- and second-year tests, respectively ($C0_F$, $C1_F$, $C2_F$, and $C3_F$ represent compaction levels zero, one, two, and three, respectively). Means labeled with different letters in a graph were significantly different at $p < 0.05$.

The emergence rate of canola was also significantly different between compaction levels for the silt clay soil (Figure 7b). The emergence rate observed for the silt clay was 100% for $C0_L$, meaning that all seeds emerged when the soil was not compacted. The $C1_L$ and $C2_L$ treatments had 65 and 60% rates, respectively, which were significantly lower than $C0_L$. Very low emergence rates were observed in the higher compaction levels ($C3_L$, and $C4_L$).

The general trend of the clay soil was that compaction decreased the emergence rate. However, the trend was not statistically significant. None of the compaction levels had an emergence rate of 80% or higher (Figure 7c).

The emergence rate of the field test was determined as the ratio of the number of emerged plants to the total seeds sown per unit area, calculated using the known seeding rate and thousand kernel weight of canola. The results revealed a general trend of an inverse relationship between emergence rate and compaction level (Figure 7d,e). The seeding rate of 1 kg/m^2 used in the first-year test resulted in a much higher and significant change in magnitude between the zero compaction ($C0_F$) and the other compaction levels (Figure 7d). No significant differences were observed among compaction levels $C1_F$, $C2_F$, and $C3_F$.

Increasing the seeding rate in the second-year test to 3.8 kg/m^2 increased the emergence rate by 24% for $C0_F$, 42.97% for $C1_F$, 60.6% for $C2_F$, and 47.54% for $C3_F$ (Figure 7e). The 80% emergence rate was achieved only in the zero compaction level, which recorded a value of 87.59% (Figure 7e).

3.2.4. Plant Height

Plant height was measured in the lab condition, not in the field, since plant structure is not stable under weather conditions (wind and rain). In the lab experiment, an increasing trend was observed in the plant height with time, which lasted until the end of the observation period (Figure 8).

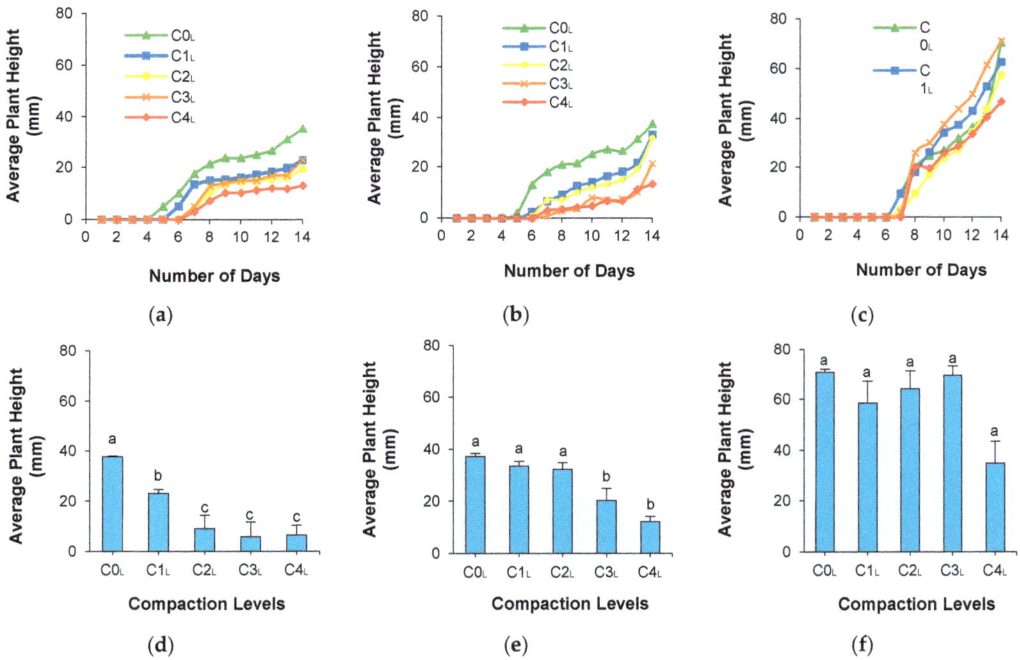

Figure 8. Lab measurement of plant height at varying compaction levels, where (**a–c**) are increasing values with time and (**d–f**) are average values. Both groups are for the sandy loam, silt clay, and clay soils, respectively. $C0_L$, $C1_L$, $C2_L$, $C3_L$, and $C4_L$ represent compaction levels zero, one, two, three, and four, respectively. Means labeled with different letters in a graph were significantly different at $p < 0.05$.

The initial growth rate for the sandy loam soil was similar for all the soil compaction levels, reflected by the similar slope of the plant height curves before the 12twelfth day (Figure 8a). Afterwards, there was a small increase in the plant height until the end of the observation period. However, higher heights were observed in the zero compaction level than the other compaction levels from the start of emergence until the end of the observation period. Very similar observation in plant height was observed between $C2_L$, $C3_L$, and $C4_L$.

Like the sandy loam soil, plant height data for the silt clay showed that the least compacted soil had a greater height at the initial growing stage (Figure 8b). Between the seventh and the twelfth day, plants at all compaction levels had a similar growth rate. The plant height experienced a rapid increase during the last two days. The $C3_L$ and $C4_L$ compaction levels had similar growth rates, as shown by the overlapped growth curves.

As compared with the other two soils, where the plant height increased gradually and non-linearly, the clay soil demonstrated a more rapid increase in plant height with a more linear increasing trend for the 14 days of growth (Figure 8c). Irrespective of the low plant count, the clay soil had higher plant height after the 14-day period.

To further demonstrate the effects of soil compaction on plant height, the data of the plant heights at the end of the observation period were averaged at each compaction level for each soil (Figure 8d–f). The sandy loam soil exhibited a decreasing trend in average plant height with increasing compaction level (Figure 8d). The maximum plant growth was observed at the non-compacted level, $C0_L$, with an average plant height of 37.7 cm. With slight compaction ($C1_L$), the average plant height reduced to 23.0 cm, a 39% reduction from $C0_L$. Additional compaction ($C2_L$) led to significantly lower plant heights. Subsequent compactions ($C3_L$ and $C4_L$) did not yield significant changes in average plant height.

In the silt clay soil, a similar average plant height to that observed in the sandy loam soil was seen for the $C0_L$ (Figure 8e). However, the change in magnitude between the compaction levels is less than that observed in the sandy loam soil. The zero compaction, $C0_L$, had the highest plant height, followed by $C1_L$ and $C2_L$ in that order, with no significant difference between the three levels. A significantly lower plant height was observed with additional compaction at $C3_L$ and $C4_L$. However, average plant height did not differ significantly between these two levels.

Compaction had no significant influence on the average height of canola plants within the clay soil (Figure 8f). The clay soil exhibited the highest plant height among all the three soils at all compaction levels. The plant emergence rate was notably less pronounced in the clay soil among all soils (Figure 7c). However, here, the clay soil had greater plant height than the other two soils.

3.2.5. Above-Ground Biomass

The sandy loam soil in the lab test showed biomass data exhibiting a similar trend to that of the plant height data. The highest biomass value was recorded for $C0_L$ (Figure 9a), which corresponded to the highest plant height. As the soil compaction level increased, significantly less biomass was produced. Biomass for the silt clay soil was much higher than the sandy loam soil and decreased with increasing compaction level (Figure 9b). The clay soil showed irregular trend with no significant difference in biomass with increasing compaction level (Figure 9c). The plant emergence data showed the worst performance in the clay soil, regardless of the soil compaction level. However, the overall amount of biomass was better than the sandy loam and as good as that of the silt clay.

In the field study, biomass was very sensitive to soil compaction in the first-year test (Figure 9d). A significant effect of compaction level on the biomass was observed. The decreasing trend of biomass with increasing compaction levels observed in both the first and second-year tests was similar between the field results (Figure 9d,e) and the lab results (Figure 9a), as the soil was sandy loam in both cases. However, compaction in the second-year test had no significant effect on biomass.

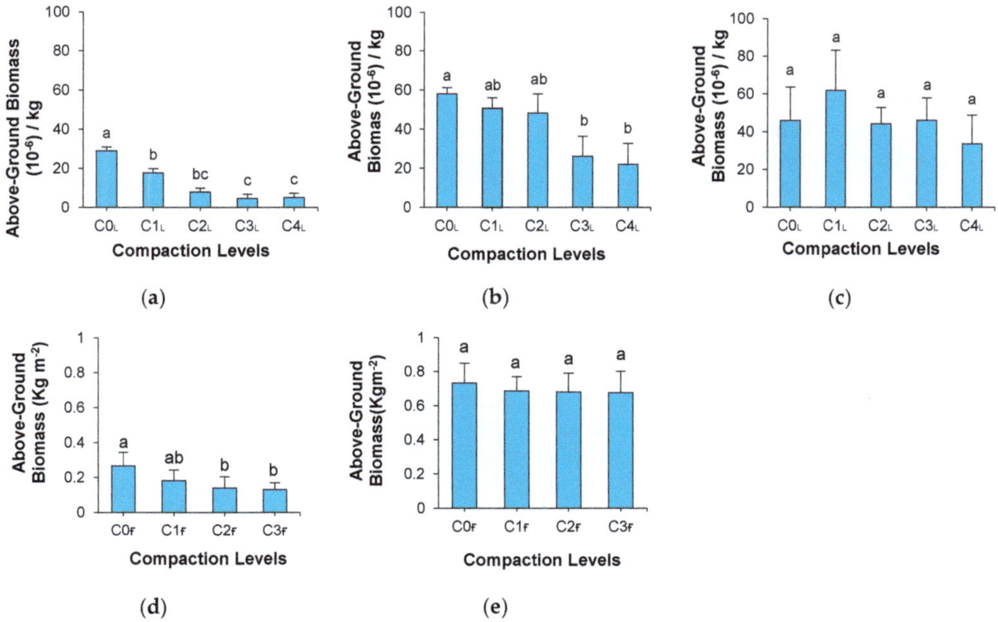

Figure 9. Canola above-ground biomass at varying soil compaction levels, where (**a–c**) are lab measurements taken 14 days post-seeding for the sandy loam, silt clay, and clay soils, respectively ($C0_L$, $C1_L$, $C2_L$, $C3_L$, and $C4_L$ represent compaction levels zero, one, two, three, and four, respectively); (**d**,**e**) are field measurements on the sandy loam soil for first- and second-year test taken 16 weeks post-seeding, respectively ($C0_F$, $C1_F$, $C2_F$, and $C3_F$, represent compaction levels zero, one, two, and three, respectively). Means labeled with different letters in a graph were significantly different at $p < 0.05$.

4. Discussion

The trend in bulk density observed in the lab test was expected, as sand particles have a higher density than silt clay and clay particles [23]. The observed increased trend in both the lab and field test was consistent with the findings of Phakdee and Suvanjumrat (2023) [24], who reported that soil compaction exhibits a positive linear relationship with soil bulk density. Lower compaction levels had less effect on soil porosity. Bulk density exhibits an inverse relationship with the porosity of soil samples of equal volume [25]. The lower values observed in the second-year test compared to the first year across all compaction levels may be due to differences in field locations, as there is no homogeneity among them (Figure 3d,e). The common lack of a further increase in soil bulk density beyond the third compaction level could likely be attributed to the limited number of soil pores which restrict further soil deformation under compaction [26].

The increase in compaction level caused a significant increase in surface resistance, as also recorded in the work carried out by Mileusnić et al. (2022) [27]. A similar trend in bulk density by soil type was translated into the trend in surface resistance, indicating that soil resistance was highly variable with soil type. The coarser the soil type, the higher its surface resistance, and vice versa [18]. Overall, different soils had different responses to the compaction efforts, in terms of soil resistance. The different behaviors of soil in response to compaction demonstrated the significance of investigating crop emergence and critical compaction levels for different soil types.

Higher soil bulk density and surface resistance were associated with a lower growth rate of canola [28,29]. Seedling emergence was delayed in clay soil due to large soil aggregates which impeded the upward movement of the canola cotyledons, as observed by

Vance et al. (2024) [30] in chickpea. As a result, some canola seeds were not able to emerge from the soil's surface. However, the clay soil's plant height was greater than sandy loam and silt clay. This observation might be due to the better water holding capacity of clay particles than sand particles in soil [31]. Comparable findings for sandy loam soil were observed in both the field and lab tests, as both used soil with the same composition. It was expected that plant count would reduce with the increasing compaction level from $C1_F$, to $C2_F$ and $C3_F$ for both tests, as observed by Kahlon et al. (2023) [32] for summer moon bean. However, the reverse was the case. Plant count decreased from $C1_F$ to $C3_F$ and then to $C2_F$ in the first-year test, and from $C2_F$ to $C3_F$, and then to $C1_F$ in the second-year test. This could be due to the heterogenous nature of field soils [33,34]. Similarly, this reverse phenomenon was also observed in the laboratory tests where the highest compaction level did not always result in the lowest number of emerged plants (Figure 6b,c).

High bulk density and surface resistance at high compaction levels adversely affected canola performance in terms of emergence rate. Specifically, the application of compaction pressure caused the reorganization of individual soil particles, leading to the closure of soil pores (reducing soil aeration), as discussed by Gong et al. (2022), Hess et al. (2019), and Otalvaro et al. (2016) [5,35,36], potentially impeding the upward movement of seedlings. This aligns with existing studies for different crops, such as corn [11], maize, sorghum, wheat [10], soybean [16], winter oilseed rape [37], and alfalfa [38]. The significantly low averaged emergence rate for sandy loam soil at $C2_L$ and $C3_L$ (Figure 7a) would significantly increase seed costs. Assuming a target emergence rate of 80%, only the $C0_L$ and $C1_L$ treatments were acceptable. Thus, the critical compaction level for the sandy loam soil is $C1_L$, which corresponded to a compaction energy of 20.2 J, bulk density of 1265 kg m^{-3}, and surface resistance of 160 k Pa. For silt clay, the only condition that had the emergence rate of 80% and greater was $C0_L$ (Figure 7b). Therefore, any form of compaction is unacceptable for seeding canola. Thus, its critical compaction level was $C0_L$, which corresponds to zero compaction energy, a soil bulk density of 848 kg m^{-3}, and a surface resistance of 9 k Pa. For clay soil, the highest emergence rate observed at $C0_L$ meant that at least 40% more seeds would need to be sown when growing canola crops in clay soil (Figure 7c). This would significantly increase production costs. The results implied that clay soil severely limited canola emergence, preventing it from reaching the 80% emergence rate, regardless of the compaction level. The reasons could be similar to those that affected plant count. The highest emergence rate of 63% in the first-year field condition was within the reported range in the literature [39], but it was far below the assumed target rate of 80% (Figure 7d). However, the second-year emergence rate (87.59%) increased beyond the target rate (Figure 7e).

For sandy loam and silt clay, the soil properties affected by compaction impacted the canola's above-ground biomass. Higher bulk densities and surface resistance of the soil resulted in significantly lower biomass (Figure 9). Compaction significantly reduced the canola emergence rate by 81.90% in sandy loam, 62.95% in silt clay and 45.70% in clay, consistent with studies of compaction effects on sugar cane [40], corn [41], and soybean [42]. Higher bulk density also influenced canola yield, causing a yield reduction, as noted in Gawęda and Haliniarz (2022) [43]. Future studies should include measurements of canola yields. These findings suggested that soil compaction from wheel traffic could be detrimental to soil health and plant growth in canola crop production.

5. Conclusions

The study highlights the need to minimize soil compaction to optimize canola emergence and growth. Lab and field tests revealed that higher compaction increased soil bulk density and surface resistance, which negatively affected canola performance, particularly in sandy loam soils. Silt clay soils showed similar sensitivity under lab conditions, while clay soil was largely unaffected by compaction. For sandy loam, no or slight compaction is recommended, and for silt clay, compaction should be avoided altogether. Clay soil did not meet the acceptable 80% emergence rate at any compaction level. In the field conditions, the target emergence rate of 80% was achieved only under no compaction treatment in one of the two studying years. Therefore, avoiding soil compaction is recommended regardless of soil type to optimize canola production. These findings provide practical guidance for agricultural producers in managing field traffic and modifying seeder equipment to prevent economic losses due to compaction-induced poor emergence. Future studies on additional soil types in field conditions, along with the application of these findings in designing and adjusting seeders, are recommended.

Author Contributions: Y.C.: Conceptualization, methodology, validation, resources, writing—review and editing, supervision, project administration, funding acquisition. E.O.-S.: Methodology, formal analysis, investigation, data curation, writing—original draft preparation, writing—review and editing, visualization. All authors have read and agreed to the published version of the manuscript.

Funding: This research was funded by Natural Sciences and Engineering Research Council of Canada (grant number: RGPIN-2019-05861).

Institutional Review Board Statement: Not applicable.

Data Availability Statement: All relevant data are within the manuscript.

Acknowledgments: The authors thank all students who helped during the test the data collection.

Conflicts of Interest: The authors declare no conflicts of interest. The funders had no role in the design of the study; in the collection, analyses, or interpretation of data; in the writing of the manuscript; or in the decision to publish the results.

References

1. Rempel, C.B.; Hutton, S.N.; Jurke, C.J. Clubroot and the importance of canola in Canada. *Can. J. Plant Pathol.* **2014**, *36*, 19–26. [CrossRef]
2. Hyatt, J.; Wendroth, O.; Egli, D.B.; TeKrony, D.M. Soil Compaction and Soybean Seedling Emergence. *Crop Sci.* **2007**, *47*, 2495–2503. [CrossRef]
3. Harker, K.N.; O'Donovan, J.T.; Blackshaw, R.E.; Johnson, E.N.; Lafond, G.P.; May, W.E. Seeding Depth and Seeding Speed Effects on No-till Canola Emergence, Maturity, Yield and Seed Quality. *Can. J. Plant Sci.* **2012**, *92*, 795–802. [CrossRef]
4. Lamb, K.E.; Johnson, B.L. Seed Size and Seeding Depth Influence on Canola Emergence and Performance in the Northern Great Plains. *Agron. J.* **2004**, *96*, 454. [CrossRef]
5. Gong, H.; Chen, Y.; Wu, S.; Tang, Z.; Liu, C.; Wang, Z.; Fu, D.; Zhou, Y.; Qi, L. Simulation of canola seedling emergence dynamics under different soil compaction levels using the discrete element method (DEM). *Soil Tillage Res.* **2022**, *223*, 105461. [CrossRef]
6. Zuo, Q.; Kuai, J.; Zhao, L.; Hu, Z.; Wu, J.; Zhou, G. The effect of sowing depth and soil compaction on the growth and yield of rapeseed in rice straw returning field. *Field Crops Res.* **2017**, *203*, 47–54. [CrossRef]
7. Sivarajan, S.; Maharlooei, M.; Bajwa, S.G.; Nowatzki, J. Impact of soil compaction due to wheel traffic on corn and soybean growth, development and yield. *Soil Tillage Res.* **2018**, *175*, 234–243. [CrossRef]
8. Schjønning, P.; Lamandé, M.; Keller, T.; Labouriau, R. Subsoil shear strength–Measurements and prediction models based on readily available soil properties. *Soil Tillage Res.* **2020**, *200*, 104638. [CrossRef]
9. Horn, R.; Fleige, H.; Mordhorst, A.; Dörner, J. Structure dependent changes in pore water pressure due to stress application and consequences on the effective stress. *Soil Tillage Res.* **2023**, *231*, 105719. [CrossRef]
10. Radford, B.; Yule, D.; McGarry, D.; Playford, C. Crop Responses to Applied Soil Compaction and to Compaction Repair Treatments. *Soil Tillage Res.* **2001**, *61*, 157–166. [CrossRef]
11. Singh, R.; Ghildyal, B.P. Influence of soil edaphic factors and their critical limits on seedling emergence of corn (*Zea mays* L.). *Plant Soil* **1977**, *47*, 125–136. [CrossRef]
12. Batey, T. Soil compaction and soil management—A review. *Soil Use Manag.* **2009**, *25*, 335–345. [CrossRef]
13. Hamza, M.A.; Anderson, W.K. Soil compaction in cropping systems: A review of the nature, causes and possible solutions. *Soil Tillage Res.* **2005**, *82*, 121–145. [CrossRef]
14. Gelder, B.; Cruse, R.; Zhang, X. Comparison of track and tire effects of planter tractors on corn yield and soil properties. *Trans. ASABE* **2007**, *50*, 365–370. [CrossRef]
15. Ishaq, M.; Hassan, A.; Saeed, M.; Ibrahim, M.; Lal, R. Subsoil compaction effects on crops in Punjab, Pakistan: I. Soil physical properties and crop yield. *Soil Tillage Res.* **2001**, *59*, 57–65. [CrossRef]
16. Buttery, B.R.; Tan, C.S.; Drury, C.F.; Park, S.J.; Armstrong, R.J.; Park, K.Y. The Effects of Soil Compaction, Soil Moisture and Soil Type on Growth and Nodulation of Soybean and Common Bean. *Can. J. Plant Sci.* **1998**, *78*, 571–576. [CrossRef]
17. Chan, K.Y.; Oates, A.; Swan, A.D.; Hayes, R.C.; Dear, B.S.; Peoples, M.B. Agronomic consequences of tractor wheel compaction on a clay soil. *Soil Tillage Res.* **2006**, *89*, 13–21. [CrossRef]

18. Lehmann, P.; Merlin, O.; Gentine, P.; Or, D. Soil texture effects on surface resistance to bare-soil evaporation. *Geophys. Res. Lett.* **2018**, *45*, 10398–10405. [CrossRef]
19. ASABE. *American Society of Agricultural and Biosystems Engineering (ASABE) Standards, Moisture Measurement, Forages*; ASABE: St. Joseph, MI, USA, 2012.
20. Manitoba Agriculture. Manitoba Ag-Weather Program. Manitoba, Agriculture, Food and Rural Development. [Online]. Available online: https://web43.gov.mb.ca/climate/DailyReport.aspx (accessed on 28 November 2023).
21. Li, B.; Chen, Y.; Chen, J. Modeling of soil-claw interaction using the discrete element method (DEM). *Soil Tillage Res.* **2016**, *158*, 177–185. [CrossRef]
22. Harker, K.N.; O'donovan, J.T.; Smith, E.G.; Johnson, E.N.; Peng, G.; Willenborg, C.J.; Gulden, R.H.; Mohr, R.; Gill, K.S.; Grenkow, L.A. Conséquences du calibre des semences et de la densité des semis sur la levée, le développement et le rendement du canola ainsi que le poids des grains. *Can. J. Plant Sci.* **2015**, *95*, 1–8. [CrossRef]
23. Campbell, G.S. Soil Physics with Basic. *Soil Sci.* **1986**, *142*, 367–368. [CrossRef]
24. Phakdee, S.; Suvanjumrat, C. Development of a tire testing machine for evaluating the performance of tractor tires based on the soil compaction. *J. Terramechanics* **2023**, *110*, 13–25. [CrossRef]
25. de Jesus Duarte, S.; Glaser, B.; Pellegrino Cerri, C.E. Effect of biochar particle size on physical, hydrological and chemical properties of loamy and sandy tropical soils. *Agronomy* **2019**, *9*, 165. [CrossRef]
26. Saljnikov, E.; Mueller, L.; Lavrishchev, A.; Eulenstein, F. *Advances in Understanding Soil Degradation Innovations in Landscape Research*; Springer: Cham, Switzerland, 2022. [CrossRef]
27. Mileusnić, Z.I.; Saljnikov, E.; Radojević, R.L.; Petrović, D.V. Soil compaction due to agricultural machinery impact. *J. Terramechanics* **2022**, *100*, 51–60. [CrossRef]
28. Chen, H.; Gao, L.; Li, M.; Liao, Y.; Liao, Q. Effect of soil surface roughness on emergence rate and yield of mechanized direct-seeded rapeseed based on 3D laser scanning. *Int. J. Agric. Biol. Eng.* **2023**, *16*, 110–119. [CrossRef]
29. Acquah, K.; Chen, Y. Soil compaction from wheel traffic under three tillage systems. *Agriculture* **2022**, *12*, 219. [CrossRef]
30. Vance, W.H.; Bell, R.W.; Johansen, C. Physical Conditions That Limit Chickpea Root Growth and Emergence in Heavy-Textured Soil. *Seeds* **2024**, *3*, 26–39. [CrossRef]
31. Yang, C.; Wu, J.; Li, P.; Wang, Y.; Yang, N. Evaluation of Soil-Water Characteristic Curves for Different Textural Soils Using Fractal Analysis. *Water* **2023**, *15*, 772. [CrossRef]
32. Kahlon, M.S.; Singh, C.B.; Dhingra, M. Effect of compaction and irrigation regimes on soil physical characteristics, emergence, growth and productivity of summer moongbean. *Legum. Res. Int. J.* **2023**, *46*, 473–481. [CrossRef]
33. Habib-ur-Rahman, M.; Raza, A.; Ahrends, H.E.; Hüging, H.; Gaiser, T. Impact of in-field soil heterogeneity on biomass and yield of winter triticale in an intensively cropped hummocky landscape under temperate climate conditions. *Precis. Agric.* **2022**, *23*, 912–938. [CrossRef]
34. Zhang, S.; Zhang, X.; Liu, Z.; Sun, Y.; Liu, W.; Dai, L.; Fu, S. Spatial heterogeneity of soil organic matter and soil total nitrogen in a Mollisol watershed of Northeast China. *Environ. Earth Sci.* **2014**, *72*, 275–288. [CrossRef]
35. Hess, M.C.; Buisson, E.; Mesléard, F. Soil compaction enhances the impact of microwave heating on seedling emergence. *Flora* **2019**, *259*, 151457. [CrossRef]
36. Otalvaro, I.F.; Neto, M.P.C.; Delage, P.; Caicedo, B. Relationship between soil structure and water retention properties in a residual compacted soil. *Eng. Geol.* **2016**, *205*, 73–80. [CrossRef]
37. Orzech, K.; Wanic, M.; Załuski, D. The effects of soil compaction and different tillage systems on the bulk density and moisture content of soil and the yields of winter oilseed rape and cereals. *Agriculture* **2021**, *11*, 666. [CrossRef]
38. Yan, M.; Yang, D.; He, Y.; Ma, Y.; Zhang, X.; Wang, Q.; Gao, J. Alfalfa Responses to Intensive Soil Compaction: Effects on Plant and Root Growth, Phytohormones and Internal Gene Expression. *Plants* **2024**, *13*, 953. [CrossRef] [PubMed]
39. Harker, K.N.; Clayton, G.W.; Blackshaw, R.E.; O'Donovan, J.T.; Stevenson, F.C. Seeding Rate, Herbicide Timing and Competitive Hybrids Contribute to Integrated Weed Management in Canola (*Brassica napus*). *Can. J. Plant Sci.* **2003**, *83*, 433–440. [CrossRef]
40. Barbosa, L.C.; Tenelli, S.; Magalhães, P.S.; Bordonal, R.O.; Cherubin, M.R.; de Lima, R.P.; Castioni, G.A.; Neto, J.R.; Carvalho, J.L.N. Linking soil physical quality to shoot and root biomass production in scenarios of sugarcane straw removal. *Eur. J. Agron.* **2024**, *152*, 127029. [CrossRef]
41. Nawaz, M.M.; Noor, M.A.; Latifmanesh, H.; Wang, X.; Ma, W.; Zhang, W. Field traffic-induced soil compaction under moderate machine-field conditions affects soil properties and maize yield on sandy loam soil. *Front. Plant Sci.* **2023**, *14*, 1002943. [CrossRef]
42. Capobiango, N.P.; Bessa, G.B.; Peris, G.C.d.O.; da Silva, F.L.; Dias, D.C.F.d.S.; Fernandes, R.B.A.; da Silva, M.F.; da Silva, L.J. Evaluation of soybean genotypes grown under soil compaction. *J. Agron. Crop Sci.* **2023**, *209*, 517–531. [CrossRef]
43. Gawęda, D.; Haliniarz, M. The Yield and Weed Infestation of Winter Oilseed Rape (*Brassica napus* L. ssp. *oleifera* Metzg) in Two Tillage Systems. *Agriculture* **2022**, *12*, 563. [CrossRef]

agriculture

MDPI

Article

Development of a Prediction Model for Specific Fuel Consumption in Rotary Tillage Based on Actual Operation

Seung-Jun Kim [1,2,†], Moon-Kyeong Jang [1,2,†], Seok-Joon Hwang [3], Won Suk Lee [4] and Ju-Seok Nam [1,2,*]

1 Department of Biosystems Engineering, Kangwon National University, 1 Kangwondaehak-gil, Chuncheon 24341, Gangwon-do, Republic of Korea; gracenav4@naver.com (S.-J.K.); moon2842@kangwon.ac.kr (M.-K.J.)

2 Interdisciplinary Program in Smart Agriculture, Kangwon National University, 1 Kangwondaehak-gil, Chuncheon 24341, Gangwon-do, Republic of Korea

3 National Academy of Agricultural Science, Rural Development Administration, 310 Nong-saengmyeong-ro, Jeonju 54875, Jeollabuk-do, Republic of Korea; human51@korea.kr

4 Department of Agricultural and Biological Engineering, University of Florida, Gainesville, FL 32611, USA; wslee@ufl.edu

* Correspondence: njsg1218@kangwon.ac.kr

† These authors contributed equally to this work.

Abstract: Tractor fuel consumption has typically been predicted using indoor test results under specific conditions. This study analyzes the factors affecting fuel consumption during rotary tillage in actual fields and develops a prediction model. The test field was divided into sections using a grid method, and rotary tillage operations were performed to measure various parameters, including soil strength, tractor's transmission and PTO gear stages, tillage pitch, travel speed, engine and PTO shaft torque and speed, and fuel consumption. Pearson correlation identified variables affecting specific fuel consumption, and regression analysis was used to develop a prediction model. The model's accuracy was analyzed using the coefficient of determination (R^2) and root mean square error (RMSE), and it was compared with the ASABE's fuel consumption prediction model. The test results showed that higher transmission and PTO gear stages, and tillage pitch decreased specific fuel consumption, while soil strength had no significant effect. Thus, operating at higher gear and PTO stages within suitable conditions enhances energy efficiency in rotary tillage. Statistical analysis showed that specific fuel consumption significantly correlated with tractor travel speed, PTO shaft power, and PTO shaft speed. The prediction model, including these variables, had the highest accuracy with R^2 of 0.91 and RMSE of 0.011 L/kW·h. The developed prediction model showed significantly improved accuracy compared to the ASABE model, indicating that it can predict specific fuel consumption based on key operational variables in actual rotary tillage operations.

Keywords: prediction model; rotary tillage; specific fuel consumption; tractor

1. Introduction

Various implements, such as rotavators, front loaders, and harvesters, are attached to agricultural tractors to perform different tasks [1]. Among them, the rotavator is attached to the rear three-point hitch of the tractor and receives power from the PTO shaft to perform tillage operations [2,3]. Tillage operations are essential for creating a suitable soil environment for crop cultivation and are classified into primary tillage, which breaks up large clods of soil, and secondary tillage, which further pulverizes the soil into smaller aggregates [4,5]. Plows are mainly used for primary tillage, whereas harrows are used for secondary tillage. In contrast, rotary tillage using a rotavator can perform both primary and secondary tillage simultaneously, offering the advantages of reducing the working time and labor [6–9]. However, rotary tillage is also one of the agricultural tasks that imposes the greatest load due to friction with soil particles during operation [10,11]. Recently, rising international oil prices have increased the burden of fuel costs in agricultural operations. Inappropriate working conditions can also lead to excessive operational loads in terms of increasing fuel consumption, and thus the efficiency analysis of rotary tillage operations [12,13] becomes necessary. The energy efficiency of rotary tillage operations can be evaluated by specific fuel consumption, which is mainly influenced by the tractor's transmission gear stages and PTO gear stages [14,15]. Specific fuel consumption is the amount of fuel consumed divided by the output of the engine, expressed as the fuel consumption per unit of output. When performing the same task, a lower specific fuel consumption indicates higher energy efficiency [16,17]. Analyzing the specific fuel consumption can enable fuel savings, maximize implementation efficiency, and suggest optimal working conditions for specific agricultural tasks [18–21]. Therefore, various studies have been conducted to analyze and predict the specific fuel consumption of tractors.

Kocher et al. [22] developed and evaluated the accuracy of five different specific fuel consumption prediction models for tractors using the Organization for Economic Cooperation and Development test report data from eight tractors ranging from 171.5 to 275.9 kW. The models used independent variables such as drawbar load, travel speed, and engine speed either individually or in combination. Model 1 considered engine speed by travel speed, whereas Model 2 only considered drawbar load. Model 3 considered travel speed and drawbar load, and Model 4 considered engine speed and drawbar load. Model 5 considered drawbar load, travel speed, and engine speed. Models that considered travel speed showed higher accuracy compared to those that did not, and the model considering all three variables (drawbar load, travel speed, and engine speed) showed the highest accuracy. Cutini et al. [23] analyzed the effects of tire slip, tractor weight, engine power, and drawbar load on fuel consumption using a 208 kW tractor. Tire slip, tractor weight, and engine power affected fuel consumption significantly, whereas drawbar load did not have a significant effect. Therefore, to increase the fuel efficiency of the tractor, they recommended operating at an optimal travel speed with appropriate slip and engine power. Park et al. [24] analyzed the effects of engine speed, travel speed, CO_2 emissions, and PTO output on fuel consumption using a 37.3 kW tractor. They found that the engine speed, CO_2 emissions, and PTO output affected fuel consumption significantly, which was proportional to the engine speed and inversely proportional to the transmission gear stage. Igoni et al. [25] developed a fuel consumption prediction model using a 53.7 kW tractor. They set the drawbar pull, travel speed, and soil moisture content as variables and identified their effects on fuel consumption through actual measurements. Fuel consumption increased proportionally with drawbar pull, travel speed, and soil moisture content. Park et al. [13] measured and compared fuel consumption during plowing and rotary tillage using a 37.3 kW tractor. Fuel consumption was higher during rotary tillage compared to plowing and showed different trends based on the combination of engine and PTO shaft rotational speeds. Nkakini et al. [26] developed a fuel consumption prediction model based on drawbar pull, plowing depth, plowing width, travel speed, soil strength, and moisture content using a 53.7 kW tractor during plowing operations. Naik and Raheman [16] developed a fuel consumption prediction model for rotary tillage using a 33.8 kW tractor. They found that fuel consumption in rotary tillage was significantly influenced by PTO output, travel speed, and PTO shaft rotational speed.

Most existing studies on predicting the specific fuel consumption of tractors have been based on tractor test reports [22,27,28]. Specific fuel consumption varies depending on the type of fuel, percent load, and engine throttle settings. The American Society of Agricultural and Biological Engineers (ASABE) developed a prediction model for specific fuel consumption that does not consider the engine throttle settings but can predict fuel consumption under various load conditions over a period of time. By inputting the rated engine rotational speed, rated PTO shaft rotational speed, required engine rotational speed during operation, and PTO shaft power requirement, one can predict the specific fuel consumption [27].

Predictions based on test reports utilize indoor test data obtained under fixed conditions, which do not account for the actual agricultural working environment [26,29]. Consequently, there could be differences between the predicted and actual specific fuel consumption in real agricultural operations. Recent studies have been conducted on predicting specific fuel consumption considering soil conditions, but they mainly consider the average soil strength for the entire test field. Soil strength is not uniform even within the same test field and varies by location [30,31], thus necessitating the consideration of soil strength at different working locations. There have been many studies on developing fuel consumption prediction formulas based on actual test data results and soil strength analysis, but there is little research has combined these two areas.

In this study, the effects of soil strength, transmission gear stage, PTO gear stage, and tillage pitch on the specific fuel consumption during rotary tillage operations in actual test fields were analyzed. To more thoroughly analyze the fuel consumption rate in relation to soil strength, the soil strength at different work locations was considered. In addition, a prediction model for specific fuel consumption was developed through statistical analysis and compared with the ASABE prediction model based on the test reports. This study is expected to accurately predict specific fuel consumption during actual rotary tillage operations.

2. Materials and Methods

2.1. Tractor and Rotavator

The shape and specifications of the selected tractor and rotavator for the experiment are shown in Figure 1 and Table 1, respectively. The tractor had a rated engine output of 42 kW and could travel at nominal speeds, as shown in Table 2, through a combination of main gear stages and sub gear stages. The PTO gears consisted of three settings: 1st (540 rpm), 2nd (720 rpm), and 3rd (1000 rpm). Considering the rated PTO output of 39 kW of the tractor, a rotavator with an appropriate rated output was selected and used.

Figure 1. Tractor–rotavator system for measurement.

Table 1. Specifications of the tractor and rotavator used in this study.

Tractor			Rotavator	
Model	TX58		Model	WJ185A
Nation/Company	Korea/TYM		Nation/Company	Korea/Woongjin
Weight (kg)	3894		Weight (kg)	370
Length × Width × Height (mm)	3695 × 1848 × 2560		Rated power (kW)	36–41
Engine rated power (kW)	42		No. of flanges	7
PTO rated power (kW)	39		No. of blades for each flange	6
Transmission	main gears	4	Tillage width (mm)	2020
	sub gears	6		
Wheelbase (mm)	2155			
Tire size	Front	11.2–20	Tillage depth (mm)	120
	Rear	14.9–30		

Table 2. Transmission gear stages and the corresponding nominal travel speeds.

Transmission Gear		Travel Speed (km/h)
Main Gear	Sub Gear	
1	L (Ultra-low speed on)	0.32
2		0.47
3		0.68
4		0.90
1	M (Ultra-low speed on)	1.02
2		1.51
3		2.16
4		2.88
1	H (Ultra-low speed on)	3.28
2		4.83
3		6.92
4		9.22
1	L (Ultra-low speed off)	1.07
2		1.57
3		2.25
4		3.00
1	M (Ultra-low speed off)	3.40
2		5.00
3		7.18
4		9.58
1	H (Ultra-low speed off)	10.90
2		16.03
3		22.98
4		30.64

2.2. *Measurement System*

A data acquisition (DAQ) system from Dewesoft X (Dewesoft 3X, Dewesoft, Trbovlje, Slovenia) was installed inside the tractor for data collection (Figure 2). This system measured the travel speed, engine torque, engine speed, PTO shaft torque, PTO shaft rotational speed, and fuel consumption of the tractor during rotary tillage operations. An RTK-GPS was mounted inside the cabin to determine the actual travel speed of the tractor. The shape and specifications of the RTK-GPS used are presented in Figure 3 and Table 3, respectively.

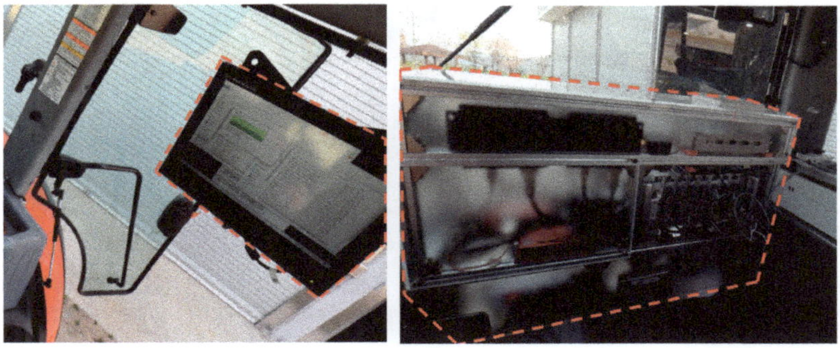

Figure 2. DAQ system used in the measurement.

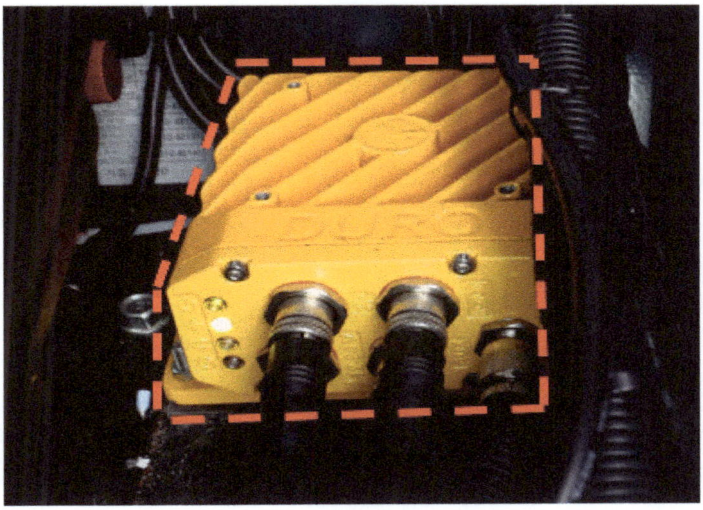

Figure 3. Shape of the RTK-GPS used for the measurement.

Table 3. Specifications of the RTK-GPS device.

Item		Specification
Model		Duro Inertial
Nation/Country/Company		USA/San Francisco/Swift Navigation
Length × Width × Height (mm)		130 × 130 × 65
Weight (kg)		0.8
RTK accuracy	Horizontal	0.01 m + 1 ppm
	Vertical	0.015 m + 1 ppm
IMU sampling rate (Hz)		25–200
GPS sampling rate (Hz)		10

Flange-type torque transducers and proximity sensors were attached to the engine and PTO shaft, respectively, of the tractor to measure the torque and rotational speed. The sampling rate was set to 1 kHz, the maximum sampling rate of the flange-type torque transducer. The shape and specifications of the flange-type torque transducer and proximity sensor are presented in Figure 4 and Table 4.

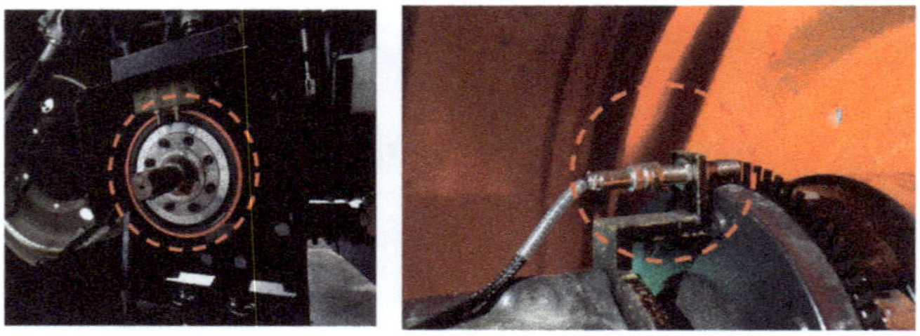

(**a**) (**b**)

Figure 4. Sensors used for torque and rotational speed measurement: (**a**) Flanged torque transducer; (**b**) Proximity sensor.

Table 4. Specifications of the flanged torque transducer and proximity sensor.

Item		Specification
Flanged torque transducer	Model	PCM 16
	Nation/Company	Germany/MANNER
	Supply voltage (V)	5
	Maximum measuring torque (kN·m)	15
	Maximum sampling rate (kHz)	1
Proximity sensor	Model	CYGTS211B-PO2
	Nation/Company	Chen Yang Technologies GmbH & Co. KG/Markt Schwaben/Germany
	Diameter/Screw size (mm)	M12×1/M16×1
	Sensing distance (mm)	<3
	Sensing object	Ferrous metal targets

The measured torque and rotational speed data of the engine and PTO shaft were inputted in Equations (1) and (2) to calculate the power requirements of the engine and PTO shaft.

$$P_e = \frac{2\pi \times T_e \times N_e}{60{,}000} \quad (1)$$

where, P_e is consumed power of engine (kW), T_e is engine torque (N·m), N_e is rotational speed of engine shaft (rpm).

$$P_{PTO} = \frac{2\pi \times T_{PTO} \times N_{PTO}}{60{,}000} \quad (2)$$

where, P_{PTO} is consumed power of PTO shaft (kW), T_{PTO} is PTO shaft torque (N·m), N_{PTO} is rotational speed of PTO shaft (rpm). In addition, flow sensors were installed at the fuel inlet and outlet to measure the fuel consumption. The shape and specifications of the flow sensors used are presented in Figure 5 and Table 5.

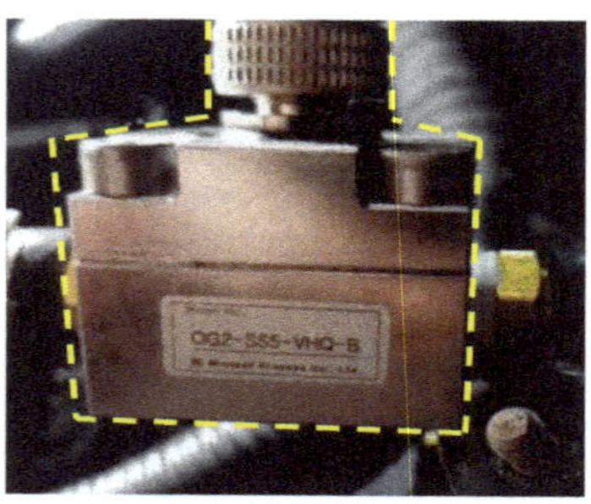

Figure 5. Flow rate sensor used in the measurement.

Table 5. Specifications of the flow rate sensor used in the measurement.

Item	Specification
Model	OG2
Nation/Company	Titan Enterprises Ltd./Sherborne/England/
Measuring oil flow range (L/min)	0.03–4
Oil Accuracy (%)	0.75
Repeatability (%)	±0.1

The specific fuel consumption was derived through Equation (3) using the measured fuel consumption and engine power requirement.

$$SFC = \frac{F_C}{P_e} \quad (3)$$

where, SFC is specific fuel consumption (L/kW·h), F_C is fuel consumption (L/h).

2.3. Test Conditions

Rotary tillage operations were performed to analyze the specific fuel consumption by the tractor under different working conditions. Considering that the typical rotary tillage speed for medium-sized tractors is 1.5–3.5 km/h, main gear stages 2, 3, 4, and sub gear stage L (ultra-low speed off) were selected as the gear conditions and PTO gear stages 1 and 2, which are commonly used for rotary tillage operations, were selected [32–34]. PTO gear stage 3 was excluded from the working conditions due to the occurrence of overload on the PTO shaft, which made accurate measurement difficult.

The length of soil tilled in the travel direction of the tractor when the rotavator blades made one rotation was defined as the tillage pitch. This pitch varies with the travel speed and the rotational speed of the PTO shaft of the tractor. The tillage pitch affects the load on the PTO shaft; a smaller tillage pitch increases the soil pulverization rate, thereby impacting operational performance [33,35]. In this study, tillage pitch was selected as one of the working variables and was calculated using Equation (4). The specific fuel consumption was then analyzed accordingly [36].

$$p = \frac{60 \times V_{act}}{Z \times N_{PTO}} \quad (4)$$

where, p is tillage pitch (cm), V_{act} is Actual travel speed of tractor (cm/s), Z is Number of rotavator blades for each flange. Field tests were conducted on two different soils in Sinbuk-eup, Chuncheon-si, Gangwon-do, Republic of Korea (Figure 6). Rotary tillage operations were performed on two fields of the same size (48 m × 30 m) by changing the transmission gear stage and PTO gear stage. The working conditions in Figure 7 represent the transmission gear stage and PTO gear stage, with the number after the hyphen (-) indicating the number of repeated tests. For example, L3P2-3 indicates the third repeated test performed with the transmission gear stage L3 and PTO gear stage 2 for rotary tillage.

The soil strength affects the tillage load and varies depending on location even within the same field [4,30]. Therefore, sampling is necessary to determine the soil strength at different working locations [31,37]. When collecting soil samples, the area was divided into a grid with uniform intervals and samples were extracted from each intersection [38,39]. In this study, as shown in Figure 8, the area was divided into a square grid with 3 m × 3 m intervals and soil strength at the intersections was measured, along with its moisture content and soil texture [40]. The cone index (CI) was used as the representative value for the soil strength. The standard test method specified by ASABE was followed, using a cone penetrometer (DIK-5532, Daiki, Saitama, Japan), which meets the test specifications [41–43]. The shape and specifications of the cone penetrometer are presented in Figure 9 and Table 6, respectively. The tillage depth of the rotavator used in this study was 120 mm. Considering that the general tillage depth of rotavators ranges from 100 to 200 mm, the cone index was measured from the soil surface (0 mm) to a depth of 200 mm in 10 mm intervals [11]. In Figure 10, the segments for each working condition consisted of two rows and nine columns, totaling 18 segments. In Figure 8, the grid intersections consisted of 15 rows and nine columns, totaling 135 intersections. To simultaneously consider the working conditions and soil strength at different locations, the segments were set such that each working condition included seven or eight soil strength measurement points (Figure 10). This allowed for the simultaneous verification of the effects of the working conditions and soil strength on the specific fuel consumption by the tractor.

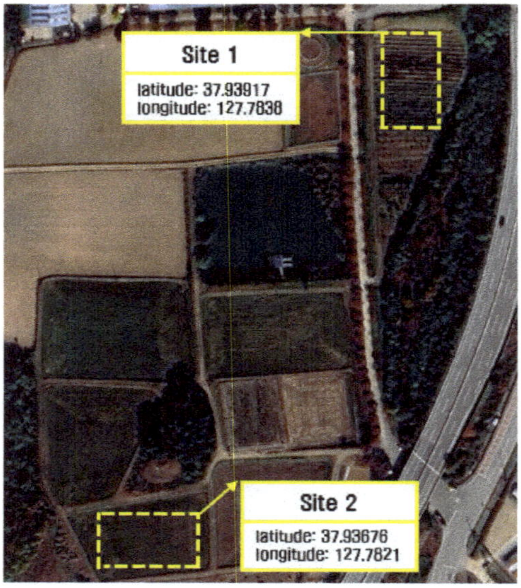

Figure 6. Location and coordinates of the test fields.

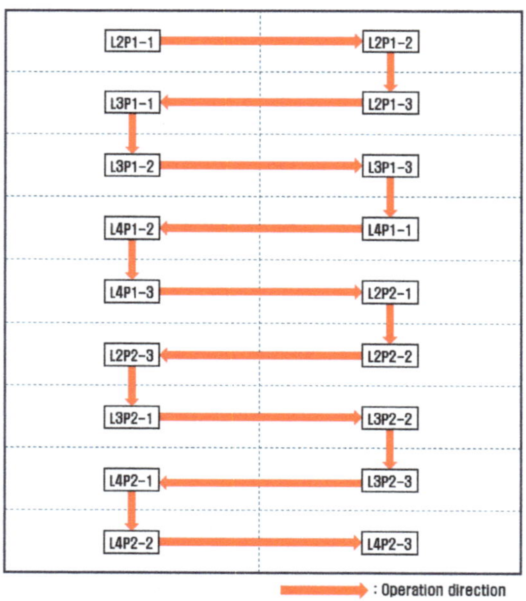

Figure 7. Rotary tillage operation sequence in the experiment.

Table 6. Specifications of the soil cone penetrometer.

Item	Specification
Model	DIK-5532
Nation/Company	Japan/Daiki
Length × Width × Height (mm)	345 × 212 × 144
Weight (kg)	4
Measuring range (kPa)	179–4903
Measuring depth (mm)	300–600

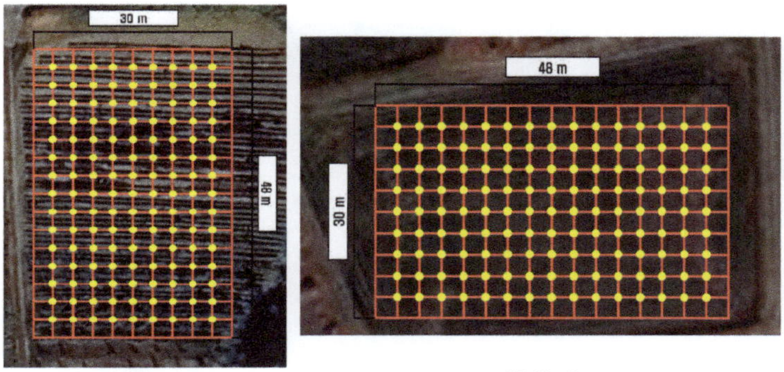

(**a**) Site 1 (**b**) Site 2

Figure 8. Grid of 3 m × 3 m for soil sampling.

Figure 9. Soil cone penetrometer.

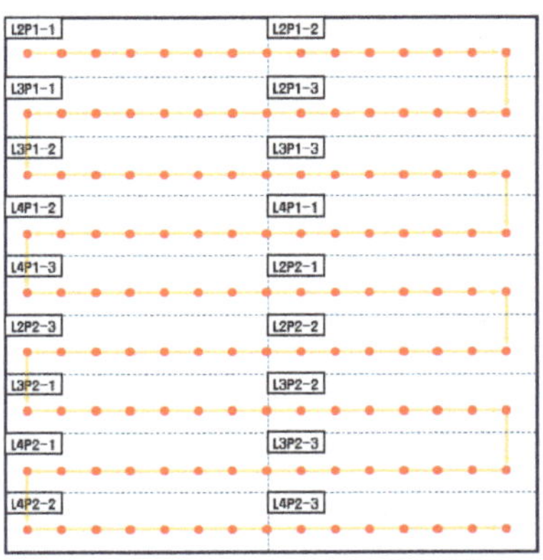

Figure 10. Matching with the sequence of the operation and grid sampling spots.

To analyze the soil texture and moisture content, soil samples were collected using a soil sampling tube (DIK-1801, Daiki Rika Kogyo Co., Ltd., Akagidai, Japan) and a soil sampling device (DIK-1815, Daiki Rika Kogyo Co., Ltd., Akagidai, Japan) (Figure 11).

Figure 11. Soil sampling process.

Soil texture was determined by the proportion of different particle sizes within the soil. The collected soil samples were dried, separated by particle size using standard sieves, and classified for each site using the USDA soil texture triangle. The weight of the soil sample was measured before and after drying at 110 °C for 24 h using an oven, and the moisture content was calculated using Equation (5).

$$W = \frac{100 \times (W_s - W_{dry})}{W_s} \quad (5)$$

where, W is soil water content (%), W_{dry} is weight of dried soil (g), W_s is weight of soil sample before drying (g).

2.4. Development of the Prediction Model for Specific Fuel Consumption

Data for each working condition were measured for 30 s to determine the specific fuel consumption of the tractor and a prediction model was developed through statistical analysis. At this time, measurements were taken only during effective working times, without any turning of the tractor, idling or other time losses. While transmission gear stages and PTO gear stages influence specific fuel consumption [14,22], they indirectly affect factors such as travel speed, PTO shaft rotational speed, PTO shaft power requirement, and tillage pitch. Considering this, the travel speed, PTO shaft rotational speed, PTO shaft power requirement, tillage pitch, and soil strength were considered as the variables influencing the specific fuel consumption by the tractor. However, the engine power requirement was excluded from the model. Since it is directly related to fuel consumption, it was excluded to avoid biasing the developed fuel consumption prediction model toward engine power. Pearson's correlation analysis was performed to verify the correlation between the specific fuel consumption and the selected working variables. Significantly correlated variables were identified and set as independent variables, and specific fuel consumption was set as the dependent variable for regression analysis to develop the prediction model.

Methods for selecting the variables for the prediction model included forward selection, backward elimination, and stepwise methods [44–46]. The stepwise method, which alternates between forward selection and backward elimination, was used to address their limitations. Starting with the second independent variable, the significance level of the F-statistic for the existing variables was checked. If its values for the existing variables remained significant, the next variable was added. If not, the added variable was removed. This process was repeated until all independent variables were considered and the final variables for the prediction model were determined. The stepwise method was used to select the suitable variables for each prediction model in this study.

The accuracy of the developed prediction model was evaluated using the coefficient of determination (R^2) and root mean square error (RMSE), which indicate the percentage of variance in the measured values compared to the predicted values and the error between the measured and predicted values, respectively. R^2 and RMSE were calculated using Equations (6) and (7). The closer R^2 is to 1 [47,48] and RMSE is to 0, the higher the accuracy of the prediction model [49–51].

$$R^2 = 1 - \frac{\sum_{i=1}^{n}(y_i - \hat{y}_i)^2}{\sum_{i=1}^{n}(y_i - \overline{y})^2} \quad (6)$$

$$RMSE = \sqrt{\frac{1}{n} \times \sum_{i=1}^{n}(y_i - \hat{y}_i)^2} \quad (7)$$

where, y_i is actual values, $\hat{y}_i$ is Predicted values, $\overline{y}$ is average of values. ASABE presented a prediction model for specific fuel consumption in PTO-based operations as shown in Equation (8) [27]. To compare it with the developed prediction model, the specific fuel consumption was calculated using the ASABE model. The rated output of the PTO shaft ($P_{PTO\ rated}$) and the rated engine speed ($N_{e\ rated}$) were set to 39 kW and 2200 rpm, respectively, for the tractor used in the experiment, whereas the PTO shaft power requirement (P_{PTO}) and engine speed (N_e) were set based on the measured data during actual operations.

$$SFC_{ASAE} = \left(0.22 + \frac{0.096 \times P_{PTO\ rated}}{P_{PTO}}\right) \times \left(1 - \left(\frac{N_e}{N_{e\ rated}} - 1\right) \times \left(0.45 \times \frac{0.096 \times P_{PTO\ rated}}{P_{PTO}} - 0.877\right)\right) \quad (8)$$

where, SFC_{ASAE} is specific fuel consumption of ASABE standard, $P_{PTO\ rated}$ is rated power of PTO shaft (kW), $N_{e\ rated}$ is rated rotational speed of engine shaft (rpm).

3. Results and Discussions

3.1. Results of the Soil Texture Analysis

The soil texture and moisture content were analyzed using the average values measured at the grid intersections of each site, and the results thus obtained are provided in Table 7. Site 1 was classified as loam and Site 2 as sandy loam. The average moisture content was 20.6% and 21.4%, respectively, showing similar levels between the two sites. The average soil strength for the entire depth of 0–200 mm and the average soil strength at 50 mm intervals for each site were derived (Table 8). Both sites showed an increase in the soil strength with increasing soil depth. In addition, the average soil strength for the entire depth was greater at Site 2 compared to Site 1. The results considering the average soil strength for the entire depth as well as the working conditions at different locations are provided in Table 9. As can be seen from the results, the soil strength varies depending on the working location within the same site.

Table 7. Results of the soil composition analysis.

Site	Water Content (%)	Silt (%)	Clay (%)	Sand (%)	Soil Texture
Site 1	20.6	34.0	20.0	46.0	Loam
Site 2	21.4	24.0	14.0	62.0	Sandy loam

Table 8. Average soil strength of the two sites.

Items		Soil Depth (mm)				
		Overall Average (0–200)	0–50	50–100	100–150	150–200
Cone Index (kPa)	Site 1	1087	482	857	1332	1676
	Site 2	1316	607	1213	1585	1856

Table 9. Soil strength for each operation condition.

Operating Conditions		Average CI (kPa)		
		1	2	3
Site 1	L2P1	993	1003	855
	L3P1	955	1051	876
	L4P1	695	1663	989
	L2P2	921	1768	953
	L3P2	1206	1101	1471
	L4P2	982	1253	1151
Site 2	L2P1	1121	1380	1411
	L3P1	1344	1761	927
	L4P1	1504	1421	1365
	L2P2	1209	1010	1623
	L3P2	917	1302	1669
	L4P2	1201	1152	1369

3.2. Results of the Specific Fuel Consumption Analysis

The fuel consumption according to the transmission gear stage, PTO gear stage, and soil strength is provided in Table 10, and the trend of the average fuel consumption according to tractor driving speed is shown in Figure 12. As the transmission gear stage and PTO gear stage increased, the tractor's driving speed also increased, and the average fuel consumption showed an increasing trend. Further, under the same PTO and transmission gear stage conditions, the average fuel consumption at Site 2 was higher than at Site 1, which is attributed to the higher soil strength at Site 2 compared to Site 1.

Table 10. Fuel consumption under different operating conditions.

Operating Conditions		Fuel Consumption (L/h)			
		1	2	3	Average
Site 1	L2P1	8.01	8.30	7.85	8.05
	L3P1	8.11	8.41	7.65	8.06
	L4P1	9.20	8.69	9.09	8.99
	L2P2	8.04	8.64	8.21	8.30
	L3P2	9.01	8.98	10.10	9.36
	L4P2	9.59	10.11	9.90	9.87
Site 2	L2P1	8.21	8.56	8.72	8.50
	L3P1	9.45	10.1	9.05	9.53
	L4P1	9.85	9.70	9.41	9.65
	L2P2	9.81	9.56	10.22	9.86
	L3P2	9.82	10.32	10.66	10.27
	L4P2	10.73	10.53	12.32	11.19

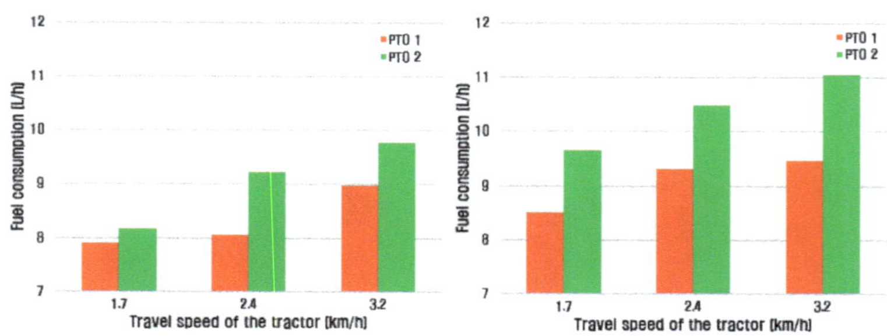

(**a**) Site 1 (**b**) Site 2

Figure 12. Average fuel consumption under different traveling speeds of the tractor.

The engine power requirement for different tractor transmission gear stages, PTO gear stages, and soil strengths is presented in Table 11 and Figure 13, with the average engine power requirement trend shown in Figure 14. As the tractor transmission gear stages and PTO gear stages increased, the average engine power requirement also increased. On the other hand, the engine power requirement did not show a specific trend with soil strength.

Table 11. Consumed power of engine for different operating conditions.

Operating Conditions		Consumed Power of Engine (kW) 1	2	3	Average
Site 1	L2P1	18.42	18.56	18.14	18.37
	L3P1	19.94	20.23	18.77	19.65
	L4P1	25.2	22.66	24.97	24.28
	L2P2	20.48	21.62	20.82	20.97
	L3P2	25.17	25.1	28.79	26.35
	L4P2	28.84	30.53	29.5	29.62
Site 2	L2P1	17.95	19.27	19.43	18.88
	L3P1	22.84	23.48	21.79	22.70
	L4P1	24.62	24.52	23.65	24.26
	L2P2	24.81	23.97	25.48	24.75
	L3P2	25.35	28.25	29.15	27.58
	L4P2	29.21	29.02	33.74	30.66

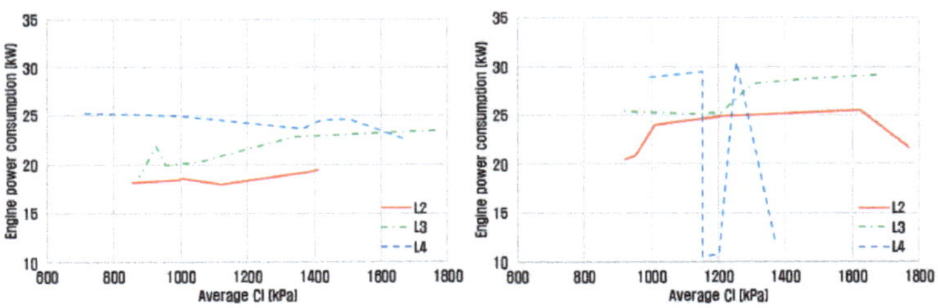

(**a**) PTO gear 1 (**b**) PTO gear 2

Figure 13. Engine power consumption based on average CI.

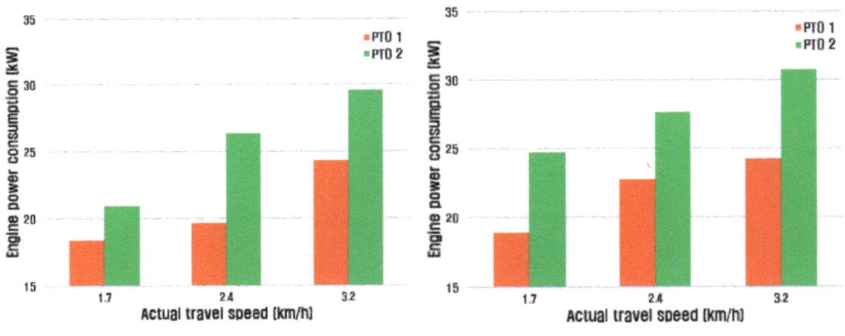

(**a**) Site 1 (**b**) Site 2

Figure 14. Average engine power under different operating conditions.

Table 12 shows the average specific fuel consumption for different values of the tractor transmission gear stage, PTO gear stage, and tillage pitch, and Figure 15 shows the trend of average specific fuel consumption as a function of these variables. As the tractor's transmission gear stages and PTO gear stages increased, the average specific fuel consumption decreased. This is because, as the tractor's transmission gear stages and PTO gear stages increase, the fuel consumption as well as engine power requirement increase, but the increase in engine power requirement is greater than that of fuel consumption.

This decreases the specific fuel consumption, which is expressed as the ratio of these two variables. Furthermore, the soil strength did not show a specific correlation with the specific fuel consumption (Figure 16). This is probably because neither fuel consumption nor engine power requirement showed a specific trend with soil strength. Soil strength often plays a critical role in determining the load on the tractor during tillage, which can affect both engine performance and fuel consumption. However, under the conditions of actual rotary tillage operations, it was found that soil strength did not significantly affect specific fuel consumption. On comparing the specific fuel consumption within the same PTO gear stage for different tillage pitches, it was found that as the tillage pitch increased, the specific fuel consumption decreased (Figure 17).

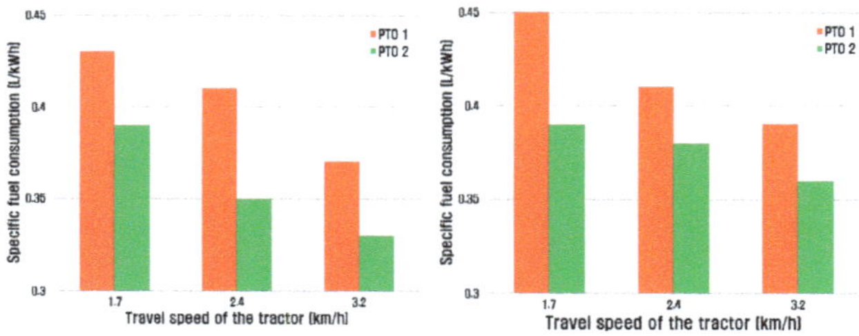

(**a**) Site 1 (**b**) Site 2

Figure 15. Average specific fuel consumption under different operating conditions.

Table 12. Specific fuel consumption according to operating conditions.

Operating Conditions		Tillage Pitch (cm)	Consumed Power of Engine (L/kW·h)			
			1	2	3	Average
Site 1	L2P1	0.82	0.44	0.45	0.43	0.44
	L3P1	1.18	0.41	0.42	0.41	0.41
	L4P1	1.57	0.37	0.38	0.36	0.37
	L2P2	0.62	0.39	0.40	0.39	0.40
	L3P2	0.89	0.36	0.36	0.35	0.36
	L4P2	1.19	0.33	0.33	0.34	0.33
Site 2	L2P1	0.83	0.46	0.44	0.45	0.45
	L3P1	1.20	0.41	0.43	0.42	0.42
	L4P1	1.61	0.40	0.40	0.40	0.40
	L2P2	0.63	0.40	0.40	0.40	0.40
	L3P2	0.90	0.39	0.37	0.37	0.37
	L4P2	1.21	0.37	0.36	0.37	0.37

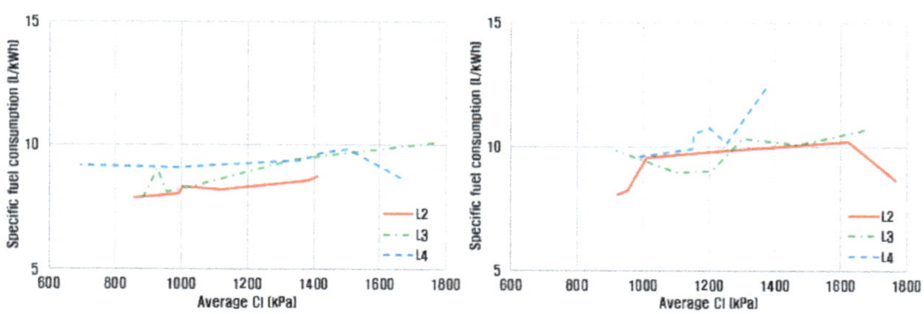

(**a**) PTO gear 1 (**b**) PTO gear 2

Figure 16. Specific fuel consumption as a function of the average CI.

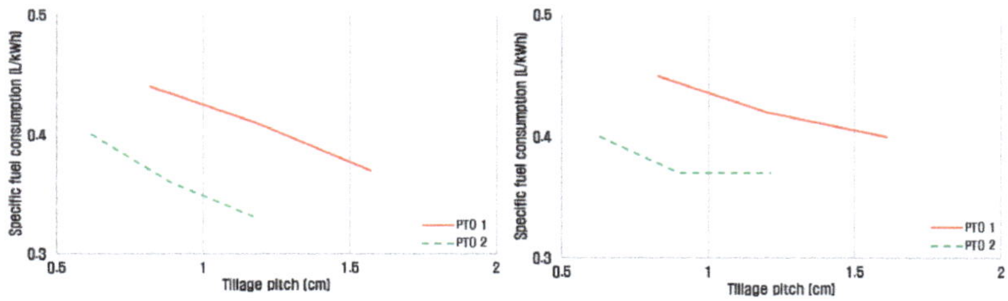

(**a**) Site 1 (**b**) Site 2

Figure 17. Average specific fuel consumption as a function of the tillage pitch.

3.3. Results of the Development of the Prediction Model for Specific Fuel Consumption

Pearson's correlation analysis was performed to verify the correlation between the specific fuel consumption and the tractor travel speed, PTO shaft power requirement, PTO shaft rotational speed, tillage pitch, and soil strength (Table 13). The specific fuel consumption showed a significant correlation with the tractor travel speed ($p < 0.01$), PTO shaft power requirement ($p < 0.01$), and PTO shaft rotational speed ($p < 0.01$). The correlation coefficients were in the order of PTO shaft rotational speed, tractor travel speed, and PTO shaft power requirement, indicating that PTO shaft rotational speed is the most influential factor on specific fuel consumption. In addition, all three variables with significant correlations showed negative correlations, indicating that lower tractor travel speed, PTO shaft power requirement, and PTO shaft rotational speed result in higher specific fuel consumption. Soil strength did not show a significant correlation with specific fuel consumption. Regression analysis was performed using the tractor travel speed, PTO shaft power requirement, and PTO shaft rotational speed as independent variables, which showed significant correlations with specific fuel consumption. The values of R^2, RMSE, types of independent variables, F-value, and *p*-value for the three prediction models derived using the stepwise method are provided in Table 14. The *p*-values for all three models were below 0.01, indicating statistical significance. The R^2 and RMSE for the third model, which included all significantly correlated independent variables, were 0.91 and 0.011 L/kW·h, respectively, indicating the highest accuracy. The regression coefficients for the third model are provided in Table 15 and the final form of the specific fuel consumption prediction model is expressed by Equation (9).

$$SFC_R = -0.0004 \times N_{PTO} - 0.043 \times V_{act} + 0.003 \times P_{PTO} + 0.67 \quad (9)$$

where, SFC_R is specific fuel consumption from this study. The R^2 and RMSE, MAPE (Mean Absolute Percentage Error) values for the prediction model developed in this study and the ASABE prediction model are provided in Table 16. The R^2 and RMSE values for the ASABE prediction model were 0.14 and 0.029 L/kW·h, respectively, indicating that the prediction model developed in this study has higher accuracy. In addition, the MAPE of the fuel consumption model developed in this study was found to be 0.02, lower than that of the ASABE prediction model, indicating its good performance. The developed prediction model is expected to accurately predict the specific fuel consumption during actual rotary tillage operations.

Table 13. Results of the Pearson's correlation analysis of specific fuel consumption.

Item		Actual Travel Speed	PTO Power	PTO Speed	Tillage Pitch	Soil Strength
Specific fuel consumption	Correlation coefficient	−0.65 *	−0.62 *	−0.66 *	−0.26	−0.01
	p-value	0.00	0.00	0.00	0.13	0.90
	N	36	36	36	36	36

* $p < 0.01$.

Table 14. Results of the stepwise method of specific fuel consumption.

Prediction Formula	Independent Variable	R^2	RMSE (L/kW·h)	F-Value	*p*-Value
1	PTO speed	0.43	0.025	25.81	0.00
2	PTO speed Actual travel speed	0.87	0.013	111.36	0.00
3	PTO speed Actual travel speed PTO power	0.91	0.011	111.17	0.00

Table 15. Regression coefficients for the prediction formula no. 3 in Table 15.

Dependent Variable	Independent Variable	Standardization Regression Coefficient (B)	Standard Error (S.E)	Standardized Coefficients (β)	t-Value	*p*-Value
Specific fuel consumption	Constant	0.67	0.017	-	39.41	0.00
	PTO speed	−0.0004	0.00	−0.93	−11.01	0.00
	Actual travel speed	−0.043	0.003	−0.80	−12.65	0.00
	PTO power	0.003	0.001	0.36	3.90	0.00

Table 16. Comparison of the predicted specific fuel consumptions.

Indicators	SFC_R	SFC_{ASABE}
R^2	0.91	0.14
RMSE	0.011 L/kW·h	0.029 L/kW·h
MAPE	0.02	0.05

4. Conclusions

In this study, the specific fuel consumption under different working conditions and soil strength during tractor rotary tillage operations was analyzed and a prediction model for specific fuel consumption was developed through regression analysis.

Rotary tillage operations were performed using a 42 kW tractor with a rotavator in two different types of soil located in Sinbuk-eup, Chuncheon-si, Gangwon-do, South Korea. Tractor travel speed, PTO shaft rotational speed, PTO shaft power requirement, tillage pitch, and soil strength were selected as the working variables, and specific fuel consumption was determined for different working conditions.

During the rotary tillage operations, as the tractor's transmission gear stages and PTO gear stages increased, the average fuel consumption and engine power requirement increased, but the specific fuel consumption decreased. This is because the increase in the engine power requirement was greater than the increase in fuel consumption as the tractor's transmission gear stages and PTO gear stages increased. As the tillage pitch increased within the same PTO gear stage, specific fuel consumption decreased. Although the fuel consumption and engine power requirement increased with higher soil strength, the specific fuel consumption did not show a specific trend. Therefore, to reduce the specific fuel consumption and increase the energy efficiency of a tractor, it is appropriate to perform operations with higher tractor transmission gear stages and PTO gear stages within suitable working conditions.

Pearson's correlation analysis performed for developing the specific fuel consumption prediction model showed that the tractor travel speed, PTO shaft power requirement, and PTO shaft rotational speed had significant correlations with specific fuel consumption. All three variables showed negative correlations with specific fuel consumption, with the PTO shaft rotational speed having the greatest influence. The prediction model, including all significant variables developed through regression analysis, had the highest accuracy, with values of R^2, RMSE and MAPE of 0.91, 0.011 L/kW·h and 0.02, respectively, outperforming the ASABE prediction model. The developed prediction model is thus expected to accurately identify the trends of specific fuel consumption depending on the working conditions.

Various mechanical factors were considered to develop a fuel consumption prediction model for a tractor performing rotary tillage operations. However, external factors such as weather and humidity, aside from soil strength, were not considered. Therefore, additional experiments are needed to develop a more comprehensive fuel consumption prediction model that includes a wider range of factors. Furthermore, we believe it is necessary to conduct experiments in diverse environments with different tractors and implements to expand the sample size of the study.

Author Contributions: Software, S.-J.K., M.-K.J. and W.S.L.; Validation, M.-K.J., S.-J.H. and W.S.L.; Investigation, S.-J.H.; Data curation, M.-K.J. and W.S.L.; Writing—original draft, S.-J.K.; Writing—review and editing, J.-S.N.; Visualization, S.-J.K. All authors have read and agreed to the published version of the manuscript.

Funding: This work was partly supported by Innovative Human Resource Development for Local Intellectualization program through the Institute of Information and Communications Technology Planning and Evaluation (IITP) grant funded by the Korea government (MSIT) (IITP-2024-RS-2023-00260267, 50) and This study was supported by the research grant of Kangwon National University in 2024(50).

Institutional Review Board Statement: Not applicable.

Data Availability Statement: The raw data supporting the conclusions of this article will be made available by the authors on request.

Conflicts of Interest: The authors have no conflicts of interest to declare.

References

1. Kim, S.J.; Gim, D.H.; Jang, M.K.; Hwang, S.J.; Kim, J.H.; Yang, Y.J.; Nam, J.S. Development of Regression Model for Predicting the Maximum Static Friction Force of Tractors with a Front-End Loader. *J. Biosyst. Eng.* **2023**, *48*, 329–338. [CrossRef]
2. Libin, Z.; Jiandong, J.; Yanbiao, L. Agricultural rotavator power requirement optimization using multi-objective probability parameter optimization. *Int. Agric. Eng. J.* **2010**, *19*, 15–22.
3. Yadav, R.K.; Singh, P.; Singh, S.K. Performance analysis of rotavator and other tillage implement driven by the tractor. *Int. J. Agric. Eng.* **2017**, *10*, 590–594. [CrossRef]
4. Myung, B.S.; Lee, H.D. Research on the actual condition of rotary tilling & rotary power requirement in the central area. *J. Korean Soc. Ind. Converg.* **2009**, *12*, 79–83.
5. Nam, J.S.; Kim, D.C.; Kim, M.H.; Kim, D.C. Tillage Characteristics Estimation of Crank-type and Rotary-type Rotavators by Motion Analysis of Tillage Blades. *J. Biosyst. Eng.* **2012**, *37*, 279–286. [CrossRef]
6. Hensh, S.; Tewari, V.K.; Upadhyay, G. An instrumentation system to measure the loads acting on the tractor PTO bearing during rotary tillage. *J. Terramech.* **2021**, *96*, 1–10. [CrossRef]
7. Al-Dosary, N.M.N.; Aboukarima, A.M.; Al-Hamed, S.A. Evaluation of Artificial Neural Network to Model Performance Attributes of a Mechanization Unit (Tractor-Chisel Plow) under Different Working Variables. *Agriculture* **2022**, *12*, 840. [CrossRef]
8. Kim, M.H.; Nam, J.S.; Kim, D.C. Comparison of tillage and loads characteristics of three types of rotavators: Rotary-type, crank-type, and plow-type. *J. Biosyst. Eng.* **2013**, *38*, 73–80. [CrossRef]
9. Kim, D.C.; Nam, J.S.; Kim, M.H.; Choe, J.S.; Inoue, E.; Okayasu, T.; Kim, D.C. Analysis of the tillage and power consumption characteristics of a crank–type rotavator according to the tillage blade shape. *J. Fac. Agric. Kyushu Univ.* **2013**, *58*, 319–328. [CrossRef]
10. Kim, W.S.; Kim, Y.J.; Baek, S.M.; Beak, S.Y.; Moon, S.P.; Lee, N.G.; Kim, T.J.; Siddique, M.A.A.; Jeon, H.H.; Kim, Y.S. Effect of the cone index on the work load of the agricultural tractor. *J. Drive Control* **2020**, *17*, 9–18.
11. Hwang, S.J.; Kim, J.H.; Jang, M.K.; Nam, J.S. Correlation between tractor variables and loan support limit in South Korea through regression analysis. *J. Biosyst. Eng.* **2022**, *47*, 402–408. [CrossRef]
12. Kumari, A.; Raheman, H. Tillage Operation with a Tractor Drawn Rotavator Using an Embedded Advisory System for Minimizing Fuel Consumption. *J. Biosyst. Eng.* **2023**, *48*, 487–502. [CrossRef]
13. Park, S.H.; Kim, Y.J.; Im, D.H.; Kim, C.K.; Jung, S.C.; Kim, H.J.; Kim, S.S. Characteristics of tractor PTO power and work loads. *J. Biosyst. Eng.* **2010**, *35*, 15–20. [CrossRef]
14. Kim, Y.S.; Kim, W.S.; Baek, S.Y.; Baek, S.M.; Kim, Y.J.; Lee, S.D.; Kim, Y.J. Analysis of tillage depth and gear selection for mechanical load and fuel efficiency of an agricultural tractor using an agricultural field measuring system. *Sensors* **2020**, *20*, 2450. [CrossRef]
15. Ranjbarian, S.; Askari, M.; Jannatkhah, J. Performance of tractor and tillage implements in clay soil. *J. Saudi Soc. Agric. Sci.* **2017**, *16*, 154–162. [CrossRef]

16. Naik, V.S.; Raheman, H. Factors affecting fuel consumption of tractor operating active tillage implement and its prediction. *Eng. Agric. Environ. Food* **2019**, *12*, 548–555. [CrossRef]
17. Moinfar, A.; Shahgholi, G.; Gilandeh, Y.A.; Gundoshmian, T.M. The effect of the tractor driving system on its performance and fuel consumption. *Energy* **2020**, *202*, 117803. [CrossRef]
18. Lovarelli, D.; Fiala, M.; Larsson, G. Fuel consumption and exhaust emissions during on-field tractor activity: A possible improving strategy for the environmental load of agricultural mechanisation. *Comput. Electron. Agric.* **2018**, *151*, 238–248. [CrossRef]
19. Sumner, H.R.; Hellwig, R.E.; Monroe, G.E. Measuring implement power requirements from tractor fuel consumption. *Trans. ASAE* **1986**, *29*, 85–89. [CrossRef]
20. Almaliki, S.; Alimardani, R.; Omid, M. Fuel consumption models of MF285 tractor under various field conditions. *Agric. Eng. Int. CIGR J.* **2016**, *18*, 147–158.
21. Fathollahzadeh, H.; Mobli, H.; Rajabipour, A.; Minaee, S.; Jafari, A.; Tabatabaie, S.M.H. Average and instantaneous fuel consumption of Iranian conventional tractor with moldboard plow in tillage. *ARPN J. Eng. Appl. Sci.* **2010**, *5*, 30–35.
22. Kocher, M.F.; Smith, B.J.; Hoy, R.M.; Woldstad, J.C.; Pitla, S.K. Fuel consumption models for tractor test reports. *Trans. ASABE* **2017**, *60*, 693–701. [CrossRef]
23. Cutini, M.; Brambilla, M.; Pochi, D.; Fanigliulo, R.; Bisaglia, C. A simplified approach to the evaluation of the influences of key factors on agricultural tractor fuel consumption during heavy drawbar tasks under field conditions. *Agronomy* **2022**, *12*, 1017. [CrossRef]
24. Park, S.H.; Kim, Y.J.; Im, D.H.; Kim, C.K.; Jang, Y.; Kim, S.S. Analysis of factors affecting fuel consumption of agricultural tractor. *J. Biosyst. Eng.* **2010**, *35*, 151–157. [CrossRef]
25. Igoni, A.H.; Ekemube, R.A.; Nkakini, S.O. Predicting tractor fuel consumption during ridging on a sandy loam soil in a humid tropical climate. *J. Eng. Technol. Res.* **2019**, *11*, 29–40. [CrossRef]
26. Nkakini, S.O.; Ekemube, R.A.; Igoni, A.H. Development of predictive model for fuel consumption during ploughing operation in agricultural soil. *Eur. J. Eng. Technol.* **2019**, *7*, 16–30.
27. *ASABE Standard ASAE D497.7*; Agricultural Machinery Management Data (R2020). American Society of Agricultural & Biological Engineers, ASABE: St. Joseph, MI, USA, 2011.
28. Ergül, İ. Türkiye'deki Tarım Traktörlerinin Enerji Verimliliği Sınıflandırmasına Yönelik bir Sistem Geliştirilmesi. Doctoral Dissertation, Al Manhal University, Abu Dhabi, United Arab Emirates, 2016.
29. Kim, W.S.; Park, S.U.; Choi, C.H.; Kim, Y.J. Analysis of Traction Performance of Agricultural Tractor according to Cone Index during Plow Tillage. *J. Drive Control* **2020**, *17*, 133–140. [CrossRef]
30. Kumar, A.; Chen, Y.; Sadek, M.A.A.; Rahman, S. Soil cone index in relation to soil texture, moisture content, and bulk density for no-tillage and conventional tillage. *Agric. Eng. Int. CIGR J.* **2012**, *14*, 26–37.
31. Mallarino, A.P.; Wittry, D.J. Efficacy of grid and zone soil sampling approaches for site-specific assessment of phosphorus, potassium, pH, and organic matter. *Precis. Agric.* **2004**, *5*, 131–144. [CrossRef]
32. Kim, Y.S.; Bae, B.M.; Kim, W.S.; Kim, Y.J.; Lee, S.D.; Kim, T.J. Working Load Analysis of a 42 kW Class Agricultural Tractor According to Tillage Type and Gear Selection during Rotary Tillage Operation. *Agriculture* **2023**, *13*, 1556. [CrossRef]
33. Kim, J.G.; Kim, Y.J.; Kim, J.H.; Shin, B.S.; Nam, J.S. Consumed-power and load characteristics of a tillage operation in an upland field in Republic of Korea. *J. Biosyst. Eng.* **2018**, *43*, 83–93. [CrossRef]
34. Ryu, M.J.; Kang, S.W.; Chung, S.O.; Kim, Y.J.; Lee, D.H.; Choi, C.H. Load characteristics of 30 and 75 kW agricultural tractors during field operations. In *2013 Kansas City, Missouri, July 21–July 24, 2013*; American Society of Agricultural and Biological Engineers: St. Joseph, MI, USA, 2013. [CrossRef]
35. Park, S.H.; Lee, D.H.; Kim, H.J.; Lee, C.S.; Cho, S.C.; Kwak, T.Y. Development of dry paddy seeder of strip tillage. *J. Biosyst. Eng.* **2002**, *27*, 25–32. [CrossRef]
36. Kim, S.S.; Lee, Y.S.; Woo, J.K. Tilling load characteristic analysis according to the shape factors of rotary blade. In Proceedings of the Korean Society for Agricultural Machinery Conference, Jeon-Ju, Republic of Korea, 6 January 1997; pp. 18–24.
37. Kerry, R.; Oliver, M.A. Average variograms to guide soil sampling. *Int. J. Appl. Earth Obs. Geoinf.* **2004**, *5*, 307–325. [CrossRef]
38. Wollenhaupt, N.C.; Wolkowski, R.P. Grid soil sampling. *Better Crop.* **1994**, *78*, 6–9.
39. Asare, E.; Segarra, E. Adoption and extent of adoption of georeferenced grid soil sampling technology by cotton producers in the southern US. *Precis. Agric.* **2018**, *19*, 992–1010. [CrossRef]
40. Tan, K.H. *Soil Sampling, Preparation, and Analysis*, 2nd ed.; CRC Press: Boca Raton, FL, USA, 2005.
41. Jabro, J.D.; Stevens, W.B.; Iversen, W.M.; Sainju, U.M.; Allen, B.L. Soil cone index and bulk density of a sandy loam under no-till and conventional tillage in a corn-soybean rotation. *Soil. Till. Res.* **2021**, *206*, 104842. [CrossRef]
42. *ASABE Standard ASAE S313.3*; Soil Cone Penetrometer. American Society of Agricultural & Biological Engineers, ASABE: St. Joseph, MI, USA, 2018.
43. *ASABE Standard ASAE EP542.1*; Procedures for Using and Reporting Data Obtained with the Soil Cone Penetrometer. American Society of Agricultural & Biological Engineers, ASABE: St. Joseph, MI, USA, 2019.
44. Blanchet, F.G.; Legendre, P.; Borcard, D. Forward selection of explanatory variables. *Ecology* **2008**, *89*, 2623–2632. [CrossRef]
45. Sutter, J.M.; Kalivas, J.H. Comparison of forward selection, backward elimination, and generalized simulated annealing for variable selection. *Microchem. J.* **1993**, *47*, 60–66. [CrossRef]

46. Thompson, B. Stepwise regression and stepwise discriminant analysis need not apply here: A guidelines editorial. *Educ. Psychol. Meas.* **1995**, *55*, 525–534. [CrossRef]
47. Rousson, V.; Goşoniu, N.F. An R-square coefficient based on final prediction error. *Stat. Methodol.* **2007**, *4*, 331–340. [CrossRef]
48. Hwang, S.J.; Kim, J.H.; Jang, M.K.; Nam, J.S. Prediction Model of Loan Support Limit of Plows and Rotavators Used in South Korea by Regression Analysis. *J. Biosyst. Eng.* **2022**, *47*, 409–416. [CrossRef]
49. Karunasingha, D.S.K. Root mean square error or mean absolute error? Use their ratio as well. *Inf. Sci.* **2022**, *585*, 609–629. [CrossRef]
50. Chai, T.; Draxler, R.R. Root mean square error (RMSE) or mean absolute error (MAE)?—Arguments against avoiding RMSE in the literature. *Geosci. Model Dev.* **2014**, *7*, 1247–1250. [CrossRef]
51. Hodson, T.O. Root-mean-square error (RMSE) or mean absolute error (MAE): When to use them or not. *Geosci. Model Dev.* **2022**, *15*, 5481–5487. [CrossRef]

agriculture

MDPI

Article

Dynamic Simulation Model of Miniature Tracked Forestry Tractor for Overturning and Rollover Safety Evaluation

Yun-Jeong Yang [1,2], Moon-Kyeong Jang [1,2] and Ju-Seok Nam [1,2,*]

[1] Department of Biosystems Engineering, Kangwon National University, 1 Kangwondaehak-gil, Chuncheon 24341, Gangwon-do, Republic of Korea; yyj000501@naver.com (Y.-J.Y.); moon2842@naver.com (M.-K.J.)
[2] Interdisciplinary Program in Smart Agriculture, Kangwon National University, 1 Kangwondaehak-gil, Chuncheon 24341, Gangwon-do, Republic of Korea
* Correspondence: njsg1218@kangwon.ac.kr; Tel.: +82-33-250-6497

Abstract: This study proposes a method to construct a dynamic simulation model to implement the lateral overturning and backward rollover characteristics of an actual tractor. Based on theoretical analysis, factors affecting these characteristics are identified, which include tractor weight, track width, wheelbase, location of mass center, weight distribution, heights of front and rear axles, and geometric shapes. The location of the mass center of the actual tractor is measured based on the standard test procedure set by the International Organization for Standardization, and the remaining influencing factors are derived through measurements. A three-dimensional (3D) model of the tractor is constructed to reflect all these factors. Additionally, a simulation model utilizing this 3D model is developed using a commercial dynamic simulation software program. The ability of the model to simulate the overturning and rollover characteristics of the actual tractor is verified by comparing the static sidelong falling angle and minimum turning radius with those of the actual tractor. The errors between the characteristics of the actual tractor and those of the 3D model and dynamic simulations are shown to be less than 5%, thus indicating that the proposed method can effectively simulate the overturning and rollover characteristics of the actual tractor.

Keywords: dynamic simulation model; miniature tracked forestry tractor; minimum turning radius; overturning; rollover; static sidelong falling angle

1. Introduction

Owing to the widespread use of agricultural machinery and the aging population of farmers, agricultural accidents have increased [1]. The rate of agricultural accidents is approximately 1.4 times higher than the average for all industries [2], with machinery-related accidents presenting the highest incidence and mortality rates among all industrial accidents [3]. In South Korea, tractors are predominantly responsible for agricultural machinery accidents [4], with overturning and rollover accidents constituting 32% of all such accidents [5]. In the United States, tractor accidents claim over 800 lives annually, with more than half of these accidents involving overturns or rollovers [6]. In Italy, 57.4% (205 cases) of 357 tractor accidents recorded between 2002 and 2012 were caused by overturning and rollover [7]. To address the global increase in tractor overturning/rollover accidents, researchers have intensively investigated tractor safety.

Previati et al. [8] analyzed the effect of ground slope angle, tire stiffness, and tractor axle position on the backward rollover of tractors on slope. They discovered that the slope angle at which backward rollover occurs increases by more than 70% for tractors equipped with front-axle suspension compared with standard tractors. Franceschetti et al. [9] analyzed the effects of tractor weight, track width, and mass center height on lateral overturning. They concluded that lateral overturning is not significantly affected by tractor weight and that safety decreases when the tractor is loaded, owing to the elevated center of mass. Lee et al. [10] theoretically analyzed the lateral overturning and backward rollover safety of small off-road vehicles to understand the manner in which these safety factors vary with the steering angle during operation.

The lateral overturning and backward rollover safety of agricultural machinery have been extensively investigated through dynamic simulation, which is a method that requires lower cost and time compared with experimental studies [11]. For instance, Shim et al. [12] analyzed the effects of driving direction and ground slope on the lateral overturning of forestry vehicles via dynamic simulation. They revealed that driving in the contour direction on slopes improved safety compared with driving along the slope direction; subsequently, they derived safe driving conditions based on the driving direction. Similarly, Kang et al. [13] performed a simulation to determine the maximum safe slope for driving a red pepper harvester and the critical angle for its backward rollover during collisions; they further suggested considering operating conditions such as crop loading and driver type in future studies. Iqbal et al. [14] performed a dynamic simulation to obtain the ground slope angle at which lateral overturning and backward rollover occur during red pepper transplanting, which provided valuable insights for safety measures. Park et al. [15] performed simulations for forestry forwarders to ascertain the critical speed for lateral overturning under different conditions, such as log loading and slope variations. They suggested reducing its driving speed by approximately 20%, as the increased mass center height decreases lateral overturning safety for a forwarder loaded with logs. Qin et al. [16] performed simulations to analyze the effects of electric power steering (EPS) on the lateral overturning and backward rollover of a tractor. They implemented an irregular road surface, which reflects actual farmland, and adjusted the ground slope angle and tractor speed as variables. The results showed that the EPS improved the lateral overturning and backward rollover safety when the tractor traveled at speeds exceeding 6 m/s on a 15° slope. They highlighted that further research is required to ensure safety on relatively harsh roads.

Whereas many studies have evaluated overturning and rollover safety through simulations, performing tests on actual machinery remains challenging owing to the risks and complexities of implementing tests under varying conditions [17]. Simulations offer a safer and more accessible alternative compared with actual experiments; however, they may not accurately reflect the physical properties of real-world scenarios, thus potentially resulting in significant inaccuracies [18]. Previous studies typically focused on the qualitative effects of key factors on safety, with few models accurately reflecting actual physical properties. Thus, this study proposes a method for constructing a dynamic simulation model to accurately represent the overturning and rollover characteristics of actual tractors. Influential factors are derived through theoretical analysis, and the model is validated via tests that consider the static sidelong falling angle and minimum turning radius. The proposed approach balances accuracy and efficiency. The results from this study are expected to serve as valuable reference data for constructing simulation models for agricultural tractors.

2. Materials and Methods

2.1. Tractor Overturning and Rollover Theory

Theoretical analysis was conducted via a literature review to determine variables that affect the lateral overturning and backward rollover of a tractor. It was conducted based on wheeled tractors, as references for tracked tractors are not available. For this analysis, the tractor was assumed to be in static equilibrium, with all elements except for the wheels regarded as rigid bodies. Owing to the typically low driving speeds of tractors, air resistance during movement was disregarded [19].

2.1.1. Lateral Overturning on Slope Under Stationary Condition

Figure 1 shows a simplified diagram of a tractor and a free-body diagram for a theoretical analysis of the slope angle at which lateral overturning occurs when the tractor is stopped on a slope [20].

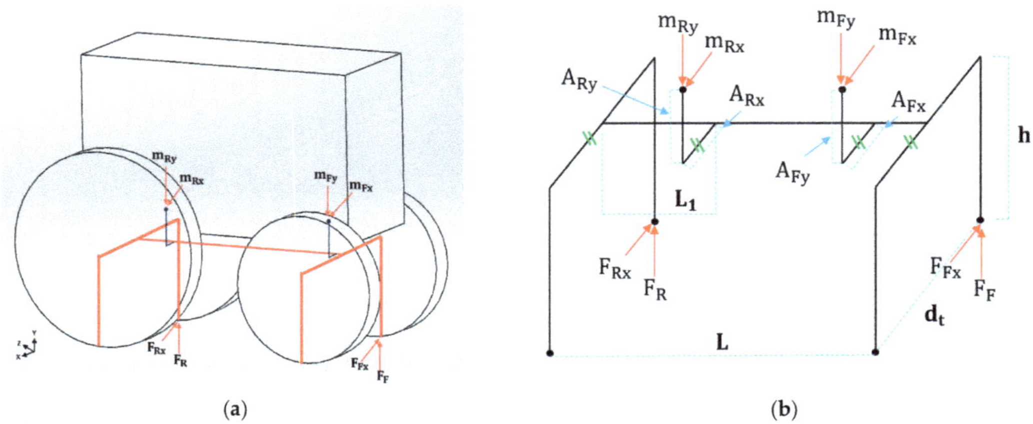

Figure 1. Diagram for lateral overturning analysis under stationary condition: (**a**) simplification of tractor model; (**b**) geometric parameters of tractor model.

When the tractor is stopped in the contour direction on a slope, the ground contact forces of the upslope rear wheel and front wheel, F_R and F_F, respectively, can be expressed, as shown in Equations (1) and (2), respectively [21].

$$F_R = \frac{m_R g}{d_t}\left[\frac{L_1}{L}\left(h\sin\theta - \frac{d_t}{2}\cos\theta\right) - (h + A_{Ry})\sin\theta + \left(\frac{d_t}{2} + A_{Rx}\right)\cos\theta\right], \quad (1)$$

where F_R is the contact force of the rear upslope wheel in the stationary state, m_R is the mass of the tractor at the rear section, g is the gravitational acceleration, d_t is the track width, L is the wheelbase, L_1 is the distance from the rear axle to the rear mass center, h is the front-axle height, θ is the ground slope angle, A_{Rx} is the downslope distance of the rear mass center with respect to the midplane, and A_{Ry} is the height of the rear mass center from the rear axle.

$$F_F = \frac{m_R\, g}{d_t}\left(\frac{L_1}{L}\right)\left[\frac{d_t}{2}\cos\theta - h\sin\theta\right] + \frac{m_F\, g}{d_t}\left[\left(\frac{d_t}{2} - A_{Fx}\right)\cos\theta - (h_1 + A_{Fy})\sin\theta\right], \quad (2)$$

where F_F is the contact force of the front upslope wheel in the stationary state, m_F is the mass of the tractor at the front section, A_{Fx} is the downslope distance of the front mass center with respect to the midplane, and A_{Fy} is the height of the front mass center from the front axle.

The contact force between the tractor tire and ground is derived based on the geometric structure of the tractor, its overall mass distribution, and the ground slope. As the ground slope increases, the upslope rear wheel lifts off the ground first, thus initiating a Phase I rollover. If the slope continues to increase, then the upslope front wheel lifts, thus resulting in lateral overturning. The slope angle at which lateral overturning begins due to the zero ground contact force of the rear wheel, θ_{R0}, can be derived by setting F_R to zero in Equation (1), as shown in Equation (3) [21].

$$\theta_{R0} = \tan^{-1}\left(\frac{d_t(L - L_1) - 2A_{Rx}L}{2\left[(h + A_{Ry})L - hL_1\right]}\right), \quad (3)$$

where θ_{R0} is the ground slope angle when the upslope rear wheel begins to lift from the ground.

In Equation (3), variables L_1, A_{Rx}, and A_{Ry} are related to the location of the mass center. The track width, wheelbase, front-axle height, and mass center of the tractor significantly affect lateral overturning under stationary conditions on a slope.

2.1.2. Lateral Overturning During Constant-Speed Turning

Figure 2 shows a diagram for the theoretical analysis of lateral overturning caused by a turning tractor [19].

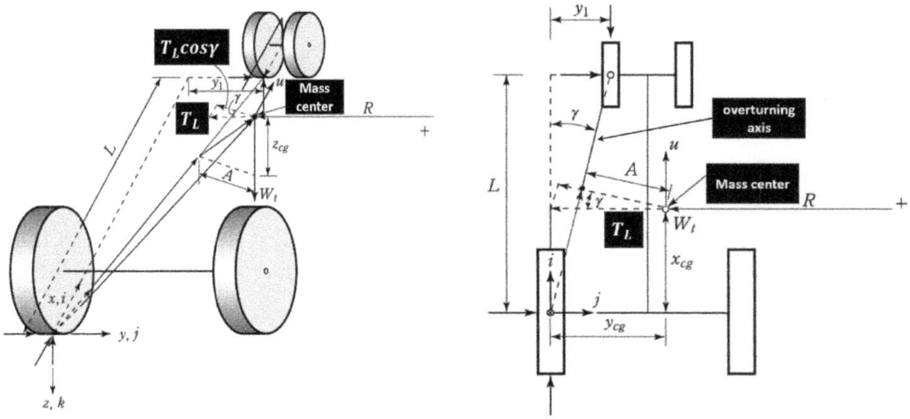

(**a**) (**b**)

Figure 2. Diagram for lateral overturning analysis under constant-speed turning: (**a**) isometric view; (**b**) top view.

When a tractor turns at a driving speed of u along a path of radius R, the lateral force exerted on the mass center of the tractor by the wheels can be obtained using Equation (4). As the driving speed increases, the lateral force at the mass center increases as well, which creates a moment about the overturning axis connecting the left front wheel to the ground contact point of the rear wheel, thereby resulting in lateral overturning. If the speed at which the right wheel begins lifting due to lateral overturning is defined as the critical speed u_s, then the moment equilibrium for the overturning axis is expressed as shown in Equation (5) [19].

$$T_L = m_t \frac{u^2}{R}, \tag{4}$$

where T_L is the lateral force of the mass center of the tractor, m_t is the tractor mass, u is the travel speed of the tractor, and R is the turning radius of the tractor.

$$m_t \frac{{u_s}^2}{R} \cos\gamma \, z_{cg} - W_t A = 0, \tag{5}$$

where γ is the angle between the lateral force of the mass center and the plane normal to the overturning axis ($\tan^{-1}\left(\frac{y_1}{L}\right)$), z_{cg} is the mass center height of the tractor, u_s is the critical travel speed of the tractor under constant-speed turning, W_t is the tractor weight, and *A* is the normal distance between the mass center and overturning axis.

In Equation (5), the variable A can be determined by considering the location of the mass center (x_{cg}, y_{cg}, and z_{cg}), the wheelbase (L), and the horizontal distance between the centers of the front and rear wheels (y_1) [19]. The horizontal distance between the centers of the front and rear wheels is affected by the track width between the front and rear wheels. Therefore, the variables affecting the critical speed of lateral overturning during a turn include the tractor weight, turning radius, mass center location, wheelbase, and track width.

2.1.3. Backward Rollover While Driving on Slope

Figure 3 shows a diagram for analyzing the backward rollover of a rear wheel drive tractor traveling on a slope [19]. The diagram captures the relative motion between the main body, including the front and rear wheels, with the dotted lines indicating the body's rotation by θ relative to the *Y*-axis.

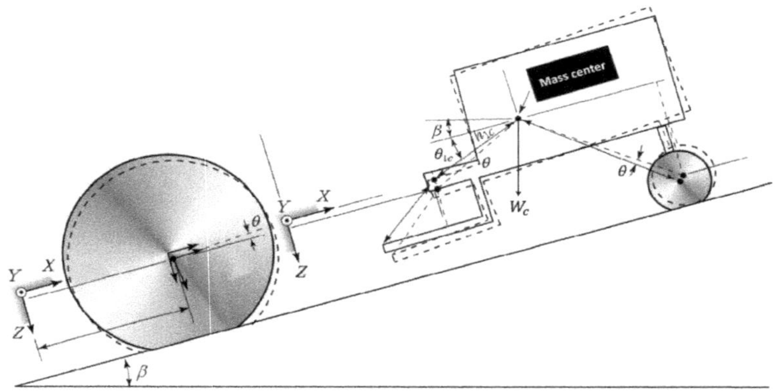

Figure 3. Diagram for backward rollover analysis.

As the angular acceleration $\ddot{\theta}$ and angular velocity $\dot{\theta}$ of the tractor increase, the risk of backward rollover caused by the rotation of the tractor's main body increases as well. If the angular velocity exceeds a certain threshold, then the tractor may rotate to a statically unstable angle, even without any driving torque or traction, thus resulting in a backward rollover. This occurs because when the angular velocity exceeds a certain value, the rotational kinetic energy promotes backward rollover. When the main body's rotation angle is θ_s, the angular velocity that causes the tractor to rotate to this unstable angle is defined as the critical angular velocity $\dot{\theta}_s$, which can be obtained using Equation (6) [19]. A higher critical angular velocity indicates improved backward rollover safety.

$$\dot{\theta}_s = \sqrt{\frac{2W_c h_{1c}(1 - \sin[\theta_{1c} + \beta + \theta_s])}{I_{yyc} + m_c {h_{1c}}^2}}, \quad (6)$$

where $\dot{\theta}_s$ is the critical angular velocity for the backward rollover of the tractor, m_c is the mass of the tractor body including the front wheels, h_{1c} is the distance between the rear wheel center and mass center of the tractor, β is the ground slope angle, θ_{1c} is the angle between the straight line of h_{1c} and the *X*-axis, θ_s is the rotation angle of the tractor body, W_c is the weight of the tractor body, and I_{yyc} is the mass moment of inertia of the tractor body about the *Y*-axis passing through the mass center.

In Equation (6), the mass moment of inertia I_{yyc} is determined by the tractor's mass center location, geometry, and weight distribution. The variables h_{1c} and θ_{1c} are determined by the rear-axle height and mass center location. Therefore, the critical factors affecting the critical angular velocity for backward rollover when the tractor is on a slope include the tractor's main body weight, ground slope angle, rear-axle height, mass center location, weight distribution, and geometric shape.

2.1.4. Analysis of Factors That Affect Lateral Overturning and Backward Rollover of Wheeled and Tracked Tractors

The theoretical analysis identified several factors influencing lateral overturning and backward rollover: tractor weight, track width, wheelbase, mass center location, weight distribution, heights of the front and rear axles, and geometric shape (Table 1).

Table 1. Factors affecting tractor overturning and rollover.

Items	Influencing Factors
Lateral overturning in stationary state	Wheelbase Track width Location of mass center Front-axle height
Lateral overturning in steady state turning	Mass Location of mass center Wheelbase Track width
Backward rollover in traveling on slopes	Mass Location of mass center Rear-axle height Weight distribution Geometric shape
Total	Mass Wheelbase Track width Location of mass center Front-axle height Rear-axle height Weight distribution Geometric shape

The theoretical analysis is discussed based on a wheeled tractor owing to the dearth of studies related to tracked tractors; however, the mass, mass center location, weight distribution, and geometric shape can be equally applied to tracked tractors. The wheelbase is defined as the straight distance between the centers of the front and rear axles for wheeled tractors but is the length of the ground contact region for tracked tractors [22]. For wheeled tractors, the wheels are elements that rotate the tires via power transmitted from the engine. In tracked tractors, components such as the sprocket, idler, and roller that rotate the track serve a similar function to the wheels. Therefore, the axle height of wheeled tractors can be equated to the height of the sprocket, idler, and roller in tracked tractors. The track width refers to the distance between the wheels on a wheeled tractor, whereas it refers to the distance between the tracks on a tracked tractor. The track width of a tracked tractor is different from that of a wheeled tractor in that the front and rear lengths of the track width are the same. Wheeled and tracked tractors have similar overall configurations except for driving devices, and the results of the theoretical analysis show that factors affecting wheeled tractors are physical properties that are applicable to tracked tractors. Therefore, theoretical analysis of wheeled tractors can be similarly applied to tracked tractors (Table 2).

Table 2. Application of influencing factors from wheeled tractors to tracked tractors.

Items	Wheeled Tractors	Tracked Tractors
Wheelbase	Straight distance between centers of front and rear axles	Length of ground contact region
Height of front and rear axles	Height of front and rear axles	Height of sprocket, idler, and roller
Track width	Different distances between front and rear regions	Same distance between front and rear regions

To accurately implement the overturning and rollover characteristics of an actual tractor, one must ensure that the influencing factors listed in Table 1 are accurately reflected in the simulation model. Among these factors, the tractor weight, track width, wheelbase, and the heights of the front and rear axles can be derived through straightforward measurements. However, determining the mass center location requires a more complex procedure that adheres to an international standard.

2.2. Miniature Tracked Forestry Tractor

A miniature tracked forestry tractor scaled down to 1:11 of the actual size was used in the tests. The model's maximum driving speed is 0.52 km/h, which is achievable in both forward and reverse motion, and it was controlled using a remote controller. The miniature tracked forestry tractor is the NST-1500VD model manufactured by Mitsubishi in Japan. Its main body was made of alloy steel and stainless steel, whereas its tracks were made of aluminum alloy. The geometry and specifications of the miniature tracked forestry tractor are shown in Figure 4 and Table 3, respectively.

Figure 4. Photograph of miniature tracked forestry tractor.

Table 3. Specifications of miniature tracked forestry tractor.

Items		Specifications
Model		NST-1500VD
Length/width/height (mm)		515/220/240
Weight (kg)		13.30
Maximum lifting weight (kg)		15
Material		Alloy steel, stainless steel, aluminum alloy
Maximum traveling speed (km/h)	Forward	0.52
	Backward	0.52

Mass Center Derivation

Determining the mass center location requires a more complex test compared with the case for the other factors. The mass center location of the miniature model was derived from an experiment conducted in accordance with the International Organization for Standardization 789-6:2019 [23], which specifies the procedure for determining the mass center of a tracked tractor.

This process involved measuring the combined reaction forces of the tractor, deck, and knife edges ($F_1 + F_2$) using a weighing machine. Subsequently, the front reaction force of the deck and knife edges alone (F_1) was measured. The tractor's specific contribution to the front section's reaction force (F_2) was derived from the difference between these two measurements. Subsequently, the horizontal coordinate of the mass center was calculated using (F_2), the total weight of the tractor (W), and the distance between the front and rear knife edges (d), as shown in Equation (7) (Figure 5).

$$\bar{x} = \frac{d \cdot F_2}{W}, \tag{7}$$

where $\bar{x}$ is the horizontal distance of the mass center from the rear knife edge, W is the total weight of the tracked tractor, d is the distance between the front and rear knife edges, and F_2 is the reaction force by the tracked tractor weight at the front knife edge.

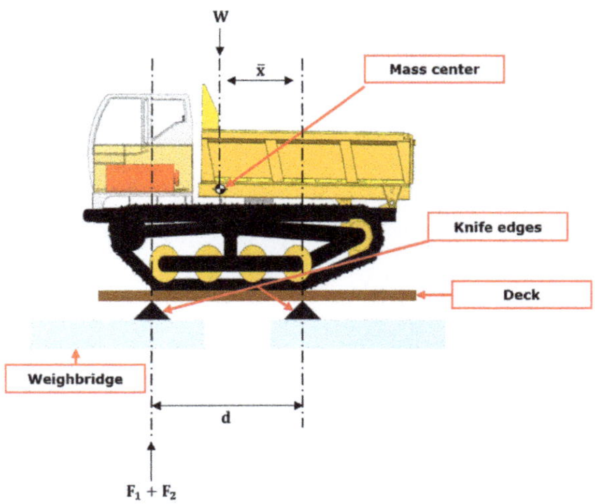

Figure 5. Diagram to determine horizontal location of mass center.

The vertical coordinate of the mass center can be determined using Equations (8) and (9), which involve the total weight of the tractor, the reaction force at the knife edge, and the horizontal distance from the knife edge to the cable (Figure 6). Initially, the rear end of the tractor track was positioned on the knife edge, and the reaction force at this point was measured, whereas the front end of the tractor was lifted to an angle between 20° and 25° using the front cable. The horizontal distance from the front cable to the knife edge (d_1) was measured, and the horizontal distance from the front cable to the mass center (c_1) was derived using Equation (8). Subsequently, a vertical line was drawn from the cable at a horizontal distance of c_1. Similarly, after locating the front end of the tractor track on the knife edge, the horizontal distance from the rear cable to the knife edge (d_2) was measured when the rear end of the tractor was lifted. The horizontal distance from the rear cable to the mass center (c_2) was derived using Equation (9). A vertical line was drawn at a horizontal distance of c_2 from the cable, and the intersection of these vertical lines from c_1 and c_2 determines the vertical coordinate of the mass center ($\bar{h}$).

$$c_1 = \frac{F_3 \cdot d_1}{W} \tag{8}$$

$$c_2 = \frac{F_4 \cdot d_2}{W}, \tag{9}$$

where c_1 is the horizontal distance of the mass center from the front end cable, c_2 is the horizontal distance of the mass center from the rear end cable, F_3 is the reaction force at the rear knife edge when lifting the front end of the tracked tractor, F_4 is the reaction force at the front knife edge when lifting the rear end of the tracked tractor, d_1 is the horizontal distance from the front cable to the rear knife edge, and d_2 is the horizontal distance from the rear cable to the front knife edge.

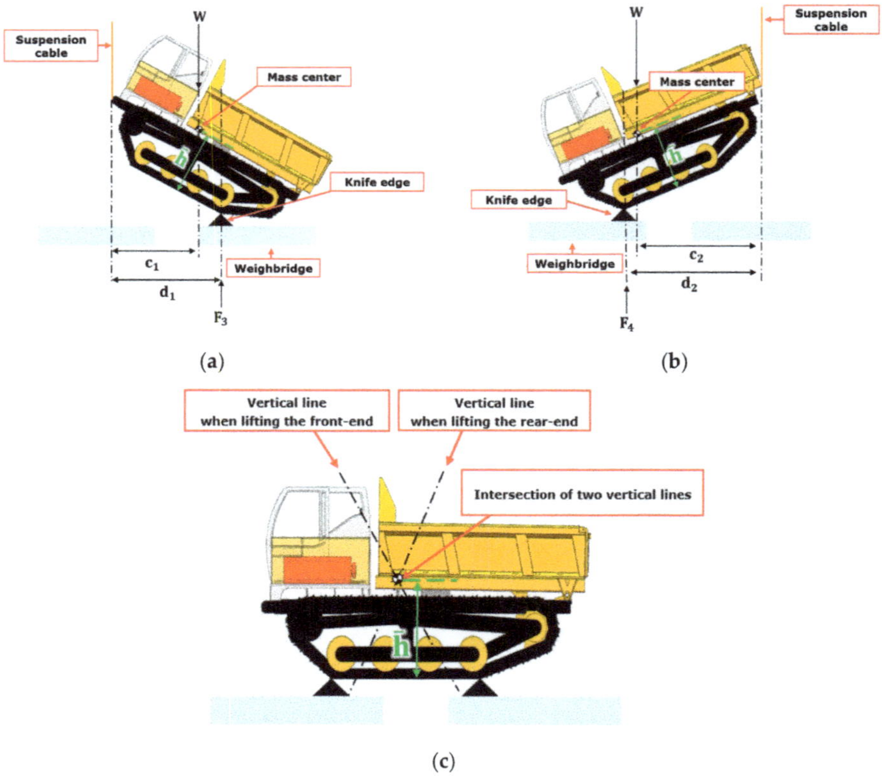

Figure 6. Diagram to determine vertical location of mass center: (**a**) lifting of front end; (**b**) lifting of rear end; (**c**) intersection of two vertical lines.

The lateral coordinate of the mass center was derived from the total weight of the tractor, the reaction force of the right track, and the track width (Figure 7). The reaction force of the right track and the track width were initially measured and then used in Equation (10) to calculate the offset of the mass center (b). Subsequently, the lateral coordinate of the mass center was derived using this offset and the track width (Equation (11)).

$$b = \frac{F_5 \cdot d_t}{W} \tag{10}$$

$$\overline{y} = \frac{d_t}{2} - b, \tag{11}$$

where $\overline{y}$ is the lateral location of the mass center from the center of two tracks, b is the lateral offset between the mass center and left track center, F_5 is the reaction force at the right-side tracks, and d_t is the track width.

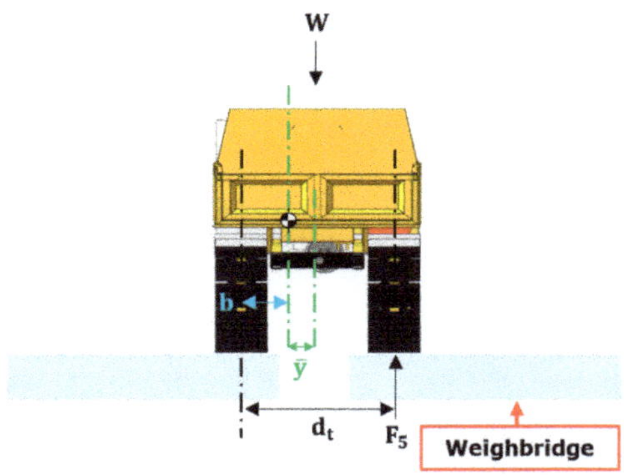

Figure 7. Diagram to determine lateral location of mass center.

2.3. Three-Dimensional (3D) Simulation Model

2.3.1. Three-Dimensional Model Construction

The 3D model of the miniature tractor was constructed by accurately reflecting all the factors affecting overturning and rollover. For this model, the measurements from the actual tractor were applied for the total weight, track width, wheelbase, height of the front and rear axles, and geometric shape. The experimentally derived value for the mass center location was incorporated as well. Additionally, the material properties of each element of the actual tractor were considered. The exterior of the miniature tracked forestry tractor was made of stainless steel, whereas the internal components, such as the hydraulic cylinder, motor, sprocket, idler, and roller, were made of alloy steel. The track was crafted from aluminum alloy. Table 4 shows the properties of the materials used [24]. To ensure an efficient and simple simulation, the smaller components inside the tractor (e.g., breadboard, wires, and MCU) were replaced with a mass cube [25,26]. The weight and position of this mass cube were adjusted to ensure that the 3D model's total weight and mass center location precisely matched those of the miniature tracked forestry tractor.

Table 4. Material properties of miniature tracked forestry tractor used in simulation.

Items		Value
Stainless steel (main body frame)	Poisson's ratio	0.3
	Elastic modulus (GPa)	180
	Density (kg/m^3)	7930
Alloy steel (hydraulic cylinder, motor, sprocket, idler, and roller)	Poisson's ratio	0.3
	Elastic modulus (GPa)	210
	Density (kg/m^3)	1900
Aluminum alloy (track)	Poisson's ratio	0.33
	Elastic modulus (GPa)	71
	Density (kg/m^3)	2700

2.3.2. Verification of Simulation Model

A simulation model utilizing the developed 3D model was constructed using commercial dynamic simulation software (Recurdyn 2023, Function Bay, Seongnam, Republic of Korea). To determine whether this model accurately represents the overturning and rollover characteristics of the actual tractor, it was verified based on the static sidelong falling angle and minimum turning radius.

Static Sidelong Falling Angle

The static sidelong falling angle test, an official method for assessing tractor safety, involves incrementally increasing the slope angle from 0° to 90°. During this test, a slope with a high friction coefficient is used to prevent slippage between the wheels and surface. The angle at which the upslope wheel or track lifts off the surface is recorded as the static sidelong falling angle [20].

To simulate the static sidelong falling angle test of an actual tractor, a test platform comprising a digital level, sloping plate, fixed plate, rubber pad, cable, and cushion (to prevent shock prevention) was set up (Figure 8). The digital level (Advanced digital level, YATO, Seoul, Republic of Korea) was used to measure the angle of the sloping plate. The rubber pad was fixed on the sloping plate to prevent track slippage. The cable, which was attached to the sloping plate through holes drilled at its upper and lower ends, was used to lift the plate quasi-statically. The cushion was installed on the fixed plate to protect against damage if the tractor overturns during the test. The sloping and fixed plates, which were made of wood and connected by hinges, allowed for the adjustment of the slope angle. During the test, the slope angle was increased gradually at increments of 0.01°, and the angle at which the upslope track lifted off the surface was measured. The test was repeated five times to ensure accuracy, and a high-speed camera (DC-GH5, Panasonic, Osaka, Japan) was used to capture the process and verify the precise angle.

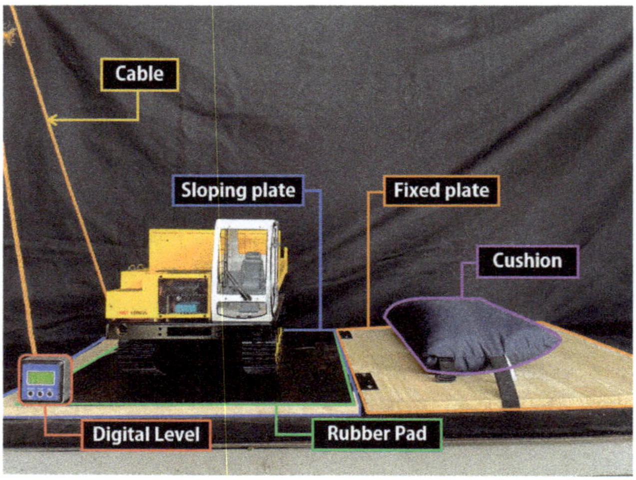

Figure 8. Platform established to perform static sidelong falling angle test.

A simulation model that simulates the static overturning characteristics of an actual tractor was developed. In this model, a 3D model tractor was placed on a slope, and the slope angle was increased at intervals of 0.01°, as performed in real-world testing, to determine the angle at which the reaction force of the upslope track becomes zero. The properties related to the track-ground contact—stiffness, friction, and damping coefficients—were selected based on a previous study [27] and are detailed in Table 5. These conditions were selected to prevent sideslip between the ground and track, thus reflecting the actual test.

Table 5. Properties used in static sidelong falling angle simulation.

Items		Value
Interaction between track and ground	Stiffness coefficient (N/mm)	10^8
	Damping coefficient (N·s/mm)	10^4
	Coefficient of dynamic friction	1.5
	Coefficient of static friction	1.85

Minimum Turning Radius

The turning radius is another critical factor affecting the overturning and rollover characteristics of a tractor. A smaller turning radius enhances work efficiency by facilitating easier steering and access to confined spaces [28]. The minimum turning radius test, which is recognized as an official test, defines the tracked tractor's minimum turning radius as the radius of the trajectory drawn by the center of the outer track when the tractor turns at a speed not exceeding 2 km/h in a counterclockwise direction, with the left track propagating backward and the right track forward.

For the miniature tracked forestry tractor, the speed was set to a minimum value of 0.04 km/h, which reflects the actual minimum turning radius used in the test. The tractor was maneuvered with the right track propagating forward and the left track backward, and the trajectory radius drawn by the center of the right track was measured.

Additionally, a simulation model capable of simulating the minimum turning radius of the actual tractor was developed. This model used the same speeds and track directions as the actual test. It calculated the trajectory radius of the right track's center during the turn.

The static sidelong falling angle test was conducted in a state where sideslip did not occur; therefore, extremely large stiffness and damping coefficients were applied to prevent sideslip from occurring in the simulation. However, sideslip occurred in the minimum turning radius test; therefore, realistic levels of stiffness and damping coefficients were applied in the simulation [29]. This aspect was integrated into the dynamic simulation, and the ground contact properties were specified accordingly (Table 6).

Table 6. Properties used in minimum turning radius simulation.

Items		Value
Interaction between track and ground	Stiffness coefficient (N/mm)	10^3
	Damping coefficient (N·s/mm)	0.5
	Coefficient of dynamic friction	1.2
	Coefficient of static friction	1.55

3. Results and Discussions

3.1. Mass Center Derivation Results for Miniature Tracked Forestry Tractor

The horizontal coordinate of the mass center was 157.9 mm from the rear knife edge of the tractor in the forward direction (Figure 9). As the distance from the rearmost section of the tractor to the rear knife edge was 125 mm, the mass center in the horizontal direction was located at 282.9 mm from the rearmost section. Because the total length of the miniature tractor was 515 mm, the mass center was located toward the front section, likely due to the placement of heavier components, such as the battery and hydraulic cylinder, in this area.

The vertical coordinate of the mass center was situated 102 mm above the ground (Figure 10). As the total height of the tractor was 240 mm, the mass center was positioned relatively low, which can be attributed to the location of heavier elements, such as the track and hydraulic cylinder, in the lower section of the tractor.

The lateral coordinate of the mass center was 1.4 mm off-center, which shifted toward the driver's seat from the midpoint between the left and right tracks (Figure 11). That the width of the tractor was 220 mm indicates that the mass center was almost centrally located within the tractor's width.

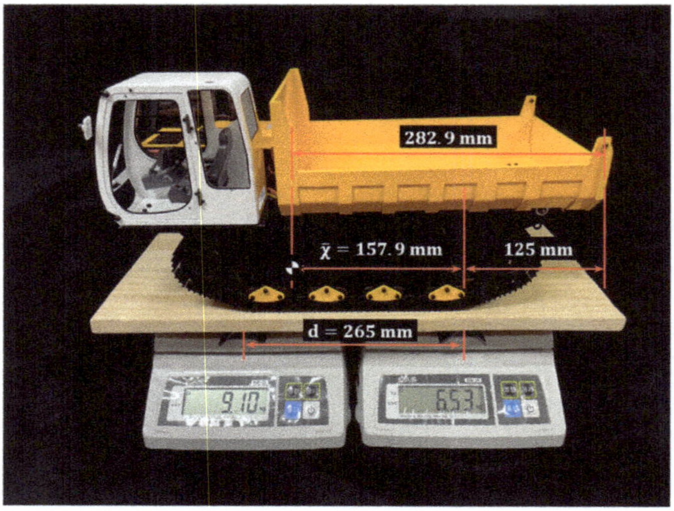

Figure 9. Horizontal location of mass center for miniature tracked forestry tractor.

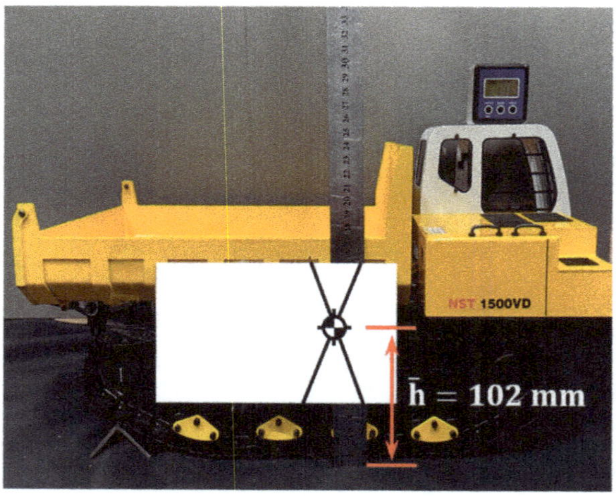

Figure 10. Vertical location of mass center for miniature tracked forestry tractor.

3.2. 3D Modeling Results for Miniature Tracked Forestry Tractor

The 3D model was constructed to replicate the geometric shape, track width, wheelbase, and axle height of the actual miniature tracked forestry tractor (Figure 12). Comparisons between the 3D model and the actual tractor showed track width and wheelbase error rates of 0.78% and 1.02%, respectively (Table 7). The error rates for the total weight and the weight distribution between the left and right tracks were both 0%, thus indicating that the developed 3D model accurately simulated these aspects of the actual tractor (Table 8).

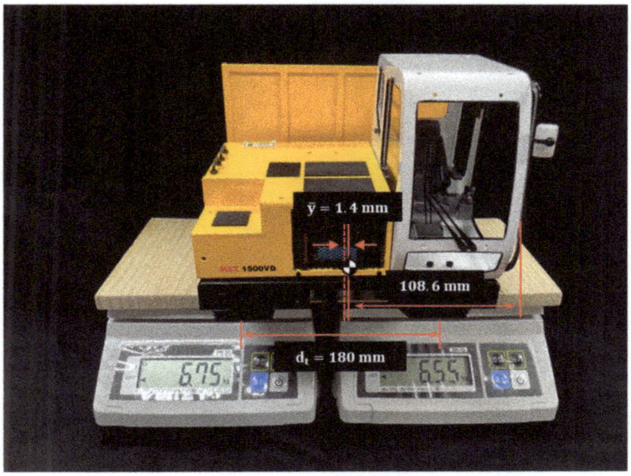

Figure 11. Lateral location of mass center for miniature tracked forestry tractor.

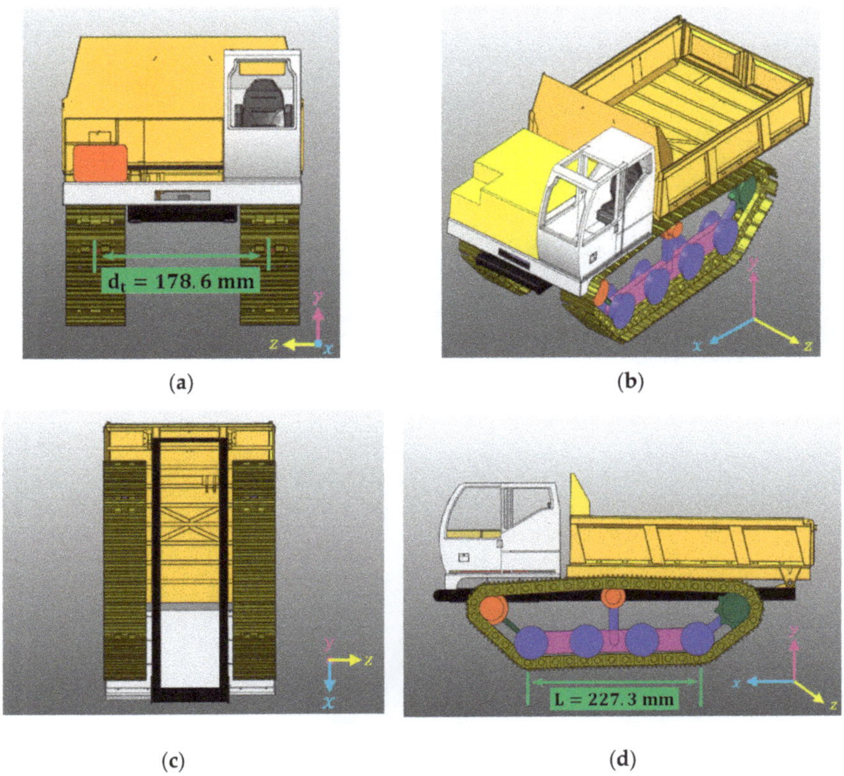

Figure 12. Three-Dimensional model of miniature tracked forestry tractor: (**a**) front view; (**b**) isometric view; (**c**) bottom view; (**d**) side view.

Table 7. Comparison of track width and wheelbase.

Items	Track Width (mm)	Wheelbase (mm)
Actual tractor	180.0	225.0
3D model	178.6	227.3
Error rate (%)	0.78	1.02

Table 8. Comparison of total weight and weight distribution.

Items	Total Weight (kg)	Weight Distribution (kg)	
		Left Track	Right Track
Actual tractor	13.30	6.55	6.75
3D model	13.30	6.55	6.75
Error rate (%)	0	0	0

The mass center location in the developed 3D model was derived (Figure 13) and compared with that of the actual tractor. The error rates for the horizontal, vertical, and lateral coordinates of the mass center were 0.2%, 0.1%, and 1.1%, respectively, thus confirming that the 3D model closely simulated the mass center of the actual tractor (Table 9).

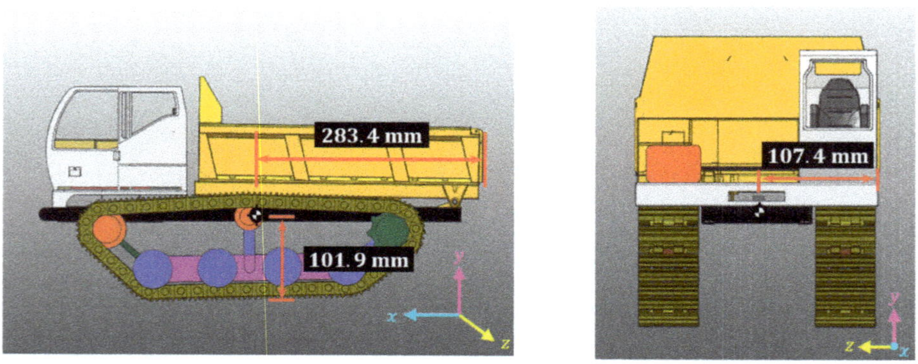

(**a**) (**b**)

Figure 13. Position of mass center in 3D model: (**a**) side view; (**b**) front view.

Table 9. Comparison of mass center positions.

Items	Coordinate (mm)		
	Horizontal	Vertical	Lateral
Actual tractor	282.9	102.0	108.6
3D model	283.4	101.9	107.4
Error rate (%)	0.2	0.1	1.1

The 3D model was designed to fully reflect the geometric shape of the actual tractor. The error rates for the track width, wheelbase, total weight, weight distribution, and mass center location were all below 5%, thus demonstrating that the model can accurately simulate the overturning and rollover characteristics of the actual tractor with high accuracy because it accurately reflected all factors affecting the overturning and rollover of the actual tractor.

3.3. Simulation Model Verification Results

Figure 14 shows the practical test and simulation results for the static sidelong falling angle of the miniature tracked forestry tractor. The average static sidelong falling angle was recorded as 50.30° based on five repeated measurements and 51.7° in the simulation, thus resulting in an error rate of 2.73% (Table 10). As the error rate was under 5%, one may conclude that the developed simulation model effectively simulated the static overturning characteristics of the actual tractor.

(**a**) (**b**)

Figure 14. Practical test and simulation of static sidelong falling angle test: (**a**) practical test; (**b**) simulation.

Table 10. Error rate of static sidelong falling angle between practical test and 3D simulation.

Item	Static Sidelong Falling Angle (°)
Practical test	50.3
Simulation	51.7
Error rate (%)	2.73

Figure 15 shows the practical test and simulation results for the minimum turning radius of the miniature tracked forestry tractor. The average minimum turning radius was 163.14 mm in the test and 162.19 mm in the simulation, which resulted in an error rate of 0.58% (Table 11). As the error rate was below 5%, one may conclude that the developed simulation model accurately simulated the turning characteristics of the actual tractor.

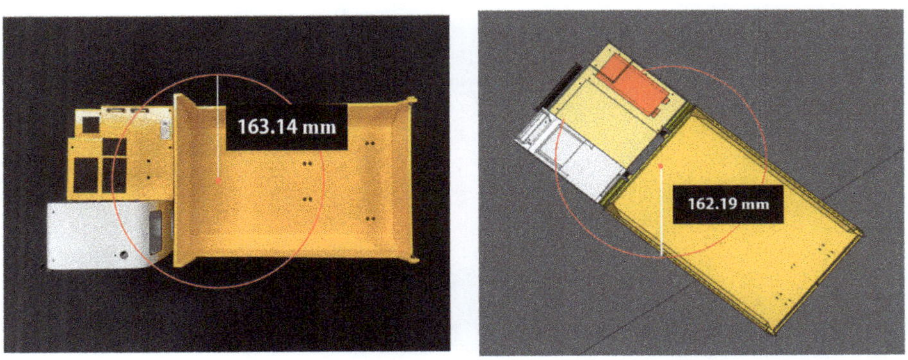

(**a**) (**b**)

Figure 15. Practical test and simulation of minimum turning radius: (**a**) practical test; (**b**) simulation.

Table 11. Error rate of minimum turning radius between practical test and 3D simulation.

Item	Minimum Turning Radius (mm)
Practical test	163.14
Simulation	162.19
Error rate (%)	0.58

The developed simulation model accurately replicated the static overturning and turning radius characteristics of the actual tractor. This high level of accuracy is attributed to the comprehensive incorporation of critical factors such as the total weight, weight distribution, and the location of the mass center into the 3D model. Therefore, if a 3D model is constructed and simulations are performed using the method proposed in this study, then one can expect the overturning and rollover characteristics of the actual tractor to be accurately predicted through simulation, thereby eliminating the necessity for physical testing. Future studies should focus on verifying the effectiveness of the simulation under varying ground and dynamic driving conditions to further enhance the model's applicability.

4. Conclusions

In this study, a simulation model that can accurately simulate the overturning and rollover characteristics of an actual tractor was developed and verified via testing. Influencing factors vital to the simulation's accuracy were derived through a theoretical analysis of tractor overturning and rollover dynamics. They were described based on wheeled tractors, as prior research pertaining to tracked tractors is not available; however, all the factors are applicable to tracked tractors. Among the factors, the mass center location was measured in accordance with the international standards, whereas the tractor's weight, track width, wheelbase, weight distribution, height of the front and rear axles, and geometric shape were derived through direct measurements. A 3D model was created to encompass all these factors. In this model, smaller components that did not affect the geometric shape were replaced with a mass cube. Comparisons between the 3D model and the actual tractor showed track width and wheelbase error rates of 0.78% and 1.02%, respectively. Errors in the weight and the distribution of weight between the left and right tracks were negligible (0%). The error rates for the coordinates of the mass center between the 3D model and the actual tractor were under 5%, thus demonstrating that the 3D model precisely represented the actual tractor.

The dynamic simulation model, which incorporated the developed 3D model, was verified based on the static sidelong falling angle and minimum turning radius through tests. For these characteristics, the error rates derived from comparisons with the actual tractor, based on the averages of five repeated measurements, were 2.73% and 0.58%, respectively. As these error rates were below 5%, the model was confirmed to effectively replicate the overturning and rollover characteristics of the actual tractor. This indicates that a simulation model constructed using this method can reliably emulate the physical attributes and behavior of actual tractors, i.e., physical tests are not necessitated.

Author Contributions: Validation, M.-K.J.; Formal analysis, Y.-J.Y. and M.-K.J.; Writing—original draft, Y.-J.Y.; Writing—review & editing, J.-S.N. All authors have read and agreed to the published version of the manuscript.

Funding: This study was partly supported by the Korea Institute of Planning and Evaluation for Technology in Food, Agriculture and Forestry (IPET) through the Machinery Mechanization Technology Development Program for Field Farming Program, which is funded by a grant from the Ministry of Agriculture, Food and Rural Affairs (MAFRA) (RS-2023-00236201, 50). Additional support was provided by the Innovative Human Resource Development for Local Intellectualization program through a grant from the Institute of Information & Communications Technology Planning & Evaluation (IITP), funded by the Korean government (MSIT) ((IITP-2024-RS-2023-00260267), 50).

Institutional Review Board Statement: Not applicable.

Data Availability Statement: The original contributions presented in the study are included in the article, further inquiries can be directed to the corresponding author.

Conflicts of Interest: The authors declare no conflicts of interest.

References

1. Shin, Y.S.; Youn, K.W.; Kim, K.S.; Choi, D.P.; Hong, S.J.; Lee, M.J. Characteristics and prevention measures of traffic accidents causing injuries to agricultural machinery occupants in the Jeonnam region. *JKDAS* **2023**, *25*, 1577–1595. [CrossRef]
2. Koh, J.W.; Kwon, S.C.; Kim, K.R.; Lee, K.S.; Jang, E.C.; Kwon, Y.J.; Ryu, S.H.; Lee, S.J.; Song, J.C. A study on the development of surveillance system for agricultural injuries in Korea. *J. Agric. Med. Community Health* **2007**, *32*, 139–153. [CrossRef]
3. Sohn, J.R.; Park, J.H.; Kim, S.P.; Kim, S.J.; Cho, S.H.; Cho, N.S. Analysis of risk factors influencing the severity of agricultural machinery related injuries. *JKSEM* **2007**, *18*, 300–306.
4. Kim, S.J.; Gim, D.H.; Jang, M.K.; Hwang, S.J.; Yang, Y.J.; Nam, J.S. Development of regression model for predicting the maximum static friction force of tractors with a front-end loader. *J. Biosyst. Eng.* **2023**, *48*, 329–338. [CrossRef]
5. Jeong, C.W.; Kim, D.J. A Study on the disaster analysis and accident prevention measures for agricultural work. *Korean J. Hazard. Mater.* **2023**, *11*, 50–57. [CrossRef]
6. Togaev, A.; Shermukhamedov, A. Tractor rollover accidents: A review of factors and safety measures. *E3S Web Conf.* **2023**, *449*, 09011. [CrossRef]
7. Facchinetti, D.; Santoro, S.; Galli, L.E.; Pessina, D. Agricultural tractor roll-over related fatalities in Italy: Results from a 12 years analysis. *Sustainability* **2021**, *13*, 4536. [CrossRef]
8. Previati, G.; Gobbi, M.; Mastinu, G. Mathematical models for farm tractor rollover prediction. *J. Veh. Design.* **2014**, *64*, 280–303. [CrossRef]
9. Franceschetti, B.; Rondelli, V.; Ciuffoli, A. Comparing the influence of roll-over protective structure type on tractor lateral stability. *Saf. Sci.* **2019**, *115*, 42–50. [CrossRef]
10. Lee, D.G.; Yoo, H.J.; Shin, M.J.; Oh, J.S.; Shim, S.B. Analysis of overturning stability of small off-road vehicle. *J. Biosyst. Eng.* **2023**, *48*, 309–318. [CrossRef]
11. Kim, Y.S.; Lee, S.D.; Kim, Y.J.; Kim, Y.J.; Choi, C.H. Effect of tractor travelling speed on a tire slip. *KJOAS* **2018**, *45*, 120–127. [CrossRef]
12. Shim, S.B.; Park, Y.J.; Kim, K.U.; Kim, J.W.; Park, M.S.; Song, T.Y. Computer simulation of sideways overturning of side-loaded mini forwarder. *J. Biosyst. Eng.* **2007**, *32*, 69–76. [CrossRef]
13. Kang, S.H.; Kim, J.H.; Kim, Y.S.; Woo, S.M.; Daniel, D.U.; Ha, Y.S. A simulation study on the dynamics characteristics of hot pepper harvester. *J. Korean Soc. Simul.* **2020**, *29*, 19–25. [CrossRef]
14. Iqbal, M.Z.; Islam, M.N.; Ali, M.; Kiraga, S.; Kim, Y.J.; Chung, S.O. Theoretical overturning analysis of a 2.6-kW two-row walking-type automatic pepper transplanter. *J. Biosyst. Eng.* **2022**, *47*, 79–91. [CrossRef]
15. Park, H.K.; Kim, K.U.; Kim, J.W.; Song, T.Y.; Park, M.S.; Cho, K.H. Sideways overturning analysis of forwarder using a multibody dynamics analysis program. *J. Biosyst. Eng.* **2002**, *27*, 185–194. [CrossRef]
16. Qin, J.; Zhu, Z.; Ji, H.; Zhu, Z.; Li, Z.; Du, Y.; Song, Z.; Mao, E. Simulation of active steering control for the prevention of tractor dynamic rollover on random road surfaces. *Biosyst. Eng.* **2019**, *185*, 135–149. [CrossRef]
17. Eom, B.G.; Kang, B.B.; Lee, H.S. A running stability test of 1/5 scaled bnogie using small-scaled derailment simulator. *KISTI* **2012**, *15*, 9–16. [CrossRef]
18. Carson, J.S. Introduction to modeling and simulation. *Johns Hopkins APL Tech. Dig.* **1995**, *12*, 6–17.
19. Shin, B.S.; Kim, D.C.; Kim, Y.J.; Kim, H.J.; Nam, J.S.; Park, Y.J.; Shim, S.B.; Lee, D.H.; Lee, J.W.; Cho, Y.J.; et al. *Tractor Engineering Principles*, 2nd ed.; Moon Woon Dang: Seoul, Republic of Korea, 2021.
20. Jang, M.K.; Hwang, S.J.; Shin, C.S.; Nam, J.S. A novel approach to determine static falling down sidelong angle of tractor using a 3D printed miniature model. *Appl. Sci.* **2021**, *12*, 43. [CrossRef]
21. Baker, V.; Guzzomi, A.L. A Model and Comparison of 4-wheel-drive fixed-chassis tractor rollover during Phase 1. *Biosyst. Eng.* **2013**, *116*, 179–189.
22. *No. 2017-668*; Performance of Automobiles and Automobile Parts and Detailed Rules for Implementation of Standards. Ministry of Land, Infrastructure and Transport: Sejong-si, Republic of Korea, 2017.
23. *No. 789-6:2019*; Agricultural Tractors-Test Procedures-Part 6: Centre of Gravity. ISO: Geneva, Switzerland, 2019.
24. Khot, S.S.; Navthar, R.R. Design and optimization of front axle of heavy truck. *IJEAST* **2019**, *4*, 183–191. [CrossRef]
25. Kim, T.J.; Jeon, H.H.; Kim, Y.J. Dynamic characteristic analysis of an autonomous tractor according to plow tillage. *PASTJ* **2019**, *1*, 56. [CrossRef]
26. Lysych, M.N. Study driving dynamics of the machine-tractor unit on a virtual stand with obstacles. *J. Phys. Conf. Ser.* **2020**, *1515*, 042079. [CrossRef]
27. Jeong, H.J.; Yu, J.W.; Lee, D.H. Track HM design for dynamic analysis of 4-tracked vehicle on rough terrain using Recurdyn. *Trans. Korean Soc. Mech. Eng. A* **2021**, *45*, 275–283. [CrossRef]

28. Kim, Y.Y.; Lim, G.K.; Shin, S.Y.; Kim, H.J.; Kim, B.G.; Kim, H.G. Development of a turning radius measurement system using DGPS for agricultural tractors. *J. Biosyst. Eng.* **2010**, *35*, 85–90. [CrossRef]
29. Wang, Y.; Yang, F.; Pan, G.; Liu, H.; Liu, Z.; Zhang, J. Design and testing of a small remote-control hillside tractor. *Trans. ASABE* **2014**, *57*, 363–370. [CrossRef]

agriculture

MDPI

Article

Reducing Grain Damage in Moist Corn Threshing via Corncob Division

Gang Wang [1,2], Chengqian Jin [1,2,*], Min Zhang [1], Chongyou Wu [1], Qing Tang [1] and Yao Yang [1]

[1] Nanjing Institute of Agricultural Mechanization, Ministry of Agriculture and Rural Affairs, Nanjing 210014, China; wanggang05@caas.cn (G.W.); zhangmin01@caas.cn (M.Z.); wuchongyou@caas.cn (C.W.); tangqing01@caas.cn (Q.T.); yangyao@caas.cn (Y.Y.)
[2] Graduate School, Chinese Academy of Agricultural Sciences, Beijing 100083, China
* Correspondence: jinchengqian@caas.cn

Abstract: For the prompt planting of subsequent crops, most of China's corn harvest must occur before full maturity, with a grain moisture content above 25%. Harvesting moist corn presents challenges due to significant grain damage during threshing. We conducted a high-speed photography observation test of moist corn threshing. It demonstrated that corn ears, when passing through the threshing cylinder, often break into pieces. Grains on divided corncobs can be threshed more easily and earlier than those on undivided ones, suggesting that pre-dividing corn ears reduces grain damage. An experiment using the Lianchuang 825 variety examined the effect of moisture content and the divided rate of corncobs (DRC) on grain damage. The results showed that as moisture content increased from 25% to 37%, grain damage to undivided ears increased from 3.75% to 37.71%. Dividing corn ears before threshing significantly reduced damage, with an eight-piece division reducing damage by approximately 70% across all moisture levels. Verification with the Jinyu 1233 variety confirmed that a higher DRC consistently reduced damage. This study provides a new approach to reducing damage in moist corn threshing and aids in the development of low-damage threshing devices.

Keywords: corn threshing; threshing cylinder; grain damage; moisture content

1. Introduction

Corn is a critical global crop and is the most widely cultivated food crop in China [1–3]. Typically, corn is harvested before full maturity to allow for the timely planting of subsequent crops, resulting in a moisture content of over 25% at harvest (moist corn). Direct harvesting of corn has become the predominant method both domestically and internationally because of its convenience and efficiency [4–6]. However, in China, harvesting moist corn presents challenges, primarily due to significant grain damage during threshing. To mitigate these challenges, corn ears are harvested rather than individual grains, which reduces harvesting efficiency and increases costs [7–9]. Damaged corn grains not only decrease market value but also contribute to heat generation, fermentation, and bacterial infections due to increased respiration. In response to the issue of high grain damage during threshing, extensive research has been conducted on the material strength of corn grains, optimization of threshing device structures and parameters, and analysis of threshing theories [10–15]. These efforts have led to notable reductions in threshing damage. Nevertheless, harvesting and threshing moist corn continue to result in substantial grain damage in China.

Corn threshing damage is influenced by the material strength of corn grains. Several studies have investigated the mechanical and physical properties of corn grains using various methods. Chen and Wassgren [16] compared the damage resistance of corn grains under compression, friction, and repeated impact. They fitted the data with statistical models and found that corn grains were particularly susceptible to compression and impact loading but exhibited high resistance to wear damage. Srivastava et al. [17] examined the mechanical properties of corn grains and identified specific impact features contributing to damage during harvesting and handling. Their measurements revealed that longitudinal shear was weaker than transverse shear when corn grains were subjected to impact forces. Shahbazi et al. [18] evaluated the effects of free-fall conditions on corn grain damage. Their study found that grains dropped onto metal surfaces experienced greater overall damage compared with those dropped onto concrete or subjected to grain-to-grain abrasion. Additionally, increased drop height caused both physical and physiological damage to the grains, while higher moisture levels reduced physical damage but increased physiological damage. Li et al. [19] analyzed the mechanical properties of corn grains with varying moisture contents and impact positions. They observed that the maximum impact force required to cause grain damage decreased as the moisture content increased. They also noted significant variations in the maximum impact force depending on the impact surface. Furthermore, different corn grain varieties exhibited varying resistance to impact rupture on the same surface, which was attributed to differences in the internal structure and shape of the grains.

Several studies have examined the structural parameters of threshing devices. Geng et al. [20] designed a transverse axial flow corn threshing test system utilizing flexible teeth and elastic short rasp bars as threshing elements arranged in four heads. They found that with a threshing cylinder speed of 450 rpm, a concave clearance of 40 mm, and a feed rate of 8 kg/s, the grain damage was 0.65% and the unthreshed grain rate was 0.59%. Yang et al. [10] developed a dual-axial flow threshing device featuring a U-shaped conveying system to transport corn ears within the threshing chamber. They observed a forward speed of 0.61 m/s, a second threshing drum speed of 500 rpm, a threshing gap of 40 mm, grain damage of 1.47%, and threshing loss of 0.89%. Qu et al. [21] designed a longitudinal axial-flow corn threshing and separating device to determine the optimal range of threshing cylinder speeds for various feed rates in a corn combine harvester. Their findings indicated that as the threshing cylinder speed increased, the corn grain damage rate also increased, while the unthreshed rate decreased. At the same cylinder speed, a higher feed rate slightly increased both the grain damage and unthreshing rates. Pickard et al. [22] investigated the effects of cylindrical and concave bar variations on corn threshing. They found that a rasp-type cylindrical bar provided better shelling efficiency and caused less grain damage compared with an angle-type cylindrical bar. Additionally, they noted that covering the cylindrical or concave bars with rubber had no impact on shelling efficiency or grain damage. To address the lack of automatic adjustment for threshing clearance in corn threshers, Li et al. [23] developed a floating corn single-panicle threshing device. This device included a clearance adjustment mechanism, feeding hopper, discrete roller, threshing roller, and differential roller. In their experiment, the discrete roller speed was set to 234 rpm, the threshing roller speed to 511 rpm, and the differential roller speed to 91 rpm. The results demonstrated a grain breakage rate of 0.25%, an unthreshing rate of 0.76%, and a 100% integrity rate of the corncob.

Several studies have theoretically analyzed corn threshing. Miu and Kutzbach [24] developed a comprehensive stochastic mathematical model to describe grain threshing and separation in both axial and tangential threshing units. They derived equations to quantify the percentages of unthreshed, free, and separated grains (both cumulative and distributed) along the length of the threshing space. Wang et al. [25] employed three-dimensional scanning technology, finite element method-based explicit dynamics simulations, and the response surface method to investigate the damage susceptibility and time-dependent non-linear mechanical behavior of corn kernels under various impact scenarios with a threshing unit. They examined contact force, Von Mises stress, and displacement characteristics to determine the critical velocity of the corn kernels. Cundall [26] introduced the discrete element method (DEM), which treats matter as a series of discrete particles and explains the motion of each particle based on classical Newtonian mechanics. This method has become widely used for studying the fundamental characteristics of corn threshing. To determine the structure and operational parameters of threshing devices, Li et al. [11] established and analyzed a crop–crop contact model and a crop-mechanical interaction system using DEM.

They derived mathematical expressions for the kinematic response of corn grains under external forces, which assisted in determining the structure and working parameters of threshing devices.

Both foreign and domestic scholars have made significant advances in the study of corn threshing, leading to a reduction in grain damage. Despite these advancements, the issue of high grain damage during threshing of high-moisture corn persists. While some researchers attribute this problem to the mutual support and squeezing of corn grains on corncobs, research on reducing grain damage through corncob division is still limited [27–29].

Li et al. [30] proposed a "pre-dispersion and post-threshing" process to address the high kernel damage during corn threshing. They designed a bionic thresher inspired by the beak's ability to insert into gaps between kernels with minimal damage and the low crack rate observed with manual threshing techniques. To study the effect of moisture content on seed corn threshing mechanisms, Gao et al. [31] conducted a series of experiments to evaluate seed damage strength, handle strength, and threshing force across different moisture levels. Their findings indicated that moisture content affects corn threshing primarily through seed damage strength and threshing force. They observed a significant increase in seed damage strength as moisture content decreased, which made the kernels less prone to breakage during threshing. The larger clearance and absence of extrusion or support among kernels were identified as key factors in this outcome.

However, research on the effect of dividing corn ears before threshing remains scarce. Therefore, we conducted a high-speed photography observation test of moist corn threshing to investigate the behavior of corn ears within threshing devices. Additionally, we performed grain damage experiments with varying moisture contents and corncob division rates (DRC) to explore methods for reducing damage during the threshing of high-moisture corn. Our analysis of these factors provides valuable insights for optimizing low-damage threshing devices and improving corn grain harvesting machinery.

2. Materials and Methods

2.1. High-Speed Photography Observation of Threshing Moist Corn Ears

The corn threshing process is complex because of the unique characteristics of corn ears as biological materials. The intricate kinematics and mechanical interactions between the threshing device and corn ears are significant but not easily identifiable. To analyze the working process of corn ears in the threshing device and assess the impact on grain damage, we conducted high-speed photography observations and analysis tests.

This study focused on the Lianchuang 825 corn variety, which is widely cultivated in the Huang-Huai-Hai region of China. We collected a sample of 500 corn ears free from insects and diseases from the field. The ears were allowed to dry naturally, and their moisture content was measured periodically to ensure precise control during the experiment. The bracts of the corn ears were then removed. Table 1 presents the main characteristics of the corn ears.

Table 1. Biological characteristics of corn ears of Lianchuang 825.

Parameters	Unit	Range	Average
Length of corn ear	mm	163–198	179
Diameter of large end	mm	46–54	50
Diameter of small end	mm	33–45	38
Mass of per corn ear	g	236–289	257
Grain moisture content	%	28–30	29

The primary equipment used in the test included a self-developed corn threshing test bench designed specifically for threshing corn ears. Table 2 presents the main structures and parameters of the test bench. Other test equipment included an LDS-1G grain moisture measuring instrument (Shanghai Tuoxi Electronic Technology Co., Ltd., Shanghai,

China), which was used to measure the moisture content in corn grains. The measurement error range of the instrument was ±0.05%. The DNC20002A electronic scale (Nanjing Bonita Scientific Instrument Co., Ltd., Nanjing, China) was used for weighing corn grains, with an indication error of ±0.5% and a measuring range of 2000 g. A Fastec HiSpec 5 high-speed camera (Hong Kong Peiport International Co., Ltd., Hong Kong, China) was used to record videos during the threshing test. A CLL-1600TDX supplementary lighting fixture (Guangzhou Falcon Eyes Stage Equipment Co., Ltd., Guangzhou, China) provided additional lighting for high-speed photography. Vernier calipers were used to measure the basic parameters of the corn ears. A tachometer was used to measure the speeds of the threshing cylinder and electric motor. A microcomputer was used for the control, shooting, and image storage in the high-speed photography tests.

Table 2. Main parameters of the corn threshing test bench.

Parameters	Value
Diameter of concave (mm)	320
Length of concave (mm)	1100
Speed of threshing cylinder (m/s)	0–23
Type of concave sieve	Punching type
Clearance of concave and threshing cylinder (mm)	25–50
Threshing cylinder rasp bars	Combination of plate teeth and pattern bars
Wrap angle of concave sieve (°)	178

The corn threshing test bench was equipped with a frequency conversion motor that provided power via a belt drive. A frequency conversion regulator was used to control the speed of the motor and threshing cylinder. The clearances of the concave and threshing cylinders can be adjusted by modifying the height of the shim beneath the bearing seat of the threshing cylinder. The feeding rate can be controlled by changing the feeding time for a certain number of corn ears. The test bench allowed for the replacement of different types of threshing cylindrical rasp bars, including patterned bars, plate teeth, and a combination of patterned bars and plate teeth. The structure and principles of the threshing test bench are shown in Figure 1a.

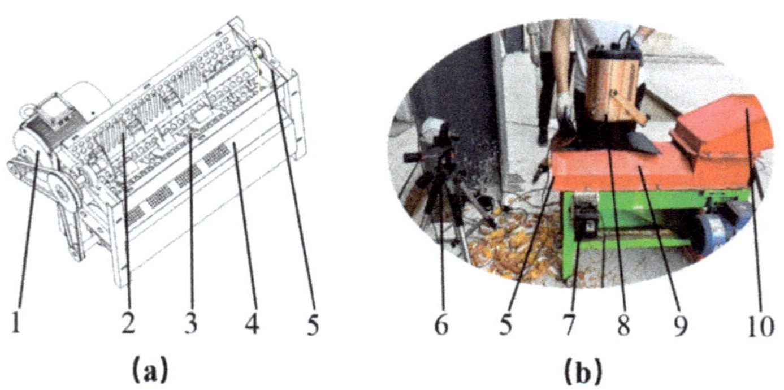

Figure 1. (**a**) Test bench of corn threshing without cover. 1—Electric motor; 2—threshing cylinder; 3—concave sieve; 4—export of corn grains; 5—export of corn cobs; (**b**) threshing test with a high-speed camera. 6—high-speed camera; 7—frequency converter; 8—supplementary lighting fixture; 9—upper cover plate of threshing cylinder; 10—feeding inlet.

To capture the threshing process more clearly, a square hole was created above the cover of the threshing cylinder. This square hole was covered with transparent glass that possessed specific qualities such as strength, impact resistance, and good light transmittance. During the test, high-speed photography equipment was positioned behind the threshing cylinder to facilitate shooting. The light fixture was placed above the transparent glass with a square hole. The arrangement of the threshing test bench and high-speed photography equipment is shown in Figure 1b.

Based on the results of single-factor tests on the clearance of the concave and threshing cylinders, speed of the threshing cylinder, and feeding rate, the following parameters were determined: clearance of 40 mm between the concave and threshing cylinders, speed of 17 m/s for the threshing cylinder, and corn ear feeding rate of 3.5 kg/s. The tested threshing bars consisted of a combination of patterned bars and plate teeth. These data were similar to those published previously [10,11,27]. Before the test, the movement parameters of the threshing test bench were adjusted to the desired level. The focus point and color temperature of the lighting fixture were adjusted, and the high-speed photography equipment was positioned and fine-tuned with optimal settings for focus, aperture, ISO, and other parameters. During the test, the power was turned on, and the threshing cylinder speed was stabilized before evenly feeding the preprocessed corn ears into the threshing test bench through the feeding inlet. High-speed photography equipment was used to capture online photographs of the threshing process of the corn ears in the threshing cylinder. After capturing, the image data were transferred to a microcomputer for storage, enabling subsequent observation and analysis.

2.2. Grain Damage Test of Threshing Moist Corn Ears

High-speed photography of threshing moist corn ears can provide clear insights into the situation of corn in the threshing device; however, the high-speed photography observation test of the corn threshing process has a subjective element. To further validate and quantify these findings, threshing experiments were conducted on corn ears with varying moisture content and different rates of corncob division.

A grain damage test for threshing the corn ears was conducted using the test equipment and materials described in Section 2.1 of this article. The Lianchuang 825 corn variety was used for the test. The factors considered in the grain damage test of threshing were the moisture content of the corn grains and the division rate of the corncobs. The concept of DRC was proposed to quantify the division rate of corncobs. This refers to the division of a complete corn ear into n evenly divided pieces using a specific method. The DRC value is represented by n. For example, if a corn ear is divided into two equal pieces, the DRC value is 2. Note that achieving complete and evenly divided corncobs is challenging; therefore, a permissible error of $\pm 10\%$ in the weight of corn ear pieces was allowed in this study. Corncob division can be achieved through external loads, such as beating, extrusion, and cutting. However, using external loads to divide corn ears may alter the connection state of the corn grain stalk and cause damage to the corn grains. To overcome this issue, the internal expansion method was employed to divide the corn ears into several pieces manually, as shown in Figure 2. This method relies on the embedding effect of the splitting cone to avoid changing the connection state of the corn grain stalk and damaging the corn grains.

Figure 2. Pretreatment of corn ears.

In the main corn-producing regions of Huang-Huai-Hai, China, the moisture content of corn during harvest ranges from 25% to 35%. For this experiment, corn grains with moisture contents of 37%, 34%, 31%, 28%, and 25% were selected, with an allowable fluctuation of $\pm 0.5\%$ from the specified levels. Corn ears with higher moisture content were manually harvested from the field and allowed to dry naturally. As the drying process progressed, the moisture content of the corn gradually decreased. The moisture content was tested every few hours until it was slightly above the desired test level. At this point, the bracts were removed, and the corncobs were manually processed according to the method described in Figure 2. The moisture content was re-measured before the experiments to ensure it was at the expected level. The test adheres to the Chinese national standards GB/T 21961-2008 [32], "Test Methods for Corn Harvest Machinery", and GB/T 21962-2008 [33], "Technical Conditions for Corn Harvest Machinery". The rate of grain damage, Z_s, was used as the evaluation index for the test, expressed as follows:

$$Z_s = \frac{W_s}{W_j} * 100\%, \quad (1)$$

where Z_s is the rate of grain damage (%), W_s is the mass of damaged grains in the sample (g), and W_j is the total mass of the grains in the sample (200 g).

During the test, the threshing cylinder was set to a speed of 17 m/s, and the corn ears were fed at a rate of 3.5 kg/s. The clearance between the threshing cylinder and the concave sieve was maintained at 40 mm. Once the threshing test bench was stabilized, the pretreated corn ears were evenly fed into the threshing cylinder through the feed inlet. After each test, the grains threshed from the cylinder were collected and mixed thoroughly. A five-point sampling method was used to obtain 200 g of grain from each point. The grains were then manually screened for damaged, cracked, or broken corn grains, and their weights were recorded.

3. Results and Discussion

3.1. *High-Speed Photography Observation of Threshing Moist Corn Ears*

The high-speed photography experiment aimed to investigate methods for reducing corn threshing damage by observing the phenomena occurring within the threshing device. After observing and analyzing the experimental video, we found that the corn ears passed through the threshing cylinder from the feeding inlet port to the export port. The photography equipment was positioned near the exit of the threshing cylinder, resulting in captured images that mainly focused on the rear section of the cylinder. These images provided an objective and accurate depiction of the state of corn ears after threshing for a certain period. They showed movement of the corn ears, corncobs, and grains, as well as damage to the grains. For the analysis, two high-speed photography videos were selected as representatives of the high-speed photography observation and analysis tests, as shown in Figure 3.

According to the photographs, the corn materials were impacted and rubbed by the threshing cylinder and concave sieve, resulting in collisions among the corn ears, threshing cylinder, concave sieve, and other corn ears. The movement trajectory and posture of each individual corn ear were irregular, but all corn ears within the annular region formed by the threshing cylinder and concave sieve exhibited various compound movements, such as tumbling and rotation, as observed in regions a1, a2, b1, and b2 in Figure 3. Towards the rear of the threshing cylinder, a significant portion of the grains separated from the corncobs because of threshing, as depicted in Figure 3(a1-2,a2-2,a2-4,b1-2,b2-2,b2-4).

After undergoing multiple impacts and rubbing from the threshing cylinder, a portion of the corncobs broke into several pieces, as shown in Figure 3(a2-2,a2-4,b2-4), whereas a portion of the corncobs did not divide (Figure 3(a1,b1)). The majority of the grains in the divided corn ear were already threshed, as shown in Figure 3(a2-4,b2-4). Only a small number of corn grains located in the central region of the divided corncobs remained unthreshed, as shown in Figure 3a2-2. Some grains of the undivided corncob remained unthreshed, and these unthreshed grains were concentrated in the central region of the corncob, as shown in Figure 3(a1,b1,b2). Additionally, some grains remained attached to the corncob despite being damaged, as shown in Figure 3(b1-5,b2-5).

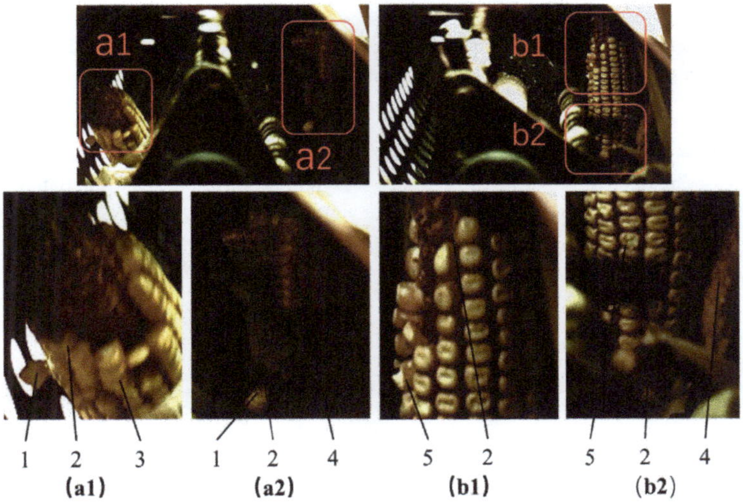

Figure 3. Photographs of threshing progress: 1—threshed grains; 2—corncob with threshed grains on the edge; 3—grains that have unthreshed after being hit; 4—corncobs without grains 5—damaged corn grains. a1, a2, b1, and b2 correspond to the enlarged red marked areas in the figures.

The purpose of corn threshing is to separate all corn grains from corncobs without damaging them. Macroscopically, separation was achieved through the contact force applied by the threshing cylinder to the corn ear. However, grain damage can occur during threshing if the strength and the number of contact forces exceed certain limits. From a microscopic perspective, if the strength and number of contact forces on a corn grain fall below a certain limit, the threshing cylinder cannot completely destroy the stalk on the grain, preventing the grain from separating from the corncob. Therefore, the objective of corn threshing is to maintain a condition in which the contact force of the threshing cylinder on the corn ears is greater than the limit for grain separation but lower than the limit for grain damage.

According to the threshing phenomenon observed in the captured images of corn ears in the threshing cylinder, after a certain period of threshing, many corn ears were divided into several pieces, whereas a small number of corn ears remained undivided. The undivided corn ears only had separated grains at the ends, and some grains located in the central region of the corncobs remained unseparated despite being damaged. However, in the divided corn ears, only a small portion of the grains in the central region of the corncobs remained unseparated. Analysis of the causes of this phenomenon showed that the grains located in the central region of the corncobs were more strongly supported than those in the edge region because they were surrounded by other grains on all sides. Considering the determined boundary conditions for grain damage, it can be concluded that the separation boundary conditions for the grains in the edge region were lower, which serves as evidence that the majority of the grains in the divided corn ear completed the threshing operations.

The corncob division increased the number of grains in the edge region. As these grains were separated, the grains in the secondary edge region next to the edge region transformed into new grains. These new edge grains were easier to separate. Therefore, for corn threshing, the division of the corn ear forms a virtuous cycle, facilitating the threshing and separation of all the grains on the corncobs.

For undivided corn ears, the grains located at the ends of the corncob had a lower supporting effect than those in the central region. Additionally, the boundary conditions for grain separation were lower in the central region, making it easier for the grains to separate from the corncobs. Under the same conditions, the grains in the central region required more strikes or a greater striking force to achieve the same threshing effect as those in the edge region. However, as the number of strikes increased, the probability of grain damage increased. This phenomenon can explain the occurrence of b1-5 and b2-5, where some corn grains failed to separate from the corncob despite being damaged by the striking force.

3.2. Grain Damage Test of Threshing Moist Corn Ears

The rate of damaged grains was calculated using Equation (1) after each grain damage test of threshed corn ears, and the average value is recorded in Table 3. Each test was repeated five times.

Table 3. Results of the grain damage test of threshing corn ears.

Moisture Content (%)	Divided Rate of Corncobs	Rate of Damaged Grains/%						Coefficient of Variation (%)
		1	2	3	4	5	Mean	
37	1	37.55	36.45	36.96	38.76	38.81	37.71	2.81
	2	27.08	27.17	27.92	27.28	26.45	27.14	1.93
	4	18.73	18.68	19.48	17.53	17.97	18.48	4.07
	6	14.22	13.42	13.66	12.42	14.15	13.57	5.35
	8	9.87	9.31	8.86	9.43	9.26	9.35	3.88
34	1	24.11	23.26	25.16	24.25	23.47	24.05	3.10
	2	18.87	18.23	19.02	19.57	18.51	18.84	2.72
	4	13.36	13.33	12.90	13.81	13.65	13.41	2.60
	6	9.93	9.36	9.67	10.28	10.55	9.96	4.75
	8	7.38	7.41	7.84	7.11	6.90	7.33	4.84
31	1	15.25	15.54	14.37	15.52	15.32	15.20	3.16
	2	11.65	11.23	11.39	12.22	12.05	11.71	3.60
	4	7.93	7.41	7.73	7.62	7.58	7.65	2.51
	6	6.14	5.98	6.02	5.76	5.83	5.95	2.56
	8	4.97	4.83	4.87	4.90	5.30	4.97	3.81
28	1	8.67	8.98	8.15	9.14	9.17	8.82	4.82
	2	6.34	6.78	6.83	6.21	5.96	6.42	5.82
	4	4.15	4.61	4.06	4.14	3.78	4.15	7.20
	6	3.34	3.47	3.33	3.09	3.16	3.28	4.64
	8	2.81	2.84	2.72	2.73	2.94	2.81	3.19
25	1	3.73	3.71	4.03	3.64	3.65	3.75	4.27
	2	2.65	2.84	2.95	2.75	2.71	2.78	4.22
	4	1.62	1.76	1.62	1.52	1.63	1.63	5.24
	6	1.51	1.33	1.36	1.34	1.22	1.35	7.68
	8	1.13	1.05	1.14	0.87	0.92	1.02	7.97

3.2.1. Corn Grain Damage in Threshing Affected by Moisture Content

The data presented in Table 3 were used to generate the change curve of corn threshing damage under varying moisture contents and corncob division rates, as illustrated in Figure 4. The coefficient of variation for each repeated experiment was less than 10%, indicating that the experimental results were relatively small in error, reliable, and representative. First, the analysis focused on grain damage during the threshing of undivided corn ears with varying moisture contents. Subsequently, the analysis was extended to corn threshing with moisture content levels of 37%, 34%, 31%, 28%, and 25% under different corncob division rates.

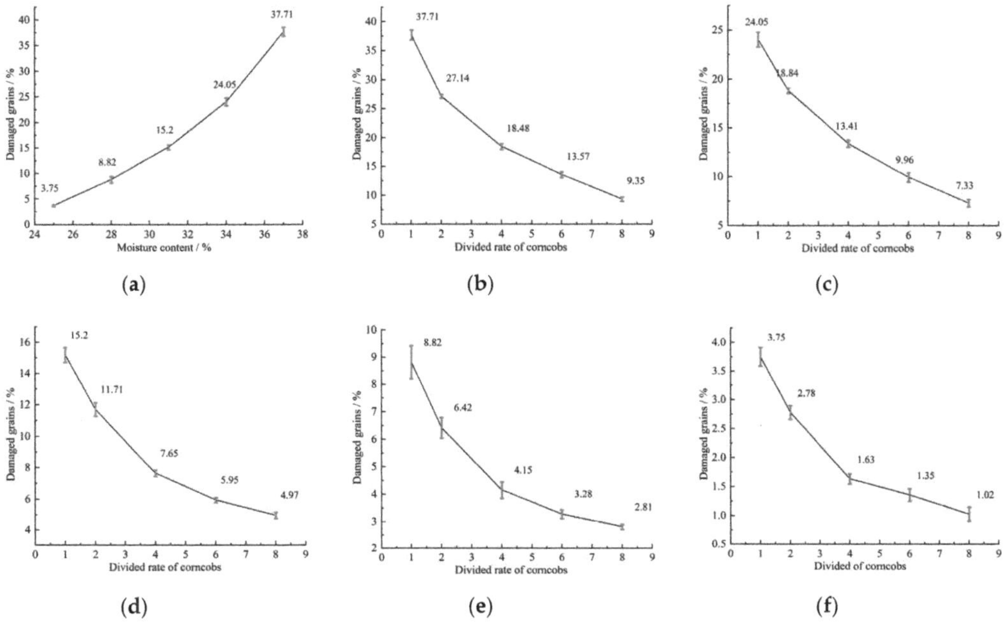

Figure 4. Rates of damaged grains (RDG). (**a**) Damaged grains of threshing undivided corn. (**b**) Damaged grains of different divided rates of corncobs with a moisture content of 37%. (**c**) Damaged grains of different divided rates of corncobs with a moisture content of 34%. (**d**) Damaged grains of different divided rates of corncobs with a moisture content of 31%. (**e**) Damaged grains of different divided rates of corncobs with a moisture content of 28%. (**f**) Damaged grains of different divided rates of corncobs with a moisture content of 25%.

Figure 4a shows that the grain damage during corn threshing decreased as the moisture content of the corn grains decreased. The percentages of grain damage at moisture contents of 37%, 34%, 31%, 28%, and 25% were 37.71%, 24.05%, 15.20%, 8.82%, and 3.75%, respectively. These findings align with previous studies [14,19,29]. Specifically, at a moisture content of 37%, the grain damage during threshing reached 37.71%, whereas at a moisture content of 25%, the grain damage decreased to 3.75%. This suggests that the structure and motion parameters of the test bench were appropriately designed to meet the requirement of corn grain damage when the threshing was less than 5% at a moisture content of 25%. The damage to the corn grains at a moisture content of 25% was only 9.94% of that at a moisture content of 37%, which was less than one-tenth. Thus, moisture content significantly influences the grain damage caused by threshing.

3.2.2. Analysis of the Effect of the Corncob Division Rate on Grain Damage under Different Moisture Levels

The results in Figure 4b–f demonstrate a consistent decrease in grain damage during threshing as DRC increases. Within the moisture content range of 25–37%, undivided corn had an average grain damage of 21.15%. However, when DRC reached 2, the average grain damage decreased to 14.72%. This trend continued with increasing DRC, reducing threshing damage to 9.10% at a DRC of 4, 5.10% at a DRC of 6, and 2.11% at a DRC of 8. When corn grains with moisture contents of 37%, 34%, 31%, 28%, and 25% were processed with a DRC of 8, the resulting grain damage decreased by 75.20%, 69.52%, 67.30%, 68.14%, and 72.8%, respectively, compared with undivided corn ears. Overall, at a DRC of 8, the average damage caused by threshing decreased by approximately 70%. The primary mechanisms of corn threshing include impact, kneading, combing, rolling, and squeezing.

Threshers that rely predominantly on impact methods tend to have a high threshing drum speed, which often results in significant grain damage. Although many researchers have optimized the motion and structural parameters of threshing devices to reduce damage, threshing corn with a high moisture content remains challenging. Currently, few threshing machines can reduce damage to less than 5% for corn with a moisture content higher than 30%.

For corn grains with a moisture content of 25% to 28%, a DRC of 4 is sufficient to meet the standard of less than 5% grain damage during threshing. Similarly, for grains with a moisture content of 31%, a DRC of 8 achieves a damage reduction of less than 5%. However, for corn with a moisture content exceeding 34%, even with a DRC of 8, grain damage during threshing still exceeds 7%. This indicates that moisture content significantly impacts grain damage during threshing.

Because of variations in planting methods, corn varieties, and harvest periods, the moisture content of corn in the field can vary. Dividing corn before threshing is a technique used to reduce threshing damage, and the extent of damage is influenced by both the corn's moisture content and the DRC. To minimize threshing damage, it is advisable to process corn at the highest feasible DRC level if conditions permit. In the experiment, it was observed that, under consistent conditions, an increase in corn moisture content results in increased threshing damage. Conversely, an increase in DRC leads to a decrease in threshing damage, provided other conditions remain unchanged. The reduction in threshing damage due to higher DRC is also dependent on the moisture content of the corn. Therefore, it can be inferred that, regardless of moisture content, a higher DRC results in reduced threshing damage, and dividing corn before threshing effectively diminishes such damage.

3.2.3. Analysis of Variance and Regression of Test Results

The data in Table 3 were analyzed using one-way analysis of variance (ANOVA) to determine the significance of the effect of DRC on corn threshing damage at different moisture content levels. Table 4 presents the results of these analyses.

Table 4. Variance analysis at different feed rates.

Moisture Content/%	Source of Variation				
	Sum	**df**	**Square**	**F**	**Sig.**
25	25.669	4	6.417	442.807	$p < 0.001$
28	125.427	4	31.357	355.672	$p < 0.001$
31	365.371	4	91.343	902.667	$p < 0.001$
34	915.436	4	228.859	874.676	$p < 0.001$
37	2571.374	4	642.843	1225.369	$p < 0.001$

Note: $p < 0.05$, significant effect; $p < 0.01$, extremely significant effect.

According to the analysis of variance in Table 4, the division rate of corncobs had a significant impact on grain damage under different moisture contents. To analyze the effect of DRC on grain damage during corn threshing at different DRC levels, the objective function (Y) was defined as the grain damage of threshing, and the independent variable (X) was set as the DRC. The test results at various grain moisture contents were analyzed using Origin 2021 software through univariate quadratic polynomial nonlinear regression analysis. The fit function for grain damage was determined as follows:

$$Y = AX^2 + BX + C \quad (2)$$

Table 5 presents the results of the objective function for different grain moisture content values. A determination coefficient R^2 closer to 1 indicates a more effective mathematical model. The significance value, $p < 0.01$, determines the meaningfulness and significance of the mathematical model. The regression fitting of the grain damage yielded a minimum determination coefficient, R^2, of 0.970, suggesting a high degree of fit between the regression equation and the experimental results.

Table 5. Results of regression analysis.

Moisture Content/%	Fitting Function	R^2	Adjusted R^2	p
37	$Y_1 = 44.24X^2 - 8.54X + 0.53$	0.978	0.974	<0.001
34	$Y_2 = 27.93X^2 - 4.7X + 0.27$	0.987	0.986	<0.001
31	$Y_3 = 18.24X^2 - 3.57X + 0.24$	0.988	0.987	<0.001
28	$Y_4 = 10.62X^2 - 2.24X + 0.16$	0.972	0.970	<0.001
25	$Y_5 = 4.56X^2 - 0.98X + 0.068$	0.973	0.971	<0.001

Note: if $p < 0.05$, the effect was significant; if $p < 0.01$, the effect was extremely significant.

3.3. *Analysis of Corn Grain Damage during Threshing*

The purpose of corn threshing is to separate all the corn grains from the corncobs without damaging them. Macroscopically, this separation is achieved through the contact force applied by the threshing device. However, if the strength and number of contact forces on a corn grain fall below the threshold required for effective threshing, the device may not fully remove the grain from the corncob. Conversely, if these forces exceed the threshold for grain damage, the grain may be damaged during threshing. Thus, the objective of corn threshing is to ensure that the contact force is sufficient to separate the grain without surpassing the damage threshold. To minimize threshing damage, it is essential to maximize the threshold for threshing damage while minimizing the threshold for effective threshing.

From a microscopic perspective, the threshold conditions for grain damage are influenced by the biological structure (BS) and chemical material composition (CMC) of the grain. In contrast, the threshing conditions are determined by the material strength of the grain stalk (MSGS) and the degree of mutual support and extrusion between the grains (DMSE). Thus, the BS, CMC, MSGS, and DMSE collectively influence whether a corn kernel will be damaged during threshing. As the moisture content of a corn kernel decreases, changes occur in its CMC, MSGS, and DMSE. This explains why threshing damage increases with higher corn moisture content, as shown in Table 3 and Figure 4a. During threshing, the BS, CMC, and DMSE remain constant if moisture content does not change. However, dividing the corncobs before threshing reduces the mutual extrusion and support among the grains on a corncob. This segmentation effectively reduces the DMSE, lowering the threshold conditions for threshing and, consequently, decreasing threshing damage, without altering other factors.

Moreover, as DRC increases, the DMSE decreases, leading to a reduction in the boundary conditions for corn threshing and, consequently, lower threshing damage. This explains why dividing the corn before threshing reduces damage; as DRC increases, threshing damage continues to decrease, regardless of the moisture content of the corn, as evidenced by the experimental results. Therefore, minimizing the degree of mutual extrusion and support among the corn grains is essential. Dividing corn ears before threshing effectively reduces grain damage caused by moist corn threshing.

Grain damage decreased as DRC increased. Regardless of the grain moisture content, the lowest level of damage was observed when DRC reached 8 under the experimental conditions. This indicates that dividing the corncob reduces the boundary conditions for separation, thereby decreasing damage during threshing. Dividing corncobs lowers the degree of mutual support and extrusion between the grains, which reduces the separation threshold and minimizes threshing damage. Further analysis using limit methods suggests that grain damage consistently decreases with an increase in DRC, regardless of the moisture content of the grains. When each divided corncob contains only one grain, threshing damage approaches zero because this scenario is effectively equivalent to having already completed the threshing process.

4. Verification Test

We conducted a threshing test on Jinyu 1233, another major Chinese corn variety, to investigate the potential reduction in grain damage caused by the division of corncobs. The test followed the experimental method described in Section 2.1. The Jinyu 1233 corn had an average corn ear length of 183 mm, an average diameter of the large end of 51 mm, an average diameter of the small end of 35 mm, an average corn ear mass of 272 g, and a weight of 1000 grains of 354.0 g.

The moisture contents of the tested corn grains were set at 30%, 27%, and 25%. Each experiment was repeated five times, and the average values are listed in Table 6.

Table 6. Results of threshing verification test.

Moisture Content (%)	Divided Rate of Corncobs	Rate of Damaged Grains/%					
		1	2	3	4	5	Mean
30	1	14.36	13.97	14.22	13.98	14.53	14.21
	2	9.94	9.23	10.15	11.03	11.41	10.35
	4	7.73	7.70	6.65	7.48	6.94	7.30
	6	5.25	5.47	4.73	5.32	5.46	5.25
	8	3.87	4.31	4.52	3.73	4.26	4.14
27	1	9.36	8.73	8.64	8.64	9.03	8.88
	2	6.17	6.78	6.27	5.87	6.06	6.23
	4	3.86	4.27	3.93	4.24	3.84	4.03
	6	3.10	3.26	3.15	3.32	3.21	3.21
	8	2.67	2.82	2.75	2.89	2.76	2.79
25	1	3.63	3.84	3.69	3.76	3.81	3.75
	2	2.72	2.72	2.68	2.66	2.80	2.72
	4	1.73	1.86	1.91	1.88	1.78	1.83
	6	1.26	1.34	1.39	1.41	1.18	1.32
	8	0.88	1.02	0.94	0.92	1.06	0.96

The research findings indicate that grain damage from threshing in the undivided corn ears of Jinyu 1233 decreased as the moisture content decreases. Specifically, at moisture contents of 30%, 27%, and 25%, the corresponding percentages of grain damage were 14.21%, 8.88%, and 3.75%, respectively. Furthermore, this study revealed that the grain damage caused by threshing consistently decreased as DRC increased. When DRC reached 8, the grain damage of the threshed Jinyu 1233 corn at moisture contents of 30%, 28%, and 25% decreased to 4.14%, 2.79%, and 0.96%, respectively. Notably, the influence of the division rate of the Jinyu 1233 corncobs on grain damage aligned with that of the Lianchuang 825 corn variety under the same experimental conditions. These findings provide further support for the previously discussed fitting function predictions.

5. Conclusions

This study presented a novel method for reducing grain damage during moist corn threshing. We conducted a high-speed photography observation test of moist corn threshing to investigate the working process of the corn ears in a threshing device. We also conducted grain damage experiments to further explore methods for reducing the damage caused by threshing high-moisture corn using different moisture contents and DRC values. Additionally, we verified the influence of DRC on grain damage in different corn varieties. The main conclusions of this study are as follows:

(1) During threshing, most corn ears broke into pieces, with edge grains separating more easily than central grains. The grains on the divided corncobs were more easily threshed than those on the undivided corncobs.

(2) Introducing DRC showed that as the grain moisture content increased from 25% to 37%, damage to intact (undivided) corn ears increased from 3.75% to 37.71%. Dividing

corn ears significantly reduced damage at all moisture content levels. An eight-piece division decreased damage by approximately 70%.

(3) During threshing, the BS, CMS of the grains, and MSGS remained unchanged; however, dividing the corncobs before threshing reduced the DMSE among the corn grains.

(4) Grain damage during the verification experiment was consistent with the Lianchuang 825 variety. Thus, dividing corn ears into pieces before threshing decreased grain damage applicable to all corn varieties.

Author Contributions: Conceptualization, C.J. and G.W.; methodology, C.W.; software, M.Z.; validation, Y.Y. and G.W.; formal analysis, G.W.; investigation, C.W. and Q.T.; resources, M.Z.; data curation, C.J.; writing—original draft preparation, G.W.; writing—review and editing, C.W., C.J., G.W. and M.Z.; visualization, C.W. and Y.Y.; supervision, Q.T. and Y.Y.; project administration, C.J.; funding acquisition, C.J. All authors have read and agreed to the published version of the manuscript.

Funding: This research was funded by the National Key R&D Project of China, Research and Development of Intelligent Control System for High Performance Combined Harvester with Large Feed Quantity (2021yfd2000503) and the General Program of the National Natural Science Foundation of China, Research on Adaptive Mechanized Harvesting Method for Soybean with Low Crushing Based on Flexible Contact (32171911).

Institutional Review Board Statement: Not applicable.

Data Availability Statement: The raw data supporting the conclusions of this study are available from the corresponding author upon reasonable request.

Acknowledgments: The authors thank the helpful feedback from anonymous reviewers, which contributed to improving the quality of this manuscript.

Conflicts of Interest: The authors declare no conflicts of interest.

References

1. Li, S.K.; Wang, K.R.; Xie, R.Z.; Li, L.L.; Ming, B.; Hou, P.; Liu, C.W. Grain breakage rate of maize by mechanical harvesting in China. *Crops* **2017**, *2*, 76–80.
2. Cui, T.; Fan, C.L.; Zhang, D.X.; Yang, L.; Li, Y.B.; Zhao, H.H. Research progress of maize mechanized harvesting technology. *Nongye Jixie Xuebao Trans. Chin. Soc. Agric. Mach.* **2019**, *50*, 1–13.
3. Wang, K.; Li, S. Progresses in research on grain broken rate by mechanical grain harvesting. *Sci. Agric. Sin.* **2019**, *50*, 2018–2026.
4. Li, X.; Zhang, W.; Xu, S.; Du, Z.; Ma, Y.; Ma, F.; Liu, J. Low-damage corn threshing technology and corn threshing devices: A review of recent developments. *Agriculture* **2023**, *13*, 1006. [CrossRef]
5. Yang, L.; Lü, Q.; Zhang, H. Experimental study on direct harvesting of corn kernels. *Agriculture* **2023**, *12*, 919. [CrossRef]
6. Geng, D.; Tan, D.; Yu, X. Design and test of corn flexible threshing cylinder element. *Jilin Daxue Xuebao (Gongxueban)* **2020**, *50*, 1923–1933.
7. Yang, L.; Wang, W.; Zhang, H.; Li, L.; Wang, M.; Hou, M. Improved design and bench test based on tangential flow-transverse axial flow maize threshing system. *Nongye Gongcheng Xuebao Trans.* **2018**, *34*, 35–43.
8. Li, X.; Gao, L. Performance test on corn thresher with different-speed threshing parts. *Nongye Gongcheng Xuebao Trans.* **2019**, *25*, 102–106.
9. Di, Z.; Cui, Z.; Zhang, H.; Zhou, J.; Zhang, M.; Bu, L. Design and experiment of rasp bar and nail tooth combined axial flow corn threshing cylinder. *Nongye Gongcheng Xuebao Trans.* **2018**, *34*, 28–34.
10. Yang, R.; Chen, D.; Zha, X.; Pan, Z.; Shang, S. Optimization design and experiment of ear-picking and threshing devices of corn plot kernel harvester. *Agriculture* **2021**, *11*, 904. [CrossRef]
11. Li, X.; Du, Y.; Mao, E.; Liu, L.; Guo, D. Design and experiment of corn low damage threshing device based on DEM. *Int. J. Agric. Biol. Eng.* **2023**, *16*, 55–63. [CrossRef]
12. Li, X.; Zhang, W.; Xu, S.; Ma, F.; Du, Z.; Ma, Y.; Liu, J. Calibration of collision recovery coefficient of corn seeds based on high-speed photography and sound waveform analysis. *Agriculture* **2023**, *13*, 1677. [CrossRef]
13. Bakharev, D.; Pastukhov, A.; Volvak, S.; Kovalev, S. Study of seed corn threshing process. *Eng. Rural Dev.* **2020**, *19*, 1036–1041.
14. Gu, R.L.; Huang, R.; Jia, G.Y.; Yuan, Z.P.; Ren, L.S.; Li, L.I.; Wang, J.H. Effect of mechanical threshing on damage and vigor of maize seed threshed at different moisture contents. *J. Integr. Agric.* **2019**, *18*, 1571–1578. [CrossRef]
15. Fan, G.C.; Wang, H.X.; Ji, J.J.; Cao, W.H.; Liu, H.X.; Hao, J.K.; Wang, S.H. Analysis of influence factor on seed damage rate and loss rate during picking corn-cob. *Nongye Gongcheng Xuebao Trans.* **2002**, *18*, 72–74.
16. Chen, Z.; Wassgren, C.; Ambrose, R.K. Measured damage resistance of corn and wheat kernels to compression, friction, and repeated impacts. *Powder Technol.* **2021**, *380*, 638–648. [CrossRef]

17. Srivastava, A.K.; Herum, F.L.; Stevens, K.K. Impact parameters related to physical damage to corn kernel. *Trans. ASABE* **1976**, *19*, 1147–1151. [CrossRef]
18. Shahbazi, R.; Shahbazi, F.; Nadimi, M.; Paliwal, J. Assessing the effects of free fall conditions on damage to corn seeds: A comprehensive examination of contributing factors. *AgriEngineering* **2023**, *5*, 1104–1117. [CrossRef]
19. Li, X.; Ma, F.; Gao, L. Dropping impact experiment on corn seeds. *Nongye Gongcheng Xuebao Trans.* **2009**, *25*, 113–116.
20. Geng, D.; He, K.; Wang, Q.; Jin, C.; Zhang, G.; Lu, X. Design and experiment on transverse axial flow flexible threshing device for corn. *Nongye Jixie Xuebao Trans. Chin. Soc. Agric. Mach.* **2019**, *50*, 101–108.
21. Qu, Z.; Zhang, D.; Yang, L.; Zhang, T.; Wan, Z.; Cui, T. Experiment on feed rate and cylinder speed of longitudinal axial flow threshing and separating device for maize. *Nongye Jixie Xuebao Trans. Chin. Soc. Agric. Mach.* **2018**, *49*, 58–65.
22. Pickard, G.E. Laboratory studies in corn combining. *Agric. Eng.* **1955**, *36*, 792–794.
23. Li, X.; Xiong, S.; Du, Z.; Geng, L.; Ji, J. Design and experiment on floating corn single panicle threshing device. *Int. J. Agric. Biol. Eng.* **2017**, *48*, 104–111.
24. Miu, P.I.; Kutzbach, H. Modeling and simulation of grain threshing and separation in threshing units—Part I. *Comput. Electron. Agric.* **2008**, *60*, 96–104. [CrossRef]
25. Wang, B.; Wang, J.; Du, D. Finite element analysis of dynamic impact damage process of maize kernel based on HyperMesh and LS-DYNA. *J. Zhejiang Univ. Agric. Life Sci.* **2018**, *44*, 465–475.
26. Cundall, P.A.; Strack, O.D. A discrete numerical model for granular assemblies. *Geotechnique* **1979**, *29*, 47–65. [CrossRef]
27. Li, X.; Wu, K. Design and experiment of bionic discrete devices based on corn threshing system. *Chem. Eng. Trans.* **2016**, *51*, 127–132.
28. Dai, F.; Zhao, Y.M.; Liu, Y.X.; Shi, R.J.; Xin, S.L.; Fu, Q.F.; Zhao, W. Analysis and performance test on dynamic seed corn threshing and conveying process with variable diameter and spacing. *Int. J. Agric. Biol. Eng.* **2023**, *16*, 259–266. [CrossRef]
29. Li, X.; Du, Y.; Guo, J.; Mao, E.R. Design, simulation, and test of a new threshing cylinder for high moisture content corn. *Appl. Sci.* **2020**, *10*, 4925. [CrossRef]
30. Li, X.; Ma, Y.; Jin, X.; Gao, L. Design and test of corn seed bionic thresher. *Trans. Chin. Soc. Agric. Mach.* **2015**, *46*, 97–101. (In Chinese)
31. Gao, L.; Li, F.; Zhang, X.; Zhang, Y. Mechanism of moisture content affect on corn seed threshing. *Trans. Chin. Soc. Agric. Mach.* **2011**, *42*, 92–96, (In Chinese with English abstract).
32. *GB/T 21961-2008*; Test Methods for Corn Harvest Machinery. National Standards of the People's Republic of China: Bejing, China, 2008.
33. *GB/T 21962-2008*; Technical Conditions for Corn Harvest Machinery. National Standards of the People's Republic of China: Bejing, China, 2008.

agriculture

MDPI

Article

Design and Fatigue Life Analysis of the Rope-Clamping Drive Mechanism in a Knotter

Jianjun Yin [1,*], Zefu Chen [1], Shiyu Lv [1], Han Wu [1], Yansu Gao [1] and Luning Wu [2]

[1] School of Agricultural Engineering, Jiangsu University, Zhenjiang 212013, China; chen_sama@163.com (Z.C.); 2212216049@stmail.ujs.edu.cn (S.L.); 18860879230@163.com (H.W.); dislike2019@outlook.com (Y.G.)

[2] Faculty of Aerospace Engineering, Jiangsu Aviation Technical College, Zhenjiang 212234, China; 2222016061@stmail.ujs.edu.cn

* Correspondence: yinjianjun@ujs.edu.cn

Abstract: A knotter is a core component for the automatic bundling of agricultural materials, and a knotter with double-fluted discs is one type. Currently, the research on knotters with double-fluted discs has gradually transitioned from structural design to reliability optimization. To address rope-clamping failures in the rope-clamping drive mechanisms in knotters, the specific failure position of the rope-clamping mechanism and the failure causes were analyzed first. The redesign of the rope-clamping drive mechanism in knotters with double synclastic fluted discs was proposed, including structure optimization and 3D modeling using the GearTrax/KISSsoft and SolidWorks software. A virtual prototype model of a knotter with a flexible rope was established by combining ANSYS with the ADAMS software. A rigid–flexible coupling dynamic simulation of the knotter was carried out using ADAMS, and the simulation results were used as the data input for the ANSYS nCode DesignLife module for the fatigue life simulation of the weak parts (the worm shaft) of the knotter. The operation test results for the rope-clamping drive mechanism indicate that the redesigned rope-clamping drive mechanism is reliable in transmission, with a rope-clamping success rate of 100%. The actual operation times for the worm shaft exceed the minimum fatigue life obtained through joint simulation. The applied joint simulation method has high simulation accuracy.

Keywords: knotter; reliability optimization design; joint simulation; fatigue life

1. Introduction

In terms of the core components of square balers, a representative product is the Rappse knotter made in Germany, which is widely used in China. Due to the complex spatial structures of knotters, the motion timing matching for each drive mechanism, and the complicated manufacturing of each part, the service life requirements of knotters are difficult to guarantee. This has become a prominent problem, restricting their practical application. Research on knotters is mainly focused on the fields of structure reconstruction, optimization and innovation, motion timing matching, and reliability. Yin et al. [1] analyzed rope-twining failures in knotter hooks and improved their curved surface shapes. Huang et al. [2] designed a hand rope-knotting device based on the knotting action of the hand rope of a packaging bag, and a dynamic simulation analysis was conducted using ADAMS. The simulation results were essentially consistent with the prototype test results, indicating this work's value as a reference for similar knotting devices. Yin et al. [3] developed a knotter driven by double-fluted discs based on the knotting principles of D-type knotters and carried out straw-bundling tests. Yu et al. [4] established a multibody dynamic simulation model for a D-type knotter, discerned the action timing for each mechanism in the knotter, and verified it using knotting experiments on a bench. Zhang et al. [5] established a numerical model of a knotter's action based on differential geometry theory and determined the manufacturing and processing technology for each component. In response to wear issues in the knotter release mechanism, Li et al. [6] optimized the cam profile of the mechanism to improve the wear. Prototype experiments showed that the cam optimization effect was significant. Ma et al. [7] conducted reliability calculation and fault diagnosis on a knotter based on Bayesian network reliability analysis technology. The accuracy of the calculation results was verified through simulation experiments, and a structural optimization plan was proposed. Naresu et al. [8] conducted a harmonic response and fatigue life analysis on a D-type knotter bracket. Based on the analysis results, the structure of the knotter bracket was optimized, and the reliability of the optimized knotter bracket was verified through orthogonal experiments. In 2021, New Holland [9] unveiled a novel knotter, designated the "Loop Master™", that was capable of tying not only double knots but also circular knots. This innovation has effectively enhanced the utilization efficiency of a bundle of rope, circumventing the waste associated with conventional double knots. In 2022, KRONE [10] unveiled a novel V-shaped knotter, which markedly enhances the knot efficiency by widening and opening the bottom of the knotting nozzle, thereby facilitating the creation of a knot without a rope head. Cen et al. [11] analyzed the failure forms of a knotter bracket and proposed a scheme for the installation of self-lubricating bearings on the knotter bracket. Through finite element software analysis, the results showed that the reliability of the bracket was improved.

In this paper, to address the problem of rope-clamping failures in the rope-clamping drive mechanism in knotters with double-fluted discs, the specific failure position of the rope clamping and the failure causes were analyzed first. The rope-clamping drive mechanism was redesigned, including the kinematic design, structure optimization, and 3D modeling using the GearTrax/KISSsoft 2022 and SolidWorks 2020 software. A virtual prototype model of a knotter with a flexible rope was established by combining ANSYS with the ADAMS 2020 software. A rigid–flexible coupling dynamics simulation of the knotter was carried out using ADAMS, and the simulation results were used as the data input for the ANSYS nCode DesignLife module for the fatigue life simulation of the weak parts of the knotter. A knotting test of the knotter prototype using a test bench and a baler was carried out, and the test results verified the feasibility and reliability of the redesign of the rope-clamping drive mechanism, which provides a reference for research on knotters.

2. Analysis of the Failure Causes of Rope Clamping and the Transmission Redesign of the Rope-Clamping Drive Mechanism in a Knotter with Double-Fluted Discs

The rope-clamping drive mechanism of a knotter with double reverse fluted discs is composed of incomplete bevel gear transmission with orthogonal intersecting axes and a worm drive with a spatially staggered shaft, as shown in Figure 1a. A rope-clamping drive mechanism is used in the knotter. Under the actual working conditions, the fatigue fracture failure of the worm shaft occurs before it reaches the expected service life, leading to the failure of rope clamping, as shown in Figure 1b. Obviously, the rope-clamping failure was caused by the inappropriate transmission design for the rope-clamping drive mechanism.

(a) (b) (c)

Figure 1. Transmission diagrams of the rope-clamping drive mechanism. (**a**) A knotter with double reverse fluted discs; (**b**) a worm shaft failure in a knotter with double reverse fluted discs; (**c**) the force acting on the worm shaft. The individual components in (**a**) are (1) the bevel gear on a small fluted disc; (2) an incomplete bevel pinion; (3) a left-hand worm; (4) a helical gear.

In worm transmission, the forces on the worm can be expressed as [12]

$$\sum = \beta_1 + \beta_1' \tag{1}$$

$$F_{a1} = F_{t1}' = 2000T_1'/d_{m1} \tag{2}$$

$$F_{r1} = F_{r1}' = F_{t1}' \tan\alpha_{n1} \cos\rho' / \cos(|\beta_1'| \pm \rho') \tag{3}$$

$$F_{t1} = F_{a1}' = F_{t1}' \tan(|\beta_1'| - \rho') \tag{4}$$

$$F_n = F_{t1} \cos\rho' / [\cos\alpha_{n1} \cos(|\beta_1| - \rho')] \tag{5}$$

$$\tan\rho' \approx \tan\rho \cos\alpha_{n1} \tag{6}$$

where β_1 is the helix angle of the worm; $\beta_1{}'$ is the helix angle of the helical gear; and Σ is the cross-axis angle. The helix angle of the right-hand helical gear and the lead angle of the right-hand worm are recorded as positive values, whereas the helix angle of the left-hand helical gear and the lead angle of the left-hand worm are recorded as negative values. T'_1 is the resistance torque applied to the helical gear; F_{a1} is the axial force of the worm; F'_{t1} is the circumferential force of the helical gear; d_{m1} is the diameter of the helical gear indexing circle; F_{r1} is the radial force of the worm; F'_{r1} is the radial force of the helical gear; F_{t1} is the circumferential force of the worm; F'_{a1} is the axial force of the helical gear; F_n is the normal force of the worm; α_{n1} is the normal pressure angle of the worm; and ρ is the friction angle of the worm pair.

Similarly, the force on the bevel pinion can be expressed as

$$F_{t2} = 2000T_1/d_{m2} = 2000T_1'/i\eta d_{m2} \tag{7}$$

$$F_{a2} = F_{t2} \tan\alpha_{n2} \cos\delta \tag{8}$$

$$F_{r2} = F_{t2} \tan\alpha_{n2} \sin\delta \tag{9}$$

where F_{t2} is the circumferential force of the bevel pinion; T_1 is the resisting torque applied to the worm shaft; i is the transmission ratio of the worm and helical gear; η is the transmission efficiency of the worm and helical gear; d_{m2} is the diameter of the bevel pinion indexing circle; F_{a2} is the axial force of the bevel pinion; F_{r2} is the radial force of the bevel pinion; δ is the taper angle corresponding to the bevel pinion; and α_{n2} is the normal pressure angle of the bevel pinion. When the influence of friction forces is ignored, the force acting on the worm shaft is as shown in Figure 1c.

As illustrated in Formulas (3) and (4), the force acting on the worm shaft is more sensitive to the helix angle $\beta_1{}'$ of the helical gear. When the cross-axis angle is held constant and the left-hand worm drive is employed, the helix angle of the worm is smaller, while the helix angle of the helical gear is larger, which results in larger circumferential and radial forces acting on the worm shaft. The larger radial forces make the worm shaft incapable of fulfilling the requirements of higher bending strength. Once the worm shaft has undergone bending fatigue deformation, the worm drive will fail, resulting in the rope-clamping failure of the knotter.

Aiming to address the above-mentioned issue of rope-clamping failure, a novel transmission scheme for the rope-clamping drive mechanism is proposed. In order to differentiate it from a knotter with double reverse fluted discs, a knotter with a novel transmission scheme for the rope-clamping drive mechanism is designated as a knotter with double synclastic fluted discs, as illustrated in Figure 2. On the premise of maintaining the limited space occupied by the rope-clamping drive mechanism, adjusting the meshing direction of the incomplete bevel gear pair, and changing the left-hand worm drive to a right-hand worm drive, the helix angle of the helical gear is reduced and the stress on the worm shaft is improved. This is expected to meet the fatigue life requirements of the worm shaft and improve the reliability of the rope-clamping drive mechanism.

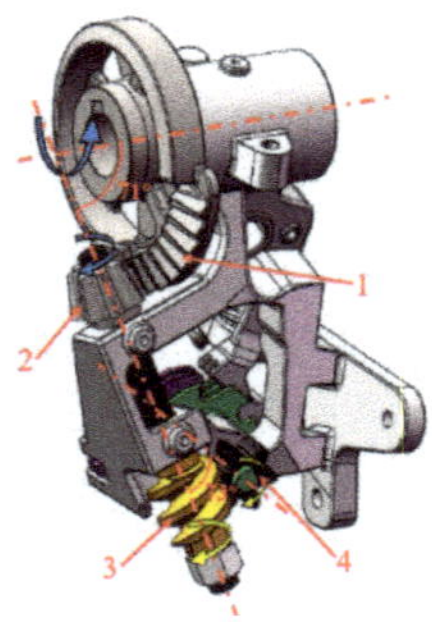

Figure 2. A novel transmission scheme for the rope-clamping drive mechanism in a knotter with double synclastic fluted discs. The individual components are (1) a bevel gear on a small fluted disc; (2) an incomplete bevel pinion; (3) a right-hand worm; (4) a helical gear.

3. Design of the Transmission Structure of the Rope-Clamping Drive Mechanism in a Knotter with Double Synclastic Fluted Discs

3.1. Design of Bevel Gear Transmission with Intersecting Axis Angle of 71°

In order to maintain the position of the involute helical gear axis, it is essential to consider the axial installation space limitations of knotters, and the optimal shaft intersection angle of the incomplete bevel gear transmission of the rope-clamping drive mechanism is preferred to be 71°. In accordance with the specific transmission requirements and general mechanical design principles [13–15], the key parameters of the bevel gear transmission with an intersecting axis angle of 71° are as shown in Table 1.

Table 1. Key parameters of bevel gear transmission with intersecting axis angle of 71°.

	Theoretical Gear Teeth	Module (mm)	Pressure Angle α (°)	Pitch Angle δ (°)	Modification Coefficient x_1 (mm)
Bevel Pinion	8	4	20	10.55	0.4396
Bevel Gear	38			60.45	−0.4396

The joint design method of GearTrax and SolidWorks can be employed to generate the solid model of an incomplete bevel gear pair. First, the bevel gear transmission parameters shown in Table 1 and the tooth profile settings were set in GearTrax to obtain an accurate three-dimensional model of the bevel gear pair. Subsequently, according to the specific transmission requirements of the knotter, a secondary structural design for the bevel gear pair generated by GearTrax was performed under SolidWorks. The completed bevel gear pair design is shown in Figure 3.

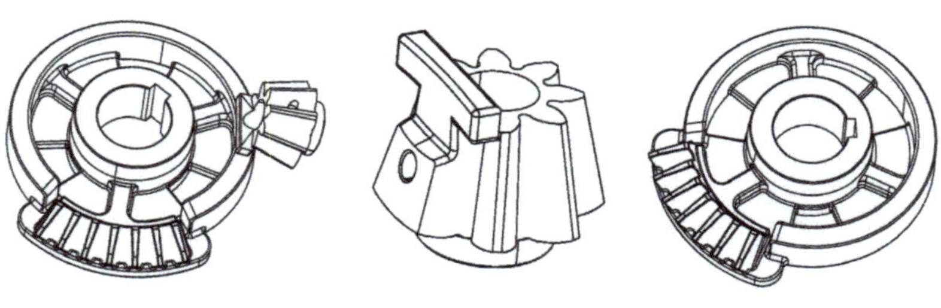

Figure 3. Bevel gear pair. (**a**) Assembly diagrams; (**b**) bevel pinion; (**c**) bevel gear on small fluted disc.

3.2. Design of Worm Drive with Spatial Staggered Shaft Angle of 73°

In order to reduce the helix angle of the helical gear, decrease the force exerted on the worm shaft by the helical gear, and improve the reliability of the rope-clamping drive mechanism, the original left-hand worm drive is modified to a spatial staggered axis right-hand involute worm drive in the new transmission scheme, with a preferred shaft intersection angle of 73° [16–18]. The main design parameters of the involute worm gear pair are shown in Table 2.

Table 2. Main design parameters of the involute worm gear pair.

	Gear Teeth	Module (mm)	Normal Pressure Angle α (°)	Helix Angle β (°)	Lead Angle γ (°)	Modification Coefficient x_2 (mm)
Worm	2	4.5	20	76.78	13.22	−0.20
Helical Gear	6			−3.78	−86.22	0.20

By using the KISSsoft parameterized design software, the precise modeling of the right-handed worm drive with a spatial staggered axis angle of 73° was achieved. First, by setting the relevant parameters of the right-handed worm drive in KISSsoft, the solid models were generated after model verification in KISSsoft. Subsequently, according to the specific transmission requirements of the worm drives in knotters, a secondary structural design for the involute worm gear pair generated by KISSsoft was performed under SolidWorks. The design of the completed involute worm gear pair is shown in Figure 4.

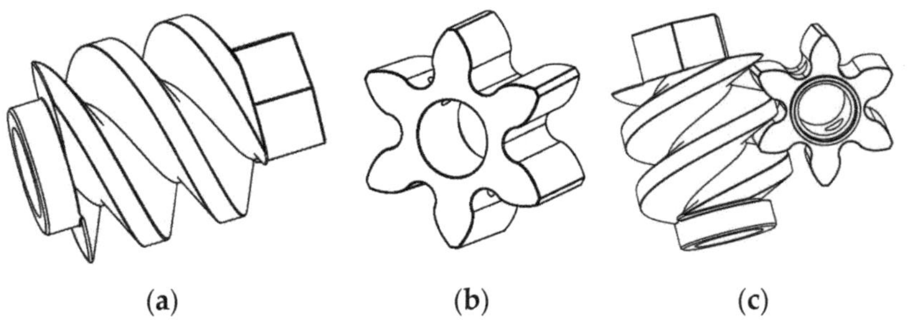

Figure 4. Involute worm gear pair. (**a**) Right-handed worm; (**b**) helical gear; (**c**) assembly diagram.

4. Dynamic Simulation of the Rope-Clamping Drive Mechanism

On the premise of ensuring that the rotation phase of the rope-clamping disc remains unchanged, with the design goal of improving the reliability of the rope-clamping drive mechanism, its kinematic design and structural optimization were carried out. In order to demonstrate the performance of the redesigned rope-clamping drive mechanism, two virtual prototype models of knotters with double-fluted discs were established using the ADAMS software [19–21], and a flexible tying rope was added to obtain the dynamic simulation model, as shown in Figure 5. The overstepping function 'STEP' of ADAMS was used to control the tension of the rope, maintaining tension of 220 Newtons from 0 to 0.48 s and rapidly decreasing to 0 Newtons between 0.48 and 0.5 s.

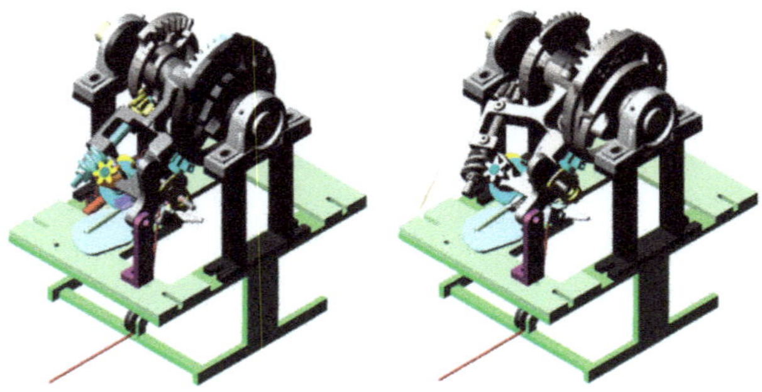

Figure 5. The dynamic simulation model of a knotter. (**a**) A knotter with double reverse fluted discs; (**b**) a knotter with double synclastic fluted discs.

4.1. *Simulation and Analysis of Incomplete Bevel Gear Transmission with Intersecting Axis Angle of 71°*

Based on a complete working cycle of 0.7 s for a knotter, the simulation time was set to 0.7 s with a step size of 300 steps in ADAMS. During the full-cycle simulation process, the meshing of the incomplete bevel gear pair and the surface fitting of the resting segment were both correct, as shown in Figure 6. The force curves of the bevel gear pair of the rope-clamping drive mechanism for the two types of knotters are depicted by the red and black curves in Figure 7.

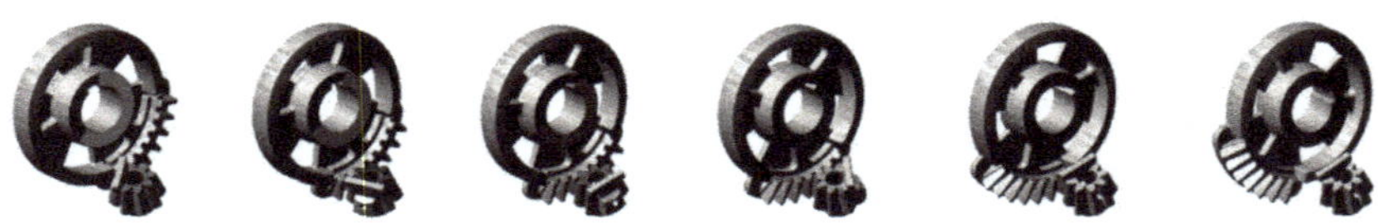

Figure 6. Meshing processes of incomplete bevel gear pair. (**a**) Start of meshing; (**b**) meshing process I; (**c**) meshing process II; (**d**) meshing process III; (**e**) resting surfaces prior to contact; (**f**) surface fitting of resting segment.

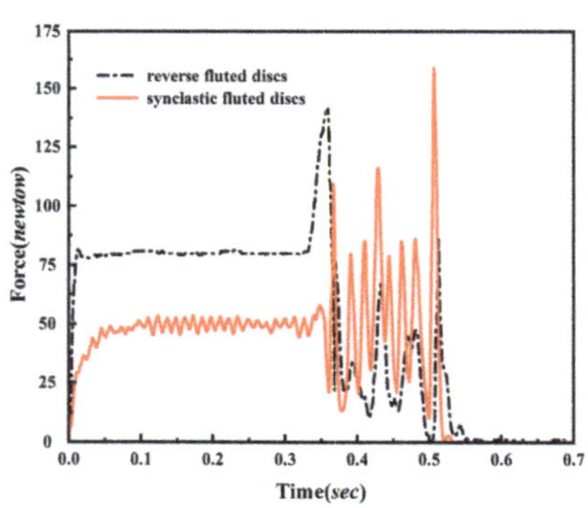

Figure 7. Force curves of the bevel gear pair of the rope-clamping drive mechanism for the two types of knotters.

From the comparison of the force curves shown in black and red in Figure 7, the meshing period of the bevel gear pair with double synclastic fluted discs is more in line with the force law of gear meshing. Moreover, during the resting segment, the mean force amplitude of the bevel gear pair in the knotter with double synclastic fluted discs is reduced by approximately one third compared to the bevel gear pair in the knotter with double reverse fluted discs, improving the force situation of the bevel gear.

4.2. Simulation and Analysis of Involute Worm Drive with Spatial Staggered Shaft Angle of 73°

According to the transmission ratio of the helical gear driven by the worm, the worm rotates once, and the helical gear drives the rope-clamping disc to rotate 120 degrees. The meshing of the involute worm drive in the knotter with double synclastic fluted discs is correct, as shown in Figure 8. The force curves of the worm pair of the rope-clamping drive mechanism for the two types of knotters are depicted by the red and black curves in Figure 9.

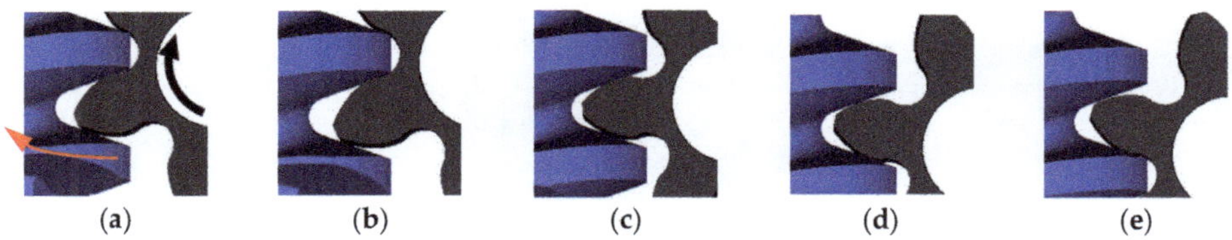

Figure 8. Meshing processes of involute worm drive. (**a**) Start of meshing; (**b**) meshing process I; (**c**) meshing process II; (**d**) meshing almost complete; (**e**) end of meshing. The illustration depicts the direction of rotation of the right-handed worm (red arrow) and the helical gear (black arrow).

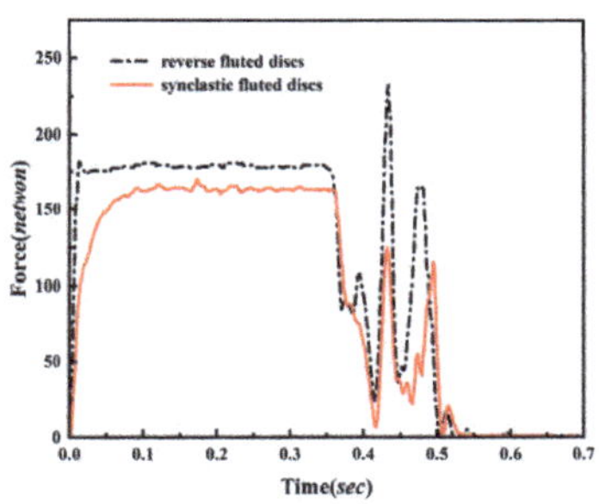

Figure 9. Force curves of the worm pair of the rope-clamping drive mechanism for the two types of knotters.

From the motion simulation of the rope-clamping drive mechanism, it can be seen that the rope gripper completes the rope-clamping action when the worm gear pair engages between 0.36 s and 0.5 s. The comparison of the force curves shown in black and red in Figure 9 reveals that the redesigned worm drive has a peak force reduction of approximately one half compared to the original left-handed worm drive, indicating that the redesigned right-handed worm drive has better transmission performance and the force applied to the worm shaft will also be reduced.

4.3. Virtual Simulation of the Rope-Clamping Drive Mechanism

In order to improve the computational efficiency and ensure the accuracy of the virtual rope-clamping simulations, the simulations used a single-strand rope model instead of the actual two-strand rope. The virtual rope-clamping processes based on the ADAMS dynamic simulation model are shown in Figure 10.

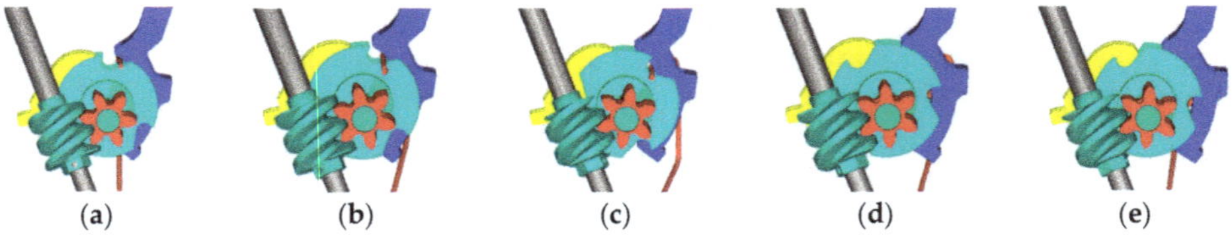

Figure 10. Simulation of rope-clamping processes. (**a**) Initial position; (**b**) start of rope clamping; (**c**) rope-clamping process I; (**d**) rope-clamping process II; (**e**) end of rope-clamping.

From Figure 10, it can be seen that the rope composed of micro-cylindrical segments is located in the groove of the rope-clamping disc in the initial position of the rope-clamping mechanism. When the rope-clamping drive mechanism acts, the rope-clamping disc rotates according to the phase relationship, and the rope follows the groove of the rope-clamping disc to move to the next phase. The results of the rope-clamping simulation are in line with the actual rope-clamping process of the rope-clamping disc.

5. Simulation of Rigid–Flexible Coupling Dynamics and Fatigue Life of the Rope-Clamp Drive Mechanism Based on the Collaborative Simulation of ADAMS and ANSYS

5.1. Simulation of Rigid–Flexible Coupling Dynamics of the Rope-Clamping Drive Mechanism

Considering that the worm shaft is susceptible to fatigue bending and fracture in the rope-clamping drive mechanism, it is necessary to simulate and predict the fatigue life of the worm shaft. Therefore, the ADAMS/ViewFlex module function was used [22–26]. The MNF flexible body neutral file, generated by ANSYS, was imported into the ADAMS rigid body dynamics model, and the original rigid model of the worm shaft was substituted. A rigid–flexible coupling dynamic simulation model of the knotter considering the flexibility of the worm shaft was established, as shown in Figure 11. The flexible worm shaft is shown in red in Figure 11.

Figure 11. Rigid–flexible coupling dynamic simulation model of the knotter.

Given that the two extremities of the worm shaft are, respectively, affixed with the bevel pinion and the worm, it is possible to approximate the force exerted on both ends of the worm shaft as the combined force acting on the bevel pinion and the worm. The simulation of the knotter's rigid–flexible coupling dynamics is similar to the multi-rigid-body dynamics simulation. Contact constraints are added between the flexible worm shaft and its connected components, and the contact type is flexible body to rigid body. The force curves of the worm and the bevel pinion obtained from the simulation of the rigid–flexible coupling dynamics are shown in Figure 12a,b, respectively.

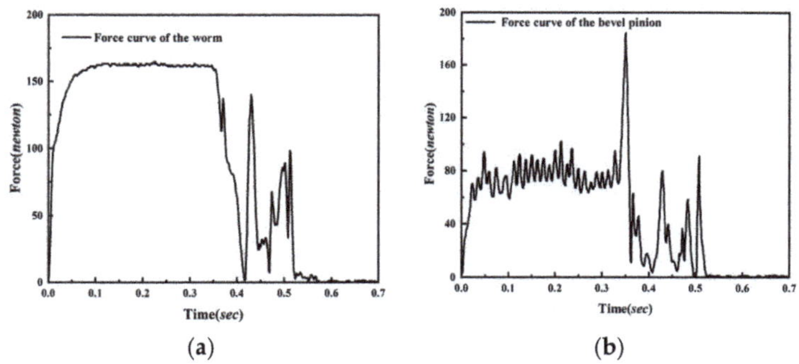

Figure 12. Force curve. (**a**) Force curve of the worm; (**b**) force curve of the bevel pinion.

In the following fatigue simulation calculation of the worm shaft, the force acting on the bevel pinion and the force acting on the worm will be taken as the forces acting on the connection point between the worm shaft and the two components. The results of the rigid–flexible coupling dynamics simulation of the worm shaft will be used as the data input for its fatigue life simulation.

5.2. Simulation of the Fatigue Life of the Rope-Clamping Drive Mechanism

In this paper, the ANSYS nCode DesignLife fatigue simulation module is used to simulate and predict the fatigue life of the worm shaft. First, an '.rst' format file was obtained for the static analysis of the worm shaft in the ANSYS WorkBench and then the static analysis results were input into ANSYS nCode DesignLife, with the load spectrum of the worm shaft obtained from Figure 12a,b used to calculate the fatigue life of the worm shaft. The detailed process is shown in Figure 13.

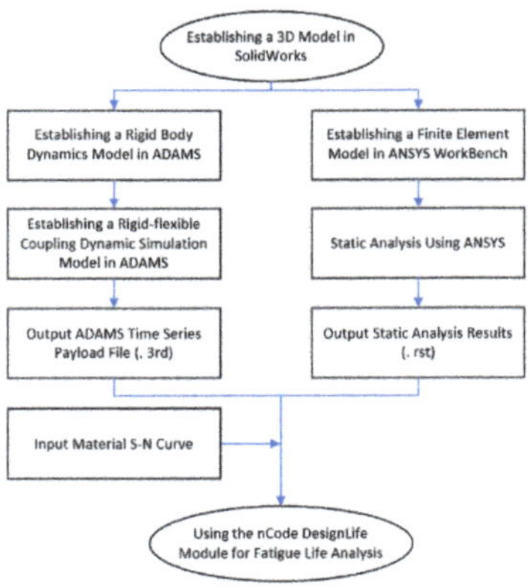

Figure 13. Flow chart for fatigue life simulation of rope-clamping drive mechanism.

The material of the worm shaft is 40 Cr alloy steel, and the heat treatment process of the worm shaft is quenching and tempering. The anticipated survival rate of the worm shaft is set to 95% in ANSYS nCode DesignLife. The cloud diagram of the fatigue life simulation of the worm shaft is shown in Figure 14.

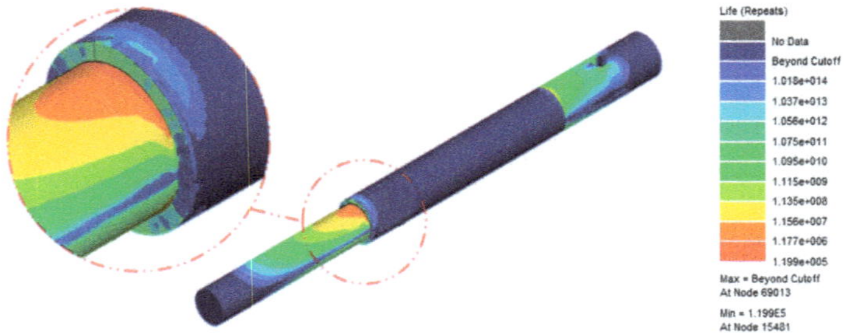

Figure 14. The cloud diagram of the fatigue simulation analysis of the worm shaft.

The minimum fatigue life of the worm shaft occurs at the shoulder of the shaft connecting the worm and the worm shaft, with significant stress concentration. The simulation results for the fatigue failure of the worm shaft are highly consistent with the actual situation. The minimum fatigue life of the worm shaft is 1.199×10^5 times, which exceeds the service life requirement of 100,000 times determined by the knotter industry standard. From the perspective of theoretical simulation, the expected lifespan goal for the reliability improvement of the rope-clamping drive mechanism was achieved.

6. Reliability Test of the Rope-Clamping Drive Mechanism

In order to verify the reliability of the rope-clamping drive mechanism and the actual minimum fatigue life of the worm shaft, two knotters with double synclastic fluted discs were trial-produced and installed on a fatigue test bench for a rope-knotting test to determine the success rate of the rope-clamping action and the service life of the worm shaft due to fatigue failure, as shown in Figure 15.

(**a**) (**b**)

Figure 15. The fatigue test bench for the knotter and rope-knotting test. (**a**) Overall structure of the fatigue test bench of the knotter; (**b**) partial enlarged view of the trial-produced knotter.

Tests of the rope-gripping action were carried out, as shown in Figure 15. From the rope-clamping process shown in Figure 16, it can be seen that the action of the rope-clamping drive mechanism is consistent with the simulation results, and it can drive the rope gripper to reliably grip the rope on the rope needle.

(a) (b) (c) (d)

Figure 16. The test of the rope-gripping action. (**a**) Before rope-gripping; (**b**) about to grip the rope; (**c**) during rope gripping; (**d**) end of rope-gripping action.

A total of 5000 knotting tests with tying ropes did not result in rope-clamping failure, which proves the effectiveness of the rope-clamping drive mechanism design. The rope knots are shown in Figure 17.

Figure 17. The rope knots of the 5000 knotting tests.

In order to reduce the testing time to verify the minimum fatigue life of the worm shaft, the fatigue test bench adopted a fast-knotting mode without considering the success rate of knotting. This allowed the rope-clamping drive mechanism to operate continuously for 210 h at a speed of 90 revolutions per minute, which is equivalent to 113,400 knots. The worm shaft did not exhibit fatigue bending failure.

After 113,400 knots, two trial-produced knotters were installed on a baler, and experiments on the bundling of rice straw were conducted at a base in Jiangsu Province, as shown in Figure 18. After the bundling of 5400 bags of rice straw, the worm shaft did not experience fatigue fracture.

(a) (b)

Figure 18. The experiments on the bundling of rice straw. (**a**) A baler installed with two trial-produced knotters; (**b**) a partial enlarged view of two trial-produced knotters installed on a baler.

Based on the above three tests, the total work number of the rope-clamping drive mechanism reached 123,800, exceeding the minimum fatigue life obtained through the simulations. This indicated that the parameters of the rope-clamping drive mechanism were correctly matched, and the worm shaft parameters and heat treatment method were appropriate.

7. Conclusions

The joint application of the GearTrax, KISSsoft, and SolidWorks software can be used to effectively design rope-clamping drive mechanisms with spatial structures, perform accurate solid modeling, and provide a reference for other complex spatial transmission designs. A rigid–flexible coupling dynamics model of a knotter with a flexible worm shaft can be established based on the collaborative use of ADAMS and ANSYS, and the fatigue life of the worm shaft can be analyzed using ANSYS nCode DesignLife, which is capable of predicting the fatigue life of weak parts. The operation test results of the rope-clamping drive mechanism indicate that the redesigned rope-clamping drive mechanism is reliable in transmission, with a rope-clamping success rate of 100%. The actual operation times of the worm shaft exceed the minimum fatigue life obtained through joint simulation. The applied joint simulation method has high simulation accuracy.

Author Contributions: Conceptualization, Z.C. and H.W.; methodology, J.Y.; writing—original draft preparation, Z.C.; writing—review and editing, S.L., Y.G. and L.W. All authors have read and agreed to the published version of the manuscript.

Funding: This research was funded by the National Natural Science Foundation of China, grant number 52375248.

Institutional Review Board Statement: Not applicable.

Data Availability Statement: The datasets supporting the conclusions of this article are included within the article.

Acknowledgments: The authors thank the anonymous reviewers and journal editor for their valuable suggestions, which helped to improve the manuscript. Moreover, they would like to thank Zhou for helping with the work.

Conflicts of Interest: The authors declare no conflicts of interest.

References

1. Yin, J.J.; Zheng, J.; Wang, X.X.; Zhu, H. Analysis of Rope-twining Failure of Knotter Hook and Its Improved Design of Curved Surface Shape. *Trans. Chin. Soc. Agric. Mach.* **2018**, *49*, 82–89. [CrossRef]
2. Huang, S.T.; Hu, C.; Lv, J.J.; Ma, L.; Ying, F. Key parameters of knotting process of new knot tying device. *J. Mech. Electr. Eng.* **2018**, *3*, 261–265.
3. Yin, J.; Zhang, W.; Chen, Y.; Gao, Q. Design and Knotting Test Analysis of Knotter Driven by Double Gear-discs. *Trans. Chin. Soc. Agric. Mach.* **2016**, *47*, 98–105.
4. Yu, X.; Chen, L.; Wu, M.; Zhang, J. Simulation and Experiment of D-bale Based on Multi-body Dynamics. Master's Thesis, Anhui Agricultural University, Hefei, China, 2017.
5. Zhang, A.; Chen, L.; Dong, H.; Zhang, S.; Li, H.; Han, L. Spatial Structure Parameter Analysis of Rope Cutting and Releasing Mechanism of D-knotter. *Trans. Chin. Soc. Agric. Mach.* **2017**, *48*, 73–80.
6. Li, H.; Xiong, Y.; Chen, L.; Zhang, S.; Li, X.; Han, L. Wear Research and Improved Design of D-knotter Wiper Mechanism. *Trans. Chin. Soc. Agric. Mach.* **2015**, *46*, 118–124.
7. Ma, W. Reliability Research on Knotter of Square Baler. Master's Thesis, Inner Mongolia Agricultural University, Hohhot, China, 2021.
8. Na, R.; Liu, Y. Random Vibration Analysis and Structural Optimization of the D-Knotter Frame. *Comput. Simul.* **2021**, *38*, 235–24
9. The Newfoundland circular knotter won the gold medal at the Edison Invention Award in 2021. *Agric. Mach.* **2021**, *8*, 36.
10. Feuerborn, B. Krone Ballenpressen neuer Knoter ohne Schnipsel [EBOL]. 25 February 2022. Available online: https://www agrarheute.com/technik/krone-ballenpressen-neuer-knoter-ohne-schnipsel-590703 (accessed on 20 October 2023). (In German
11. Shuang, Z.; Cen, H.T.; Si, X.K. Frame body structural design and modal analysis of D-type double knotter. *J. Mach. Des.* **2021**, 3 (Suppl. S2), 118–122.
12. Zhang, Y.; Zhu, L.Y.; Gou, X.F. Calculation methods of load distribution ratio for spiral bevel gear. *Int. J. Mech. Sci.* **202** *257*, 108531. [CrossRef]

13. Song, C.S.; Zhu, C.C.; Lim, T.C.; Luo, J.Y.; Xu, X.Y. Time-varying mesh characteristics of beveloid gears with crossed axes and small shaft angle. *J. Vib. Shock.* **2012**, *31*, 153–157.
14. Chen, Z.; Zhou, J.; Liu, B.; Fu, H.; Meng, X.; Ji, J.; Zhang, Y.; Hua, X.; Xu, X.; Fu, Y. Meshing frictional characteristics of spur gears under dry friction and heavy loads: Effects of the preset pitting-like micro-textures array. *Tribol. Int.* **2023**, *180*, 108296. [CrossRef]
15. Ma, L.Z.; Xie, J.; Yin, X.Q. *Principle and Design of Mechanism*, 2nd ed.; China Machine Press: Beijing, China, 2015.
16. Xu, X.Y.; Li, L.; Ren, B.; Zhu, C. Study on the Dynamic Characteristics of Herringbone Planetary Gear System with Space Shafts Angle. *J. Mech. Eng.* **2023**, *59*, 141–150. [CrossRef]
17. Zhou, M.L.; Yang, J.J.; Xu, T.B.; Ying, J.; Wang, X. Optimal design of transplanting mechanism with differential internal engagement non-circular gear trains. *J. Agric. Eng.* **2022**, *53*, 1–10. [CrossRef]
18. Guo, E.K.; Hu, L.L.; Zhang, E.Z.; Liu, C.; Xu, J.; He, W. A Cylindrical Skiving Tool Design Method Based on a Conjugate Surface for Internal Gear Manufacture. *J. Manuf. Process.* **2023**, *101*, 1538–1550. [CrossRef]
19. Qiao, G.; Liao, R.; Guo, S.J.; Shi, Z.; Ma, S. Design and Dynamic Analysis of the Recirculating Planetary Roller Screw Mechanism. *Chin. J. Mech. Eng.* **2022**, *35*, 87. [CrossRef]
20. Huang, J.C.; Tan, L.; Tian, K.P.; Zhang, B.; Ji, A.; Liu, H.; Shen, C. Formation mechanism for the laying angle of hemp harvester based on ANSYS-ADAMS. *Int. J. Agric. Biol. Eng.* **2023**, *16*, 109–115. [CrossRef]
21. Li, X.F.; Wang, Y.B. Analysis of mixed model in gear transmission based on ADAMS. *Chin. J. Mech. Eng.* **2012**, *25*, 968–973. [CrossRef]
22. Huang, T.; Zhao, J.X.; Song, K.X.; Zhang, X.; Jiang, Y.; Hao, L. Fatigue Life Research of Bellows under Coupled Loading of Velocity and External Pressure Based on S-N Curve of Components. *J. Mech. Eng.* **2022**, *58*, 269–282. [CrossRef]
23. Fu, Q.; Wang, D.D.; Zhang, R.Y.; Lu, Y.; Wang, R.; Zhu, R. Stress Analysis and Stress Fatigue Life Prediction of Rcp Impeller Based on Fluid-thermal-solid Coupling. *Nucl. Eng. Des.* **2023**, *414*, 112596. [CrossRef]
24. Li, F.Z.; Tong, S.G.; Wang, X.B.; Cheng, X.M. Rigid-flexible coupling dynamics and characteristics of marine excavator's mechanical arm. *J. Vib. Shock.* **2014**, *33*, 157–163.
25. Zheng, T.; Zhang, D.G. Limitations of MSC.ADAMS for High speed and Large Deformation Problems of Rigid flexible Coupling Systems. *J. Nanjing Univ. Ence Technol.* **2012**, *36*, 993–998.
26. Wu, D.; Ding, D.; Cui, B.; Jiang, S.; Zhao, E.; Liu, Y.; Cao, C. Design and experiment of vibration plate type camellia fruit picking machine. *Trans. Chin. Soc. Agric. Eng.* **2022**, *15*, 130–138. [CrossRef]

agriculture

MDPI

Article

Experimental Structural Safety Analysis of Front-End Loader of Agricultural Tractor

Jeong-Hun Kim [1,2,†], Dong-Hyeon Gim [3,†] and Ju-Seok Nam [1,2,*]

1 Department of Biosystems Engineering, College of Agriculture and Life Sciences, Kangwon National University, Chuncheon 24341, Gangwon-do, Republic of Korea; jhkim1995@kangwon.ac.kr
2 Interdisciplinary Program in Smart Agriculture, Kangwon National University, 1 Kangwondaehak-gil, Chuncheon 24341, Gangwon-do, Republic of Korea
3 CAE Solution Department of TYMICT, Gongju 32530, Chungcheongnam-do, Republic of Korea; dhkim9075@tymict.com
* Correspondence: njsg1218@kangwon.ac.kr; Tel.:+82-33-250-6497
† These authors contributed equally to this work.

Abstract: The agricultural front-end loader is an implement attached to the front of tractors to transport various agricultural materials, including soil. Since they are subjected to various loads due to the working environment, their safety analysis in consideration of actual working conditions is required. However, there are no official standardized test codes to consider various actual working environments currently. In this study, the structural safety of a front-end loader for static and fatigue failures was evaluated using new test code reflecting actual working environments. Thirty-four measurement locations were determined as the stress concentration spots of each component of the front-end loader derived through multibody dynamic simulation. The total testing time was set to 1 h, and the test time for each task was determined considering the duty percentage of the actual loader work. The measurement results showed that the maximum stress that exceeds the material's yield strength occurred at two locations of the mount, which is the connection to the tractor body, resulting in static yielding. For tasks, the pulling and dumping exhibited the highest stress. The task that had the largest impact on fatigue damage was the dumping. The static safety factor was found to be over 1.93 and the fatigue life met the required lifespan at all measurement locations except for those exhibiting static yielding. Therefore, the most vulnerable part of the front-end loader is the mount, and it is necessary to secure the overall structural integrity by robust design for the mount.

Keywords: failure; fatigue life; front-end loader; static safety factor; structural safety

1. Introduction

Various implements have been mounted on prime movers to improve the efficiency and productivity of agricultural work [1]. Among them, the front-end loader is attached to the front of the agricultural tractors and controlled through tractor hydraulic systems to transport soil, agricultural products, and byproducts [2,3]. The front-end loader can fail or become deformed under large loads when it performs loading or unloading and is exposed to irregular road surfaces or obstacles due to the nature of the working environment [4,5]. Such failure and deformation significantly affect the productivity and work efficiency as well as the safety of workers [6,7]. Therefore, the front-end loader needs to be designed to have a safe structure and its safety must be verified through tests [8,9].

Only a few studies have been conducted on the agricultural front-end loader. Cho et al. analyzed the impact reduction effect of the boom suspension through hydraulic simulation to reduce the vibration and impact generated by the bumps on the ground. They improved the hydraulic system of the front-end loader based on simulation results and verified the impact reduction effect by performing an actual vehicle test [10]. Ahn et al. conducted research for reducing the vibration and impact from the front-end loader using an accumulator [11]. Latoree et al. developed an automatic leveling system by controlling the posture of the front-end loader during work on slopes or irregular surfaces. They expected this to reduce the labor of farmers and improve their productivity [12]. Kwon and Shin conducted structural analysis by using the load derived through hydraulic system analysis and evaluated the structural safety of the front-end loader. They also performed design optimization based on the structural analysis results [13]. In addition to structural analysis, studies evaluating the safety of front-end loaders based on measurements have also been conducted. In these studies, most instrumentation systems use strain gauges to reduce costs and enable miniaturization [14]. Malon et al. measured the stress of front-end loader components under static and dynamic loadings to the bucket for the development of a structural analysis model of the front-end loader. They verified the accuracy of the model by comparing the stress measurements with the structural analysis results [15]. Park et al. evaluated the structural safety of a front-end loader by measuring the stress generated during the drop and impact tests for the bucket [16]. Thus far, the structural safety of front-end loaders mainly considered the effects of static or dynamic loads from the weight on the bucket, as well as simple operations with a few impact conditions. However, when using a front-end loader for various tasks, diverse loads can occur in various working environments. There are no official standardized test codes to consider these various working conditions, and front-end loader manufacturers have tested their product by considering some simple impact conditions. Therefore, it is necessary to develop a test code to reflect various working conditions and evaluate the front-end loader safety based on the test code.

In this study, we suggest new test code for the front-end loader considering the actual working environments, and evaluate the structural safety based on the measurement following the code. The following conditions were included in the test code: the occurrence of static and dynamic loads caused by the payload of bucket, the occurrence of an impact load due to an obstacle, traveling on an irregular road surface with bumps, and the operation of loading, transportation, and unloading of soil. Based on the stress measurement results, the vulnerable part of the front-end loader was identified. The results of this study are expected to be used as foundational data for front-end loader design and verification.

2. Materials and Methods

2.1. Tractor and Front-End Loader

In this study, tests were conducted by attaching a front-end loader to an 18 kW-class agricultural tractor (TYM, Gongju, Korea). The tractor used for the tests weighs 8.19 kN, with a wheelbase of 1670 mm and a ground clearance of 240 mm. The front-end loader used is a product of the parallel link type with two boom cylinders and two bucket cylinders. The front-end loader weighs 4.5 kN, and the capacity and maximum allowable payload of the bucket are 0.23 m^3 and 5.96 kN, respectively. Tables 1 and 2 list the specifications of the tractor and the front-end loader, respectively. Figure 1 shows the shape of the tractor and the front-end loader. The frame of the front-end loader consists of a bucket, booms, posts, and mounts and rear frames for connecting to a tractor. The characteristics of each component are as follows:

Table 1. Specification of the tractor used.

Item	Specification
Model/Company/City/Nation	T2515/TYM/Gongju/Korea
Engine rated power (kW)	18
Engine rated speed (rpm)	2100
Total weight (kN)	8.19
Overall length × width × height (mm)	2395 × 1170 × 2500
Wheelbase (mm)	1670
Ground clearance (mm)	240

Table 2. Specifications of the front-end loader used.

Item	Specification
Model/Company/City/Nation	BL150/TYM/Gongju/Korea
Bucket capacity (m^3)	0.23
Total weight (kN)	4.5
Maximum lift height (mm)	2665
Maximum allowable payload (kN)	5.96

Figure 1. View of the tractor and front-end loader used: (**a**) tractor; (**b**) front-end loader; (**c**) front-end loader attached to the tractor.

A. Bucket: It is the loading part that directly loads soil, agricultural products, and byproducts, and its movement is controlled by bucket cylinders.

B. Boom: It is the main frame of the front-end loader equipped with the bucket at the end, and its movement is controlled using boom cylinders.

C. Post: It is the connecting frame of the boom, boom cylinder, and mount. It supports the reaction force generated by the operation of boom cylinders.

D. Mount: It connects the post to the tractor body. It is fastened to the bottom of the engine room of the tractor to support the weight of the loader and the impact and vibration generated during operation.

E. Rear frame: It connects the mount to the rear axle of the tractor to share the impact and vibration from the mount.

2.2. Test Condition

The test conditions were determined based on a newly proposed test code by the front-end loader manufacturer. They included five tasks: drop test, impact test, pulling test, driving test, and dumping test. The drop test considers the case where the load is generated by the payload when the bucket of the front-end loader rises and falls. The impact and pulling tests consider cases where an impact occurs due to an obstacle or an excessive payload is applied to the bucket. The driving test considers the vibration caused by the irregular road surface while traveling, and the dumping test considers the work of loading, transportation, and unloading of soil. In all test conditions, a rear ballast load of 3.0 kN was applied for safe operation. The drop, impact, and driving tests were conducted by applying a payload of 2.5 kN to the bucket and a ballast weight of 3.0 kN to the rear of the tractor. In the pulling and dumping tests, only a rear ballast weight of 3.0 kN was applied with no payload on the bucket. Figure 2 shows the photographs of the tests. All the tests were conducted by skilled workers. The steps performed for the different tests were as follows:

A. Drop test: The bucket was first lifted to a maximum height of 2665 mm and then dropped until its bottom reached a height of 300 mm from the ground. In this instance, the boom cylinder was controlled so that the boom could move at maximum speed.

B. Impact test: The left and right ends of the bucket collided with the upper and lower parts of the impact test structure at a traveling speed of 1 km/h. The structure was made of steel and was fixed using anchors to prevent shaking at the time of the impact.

C. Pulling test: After fixing the left and right end, and the center of the bucket to the ground using a 300 mm long chain, the bucket was lifted repeatedly. In this case, the boom cylinder was controlled so that the boom could operate at maximum speed.

D. Driving test: The tractor was driven at a speed of 1 km/h on a circular paved road equipped with various bumps provided by the manufacturer.

E. Dumping test: Loading was performed to fill the bucket with soil, following which a straight-line distance of 20 m was traveled, and, subsequently, unloading was performed. When the transport of the piled soil was completed, the tractor was driven backwards with the bottom of the bucket touching the ground to level the soil.

The total testing time was set to 1 h, and the testing time for each test condition was determined considering the duty percentage. The driving and dumping tests were continuous tests, whereas the drop, impact, and pulling tests were discontinuous tests. For the latter, the number of tests was determined considering the testing time to be performed and the time required for one test. Table 3 shows the duty percentage provided by the manufacturer and the testing time for each condition based on 1 h of total time.

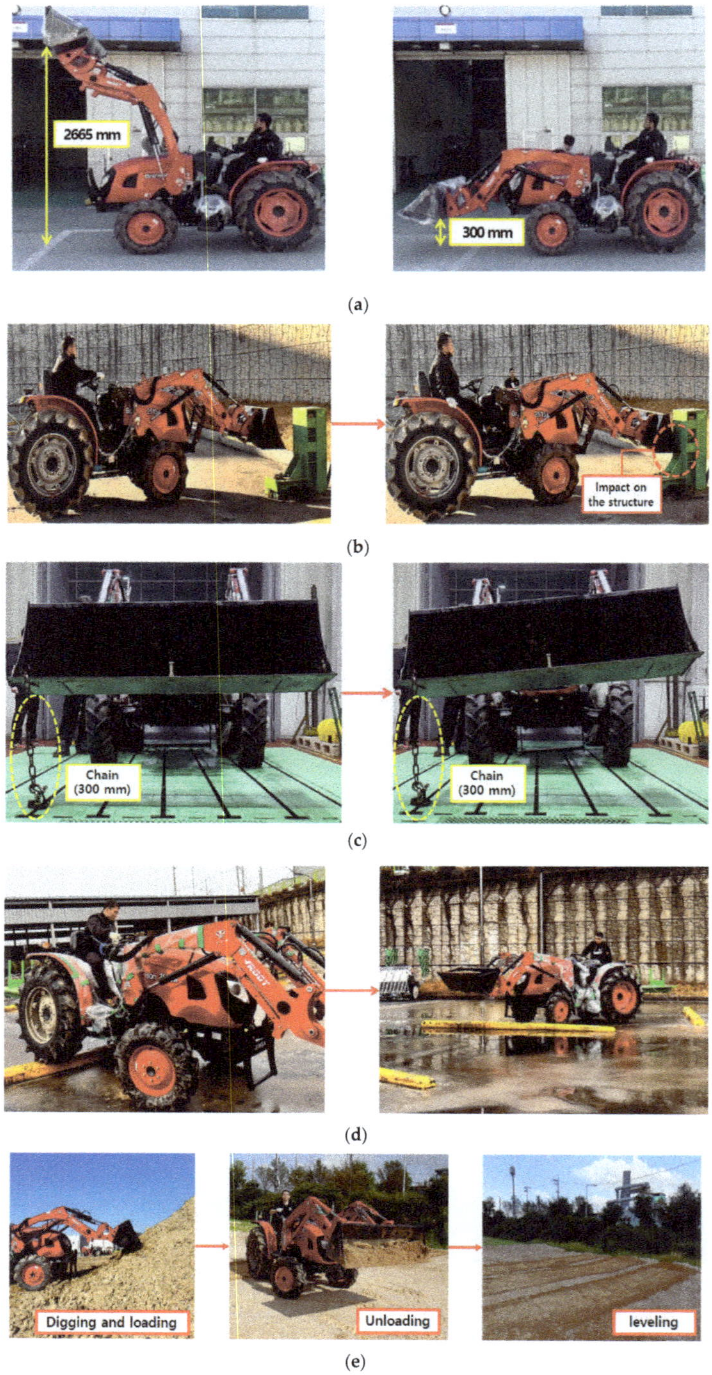

Figure 2. Test conditions of the front-end loader: (**a**) drop test; (**b**) impact test; (**c**) pulling test; (**d**) driving test; (**e**) dumping test.

Table 3. Duty percentage and test time for each test.

Test Type		Duty Percentage (%)	Test Time (s)
Drop test		3.4	122.4
Impact test	Left-end of bucket to lower part of structure	1.0	7.2
	Left-end of bucket to upper part of structure		10.8
	Right-end of bucket to lower part of structure		10.8
	Right-end of bucket to upper part of structure		7.2
Pulling test	Chain on center of bucket	2.8	33.6
	Chain on left-end of bucket		33.6
	Chain on right-end of bucket		33.6
Driving test		46.4	1670.4
Dumping test		46.4	1670.4
Total		100	3600

2.3. *Stress Measurement*

2.3.1. Determination of the Stress Concentration Spots in the Front-End Loader

The stress concentration spots of the front-end loader components and the principal stress directions there were analyzed using commercial multibody dynamic analysis software (Recurdyn V9R5, Functionbay, Seongnam, Korea). Figure 3 shows the 3D model of the front-end loader used in the dynamic simulation. The 3D model was developed by reflecting the actual geometry and material properties of the front-end loader. The weight difference between the 3D model and the actual front-end loader was approximately 1%, indicating that the developed 3D model was valid (Table 4).

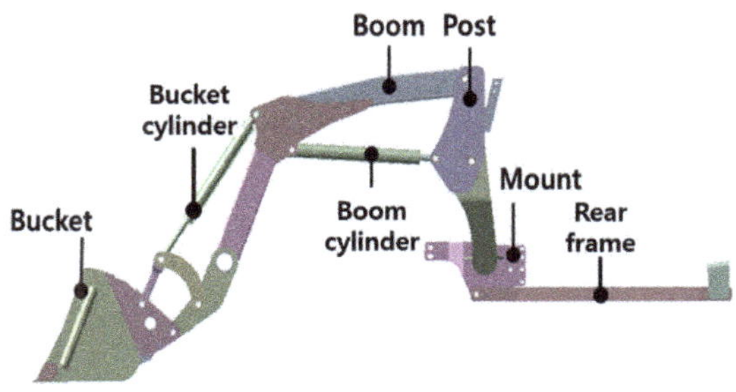

(a)

(b)

Figure 3. 3D model of the front-end loader: (**a**) side view; (**b**) ISO metric view.

Dynamic simulation was performed for the drop, pulling, and impact tests in which the occurrence of large instantaneous loads was expected. Figure 4 shows the simulation models created for each test condition. For the simulation of the drop test, a maximum allowable payload of 5.96 kN was added to the bucket and reciprocating lifting and dropping motion was applied within a height from 300 mm to 2655 mm as the actual test. The actuating force of the boom cylinder for moving the boom at maximum speed was determined to be 24.03 kN by an experiment. In the impact test simulation, a rigid body was created as an impact structure and this structure was allowed to hit the bucket at a speed of 1 km/h. In this case, a payload of 5.96 kN was applied to the bucket. In the pulling test simulation, the bucket was located at a height of 300 mm from the rigid ground. While the ground and the bucket were connected using a rigid chain, the boom was raised at maximum speed. In all simulations, the mount and rear frame were fixed with a fixed support constraint. In the simulations of the drop and impact tests, the maximum allowable payload was applied to the bucket, unlike the actual test conditions, to clearly reveal stress concentration spots by the loads.

Table 4. Weight of the 3D model and actual front-end loader.

Item	Weight (kN)	
	3D Model	Actual Product
Bucket	1.27	4.5
Link	0.14	
Boom cylinder	0.22	
Boom	0.90	
Bucket cylinder	0.20	
Post	0.30	
Mount	0.92	
Rear frame	0.50	
Total	4.45	4.5

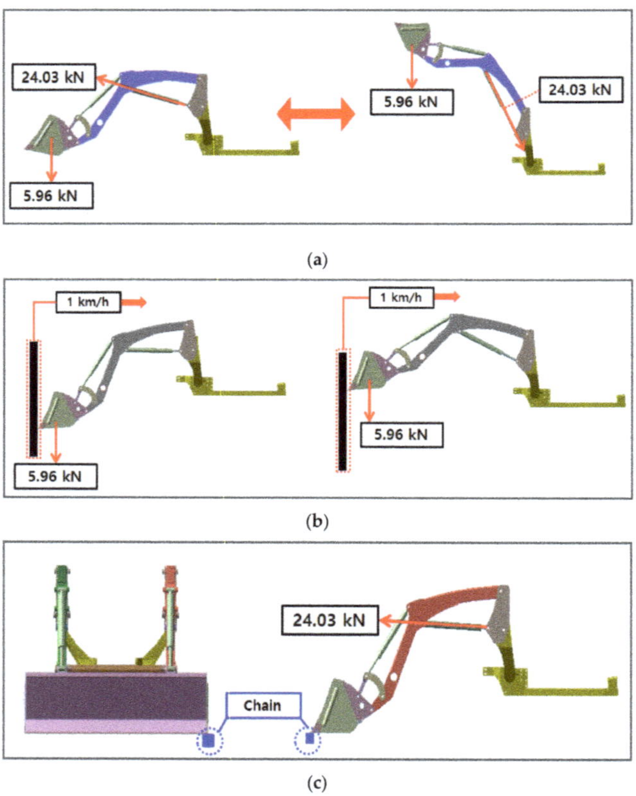

Figure 4. Multibody dynamic simulation models of the front-end loader: (**a**) drop test; (**b**) impact test; (**c**) pulling test.

The stress levels of the boom, post, mount, and rear frame, which are load-bearing frames among the components of the front-end loader and whose design can be changed relatively easily, were obtained from simulations. To facilitate analysis and save time, only the boom, post, mount, and rear frame were set as flexible bodies, while the other elements were constructed to rigid bodies. A square-shaped mesh was chosen for the flexible body analysis and its size ranged from 10 to 14.5 mm depending on the location. In the simulation results, a total of 34 stress concentration spots were derived with the principal stress directions, including 12 for the boom, 6 for the post, 8 for the mount, and 8 for the rear frame (Figure 5 and Table 5). In Figure 5, higher stress levels are represented in red, while lower stress levels are represented in blue. Since the front-end loader used in this study was horizontally symmetrical, the left side was marked with L and the right side with R based on the traveling direction of the tractor (Table 5).

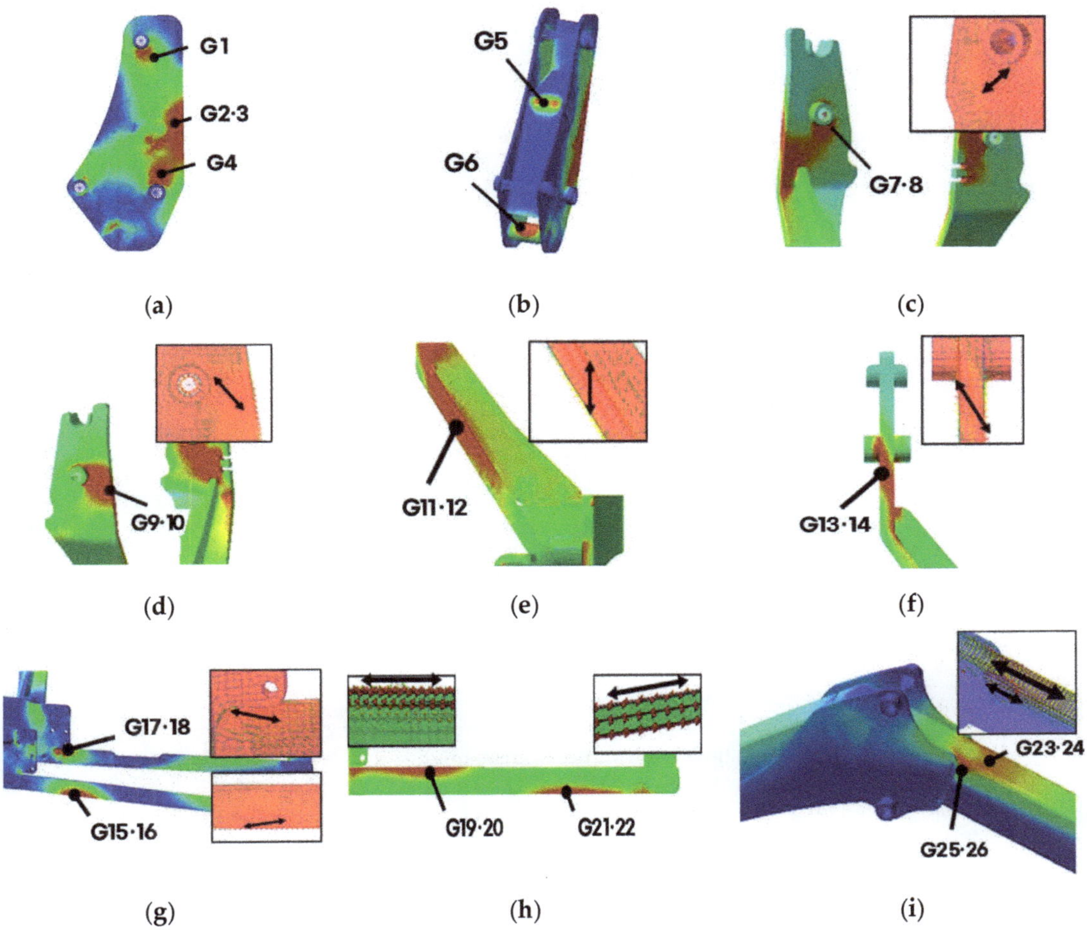

Figure 5. *Cont.*

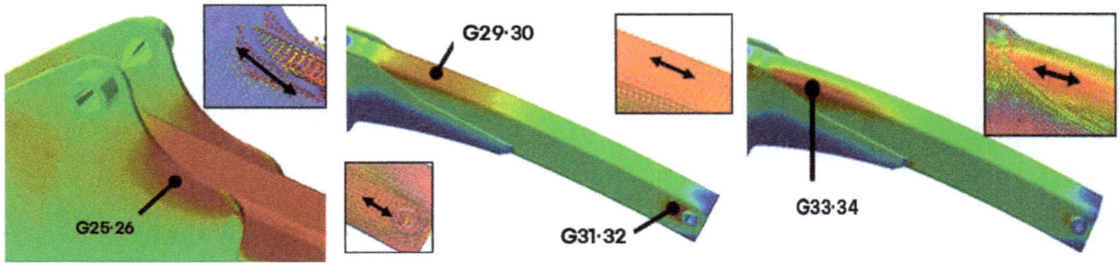

(j) (k) (l)

Figure 5. Stress of the front-end loader components from dynamic simulation: (**a**) post rear side; (**b**) post inside; (**c**) mount inside; (**d**) mount outside; (**e**) mount upper back side; (**f**) mount lower back side; (**g**) rear frame outside and upper inside; (**h**) rear frame upper and bottom side; (**i**) boom front top and inside; (**j**) boom side plate; (**k**) boom rear top and inside; (**l**) boom rear inside.

Table 5. Result of stress concentration location of each front-loader component through dynamic simulation.

Component	Symbol	Location	Component	Symbol	Location
Post	G1	Post upper collar	Rear frame	G18	Rear frame inside (R)
	G2	Post back side (L)		G19	Rear frame outside upper (L)
	G3	Post upper collar (R)		G20	Rear frame outside upper (R)
	G4	Post lower collar		G21	Rear frame rear bottom (L)
	G5	Post plate		G22	Rear frame rear bottom (R)
	G6	Post bar	Boom	G23	Front boom top side (L)
Mount	G7	Mount collar inside (L)		G24	Front boom top side (R)
	G8	Mount collar inside (R)		G25	Front boom inside (L)
	G9	Mount collar outside (L)		G26	Front boom inside (R)
	G10	Mount collar outside (R)		G27	Boom side plate (L)
	G11	Mount back side lower (L)		G28	Boom side plate (R)
	G12	Mount back side lower (R)		G29	Rear boom top side (L)
	G13	Mount back side upper (L)		G30	Rear boom top side (R)
	G14	Mount back side upper (R)		G31	Rear boom collar (L)
Rear frame	G15	Rear frame outside lower (L)		G32	Rear boom collar (R)
	G16	Rear frame outside lower (R)		G33	Rear boom inside (L)
	G17	Rear frame inside (L)		G34	Rear boom inside (R)

2.3.2. Measurement System

Uniaxial strain gauges were attached to the 34 stress concentration spots, determined from the dynamic simulation, in the principal stress direction (Figure 6). The strain gauges were connected to a bridge box (CANHEAD) and the data were transmitted to a data acquisition (DAQ) system via controller area network (CAN) communication. The data collected in the DAQ system were checked and stored in real time through a laptop. Figure 7 shows the outline of the measurement system and Table 6 lists the specifications of the strain gauge and bridge box. The strain measured using the strain gauge was converted into stress using Equation (1).

$$\sigma = E \times \varepsilon \quad (1)$$

where σ is the measured normal stress, MPa; E is modulus of elasticity, MPa; ε is measured normal strain, mm/mm.

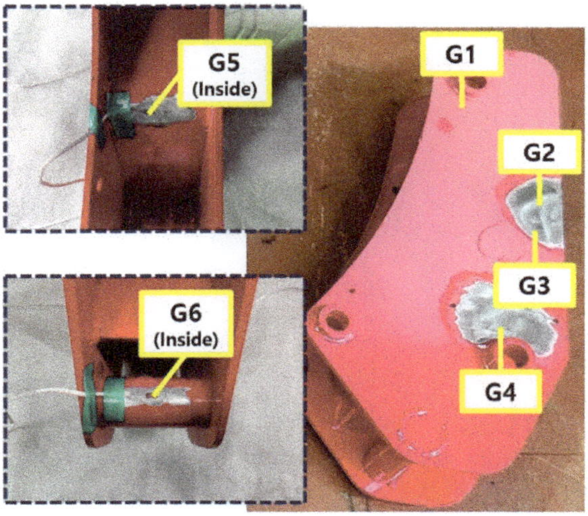

(a)

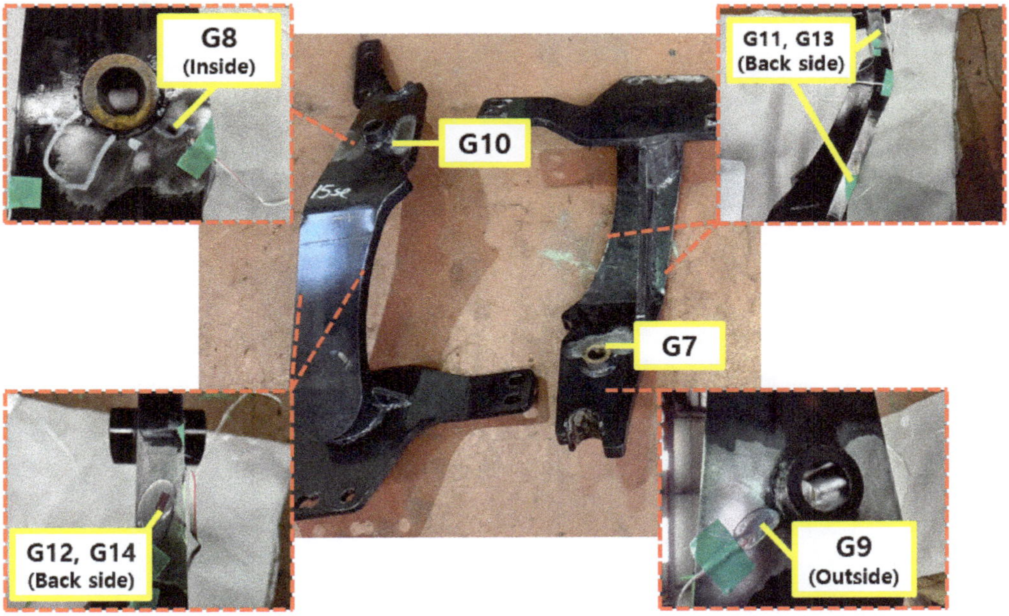

(b)

Figure 6. *Cont.*

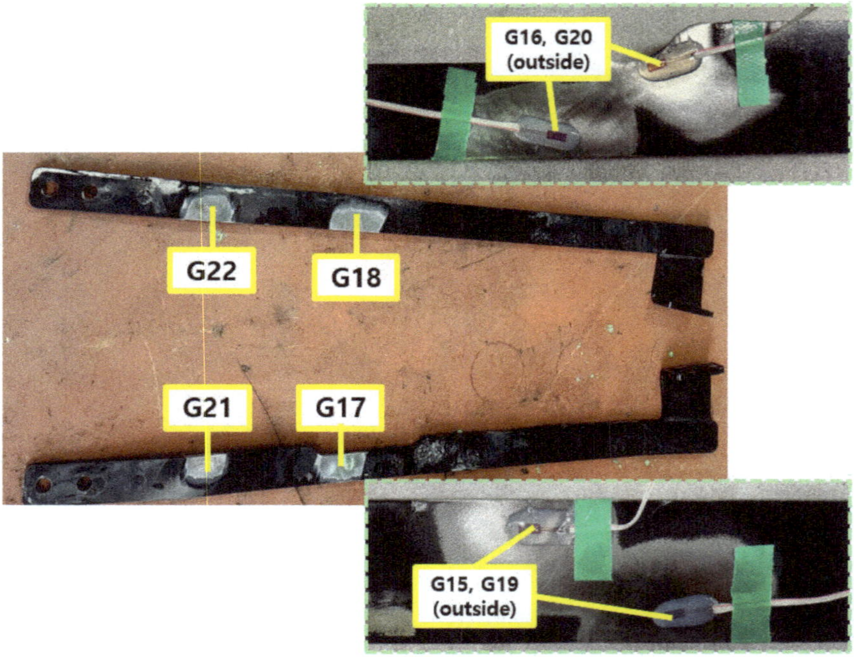

(c)

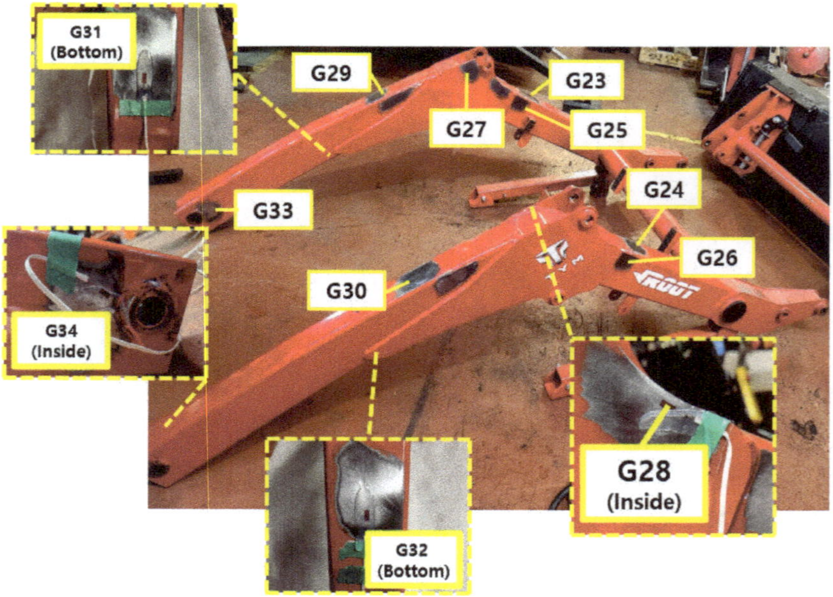

(d)

Figure 6. Attachment location of strain gauges: (**a**) post; (**b**) mount; (**c**) rear frame; (**d**) boom.

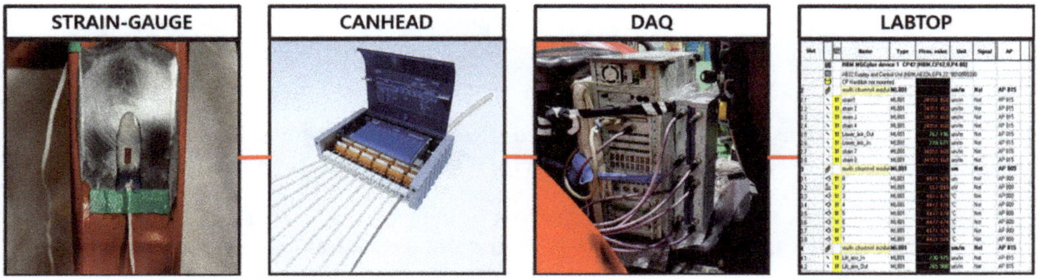

Figure 7. Diagram of the stress measurement system.

Table 6. Specification of the strain gauge and bridge box.

Item		Value
Strain gauge	Model/Company/City/Nation	KFGS−5−350−C1−11 /KYOWA/Tokyo/Japan
	Gauge factor	2.14 ± 1.0%
	Gauge length (mm)	5
	Gauge resistance (Ω)	350.0 ± 2.4
Bride box	Model/Company/City/Nation	CANHEAD/HBK/Seongnam/Korea
	Bridge excitation voltage (V)	2.5
	Measuring ranged (mV/V)	4
	Maximum sampling rate (kHz)	300
	Applicable temperature range (°C)	−30~70

2.4. *Safety Analysis*

2.4.1. Static Safety Factor

The static safety factor is an indicator of how safely agricultural machinery can respond to instantaneous high loads during agricultural work [17]. It is used to evaluate and improve the safety of the agricultural machinery in the design stage considering the strength of the materials and the working environment. The stress-based static safety factor is defined as the ratio of the allowable strength of the material to the maximum stress acting on it. If the static safety factor is higher than 1.0, the material can be judged to be safe for static loads. However, if it is less than 1.0, static failure can occur to the part or the structure [18]. In this study, the static safety factor was calculated using Equation (2) and the static safety of the front-end loader components was analyzed.

$$\mathrm{S.F.} = \left| \frac{S_y}{\sigma_{max}} \right| \tag{2}$$

where S.F. is static safety factor, S_y is tensile yield strength of the material, MPa, and σ_{max} is measured maximum normal stress during the operation, MPa.

2.4.2. Fatigue Life

Failure to the mechanical parts and structures can occur due to the fatigue caused by repeated small loads as well as the high stress that exceeds the yield strength of the material [19,20]. Since the front-end loader is subjected to repeated loads while performing various tasks, safety against fatigue failure must also be evaluated.

The stress data measured in the time domain by the strain gauge were converted into the stress data in the frequency domain using the rain-flow counting method, thus making it possible to obtain the mean stress, the stress amplitude, and the frequency of that stress occurring [21]. The stress with the mean and amplitude were converted into the equivalent completely reversed stress with a mean of zero using the Goodman's equation given in Equation (3).

$$\sigma_{eq} = \frac{S_u \times \sigma_a}{S_u - \sigma_m} \quad (3)$$

where σ_{eq} is equivalent completely reserved stress, MPa; S_u is ultimate tensile strength of the material, MPa; σ_a is measured stress amplitude, MPa; σ_m is measured mean stress, MPa.

It is possible to obtain the fatigue life cycle by applying the derived equivalent completely reversed stress to the stress–life (S–N) curve of the material. Fatigue damage was calculated using the actual frequency of the stress from rain-flow counting and the life cycle from S–N curve, by the Palmgren–Miner's linear cumulative damage rule, as shown in Equation (4).

$$D_t = \sum_{i=1}^{k} \frac{n_i}{N_i} \quad (4)$$

where D_t is total damage sum, n_i is number of actually applied cycles for the ith stress, and N_i is life cycles for ith stress.

The linear cumulative damage rule provides the total damage sum by adding the partial damage caused by all applied loads under the assumption that fatigue damage was linearly accumulated. It assumes that the fatigue failure of the material occurs when the total damage sum becomes 1.0 [22]. The fatigue life based on the total damage sum was calculated using Equation (5):

$$L_f = \frac{1}{D_t} \times t \quad (5)$$

where L_f is fatigue life and t is working time which generate damage sum.

In this study, commercial fatigue analysis software (nCode 2022, HBM Prenscia, Detroit, MI, USA) was used to calculate the fatigue damage and fatigue life. The calculation process was constructed by applying rain-flow counting, Goodman's equation, and Palmgren–Miner's linear cumulative damage rule to the measured time-domain stress data (Figure 8).

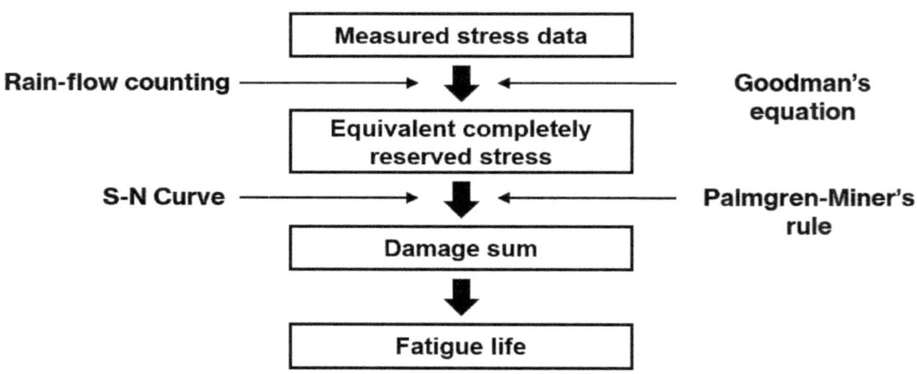

Figure 8. Process diagram to calculate the fatigue life.

The boom, post, mount, and rear frame of the front-end loader were made of the same material, and the S–N curve was obtained by reflecting the actual properties of the material (Figure 9 and Table 7). The components of the front-end loader are made from the same material, simplifying the manufacturing process, and reducing costs.

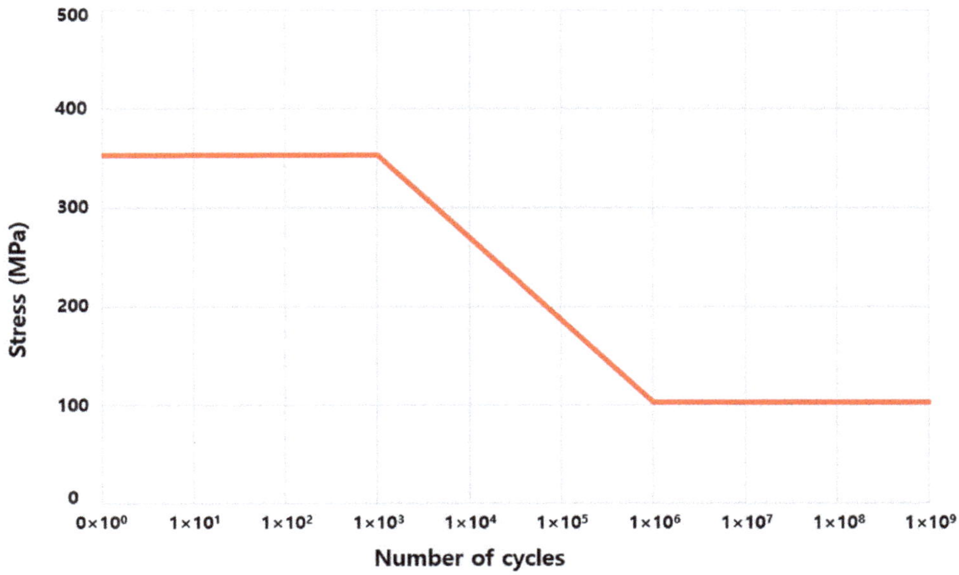

Figure 9. S–N curve of the front-end loader components.

Table 7. Material properties of the front-end loader components.

Item	Material Properties
Material	SS400
Tensile yield strength (MPa)	331
Ultimate tensile strength (MPa)	472
Elastic modulus (GPa)	202

3. Results and Discussion

3.1. Stress Measurement Results

The shape of the total measurement data at the gauge G1 considering the duty percentage by the test condition is shown in Figure 10. The stress trend was similar at all locations. In the drop test, the stress tended to increase when the bucket ascended compared to when it descended. In the impact test, lower stress occurred compared to the drop test. This indicates that a larger amount of damage occurs to the front-end loader when it rapidly raises and lowers the bucket compared to when it collides with obstacles at a speed of 1 km/h. The average stress during the driving test ranged from 5 to 60 MPa, depending on the measurement location, which was found to be the level that has no significant impact on the front-end loader damage. The maximum stress ranged from 28.86 to 3015 MPa depending on the measurement location, and it occurred in the pulling or dumping test (Table 8). In the pulling test, when the bucket was raised, the front-end loader acted as a cantilever due to the chain fixed to the ground. This appears to have caused a large stress on the components due to bending. In addition, it is thought that a large stress occurred in the dumping test because large bending loads were applied to the front-end loader by the payload while performing loading and unloading.

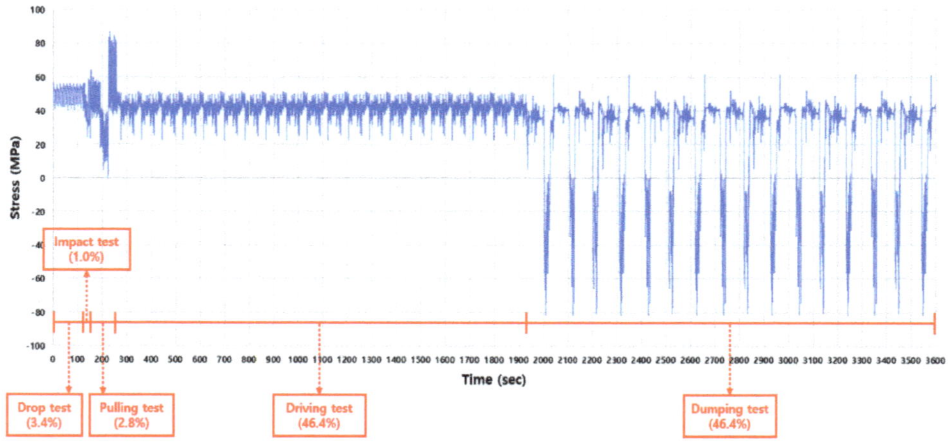

Figure 10. Shape of measure time-domain stresses at the gauge G1.

Table 8. Maximum stresses for each measurement location.

Component	Location	Maximum Stress (MPa)	Test Condition	Component	Location	Maximum Stress (MPa)	Test Condition
Post	G1	87.36	Pulling test	Rear frame	G18	36.42	Dumping test
	G2	−68.13	Dumping test		G19	−60.22	Pulling test
	G3	126.50	Dumping test		G20	−37.60	Pulling test
	G4	−116.20	Dumping test		G21	−171.70	Dumping test
	G5	−29.97	Dumping test		G22	−129.70	Dumping test
	G6	157.90	Dumping test	Boom	G23	83.26	Pulling test
Mount	G7	35.28	Dumping test		G24	96.83	Dumping test
	G8	−80.26	Pulling test		G25	95.68	Pulling test
	G9	68.00	Pulling test		G26	85.46	Pulling test
	G10	−54.57	Dumping test		G27	46.69	Pulling test
	G11	581.50	Pulling test		G28	47.50	Dumping test
	G12	−3015	Dumping test		G29	133.10	Dumping test
	G13	92.47	Pulling test		G30	112.00	Dumping test
	G14	106.60	Pulling test		G31	−81.01	Pulling test
Rear frame	G15	58.91	Dumping test		G32	−53.62	Dumping test
	G16	38.78	Pulling test		G33	63.31	Dumping test
	G17	28.86	Pulling test		G34	46.12	Pulling test

At the G12 and G11 locations of the mount, maximum stresses of 3015 and 581.5 MPa that exceeded the yield strength of the material occurred in the dumping and pulling tests, respectively. They were followed by maximum stresses of 171.70 and 157.90 MPa at the G21 location of the rear frame (dumping test) and the G6 location of the post (dumping test), respectively (Figure 11). The mount is a fastening element that fixes the front-end loader to the tractor body. It is the element to which the weight and working load of the front-end loader is first transferred. In particular, the G11 and G12 locations of the mount were located in the middle of the post–mount joint and mount–tractor body joint. Therefore, it appears that the application of an external load while both ends were fixed acted as a large bending load at the G11 and G12 locations, resulting in the maximum stress that exceeded the yield strength of the material. Relatively large stress also occurred at the G6 location of the post and the G21 location of the rear frame, even though it was lower than the yield strength of the material. It is believed that high stress occurred at the G6 location of the post because it was located in the transfer path of the load of the front-end loader at the post–mount joint. In addition, the G21 location of the rear frame was close to the mount–rear frame joint, and it was inferred that a high stress occurred because it was located in the path where the load of the mount was transferred to the rear frame (Figure 12).

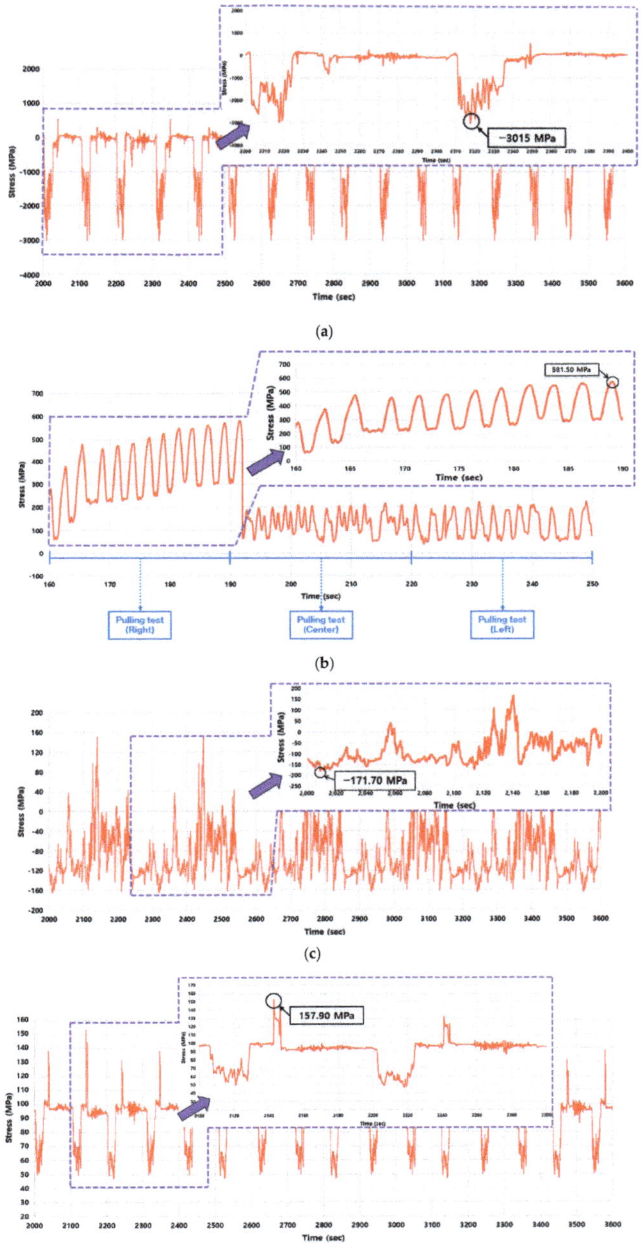

Figure 11. The location where the highest maximum stress occurred: (**a**) the 1st highest stress spot (G12, mount back side lower (R)); (**b**) the 2nd highest stress spot (G11, mount back side lower (L)); (**c**) the 3rd highest stress spot (G21, rear frame rear bottom (L)); (**d**) the 4th highest stress spot (G6, post bar).

Figure 12. Measurement location of G6 and G21.

3.2. Safety Evaluation of the Front-End Loader

3.2.1. Static Safety Factor

The static safety factor at each location was obtained using the maximum stress measured during the total testing time and the yield strength of the front-end loader components (Table 9). The G11 and G12 locations of the mount had a static safety factor of less than 1.0 and the yielding of the material occurred because the maximum stress exceeded a yield strength of 331 MPa. At the other measurement locations, the static safety factor values ranged from 2.10 to 11.04 in the post, 3.11 to 9.38 in the mount, 1.93 to 11.47 in the rear frame, and 2.49 to 7.09 in the boom, indication that static yielding did not occur. These results indicate that it is necessary to modify the design of the mount, which primarily supports the weight and working load of the front-end loader. It is possible to increase the static safety factor by utilizing a material with a higher strength or by modifying the thickness or geometry to reduce the maximum stress level. In the case of the front-end loader used in this study, a material with a yield strength of 3015 MPa or higher must be used to apply a design modification method of increased strength. This is approximately nine times the yield strength of the material used, and considerable cost will be required to use such high-strength materials. Therefore, it is deemed proper to modify the design in the direction of mainly changing the thickness or geometry of the mount frame and using a material with a slightly higher strength as a complement.

Table 9. Static safety factor of each measurement location.

Component	Location	Static Safety Factor	Component	Location	Static Safety Factor
Post	G1	3.79	Rear frame	G18	9.09
	G2	4.86		G19	5.50
	G3	2.62		G20	8.80
	G4	2.85		G21	1.93
	G5	11.04		G22	2.55
	G6	2.10	Boom	G23	3.98
Mount	G7	9.38		G24	3.42
	G8	4.12		G25	3.46
	G9	4.87		G26	3.87
	G10	6.07		G27	7.09
	G11	0.57 (Static failure)		G28	6.97
	G12	0.10 (Static failure)		G29	2.49
	G13	3.58		G30	2.96
	G14	3.11		G31	4.09
Rear frame	G15	5.62		G32	6.17
	G16	8.54		G33	5.23
	G17	11.47		G34	7.18

3.2.2. Fatigue Life

The G11 and G12 locations of the mount were excluded from the fatigue life analysis due to the occurrence of static yielding. Table A1 (Appendix A) shows the damage sum and the damage ratio of each test condition at the measurement spots except for G11 and G12. At all measurement locations, the damage by the dumping test accounted for 72.5 to 99.9% of the total damage, showing that the dumping test is the condition that has the largest impact on fatigue damage. The damage that occurred in the pulling test represented 0 to 26.6% of the total damage. The sum of the damage that occurred in the drop, impact, and driving tests was less than 8% of the total damage, indicating that these three tests hardly affected fatigue damage. These results indicate that repeating the loading and unloading tasks using the bucket is the working condition that has the largest impact on fatigue damage among the tasks using the front-end loader.

The total damage sum at each measurement location ranged from 5.938×10^{-5} to 2.946×10^{-11}. Five measurement locations that exhibited the highest total damage were G21 of the rear frame, G3 of the post, G14 of the mount, G24 of the boom, and G23 of the boom, with values of 5.938×10^{-5}, 5.210×10^{-6}, 2.731×10^{-6}, 2.605×10^{-6}, and 1.868×10^{-6}, respectively. Table 10 shows the stress conditions that exhibited the largest impact on fatigue damage for the five locations. The number of cycles of the stress that caused the maximum damage was four to five, depending on the location, and this one stress condition represented 49.3 to 66.2% of the total damage.

At each measurement location, the fatigue life ranged from 1.919×10^5 to 1.394×10^9 h in the post, 3.661×10^5 to 7.312×10^8 h in the mount, 1.684×10^4 to 3.394×10^{10} h in the rear frame, and 3.839×10^5 to 2.461×10^7 h in the boom (Table 11). This satisfied the life span required by the front-end loader manufacturer. Therefore, it was concluded that fatigue life was satisfied at all measurement locations except for G11 and G12 of the mount where static yielding occurred. The five measurement locations that showed the shortest fatigue life were G21 of the rear frame (1.684×10^5 h), G3 of the post (1.919×10^5 h), G14 of the mount (3.661×10^5 h), G24 of the boom (3.839×10^5 h), and G23 of the boom (5.352×10^5 h). In the mount and rear frame, the fatigue life was also lowest at the spots where the highest maximum stress occurred among measurement locations. However, in the case of the post and boom, the spots with the highest maximum stress (G6 for the post and G29 for the boom) and those with the shortest fatigue life were different. This could be because the stress amplitude and the equivalent completely reversed stress were also high at the spots with the higher maximum stress for the mount and rear frame, but they were relatively low at the spots with the highest maximum stress for the post and boom.

Table 10. Stress condition where higher damage occurred for five highest damage location.

Rank	Location	Mean Stress (MPa)	Stress Amplitude (MPa)	Equivalent Completely Reversed Stress (MPa)	Damage	Damage Ratio (%)	Number of Cycles
1	G21 (Rear frame rear bottom (L))	−6.89	157.94	155.67	3.93×10^{-5}	66.2	4
		−10.31	161.36	157.91	1.06×10^{-5}	17.9	1
		−1.14	98.43	98.20	2.94×10^{-6}	5.0	5
		−59.48	102.60	91.12	1.93×10^{-6}	3.3	5
		−60.91	98.13	86.91	1.48×10^{-6}	2.5	5
			Sum			94.9	
2	G3 (Post upper collar (R))	−15.68	52.46	50.77	3.30×10^{-6}	63.3	4
		−3.6	64.53	64.04	6.85×10^{-7}	13.2	1
			Sum			76.5	
3	G14 (Mount back side upper (L))	11.31	82.15	84.17	1.54×10^{-6}	56.4	5
		17.90	88.74	92.24	5.16×10^{-7}	18.9	1
		−11.28	58.70	57.33	1.78×10^{-7}	6.5	5
			Sum			81.8	
4	G24 (Front boom top side (R))	17.21	157.94	155.67	1.39×10^{-6}	53.7	5
		−10.31	161.36	157.91	7.38×10^{-7}	28.3	5
		−1.14	98.43	98.2	2.30×10^{-7}	8.8	5
		−59.48	102.6	91.12	1.87×10^{-8}	7.2	1
			Sum			98	
5	G23 (Front boom top side (L))	4.83	76.01	76.80	9.21×10^{-7}	49.3	4
		−3.41	67.69	67.21	3.48×10^{-7}	18.6	1
		−6.86	60.94	60.06	2.31×10^{-7}	12.4	5
		6.04	77.22	78.22	2.04×10^{-7}	10.9	5
		−1.75	66.04	65.80	7.72×10^{-8}	4.1	5
			Sum			95.3	

Table 11. Fatigue life for each measurement location.

Component	Location	Fatigue Life (hours)	Component	Location	Fatigue Life (hours)
Post	G1	6.447×10^5	Rear frame	G18	5.314×10^9
	G2	4.922×10^6		G19	1.675×10^9
	G3	1.919×10^5		G20	1.370×10^9
	G4	1.006×10^7		G21	1.684×10^4
	G5	1.394×10^9		G22	4.699×10^8
	G6	5.696×10^5	Boom	G23	5.352×10^5
Mount	G7	7.312×10^8		G24	3.839×10^5
	G8	2.962×10^8		G25	7.886×10^5
	G9	4.461×10^6		G26	6.585×10^5
	G10	5.714×10^6		G27	1.825×10^7
	G11	Static failure		G28	1.103×10^7
	G12	Static failure		G29	1.785×10^6
	G13	1.957×10^6		G30	9.943×10^5
	G14	3.661×10^5		G31	1.578×10^6
Rear frame	G15	1.225×10^8		G32	2.461×10^7
	G16	3.394×10^{10}		G33	8.922×10^6
	G17	3.025×10^9		G34	1.535×10^7

4. Conclusions

In this study, stress was measured for each component of an agricultural front-end loader to evaluate its structural safety. For that, we suggested new test code considering actual operational environments of the front-end loader. They include five test conditions: drop test, impact test, pulling test, driving test, and dumping test. The drop test considers the load caused by the rise and fall of the bucket. The impact and pulling tests consider cases where an impact occurs due to an obstacle or when an excessive weight is applied to the bucket. The driving test considers the load caused by the irregular road surface while traveling, and the dumping test considers the operation of loading, transportation, and unloading of soil. A total of 1 h of testing was performed, considering the duty percentage for each test condition provided by the front-end loader manufacturer. Thirty-four stress measurement locations were determined as the stress concentration spots of the front-end loader components from a dynamic simulation.

As results, the pulling and dumping tests exhibited the highest stress among the test conditions. In this case, the maximum stress that exceeded the yield strength of the material occurred at some locations of the mount. This could be because the mount is an element that primarily supports various loads generated by the weight and working load of the front-end loader. Except for the locations of the mount where static yielding occurred, the static safety factor was 1.93 or higher at all measurement locations. The fatigue damage and fatigue life were calculated by applying the rain-flow counting, Goodman's equation, and Palmgren–Miner's linear cumulative damage rule to the measured time-series stress data. The stress conditions that cause fatigue damage were also identified. Most of the fatigue damage occurred in the dumping test, followed by the pulling test. The drop, impact, and driving tests negligibly affected fatigue damage. Fatigue life at all measurement locations, except for the locations of the mount where static yielding occurred, was within the range that satisfied the life span required by the front-end loader manufacturer. Therefore, the mount is structurally the weakest point of the front-end loader. A robust design for the mount is necessary to secure the structural integrity of the entire front-end loader.

Author Contributions: Conceptualization, formal analysis, methodology, validation, visualization, writing—original draft, J.-H.K.; Conceptualization, formal analysis, methodology, validation, visualization, writing—original draft, D.-H.G.; Conceptualization, formal analysis, funding acquisition, methodology, writing—review and editing, J.-S.N. All authors have read and agreed to the published version of the manuscript.

Funding: This work was partly supported by Korea Institute of Planning and Evaluation for Technology in Food, Agriculture and Forestry (IPET) through Machinery Mechanization Technology Development Program for Field Farming Program, grant funded by Ministry of Agriculture, Food and Rural Affairs (MAFRA) (RS-2023-00235957, 50) and Innovative Human Resource Development for Local Intellectualization program through the Institute of Information & Communications Technology Planning & Evaluation (IITP) grant funded by the Korea government (MSIT) (IITP-2024-RS-2023-00260267, 50).

Institutional Review Board Statement: Not applicable.

Data Availability Statement: Data is contained within the article.

Conflicts of Interest: The authors declare no conflict of interest.

Appendix A

Table A1. Total damage and damage ratio by test condition at each measurement location.

Location	Test Condition	Damage	Damage Ratio (%)	Total Damage	Location	Test Condition	Damage	Damage Ratio (%)	Total Damage
	Drop	5.752×10^{-12}	0.0			Drop	2.284×10^{-14}	0.0	
	Impact	2.862×10^{-10}	0.0			Impact	1.923×10^{-12}	1.0	
G1	Pulling	1.990×10^{-8}	12.8	1.551×10^{-6}	G18	Pulling	3.195×10^{-11}	17.0	1.882×10^{-10}
	Driving	8.161×10^{-10}	0.0			Driving	8.562×10^{-13}	0.5	
	Dumping	1.351×10^{-6}	87.2			Dumping	1.535×10^{-10}	81.5	
	Sum		100			Sum		100	
	Drop	5.767×10^{-12}	0.0			Drop	1.640×10^{-13}	0.0	
	Impact	2.870×10^{-10}	0.0			Impact	9.131×10^{-14}	0.0	
G2	Pulling	2.224×10^{-7}	14.3	1.555×10^{-6}	G19	Pulling	6.972×10^{-11}	11.7	5.969×10^{-10}
	Driving	8.183×10^{-10}	0.1			Driving	3.882×10^{-12}	0.7	
	Dumping	1.332×10^{-6}	85.6			Dumping	5.237×10^{-10}	87.6	
	Sum		100			Sum		100	
	Drop	7.234×10^{-9}	0.1			Drop	8.039×10^{-14}	0.0	
	Impact	2.870×10^{-10}	0.0			Impact	6.874×10^{-13}	0.1	
G3	Pulling	5.865×10^{-7}	11.3	5.210×10^{-6}	G20	Pulling	1.286×10^{-10}	17.6	7.300×10^{-10}
	Driving	4.122×10^{-7}	7.9			Driving	1.535×10^{-12}	0.2	
	Dumping	4.204×10^{-6}	80.7			Dumping	5.990×10^{-10}	82.1	
	Sum		100			Sum		100	
	Drop	3.224×10^{-12}	0.0			Drop	3.789×10^{-13}	0.0	
	Impact	1.781×10^{-9}	1.8			Impact	1.369×10^{-9}	0.1	
G4	Pulling	1.177×10^{-8}	11.9	9.937×10^{-8}	G21	Pulling	6.559×10^{-10}	0.0	5.938×10^{-5}
	Driving	2.490×10^{-11}	0.0			Driving	5.054×10^{-10}	0.0	
	Dumping	8.579×10^{-8}	86.3			Dumping	5.937×10^{-5}	99.9	
	Sum		100			Sum		100	
	Drop	6.125×10^{-12}	0.9			Drop	8.048×10^{-13}	0.0	
	Impact	4.683×10^{-14}	0.0			Impact	2.430×10^{-13}	0.0	
G5	Pulling	9.700×10^{-11}	13.5	7.173×10^{-10}	G22	Pulling	2.661×10^{-10}	12.4	2.128×10^{-9}
	Driving	3.070×10^{-12}	0.4			Driving	2.549×10^{-11}	1.2	
	Dumping	6.111×10^{-10}	85.2			Dumping	1.855×10^{-9}	86.4	
	Sum		100			Sum		100	

Table A1. Cont.

Location	Test Condition	Damage	Damage Ratio (%)	Total Damage	Location	Test Condition	Damage	Damage Ratio (%)	Total Damage
G6	Drop	1.654×10^{-9}	0.1	1.756×10^{-6}	G23	Drop	1.246×10^{-11}	0.0	1.868×10^{-6}
	Impact	1.661×10^{-9}	0.1			Impact	9.269×10^{-9}	0.5	
	Pulling	1.876×10^{-7}	10.7			Pulling	1.874×10^{-7}	10.0	
	Driving	6.932×10^{-8}	3.9			Driving	1.282×10^{-9}	0.1	
	Dumping	1.496×10^{-6}	85.2			Dumping	1.670×10^{-6}	89.4	
	Sum		100			Sum		100	
G7	Drop	1.137×10^{-13}	0.0	1.368×10^{-9}	G24	Drop	1.332×10^{-11}	0.0	2.605×10^{-6}
	Impact	8.327×10^{-11}	6.2			Impact	4.427×10^{-9}	0.2	
	Pulling	2.804×10^{-10}	20.5			Pulling	4.540×10^{-7}	17.4	
	Driving	3.376×10^{-12}	0.2			Driving	2.168×10^{-9}	0.1	
	Dumping	9.995×10^{-10}	73.1			Dumping	2.144×10^{-6}	82.3	
	Sum		100			Sum		100	
G8	Drop	7.747×10^{-13}	0.0	3.376×10^{-9}	G25	Drop	1.225×10^{-11}	0.0	1.268×10^{-6}
	Impact	1.001×10^{-12}	0.0			Impact	6.826×10^{-9}	0.5	
	Pulling	8.982×10^{-10}	26.6			Pulling	1.955×10^{-7}	15.5	
	Driving	2.898×10^{-11}	0.9			Driving	1.211×10^{-9}	0.1	
	Dumping	2.447×10^{-9}	72.5			Dumping	1.064×10^{-6}	83.9	
	Sum		100			Sum		100	
G9	Drop	6.745×10^{-11}	0.0	2.241×10^{-7}	G26	Drop	8.734×10^{-12}	0.0	1.519×10^{-6}
	Impact	4.773×10^{-9}	2.1			Impact	3.989×10^{-9}	0.3	
	Pulling	3.125×10^{-8}	13.9			Pulling	2.188×10^{-7}	14.6	
	Driving	4.052×10^{-9}	1.8			Driving	1.903×10^{-9}	0.1	
	Dumping	1.842×10^{-7}	82.2			Dumping	1.272×10^{-6}	85.0	
	Sum		100			Sum		100	
G10	Drop	3.495×10^{-11}	0.0	1.750×10^{-7}	G27	Drop	8.392×10^{-13}	0.0	5.479×10^{-8}
	Impact	4.502×10^{-11}	0.0			Impact	1.769×10^{-10}	0.3	
	Pulling	2.113×10^{-8}	12.2			Pulling	8.085×10^{-9}	14.3	
	Driving	5.800×10^{-10}	0.3			Driving	4.468×10^{-11}	0.1	
	Dumping	1.528×10^{-7}	87.5			Dumping	4.826×10^{-8}	85.3	
	Sum		100			Sum		100	

Table A1. *Cont.*

Location	Test Condition	Damage	Damage Ratio (%)	Total Damage	Location	Test Condition	Damage	Damage Ratio (%)	Total Damage
G11	Static failure				G28	Drop	3.215×10^{-13}	0.0	9.064×10^{-8}
						Impact	9.897×10^{-11}	0.1	
						Pulling	4.989×10^{-9}	5.5	
						Driving	4.128×10^{-11}	0.0	
						Dumping	8.495×10^{-8}	94.4	
						Sum		100	
G12	Static failure				G29	Drop	1.656×10^{-13}	0.0	5.602×10^{-7}
						Impact	7.123×10^{-9}	1.2	
						Pulling	4.274×10^{-8}	7.5	
						Driving	7.500×10^{-11}	0.0	
						Dumping	5.242×10^{-7}	91.3	
						Sum		100	
G13	Drop	3.545×10^{-10}	0.1	5.110×10^{-7}	G30	Drop	7.385×10^{-14}	0.0	1.006×10^{-6}
	Impact	7.731×10^{-10}	0.2			Impact	2.067×10^{-9}	0.1	
	Pulling	8.645×10^{-8}	16.8			Pulling	8.669×10^{-8}	8.2	
	Driving	1.260×10^{-8}	2.5			Driving	5.701×10^{-11}	0.0	
	Dumping	4.108×10^{-7}	80.4			Dumping	9.797×10^{-7}	91.7	
	Sum		100			Sum		100	
G14	Drop	1.116×10^{-9}	0.0	2.731×10^{-6}	G31	Drop	1.175×10^{-14}	0.0	6.337×10^{-7}
	Impact	1.440×10^{-10}	0.0			Impact	2.046×10^{-10}	0.0	
	Pulling	3.285×10^{-7}	11.9			Pulling	5.167×10^{-8}	8.5	
	Driving	1.509×10^{-7}	5.5			Driving	7.456×10^{-13}	0.0	
	Dumping	2.287×10^{-6}	82.6			Dumping	5.571×10^{-7}	91.5	
	Sum		100			Sum		100	
G15	Drop	1.011×10^{-15}	0.0	8.164×10^{-9}	G32	Drop	3.554×10^{-14}	0.0	4.063×10^{-8}
	Impact	1.054×10^{-10}	1.3			Impact	2.448×10^{-11}	0.1	
	Pulling	7.194×10^{-10}	8.8			Pulling	2.431×10^{-9}	9.0	
	Driving	3.432×10^{-11}	0.4			Driving	3.220×10^{-12}	0.0	
	Dumping	7.305×10^{-9}	89.5			Dumping	2.443×10^{-8}	90.9	
	Sum		100			Sum		100	

Table A1. *Cont.*

Location	Test Condition	Damage	Damage Ratio (%)	Total Damage	Location	Test Condition	Damage	Damage Ratio (%)	Total Damage
G16	Drop	1.341×10^{-14}	0.0	2.946×10^{-11}	G33	Drop	3.532×10^{-14}	0.0	1.121×10^{-7}
	Impact	8.758×10^{-14}	0.3			Impact	1.949×10^{-9}	1.8	
	Pulling	5.027×10^{-12}	17.1			Pulling	5.268×10^{-10}	0.5	
	Driving	5.180×10^{-13}	1.8			Driving	2.800×10^{-12}	0.0	
	Dumping	2.379×10^{-11}	80.8			Dumping	1.073×10^{-7}	97.7	
	Sum		100			Sum		100	
G17	Drop	4.663×10^{-14}	0.0	3.306×10^{-10}	G34	Drop	2.493×10^{-14}	0.0	6.516×10^{-8}
	Impact	1.518×10^{-13}	0.1			Impact	6.084×10^{-10}	1.0	
	Pulling	4.029×10^{-11}	13.3			Pulling	4.915×10^{-10}	0.8	
	Driving	2.021×10^{-12}	0.7			Driving	4.588×10^{-12}	0.0	
	Dumping	2.601×10^{-10}	85.9			Dumping	6.169×10^{-8}	98.2	
	Sum		100			Sum		100	

References

1. Glancey, J.; Upadhyaya, S.; Chancellor, W.; Rumsey, J. Prediction of Agricultural Implement Draft using an Instrumented Analog Tillage Tool. *Soil Tillage Res.* **1996**, *37*, 47–65. [CrossRef]
2. Lee, C.J.; Ha, J.W.; Choi, D.S.; Kim, H.J. Development of a Self-leveling System for the Bucket of an Agricultural Front-end Loader Using an Electro Hydraulic Proportional Valve and a Tilt Sensor. *J. Drive Control* **2015**, *12*, 60–70. [CrossRef]
3. Kim, S.J.; Gim, D.H.; Jang, M.K.; Hwang, S.J.; Kim, J.H.; Yang, Y.J.; Nam, J.S. Development of Regression Model for Predicting the Maximum Static Friction Force of Tractors with a Front-End Loader. *J. Biosyst. Eng.* **2023**, *48*, 329–338. [CrossRef]
4. Jaiswal, S.; Korkealaakso, P.; Åman, R.; Sopanen, J.; Mikkola, A. Deformable Terrain Model for the Real-Time Multibody Simulation of a Tractor with a Hydraulically Driven Front-Loader. *IEEE Access* **2019**, *7*, 172694–172708. [CrossRef]
5. Han, G.H.; Kim, K.D.; Ahn, D.V.; Park, Y.J. Comparative Analysis of Tractor Ride Vibration According to Suspension System Configuration. *J. Biosyst. Eng.* **2023**, *48*, 69–78. [CrossRef]
6. Nasarwanji, M.; Pollard, J.; Porter, W. An Analysis of Injuries to Front-end Loader Operators during Ingress and Egress. *Int. J. Ind Ergon.* **2018**, *65*, 84–92. [CrossRef]
7. Modak, S.; Raheman, H. Effects of Various Machine Parameters on Cutting Performance for High-speed Cutting of Paddy Crop *J. Biosyst. Eng.* **2022**, *47*, 181–196. [CrossRef]
8. Bhola, M.; Kumar, A.; Kumar, N. Energy-efficient Control of Hydrostatic Transmission of a Front-end Loader Machine Using Machine Learning Algorithm and its Sensitivity Analysis. *Proc. Inst. Mech. Eng. Part D J. Automobile. Eng.* **2022**, *13*, 237. [CrossRef]
9. Lim, G.S.; Lee, B.Y. Study on the Impact Analysis of Front Loader for Tractor. *J. Korean Soc. Manuf. Technol.* **2015**, *16*, 5051–5059 [CrossRef]
10. Cho, B.J.; Ahn, S.W.; Lee, C.J.; Yoon, Y.H.; Lee, S.S.; Kim, H.J. Improved Design of Hydraulic Circuit of Front-end Loader for Bump Shock Reduction of an Agricultural Tractor. *J. Drive Control* **2016**, *13*, 10–18. [CrossRef]
11. Ahn, S.W.; Kim, H.J.; Lee, S.S.; Choi, D.S. Study on Driving Shock Reduction of a Front End Loader by Accumulator. *Proc. KSAM* **2014**, *19*, 91–92.
12. Latoree, J.I.; Arana, I.; Ballesteros, T.; Pintor, J.M.; Alfaro, J.R. Front End Loader with Automatic Levelling for Farm Tractors *Biosyst. Eng.* **2016**, *148*, 111–126. [CrossRef]
13. Kwon, G.B.; Shin, D.Y. Optimum Design of Mini Loader Based on the Design of Experiment. *Trans. Korean Soc. Mech. Eng. A* **2011** *35*, 693–699. [CrossRef]
14. Hensh, S.; Tewari, V.K.; Upadhyay, G. An instrumentation system to measure the loads acting on the tractor PTO bearing during rotary tollage. *J. Terramech.* **2021**, *96*, 1–10. [CrossRef]
15. Malon, H.; Ayuda, A.; Garcia, F.; Vidal, M.; Cuartero, J. Application of Low-Cost Sensors for the Development of a Methodology to Design Front-End Loaders for Tractors. *J. Sens.* **2020**, *2020*, 1–13. [CrossRef]
16. Park, Y.J.; Shim, S.B.; Nam, J.S. Experimental Study on the Structural Safety of the Tractor Front-End Loader against Impact Load *J. Biosyst. Eng.* **2016**, *41*, 153–160. [CrossRef]
17. Yilmaz, D.; Hamamci, E.; Salik, D.; Ahiskali, Y. Structural Analysis of Agricultural Machinery: A Case Study for a Transport Chassis of a Spraying Machine. *J. Agric. Mach. Sci.* **2011**, *7*, 405–409.
18. Budynas, R.G.; Nisbett, J.K. *Shigley's Mechanical Engineering Design*, 9th ed.; Mcgraw-Hill: New York, NY, USA, 2011; pp. 427–436
19. Mourad, A.; Alomomani, A.; Seikh, I.; Elsheikh, A. Failure Analysis of Gas and Wind Turbine Blades: A review, Eng. *Fail. Anal* **2023**, *146*, 107107. [CrossRef]
20. Lee, G.H.; Moon, B.E.; Basak, J.; Kim, N.E.; Paudel, B.; Jeon, S.W.; Kook, J.H.; Kang, M.Y.; Ko, H.J.; Kim, T.H. Assessment of Load on Threshing Bar During Soybean Pod Threshing. *J. Biosyst. Eng.* **2023**, *48*, 478–486. [CrossRef]
21. Eldogan, Y.; Cigeroglu, E. Vibration Fatigue Analysis of a Cantilever Beam Using Different Fatigue Theories. In *Topics in Modal Analysis, Volume 7: Proceedings of the 31st IMAC, A Conference on Structural Dynamics*; Springer: New York, NY, USA, 2013 Volume 7, pp. 471–478. [CrossRef]
22. Lee, W.S.; Lee, H.W. A Study on the Prediction of the Fatigue Life of a Lug through the Finite Element Analysis. *J. Korean Soc Precis. Eng.* **1998**, *15*, 88–95.

MDPI AG
Grosspeteranlage 5
4052 Basel
Switzerland
Tel.: +41 61 683 77 34

Agriculture Editorial Office
E-mail: agriculture@mdpi.com
www.mdpi.com/journal/agriculture

www.ingramcontent.com/pod-product-compliance
Lightning Source LLC
Chambersburg PA
CBHW061511220926
51850CB00004B/226
* 9 7 8 3 7 2 5 8 6 1 6 5 1 *